National
Fire Alarm Code
Handbook

National Fire Alarm Code Handbook

Second Edition

Based on the 1996 edition of NFPA 72,
National Fire Alarm Code

Edited by
Merton W. Bunker, JR., P.E.
Wayne Moore, P.E.

National Fire Protection Association
1 Batterymarch Park, P.O. Box 9101
Quincy, Massachusetts 02269-9101

This second edition of the *National Fire Alarm Code Handbook* provides an essential reference for everyone associated with the production, distribution, or use of flammable and combustible liquids. The handbook explains the requirements found in NFPA 72, *National Fire Alarm Code*. The handbook contains the complete text of the codes, along with explanatory commentary, which is integrated between requirements.

All NFPA codes and standards are processed in accordance with NFPA's *Regulations Governing Committee Projects*.

Project Manager: Lauren MacCarthy

Project Editor: Jennifer Falotico

Composition: Argosy

Illustrations: Todd Bowman and George Nichols

Cover Design: Betsy Greenwood for Greenwood Associates

Production: Ellen Glisker, Stephen Dornbusch, and Debra Rose

NFPA No. 72HB96

ISBN No. 0-87765-410-7

Library of Congress Catalog Card Number 96-87464

Printed in the United States of America

First Printing, February 1997

99 98 97 6 5 4 3 2 1

Dedication

This edition of the Handbook is dedicated to all of the men and women in the fire alarm profession, past and present, who have served on the NFPA fire alarm systems Technical Committees.

Contents

Preface

National Fire Alarm Code Handbook, 2nd Edition

The National Fire Alarm Code has been in print for approximately three years and most of the fire protection community have had the opportunity to use the Code in its new recombined format. The 1996 edition of the Code has been revised to incorporate new technology and further reorganized in the hope that it will be easier to use.

We have also had the opportunity to incorporate comments received on the first edition of this handbook and to include other useful information we felt would assist the users of the handbook in performing their job. As was stated in the first edition, neither the Code nor this handbook contain every fact and/or concept that may be relevant to the design of a fire alarm system for every specific situation. The design of fire protection systems, including fire alarm systems covered by this Code, is an engineering discipline requiring knowledge of special concepts. This knowledge is not normally at the command of engineers or technicians who have not been specifically trained in the field. In addition, both technology and new research are changing the ways we apply and design systems, almost on a daily basis.

The National Fire Alarm Code provides the **minimum** requirements for the application, location, testing, maintenance, spacing and general installation of all components and devices associated with a fire alarm system. Whether the fire alarm system is required by a building code, the *Life Safety Code®*, insurance provider or simply a result of the owner's desire to protect his or her property, the *National Fire Alarm Code* requirements apply to each and every system installed.

We wish to acknowledge the individuals who contributed to the Handbook, both in content and direction: Dean K. Wilson, currently the Technical Correlating Committee Chairman for NFPA 72, provided the commentary for Chapter 4, as well as technical assistance and moral support; John Cholin, a member of the Technical Committee on Initiating Devices provided extensive commentary in Chapter 5; Christopher Marrion, also a member of the Technical Committee on Initiating Devices, provided commentary that will benefit the users of Appendix B; Robert P. Schifiliti, current Chairman of the Technical Committee on Notification Appliances is responsible for the commentary in Chapter 6; and Charlie Zimmerman's continuous support, comments and advice has, as always, been of enormous value.

We also want to acknowledge all of the individuals who gave of their time and talents to make the *National Fire Alarm Code* a professional and usable document, including the consulting engineers, manufacturers and their representatives who supplied artwork, data, and photographs for this edition. Their efforts have helped make this document more informative and provided a solid foundation on we were able to build.

We would like to thank NFPA staff members, Mark Earley, Mark Ode, and Joe Sheehan. Their insight, guidance and very generous assistance during this project is appreciated. We have had a great deal of assistance from NFPA managers, project editors and other staff, specifically Lauren MacCarthy, Jenn Falotico, Ellen Gisker, Angela Murphy, Kimberly Shea, and Mary Warren-Pilson, whose tireless efforts and guidance have helped us to complete this project.

We also both thank our wives, Joan and Maureen for their love, support and understanding while we were on the phone, attending meetings, or at the computer working on our project.

Merton W. Bunker, Jr., P.E.   Wayne D. Moore, P.E.

About the Editors

Merton Bunker is the NFPA senior signaling systems engineer and the staff liaison for signaling documents , including NFPA 72, *National Fire Alarm Code®*. He is currently the Secretary of the Technical Correlating Committee on the National Fire Alarm Code.

Mr. Bunker is also a member of the ASME A17.1 Technical Committee on Emergency Operations for Elevators and Escalators, and is the Technical Advisor for the International Electrotechnical Commission (IEC) Technical Committee 79.

He received his Bachelor's Degree in Electrical Engineering from the University of Maine in 1985, and is a registered professional electrical engineer, with more 11 years of experience in the field.

Currently, in addition to his other duties, Mr. Bunker is an instructor of the NFPA Fire Alarm Systems Seminar and has contributed to its development. He has also written articles for *NFPA Journal* and is a member of the IEEE, AFAA, SFPE, IAEI and IMSA.

Wayne Moore is a registered professional fire protection engineer with more than 25 years of practical experience in the fire detection system field. Currently, he is the engineering and marketing vice president for MBS Fire Technology, Inc. and manages the company's Atlanta, Ga. office. MBS is a fire protection consulting and engineering firm headquartered in Worcester, MA.

Mr. Moore is the chairman of the NFPA 72-1993 technical committee on Protected Premises Fire Alarm Systems, and is a member of the NFPA Standards Council. He is also chairman of the Fire Detection Institute.

He is the author of numerous articles and is a Section editor for the 18th edition of the NFPA Fire Protection Handbook. He has contributed to the NFPA Fire Alarm Signaling Systems Handbook and was the editor of the 1993 editor on the National Fire Alarm Code Handbook. He is co-editor of the national newsletter, " The Moore-Wilson Signaling Report."

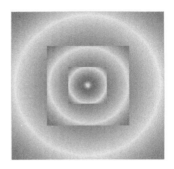

How to Use this Handbook

The text and illustrations that make up the commentary on the various sections of NFPA 72 are printed in an easy-to-read brick red. The text of the Code itself is printed in black to make this edition of the handbook easier to use.

The Formal Interpretations included in this handbook were issued as a result of questions raised on specific editions of the Code. They apply to all previous and subsequent editions in which the test in question remains substantially unchanged. Formal Interpretations are not part of the Code and therefore are printed in a shaded red box.

Paragraphs that begin with the letter "A" are extracted from Appendix A of the Code. Appendix A material is not mandatory. It is designed to help users apply the provisions of the Code. In this handbook, material from Appendix A is integrated with the text so that it follows the paragraph it explains. Appendix figures are also easily identified by the "A" and these figures follow material in the Code that they illustrate. Paragraphs that have accompanying Appendix A text or figures are marked with an asterisk (*).

1

Fundamentals of Fire Alarm Systems

Contents, Chapter 1

1-1   Scope

This code covers the application, installation, performance, and maintenance of fire alarm systems and their components.

This Code provides the minimum test, maintenance, and performance requirements for fire alarm systems. The application, location, and limitations of fire alarm components such as manual fire alarm boxes, detectors, and notification appliances are also provided.

Equipment or systems required by this Code or other codes must be installed to satisfy the requirements contained within this document. Partial systems, or equipment not necessarily required by this Code or other codes for use in fire alarm systems, should also be installed to satisfy the requirements of this Code. For example, a building owner wants a specific room to be protected by automatic smoke detection. However, separate detection for that space is not required by this Code or by other applicable codes. Nonetheless, the detection equipment that is installed in that space must satisfy the spacing requirements for the use of that type of equipment, as described in Chapter 5. The same holds true for other chapters of the Code, including notification appliances and off-premises transmission equipment.

1-2   Purpose

1-2.1* The purpose of this code is to define the means of signal initiation, transmission, notification, and annunciation; the levels of performance; and the reliability of the various types of fire alarm systems. This code defines the features associated with these systems and also provides the information necessary to modify or upgrade an existing system to meet the requirements of a particular system classification. It is the intent of this code to establish the required levels of performance, extent of redundancy, and quality of installation but not to establish the methods by which these requirements are to be achieved.

This Code describes the various types of initiating, supervisory, visible and audible notification devices, and how and where each should be used. The types of systems and methods of system transmission acceptable for each type of system are described, as are system reliability and performance requirements. However, the Code is not an installation specification or an approval guide or training manual.

A-1-2.1 In determining the performance criteria of circuits, the performance and capacity tables in Chapters 3 and 4 should be consulted. Where modifying an existing system, the system should be tested to determine the style of each circuit for the proper description and understanding of the system.

Chapter 3 describes the system capabilities required for protected premises fire alarm systems. Chapter 4 contains loading capacity tables for each of the available supervising station system transmission methods. Whenever a system is modified, it is vital for the system designer to have a thorough understanding of the existing equipment, including its capabilities and the system's wiring style, type, and configuration. In many cases, the existing equipment is older and may not interface easily with the newer technology used in the planned additional equipment. This existing equipment may or may not be able to be modified to conform to current code requirements.

1-2.2 Any reference or implied reference to a particular type of hardware is for the purpose of clarity and shall not be interpreted as an endorsement.

NFPA does not manufacture, test, distribute, endorse, approve, or list services, products, or components. See A-1-4 definition of Approved.

1-2.3 Unless otherwise noted, it is not intended that the provisions of this document be applied to facilities, equipment, structures, or installations that were existing or approved for construction or installation prior to the effective date of the document, except in those cases where it is determined by the authority having jurisdiction that the existing situation involves a distinct hazard to life or property.

With the exception of Chapter 7, the Code is not intended to be applied retroactively to existing installations. See Commentary following Section 7-1.1.1.

1-3 General

1-3.1 This code classifies fire alarm systems as follows:

(a) Household fire warning systems;

Household fire warning systems are installed to warn the occupants of a fire emergency so they may immediately evacuate the building. This system may be comprised of a control unit with connected initiating devices and notification appliances powered by the control unit. Alternatively, the system may consist of single smoke alarms or multiple-station smoke alarms located in specific areas of the family living unit. Single and multiple station smoke alarms may also be used in hotel rooms and apartment buildings. The requirements for

household fire warning systems are offered in detail in Chapter 2 of this Code.

(b) Protected premises fire alarm systems;

The primary purpose of a protected premises fire alarm system is to warn building occupants to evacuate the premises. Other purposes of protected premises fire alarm systems are to activate the building fire protection features, provide mission protection and continuity of operations. Chapter 3 of this Code describes the requirements for protected premises fire alarm systems.

(c) Supervising station fire alarm systems:

Supervising station fire alarm systems, described in Chapter 4, provide requirements for means of communication between the protected premises and a location called a supervising station. The types of supervising station fire alarm systems are enumerated below.

1. Auxiliary fire alarm systems

These systems provide a direct means of communicating between the protected premises and the fire department using fire-department-maintained transmission lines. Auxiliary fire alarm systems require a municipal fire alarm system in order to transmit the fire alarm signal.

a. Local energy type

Contacts in the fire alarm control unit activate a city fire alarm box mounted on the inside or outside of the building which, in turn, initiates an alarm at the fire department. Power to activate this system is provided through the fire alarm control unit. The fire alarm control unit circuit is isolated from the city circuit.

b. Parallel telephone type

In one version of this system type, telephone lines are used to communicate between each protected premises fire alarm box and the fire department. In another version, the leased telephone circuits interconnect between the municipal communication center and a supervised circuit of the fire alarm system.

c. Shunt type

A closed contact at the protected premises is electrically connected to, and is an integral part of, the city circuit.

Shunt circuits are normally connected to waterflow initiating devices and a manual fire alarm box. An open circuit condition in the reporting circuit will send an alarm signal to the fire department. A shunt-type system has very specific requirements and is not allowed to be interconnected to a protected premises system unless the city circuits entering the protected premises are installed in rigid conduit. This helps to prevent faults in one premises from disabling the circuit. Faults in the circuit may prevent transmission from other protected premises, leaving them unprotected.

2. Remote supervising station fire alarm systems

These systems provide a means of transmitting alarm and trouble signals from the protected premises to a remote supervising station. Typically, these systems are used in remote locations where no municipal system is available. The local fire department has the option of installing the remote supervising station to receive fire alarm signals where no municipal fire alarm system is available. The remote station is staffed 24 hours a day and properties under one or more ownership may be monitored by this station. When located at, and operated by, the public fire department, supervisory and trouble signals are generally not transmitted to the remote supervising station.

3. Proprietary supervising station systems

These systems require trained, competent personnel in constant attendance at a supervising station. The station personnel monitor the protected premises and take appropriate action when necessary. The proprietary supervising station is located at the protected property and is owned and operated by the property owner. Proprietary supervising stations monitor properties under a single ownership. The property may consist of a single building, such as a high-rise building, or several buildings, such as a college campus where the dormitories and other buildings report to a single supervising station on campus. The property may be contiguous or non-contiguous. If non-contiguous, it consists of protected property at remote locations, such as across town or across the country.

4. Central station fire alarm systems

These are privately owned systems or groups of systems with operators who, upon receipt of a signal from the protected premises, take appropriate action. Central station service is controlled and operated by an individual or a group whose business is to furnish, maintain, and monitor supervised fire alarm systems. Central station service is provided by a listed central station under contract to the subscriber. The central station provides all elements of installation, maintenance, testing, record keeping, retransmission and runner service. See Section 4-2.

5. Municipal fire alarm systems.

These are systems used to transmit alarms to the fire department from street locations, using street fire alarm boxes, or from a building, using master fire alarm boxes.

1-3.2   A device or system having materials or forms that differ from those detailed in this code shall be permitted to be examined and tested according to the intent of the code and, if found equivalent, shall be approved.

A device or system that does not meet the specific requirements of this Code may be submitted to a testing laboratory for listing by the laboratory to determine if the device or system meets the intent of the Code. The authority having jurisdiction has the responsibility to determine whether or not a product or device is suitable.

1-3.3   The intent and meaning of the terms used in this code are, unless otherwise defined herein, the same as those of NFPA 70, *National Electrical Code®*.

1-4   Definitions

For the purposes of this code, the following terms are defined as follows:

All definitions that apply to subjects covered throughout the Code have been relocated to Section 1-4 to make this edition of the Code easier to use.

Acknowledge.   To confirm that a message or signal has been received, such as by the pressing of a button or the selection of a software command.

Active Multiplex System.   A multiplexing system in which signaling devices such as transponders are employed to transmit status signals of each initiating device or initiating device circuit within a prescribed time interval so that lack of receipt of such signal may be interpreted as a trouble signal.

Active multiplex circuits utilize an interrogation (call and response) routine to determine the status of a device. Failure to receive a status signal from a device is interpreted as a trouble signal. This is one form of monitoring the wiring for integrity. See Section 1-5.8.

Addressable Device. A fire alarm system component with discrete identification that can have its status individually identified or that is used to individually control other functions.

Addressable devices may be either initiating or control devices. An initiating device will provide the control panel with its status and actual location. The control device will operate specific circuits or devices. Device addresses may be assigned by the system hardware or software. See Figures 1.1 and 1.2.

Adverse Condition. Any condition occurring in a communications or transmission channel that interferes with the proper transmission or interpretation, or both, of status change signals at the supervising station. (*See also Trouble Signal.*)

Open circuits, short-circuits, or electrical interference on the transmission line that prevents the normal operation of the system are examples of adverse conditions.

Air Sampling-Type Detector. A detector that consists of a piping or tubing distribution network that runs from the detector to the area(s) to be protected. An aspiration fan in the detector housing draws air from the protected area back to

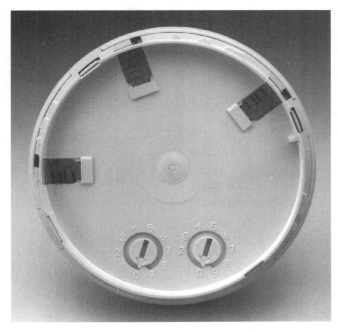

Figure 1.2   Addressable smoke detector showing programming switches. Photograph courtesy of System Sensor Corp., St. Charles, IL.

the detector through air sampling ports, piping, or tubing. At the detector, the air is analyzed for fire products.

There are passive and active air sampling-type detectors. Duct smoke detectors are typically considered passive detection devices. Active sampling requires aspirating equipment, such as a pump, to draw an air sample from the protected area or space. See Figures 1.3 and 1.4.

Figure 1.1   Addressable device programmer. Photograph courtesy of Cerberus Pyrotronics, Cedar Knolls, NJ.

Figure 1.3   Duct smoke detector (passive). Photograph courtesy of System Sensor Corp., St. Charles, IL.

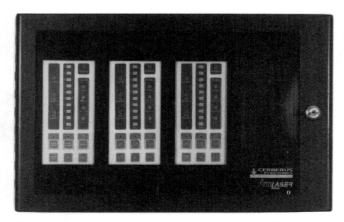

Figure 1.4   Air sampling smoke detector (active). Photograph courtesy of Cerberus Pyrotronics, Cedar Knolls, NJ.

Alarm.   A warning of fire danger.

Only the word "alarm" is used to indicate a fire alarm condition. The phrases "trouble alarm" or "supervisory alarm" are not acceptable references in cases of fire. See definitions for trouble and supervisory signals.

Alarm Service.   The service required following the receipt of an alarm signal.

Alarm service provides immediate retransmission of the alarm to the fire department, dispatching of a runner to the protected premises for either an alarm, supervisory, or trouble signal, notification of the subscriber, and notification of the authority having jurisdiction (if the signal received was not an alarm).

Alarm Signal.   A signal indicating an emergency requiring immediate action, such as a signal indicative of fire.

Alarm Verification Feature.   A feature of automatic fire detection and alarm systems to reduce unwanted alarms wherein smoke detectors report alarm conditions for a minimum period of time, or confirm alarm conditions within a given time period after being reset, in order to be accepted as a valid alarm initiation signal.

This feature has been used in facilities with a history of unwanted alarms from smoke detectors. It is available as a listed component of most fire alarm control units. The timing sequence and operation is defined by the listing laboratory.

Alert Tone.   An attention-getting signal to alert occupants of the pending transmission of a voice message.

The alert tone is used in voice communication systems to warn the occupants that a voice announcement is about to be made. The alert tone is not considered an alarm. See Section 3-12.6.3.1(a).

Analog Initiating Device (Sensor).   An initiating device that transmits a signal indicating varying degrees of condition as contrasted with a conventional initiating device, which can only indicate an on/off condition.

Analog devices are designed to measure and transmit smoke density, temperature variation, water level, water pressure changes, etc. to a fire alarm control unit. Smoke detector technology uses the analog feature to provide a warning signal to the owner when the detector is dirty or when the detector drifts outside of its listed sensitivity range. Section 7-3.2.1(d) permits analog technology to be used as a substitute for sensitivity testing.

Annunciator.   A unit containing one or more indicator lamps, alphanumeric displays, or other equivalent means in which each indication provides status information about a circuit, condition, or location.

Approved.*   Acceptable to the authority having jurisdiction.

A-1-4   Approved.   The National Fire Protection Association does not approve, inspect, or certify any installations, procedures, equipment, or materials; nor does it approve or evaluate testing laboratories. In determining the acceptability of installations, procedures, equipment, or materials, the authority having jurisdiction may base acceptance on compliance with NFPA or other appropriate standards. In the absence of such standards, said authority may require evidence of proper installation, procedure, or use. The authority having jurisdiction may also refer to the listings or labeling practices of an organization concerned with product evaluations that is in a position to determine compliance with appropriate standards for the current production of listed items.

The authority having jurisdiction may require a product to be listed or labeled, but the listing or label alone does not constitute approval. The authority having jurisdiction has the right to approve products or systems that are not labeled or listed.

Authority Having Jurisdiction.*   The organization, office, or individual responsible for approving equipment, an installation, or a procedure.

A-1-4 Authority Having Jurisdiction. The phrase "authority having jurisdiction" is used in NFPA documents in a broad manner, since jurisdictions and approval agencies vary, as do their responsibilities. Where public safety is primary, the authority having jurisdiction may be a federal, state, local, or other regional department or individual such as a fire chief; fire marshal; chief of a fire prevention bureau, labor department, or health department; building official; electrical inspector; or others having statutory authority. For insurance purposes, an insurance inspection department, rating bureau, or other insurance company representative may be the authority having jurisdiction. In many circumstances, the property owner or his or her designated agent assumes the role of the authority having jurisdiction; at government installations, the commanding officer or departmental official may be the authority having jurisdiction.

Automatic Extinguishing System Supervisory Device. A device that responds to abnormal conditions that could affect the proper operation of an automatic sprinkler system or other fire extinguishing system(s) or suppression system(s) including, but not limited to, control valves; pressure levels; liquid agent levels and temperatures; pump power and running; engine temperature and overspeed; and room temperature.

When an abnormal condition is detected, a supervisory signal, different from the alarm or trouble signal, is activated to warn the owner/attendant that the extinguishing system requires attention.

Automatic Fire Detector. A device designed to detect the presence of a fire signature and to initiate action. For the purpose of this code, automatic fire detectors are classified as follows:

Automatic Fire Extinguishing or Suppression System Operation Detector. A device that automatically detects the operation of a fire extinguishing or suppression system by means appropriate to the system employed.

Examples of alarm initiating devices are agent flow switches and agent pressure switches.

Fire-Gas Detector. A device that detects gases produced by a fire.

Examples of these fire gasses are hydrogen chloride (HCl) and carbon monoxide (CO). However, this type of CO detection should not be confused with household CO detectors designed to prevent CO poisoning by alerting occupants to the presence of CO gas in the home.

Figure 1.5 Typical heat detector. Photograph courtesy of Edwards Systems Technology/EST, Cheshire, CT.

Heat Detector. A fire detector that senses heat produced by burning substances. Heat is the energy produced by combustion that causes substances to rise in temperature.

See Figure 1.5.

Other Fire Detectors. Devices that detect a phenomenon other than heat, smoke, flame, or gases produced by a fire.

Radiant Energy-Sensing Fire Detector. A device that detects radiant energy (such as ultraviolet, visible, or infrared) that is emitted as a product of combustion reaction and obeys the laws of optics.

Smoke Detector. A device that detects visible or invisible particles of combustion.

See Figure 1.6.

Auxiliary Box. A fire alarm box that can be operated from one or more remote actuating devices.

Figure 1.6 Typical spot-type smoke detector. Photograph courtesy of System Sensor Corp., St. Charles, IL.

In an auxiliary fire alarm system, the auxiliary fire alarm box (commonly called a "master box") is normally located on the outside of a building. The protected premises fire alarm system is connected to this box and will automatically activate the box when an alarm occurs. The auxiliary box may also be manually activated.

Auxiliary Fire Alarm System.   A system connected to a municipal fire alarm system for transmitting an alarm of fire to the public fire service communications center. Fire alarms from an auxiliary fire alarm system are received at the public fire service communications center on the same equipment and by the same methods as alarms transmitted manually from municipal fire alarm boxes located on streets.

Local Energy Type.   An auxiliary system that employs a locally complete arrangement of parts, initiating devices, relays, power supply, and associated components to automatically trip a municipal transmitter or master box over electrical circuits that are electrically isolated from the municipal system circuits.

Parallel Telephone Type.   An auxiliary system connected by a municipally controlled individual circuit to the protected property to interconnect the initiating devices at the protected premises and the municipal fire alarm switchboard.

Shunt Auxiliary Type.   An auxiliary system electrically connected to an integral part of the municipal alarm system extending the municipal circuit into the protected premises to interconnect the initiating devices, which, when operated, open the municipal circuit shunted around the trip coil of the municipal transmitter or master box, which is thereupon energized to start transmission without any assistance whatsoever from a local source of power.

Average Ambient Sound Level.   The root mean square, A-weighted sound pressure level measured over a 24-hour period.

Box Battery.   The battery supplying power for an individual fire alarm box where radio signals are used for the transmission of box alarms.

Carrier.   High frequency energy that can be modulated by voice or signaling impulses.

Carrier System.   A means of conveying a number of channels over a single path by modulating each channel on a different carrier frequency and demodulating at the receiving point to restore the signals to their original form.

Ceiling.   The upper surface of a space, regardless of height. Areas with a suspended ceiling have two ceilings, one visible from the floor and one above the suspended ceiling.

Ceiling Height.   The height from the continuous floor of a room to the continuous ceiling of a room or space.

Ceiling Surfaces.   Ceiling surfaces referred to in conjunction with the locations of initiating devices are defined as follows:

Beam Construction.   Ceilings having solid structural or solid nonstructural members projecting down from the ceiling surface more than 4 in. (100 mm) and spaced more than 3 ft (0.9 m), center to center.

Girder.   A support for beams or joists that runs at right angles to the beams or joists. Where the top of girders are within 4 in. (100 mm) of the ceiling, they are a factor in determining the number of detectors and are to be considered as beams. Where the top of the girder is more than 4 in. (100 mm) from the ceiling, it is not a factor in detector location.

Central Station.   A supervising station that is listed for central station service.

The listed or approved central station is the attended location where the central station fire alarm system is located, where alarm signals are received, and which provides retransmission and runner service.

Central Station Fire Alarm System.   A system or group of systems in which the operations of circuits and devices are transmitted automatically to, recorded in, maintained by, and supervised from a listed central station having competent and experienced servers and operators who, upon receipt of a signal, take such action as required by this code. Such service is to be controlled and operated by a person, firm, or corporation whose business is the furnishing, maintaining, or monitoring of supervised fire alarm systems.

See Figure 1.7.

Central Station Service.   The use of a system or a group of systems in which the operations of circuits and devices at a protected property are signaled to, recorded in, and supervised from a listed central station having competent and experienced operators who, upon receipt of a signal, take such action as required by this code. Related activities at the protected property such as equipment installation, inspection, testing, maintenance, and runner service are the responsibility of the central station or a listed fire alarm service-local company. Central station service is controlled and operated by a person, firm, or corporation whose business is the furnishing of such contracted services or whose properties are the protected premises.

Figure 1.7 Central station fire alarm system. Photograph courtesy of AFA Protective Systems, Inc., Syosset, NY.

Certification. A systematic program using randomly selected follow-up inspections of the certified systems installed under the program that allows the listing organization to verify that a fire alarm system complies with all the requirements of this code. A system installed under such a program is identified by the issuance of a certificate and is designated as a certificated system.

Certification of Personnel. A formal program of related instruction and testing as provided by a recognized organization or the authority having jurisdiction.

NOTE: This definition applies only to municipal fire alarm systems.

Channel. A path for voice or signal transmission utilizing modulation of light or alternating current within a frequency band.

Circuit Interface. A circuit component that interfaces initiating devices or control circuits, or both, notification appliances or circuits, or both, system control outputs, and other signaling line circuits to a signaling line circuit.

Cloud Chamber Smoke Detection. The principle of using an air sample drawn from the protected area into a high humidity chamber combined with a lowering of chamber pressure to create an environment in which the resultant moisture in the air condenses on any smoke particles present, forming a cloud. The cloud density is measured by a photoelectric principle. The density signal is processed and used to convey an alarm condition when it meets preset criteria.

Cloud chamber smoke detection is another form of an active air sampling-type smoke detector. This type of detector is extremely sensitive to low levels of combustion products and is frequently used to detect very small fires in vital equipment. Please also see the definition of active air sampling-type smoke detector, described earlier in this section.

Coded. An audible or visible signal conveying several discrete bits or units of information. Notification signal examples are numbered strokes of an impact-type appliance and numbered flashes of a visible appliance.

Coded textual signals may use words familiar only to those concerned with response to the signal, in order to avoid general alarm notification disruption of the occupants. This type of signal is often used in hospitals. Appendix Subsection A-1-5.4.2.1 provides recommended assignment for simple coded signals.

Combination Detector. A device that either responds to more than one of the fire phenomenon or employs more than one operating principle to sense one of these phenomenon. Typical examples are a combination of a heat detector with a smoke detector or a combination rate-of-rise and fixed-temperature heat detector.

See Figures 1.8 and 1.9.

Figure 1.8 Combination rate-of-rise and fixed-temperature heat detector. Photograph courtesy of Mammoth Fire Alarms, Inc., Lowell, MA.

Figure 1.9 *Combination photoelectric smoke and heat detector. Photograph courtesy of Edwards Systems Technology/EST, Cheshire, CT.*

Combination Fire Alarm and Guard's Tour Box. A manually operated box for separately transmitting a fire alarm signal and a distinctive guard patrol tour supervisory signal.

Combination System. A fire alarm system whose components might be used, in whole or in part, in common with a nonfire signaling system, such as a paging system, a security system, a building automatic system, or a process monitoring system.

See Figure 1.10.

Communications Channel. A circuit or path connecting a subsidiary station(s) to a supervising station(s) over which signals are carried.

Compatibility Listed. A specific listing process that applies only to two-wire devices (such as smoke detectors) designed to operate with certain control equipment.

Two-wire, circuit-powered dectectors must be listed for compatibility with a fire alarm control unit. Detectors that are not listed for this purpose may appear to operate properly under normal operating conditions but may not work under adverse conditions such as a low voltage situation.

Compatible (Equipment). Equipment that interfaces mechanically or electrically as manufactured without field modification.

Contiguous Property. A single-owner or single-user protected premises on a continuous plot of ground, including any buildings thereon, that is not separated by a public thoroughfare, transportation right-of-way, property owned or used by others, or body of water not under the same ownership.

Control Unit. A system component that monitors inputs and controls outputs through various types of circuits.

This definition includes control unit equipment used for sprinkler supervisory service, transponders, and control units with integral digital alarm communicator transmitters (DACTs).

Delinquency Signal. A signal indicating the need for action in connection with the supervision of guards or system attendants.

This definition applies to guard's tour systems. If the guard fails to activate a tour station or fails to make a tour in a prescribed amount of time, the supervising station is notified.

Derived Channel. A signaling line circuit that uses the local leg of the public switched network as an active multiplex channel while simultaneously allowing that leg's use for normal telephone communications.

Detector. A device suitable for connection to a circuit having a sensor that responds to a physical stimulus such as heat or smoke.

Figure 1.10 *Combination burglary and fire alarm system control panel with keypad. Photograph courtesy of Ademco, Syosett, NY.*

Figure 1.11 *Digital alarm communicator receiver (DACR). Photograph courtesy of AFA Protective Systems, Inc. (Radionics), Syosset, NY.*

Digital Alarm Communicator Receiver (DACR). A system component that accepts and display signals from digital alarm communicator transmitters (DACTs) sent over the public switched telephone network.

See Figure 1.11.

Digital Alarm Communicator System (DACS). A system in which signals are transmitted from a digital alarm communicator transmitter (DACT) located at the protected premises through the public switched telephone network to a digital alarm communicator receiver (DACR).

Digital Alarm Communicator Transmitter (DACT). A system component at the protected premises to which initiating devices or groups of devices are connected. The DACT seizes the connected telephone line, dials a preselected number to connect to a DACR, and transmits signals indicating a status change of the initiating device.

Figure 1.12 *Digital alarm communicator transmitter (DACT) and programmer. Photograph courtesy of ESL Sentrol, Inc., Tualitin, OR.*

See Figure 1.12.

Digital Alarm Radio Receiver (DARR). A system component composed of two subcomponents: one that receives and decodes radio signals, and another that annunciates the decoded data. These two subcomponents can be coresident at the central station or separated by means of a data transmission channel.

Digital Alarm Radio System (DARS). A system in which signals are transmitted from a digital alarm radio transmitter (DART) located at a protected premises through a radio channel to a digital alarm radio receiver (DARR).

See Figure 1.13.

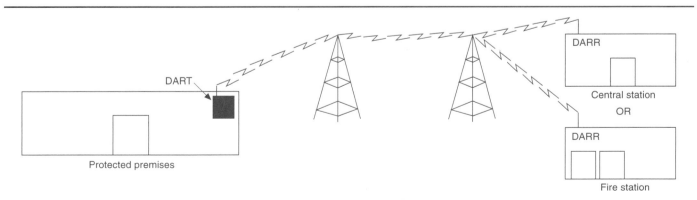

Figure 1.13 *Digital alarm radio system (DARS). Courtesy of AFA Protective Systems, Inc., (Ademco AlarmNet) North Brunswick, NJ.*

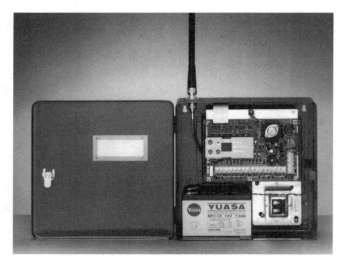

Figure 1.14   Digital alarm radio transmitter (DART). Photograph courtesy of AFA Protective Systems, Inc., (Ademco AlarmNet) Syosset, NY.

Digital Alarm Radio Transmitter (DART).   A system component that is connected to, or an integral part of, a DACT that is used to provide an alternate radio transmission channel.

See Figure 1.14.

Display.   The visual representation of output data as opposed to printed copy.

Double Doorway.   A single opening that has no intervening wall space or door trim separating the two doors. *See Figure 5-10.7.4.3.1.*

Dual Control.   The use of two primary trunk facilities over separate routes or different methods to control one communications channel.

Ember.*   A particle of solid material that emits radiant energy due either to its temperature or the process of combustion on its surface. (*See definition of Spark.*)

A-1-4   **Ember.**   Class A and Class D combustibles burn as embers under conditions where the flame typically associated with fire does not necessarily exist. This glowing combustion yields radiant emissions in parts of the radiant energy spectrum that are radically different from those parts affected by flaming combustion. Specialized detectors, specifically designed to detect those emissions, should be used in applications where this type of combustion is expected. In general, flame detectors are not intended for the detection of embers.

Emergency Voice/Alarm Communications.   Dedicated manual or automatic facilities for originating and distributing voice instructions, as well as alert and evacuation signals pertaining to a fire emergency, to the occupants of a building.

This type of system is generally used where relocation of occupants is planned and where total evacuation of a building, such as a high-rise apartment or office building or a multi-story hospital, is not necessary or practical.

Evacuation.   The withdrawal of occupants from a building.

NOTE: Evacuation does not include relocation of occupants within a building.

Evacuation Signal.   A distinctive signal intended to be recognized by the occupants as requiring evacuation of the building.

This signal meets the requirements of the *American National Standard Audible Emergency Evacuation Signal* (ANSI S3.41-1990). On July 1, 1996, the ANSI evacuation signal became the requirement for household fire warning systems and for all protected premises fire alarm systems.

Exit Plan.   A plan for the emergency evacuation of the premises.

Family Living Unit.   One or more rooms in a single-family detached dwelling, single-family attached dwelling, multifamily dwelling, or mobile home for the use of one or more persons as a housekeeping unit with space for eating, living, and sleeping and permanent provisions for cooking and sanitation. This definition covers living areas only and not common usage areas in multifamily dwellings such as corridors, lobbies, or basements.

Field of View.   The solid cone extending out from the detector within which the effective sensitivity of the detector is at least 50 percent of its on-axis, listed, or approved sensitivity.

This definition applies to Radiant Energy Fire detectors.

Fire Alarm Control Unit (Panel).   A system component that receives inputs from automatic and manual fire alarm devices and might supply power to detection devices and a transponder(s) or an off-premises transmitter(s). The control unit might also provide transfer of power to the notification appliances and transfer of condition to relays or devices connected to the control unit. The fire alarm control unit can be a local fire alarm control unit or master control unit.

See Figure 1.15.

Figure 1.15 Typical fire alarm control unit. Photograph courtesy of Fire Control Instruments, Newton, MA.

Fire Alarm/Evacuation Signal Tone Generator. A device that, upon command, produces a fire alarm/evacuation tone.

Fire Alarm Signal. A signal initiated by a fire alarm-initiating device such as a manual fire alarm box, automatic fire detector, waterflow switch, or other device whose activation is indicative of the presence of a fire or fire signature.

Fire Alarm System. A system or portion of a combination system consisting of components and circuits arranged to monitor and annunciate the status of fire alarm or supervisory signal-initiating devices and to initiate the appropriate response to those signals.

Fire Command Center. The principal attended or unattended location where the status of the detection, alarm communications, and control systems is displayed and from which the system(s) can be manually controlled.

See Figure 1.16.

Fire Rating. The classification indicating in time (hours) the ability of a structure or component to withstand fire conditions.

Fire Safety Function Control Device. The fire alarm system component that directly interfaces with the control system that controls the fire safety function.

These devices are used to control safety functions that increase the level of life safety and property protection. They may be operated manually or automatically by the fire alarm control unit. These devices may control door locking and release systems, suppression system activation, and HVAC (heating, ventilating, and air conditioning) systems.

Fire Safety Functions. Building and fire control functions that are intended to increase the level of life safety for occupants or to control the spread of the harmful effects of fire.

These functions include activation of duct smoke or fire dampers, elevator recall, HVAC shutdown, and door-release features.

Fire Warden. A building staff member or a tenant trained to perform assigned duties in the event of a fire emergency.

Fixed Temperature Detector.* A device that responds when its operating element becomes heated to a predetermined level.

See Figure 1.17.

A-1-4 Fixed Temperature Detector. The difference between the operating temperature of a fixed temperature device and the surrounding air temperature is proportional to the rate at which the temperature is rising and is commonly referred to as "thermal lag." The air temperature is always higher than the operating temperature of the device.

Typical examples of fixed temperature-sensing elements follow.

(a) *Bimetallic.* A sensing element comprised of two metals having different coefficients of thermal expansion arranged so that the effect is deflection in one direction when heated and in the opposite direction when cooled.

(b) *Electrical Conductivity.* A line-type or spot-type sensing element whose resistance varies as a function of temperature.

(c) *Fusible Alloy.* A sensing element of a special composition (eutectic) metal that melts rapidly at the rated temperature.

(d) *Heat-Sensitive Cable.* A line-type device whose sensing element comprises, in one type, two current-carrying wires separated by heat-sensitive insulation that softens at the rated temperature, thus allowing the wires to make electrical contact. In another type, a single wire is centered in a metallic tube, and the intervening space is filled with a substance that, at a critical temperature, becomes conductive, thus establishing electrical contact between the tube and the wire.

(e) *Liquid Expansion.* A sensing element comprising a liquid capable of marked expansion in volume in response to temperature increase.

Figure 1.16   A fire command center. Photograph courtesy of Simplex Time Recorder Co., Gardner, MA.

Flame.   A body or stream of gaseous material involved in the combustion process and emitting radiant energy at specific wavelength bands determined by the combustion chemistry of the fuel. In most cases, some portion of the emitted radiant energy is visible to the human eye.

Flame Detector.   A radiant energy-sensing fire detector that detects the radiant energy emitted by a flame. (*See A-5-4.2.*)

NOTE: Flame detectors are categorized as ultraviolet, single wavelength infrared, ultraviolet infrared, or multiple wavelength infrared.

Flame Detector Sensitivity.   The distance along the optical axis of the detector at which the detector can detect a fire of specified size and fuel within a given time frame.

Guard Signal.   A supervisory signal monitoring the performance of guard patrols.

Guard's Tour Reporting Station.   A device that is manually or automatically initiated to indicate the route being followed and the timing of a guard's tour.

Heat Alarm.   A single or multiple station alarm responsive to heat.

Household.   The family living unit in single-family detached dwellings, single-family attached dwellings, multifamily buildings, and mobile homes.

Household Fire Alarm System.   A system of devices that produces an alarm signal in the household for the purpose of notifying the occupants of the presence of a fire so that they will evacuate the premises.

Figure 1.17 Nonrestorable fixed-temperature heat detector. Photograph courtesy of Mammoth Fire Alarms, Lowell, MA.

Hunt Group. A group of associated telephone lines within which an incoming call is automatically routed to an idle (not busy) telephone line for completion.

Initiating Device. A system component that originates transmission of a change of state condition, such as in a smoke detector, manual fire alarm box, or supervisory switch.

Initiating Device Circuit. A circuit to which automatic or manual initiating devices are connected where the signal received does not identify the individual device operated.

This type of circuit is used with conventional fire alarm systems when individual devices are not identified at the control unit. In contrast each device on a signaling line circuit is usually identified by means of an address. See definition of Signaling Line Circuit in this section.

Intermediate Fire Alarm or Fire Supervisory Control Unit. A control unit used to provide area fire alarm or area fire supervisory service that, where connected to the proprietary fire alarm system, becomes a part of that system.

A fire alarm control unit at a protected premises connected to a proprietary system is part of the proprietary system and designated as an intermediate fire alarm control unit.

Ionization Smoke Detection.* The principle of using a small amount of radioactive material to ionize the air between two differentially charged electrodes to sense the presence of smoke particles. Smoke particles entering the ionization volume decrease the conductance of the air by reducing ion mobility. The reduced conductance signal is processed and used to convey an alarm condition when it meets preset criteria.

A-1-4 Ionization Smoke Detection. Ionization smoke detection is more responsive to invisible particles (smaller than 1 micron in size) produced by most flaming fires. It is somewhat less responsive to the larger particles typical of most smoldering fires. Smoke detectors utilizing the ionization principle are usually of the spot type.

Ionization-type smoke detectors are typically used when the designer expects a greater risk of a flaming rather than a smoky fire scenario. Ionization detectors are more sensitive to the smaller particles of smoke produced in the case of a flaming fire. See Figure 1.18.

Labeled. Equipment or materials to which has been attached a label, symbol, or other identifying mark of an organization that is acceptable to the authority having jurisdiction and concerned with product evaluation that maintains periodic inspection of production of labeled equipment or materials and by whose labeling the manufacturer indicates compliance with appropriate standards or performance in a specified manner.

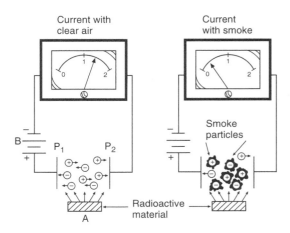

Figure 1.18 Principle of operation for ionization smoke detector.

Leg Facility.   The portion of a communications channel that connects not more than one protected premises to a primary or secondary trunk facility. The leg facility includes the portion of the signal transmission circuit from its point of connection with a trunk facility to the point where it is terminated within the protected premises at one or more transponders.

Level Ceilings.   Ceilings that are actually level or have a slope of 1½ in./ft (41.7 mm/m) or less.

Line-Type Detector.   A device in which detection is continuous along a path. Typical examples are rate-of-rise pneumatic tubing detectors, projected beam smoke detectors, and heat-sensitive cable.

Listed.*   Equipment, materials, or services included in a list published by an organization acceptable to the authority having jurisdiction and concerned with evaluation of products or services that maintains periodic inspection of production of listed equipment or materials or periodic evaluation of services and whose listing states either that the equipment, material, or service meets identified standards or has been tested and found suitable for a specified purpose.

A-1-4   Listed.   The means for identifying listed equipment may vary for each organization concerned with product evaluation, some of which do not recognize equipment as listed unless it is also labeled. The authority having jurisdiction should utilize the system employed by the listing organization to identify a listed product.

Loading Capacity.   The maximum number of discrete elements of fire alarm systems permitted to be used in a particular configuration.

The loading capacity of a system is dependent on the type of system and the style and class of the circuit used. Chapter 4 provides the loading capacities for various types of systems.

Loss of Power.   The reduction of available voltage at the load below the point at which equipment can function as designed.

Low Power Radio Transmitter.   Any device that communicates with associated control/receiving equipment by low power radio signals.

Maintenance.   Repair service, including periodic inspections and tests, required to keep the fire alarm system and its component parts in an operative condition at all times, together with replacement of the system or its components when they become undependable or inoperable for any reason.

Figure 1.19   Manual fire alarm box.

Manual Fire Alarm Box.   A manually operated device used to initiate an alarm signal.

See Figure 1.19.

Master Box.   A municipal fire alarm box that can also be operated by remote means.

See Figure 1.20.

Figure 1.20   Auxiliary (Master) box. Photograph courtesy of Mammoth Fire Alarms (Gamewell), Lowell, MA.

Master Control Unit (Panel). A control unit that serves the protected premises or portion of the protected premises as a local control unit and accepts inputs from other fire alarm control units.

Where more than one fire alarm control unit has been installed in a facility, one of the control units must act as the master control unit to monitor alarm, trouble, and supervisory conditions from the other satellite control unit panels.

Multiple Station Alarm. A single station alarm capable of being interconnected to one or more additional alarms so that the actuation of one causes the appropriate alarm signal to operate in all interconnected alarms.

This definition was added to the 1996 edition of the Code to help differentiate between detectors powered by a panel and single- and multiple-station smoke detectors powered by either AC power or by a battery. This definition corresponds with the definition used in international standards terminology.

Multiple Station Alarm Device. Two or more single station alarm devices that can be interconnected so that actuation of one causes all integral or separate audible alarms to operate. It also can consist of one single station alarm device having connections to other detectors or to a manual fire alarm box.

Multiplexing. A signaling method characterized by simultaneous or sequential transmission, or both, and reception of multiple signals on a signaling line circuit, a transmission channel, or a communications channel, including means for positively identifying each signal.

There are two types of multiplexing: active and passive. Active multiplexing devices require a command signal prior to transmission, such as the interrogation signal to a device on an addressable signaling line circuit. Passive devices, such as the simple coded system, do not require a command signal in order to transmit the signal to the receiver.

Municipal Fire Alarm Box (Street Box). An enclosure housing a manually operated transmitter used to send an alarm to the public fire service communications center.

See Figure 1.21.

Municipal Fire Alarm System. A system of alarm-initiating devices, receiving equipment, and connecting circuits (other than a public telephone network) used to transmit alarms from street locations to the public fire service communications center.

Figure 1.21   Municipal fire alarm box (street box). Photograph courtesy of Mammoth Fire Alarms (Gamewell), Lowell, MA.

Municipal Transmitter. A transmitter that can only be tripped remotely that is used to send an alarm to the public fire service communications center.

Noncoded. An audible or visible signal conveying one discrete bit of information.

Noncontiguous Property. An owner- or user-protected premises where two or more protected premises, controlled by the same owner or user, are separated by a public thoroughfare, body of water, transportation right-of-way, or property owned or used by others.

Nonrestorable Initiating Device. A device whose sensing element is designed to be destroyed in the process of operation.

One example of this type of device is the fixed-temperature heat detector that uses a fusible element that melts when subjected to heat. When the element melts, the electrical contacts are shorted together and the alarm signal is activated. See Figure 1.17.

Notification Appliance.   A fire alarm system component such as a bell, horn, speaker, light, or text display that provides audible, tactile, or visible outputs, or any combination thereof.

See Figures 1.22 and 1.23.

Audible Notification Appliance.   A notification appliance that alerts by the sense of hearing.

Audible Textual Notification Appliance.   A notification appliance that conveys a stream of audible information. An example of an audible textual appliance is a speaker that reproduces a voice message.

Olfactory   Notification   Appliance. A notification appliance that alerts by the sense of smell.

Tactile Notification Appliance.   A notification appliance that alerts by the sense of touch or vibration.

This type of notification appliance includes vibrating pagers and bed shakers used by disabled persons who are not be able to respond to an audible or visual alarm.

Visible Notification Appliance.   A notification appliance that alerts by the sense of sight.

Figure 1.23   *Typical visible notification appliance. Photograph courtesy of Gentex Corp., Zeeland, MI.*

Visible Textual Notification Appliance.   A notification appliance that conveys a stream of visible information. An example of a visible textual appliance is a monitor that displays an alphanumeric or pictorial message.

Notification Appliance Circuit.   A circuit or path directly connected to a notification appliance(s).

Notification Zone.   An area covered by notification appliances that are activated simultaneously.

Nuisance Alarm.   Any alarm caused by mechanical failure, malfunction, improper installation, or lack of proper maintenance, or any alarm activated by a cause that cannot be determined.

Off-Hook.   To make connection with the public switched telephone network in preparing to dial a telephone number.

When a telephone is raised from the receiver, it is considered off-hook. Some transmission technologies use equipment to access the public switched network and automatically provide an off-hook condition prior to transmission.

Figure 1.22   *Typical audible notification appliance. Photograph courtesy of Edwards Systems Technology/EST, Cheshire, CT.*

On-Hook. To disconnect from the public switched telephone network.

When a telephone is returned to the receiver, it is considered on-hook. Some transmission technologies accomplish this function automatically.

Open Area Detection (Protection). Protection of an area such as a room or space with detectors to provide early warning of fire.

Operating Mode, Private. Audible or visible signaling only to those persons directly concerned with the implementation and direction of emergency action initiation and procedure in the area protected by the fire alarm system.

At some locations, the private operating mode is used to alert individuals in supervising stations, such as a switchboard operators' area, and the offices of the building engineer, plant manager, boiler room operators, fire brigade leaders or other specially trained personnel, rather than the public-at- large. This alarm mode may also be used in the nursing stations of a health care facility or the guard station in either the lobby of an office building or at the main gate of an industrial facility.

Operating Mode, Public. Audible or visible signaling to occupants or inhabitants of the area protected by the fire alarm system.

This type of signaling might be used in locations such as offices, conference rooms, classrooms, lobbies, hallways, shop areas and gyms where total evacuation of or relocation from an area or building is necessary.

Operating System Software. The basic operating system software that is alterable only by the equipment manufacturer or its authorized representative. This software is sometimes referred to as "firmware," "BIOS," or "executive program."

Operating system software is similar to the main system operating software used in computers. This definition was added in the 1996 edition to provide a better understanding of the different software types. See Section 7-1.6.2.1.

Ownership. Any property or building or its contents under legal control by the occupant, by contract, or by holding of a title or deed.

Paging System. A system intended to page one or more persons by such means as voice over loudspeaker, coded audible signals or visible signals, or lamp annunciators.

See Figures 1.24 and 1.25.

Parallel Telephone System. A telephone system in which an individually wired circuit is used for each fire alarm box.

Path (Pathways). Any conductor, optic fiber, radio carrier, or other means for transmitting fire alarm system information between two or more locations.

Permanent Visual Record (Recording). An immediately readable, not easily alterable, print, slash, or punch record of all occurrences of status change.

Photoelectric Light Obscuration Smoke Detection.* The principle of utilizing a light source and a photosensitive sensor onto which the principal portion of the source emissions is focused. When smoke particles enter the light path, some of the light is scattered and some is absorbed, thereby reducing the light reaching the receiving sensor. The light reduction signal is processed and used to convey an alarm condition when it meets preset criteria.

See Figure 1.26.

A-1-4 Photoelectric Light Obscuration Smoke Detection.
The response of photoelectric light obscuration smoke detectors is usually not affected by the color of smoke.
Smoke detectors utilizing the light obscuration principle are usually of the line type. These detectors are commonly referred to as "projected beam smoke detectors."

Photoelectric-type smoke detectors react only to visible smoke. These devices are used by designers when a smoldering fire scenario is expected. This type of detector is slightly better at detecting the larger particles produced by smoldering fires. In locations where flaming fires are more likely to occur, better protection may be offered by ionization smoke detectors.

Photoelectric Light-Scattering Smoke Detection.* The principle of utilizing a light source and a photosensitive sensor arranged in a manner so that the rays from the light source do not normally fall onto the photosensitive sensor. When smoke particles enter the light path, some of the light is scattered by reflection and refraction onto the sensor. The light signal is processed and used to convey an alarm condition when it meets preset criteria.

See Figure 1.27.

Figure 1.24   *Fire fighter using paging system. Photograph courtesy of Simplex Time Recorder Co., Gardner, MA.*

A-1-4 **Photoelectric Light-Scattering Smoke Detection.** Photoelectric light-scattering smoke detection is more responsive to visible particles (larger than 1 micron in size) produced by most smoldering fires. It is somewhat less responsive to the smaller particles typical of most flaming fires. It is also less responsive to black smoke than to lighter colored smoke. Smoke detectors utilizing the light-scattering principle are usually of the spot type.

Plant.   One or more buildings under the same ownership or control on a single property.

Positive Alarm Sequence.   An automatic sequence that results in an alarm signal, even when manually delayed for investigation, unless the system is reset.

Power Supply.   A source of electrical operating power including the circuits and terminations connecting it to the dependent system components.

Primary Battery (Dry Cell).   A nonrechargeable battery requiring periodic replacement.

Primary Trunk Facility.   That part of a transmission channel connecting all leg facilities to a supervising or subsidiary station.

Prime Contractor.   The one company contractually responsible for providing central station services to a subscriber as required by this code. This can be either a listed central station or a listed fire alarm service-local company.

Private Radio Signaling.   A radio system under control of the proprietary supervising station.

Projected Beam-Type Detector.   A type of photoelectric light obscuration smoke detector wherein the beam spans the protected area.

See Figure 1.28.

Proprietary Supervising Station.   A location to which alarm or supervisory signaling devices on proprietary fire alarm

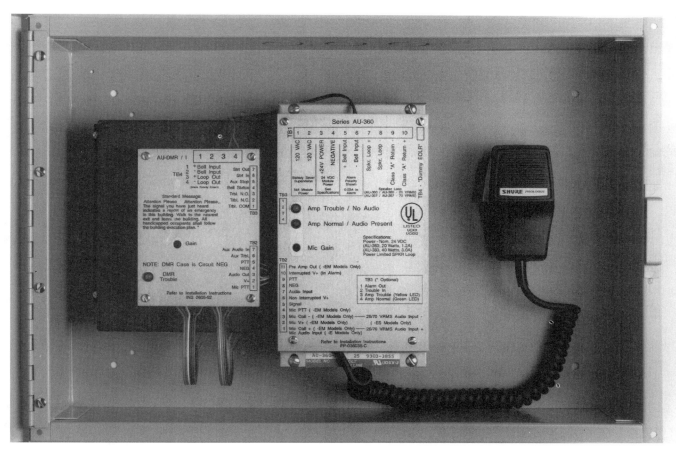

Figure 1.25   *Typical single channel paging system. Photograph courtesy of Audiosone Corp., Stratford, CT.*

systems are connected and where personnel are in attendance at all times to supervise operation and investigate signals.

See Figure 1.29.

Proprietary Supervising Station Fire Alarm System.   An installation of fire alarm systems that serves contiguous and noncontiguous properties, under one ownership, from a proprietary supervising station located at the protected property, at which trained, competent personnel are in constant attendance. This includes the proprietary supervising station; power supplies; signal-initiating devices; initiating device circuits; signal notification appliances; equipment for the automatic, permanent visual recording of signals; and equipment for initiating the operation of emergency building control services.

Protected Premises.   The physical location protected by a fire alarm system.

Protected Premises (Local) Control Unit (Panel).   A control unit that serves the protected premises or a portion of

the protected premises and indicates the alarm via notification appliances inside the protected premises.

Protected Premises (Local) Fire Alarm System.   A protected premises system that sounds an alarm at the protected premises as the result of the manual operation of a fire alarm box or the operation of protection equipment or systems, such as water flowing in a sprinkler system, the discharge of carbon dioxide, the detection of smoke, or the detection of heat.

Public Fire Service Communications Center.   The building or portion of the building used to house the central operating part of the fire alarm system; usually the place where the necessary testing, switching, receiving, transmitting, and power supply devices are located.

In a municipal system, the public fire service communication center is usually at the main fire station.

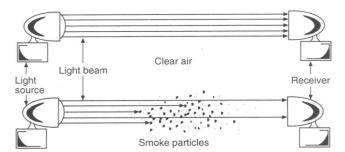

Figure 1.26 *Principle of operation for photoelectric light obscuration smoke detector.*

Figure 1.28 *Projected Beam Detector. Photograph courtesy of System Sensor, St. Charles, IL.*

Public Switched Telephone Network. An assembly of communications facilities and central office equipment operated jointly by authorized common carriers that provides the general public with the ability to establish communications channels via discrete dialing codes.

Radio Alarm Repeater Station Receiver (RARSR). A system component that receives radio signals. This component resides at a repeater station that is located at a remote receiving location.

Radio Alarm Supervising Station Receiver (RASSR). A system component that receives data and annunciates that data at the supervising station.

Radio Alarm System (RAS). A system in which signals are transmitted from a radio alarm transmitter (RAT) located at a protected premises through a radio channel to two or more radio alarm repeater station receivers (RARSR) and are annunciated by a radio alarm supervising station receiver (RASSR) located at the central station.

Radio Alarm Transmitter (RAT). A system component at the protected premises to which initiating devices or groups of devices are connected. The RAT transmits signals indicating a status change of the initiating devices.

Radio Channel. A band of frequencies of a width sufficient to allow its use for radio communications.

NOTE: The width of the channel depends on the type of transmissions and the tolerance for the frequency of emission. Channels normally are allocated for radio transmission in a specified type for service by a specified transmitter.

Rate Compensation Detector.* A device that responds when the temperature of the air surrounding the device reaches a predetermined level, regardless of the rate of temperature rise.

See Figure 1.30.

A-1-4 **Rate Compensation Detector.** A typical example is a spot-type detector with a tubular casing of a metal that tends to expand lengthwise as it is heated and an associated contact mechanism that closes at a certain point in the elongation. A second metallic element inside the tube exerts an opposing force on the contacts, tending to hold them open. The forces are balanced in such a way that, on a slow rate-of-temperature rise, there is more time for heat to penetrate to the inner element, which inhibits contact closure until the total device has been heated to its rated temperature level. However, on a fast rate-of-temperature rise, there is not as much time for heat to penetrate to the inner element, which exerts less of an inhibiting effect so that contact closure is achieved when the total device has been heated to a lower temperature. This, in effect, compensates for thermal lag.

Rate-of-Rise Detector.* A device that responds when the temperature rises at a rate exceeding a predetermined value.

A-1-4 **Rate-of-Rise Detector.** Typical examples of rate-of-rise detectors follow.

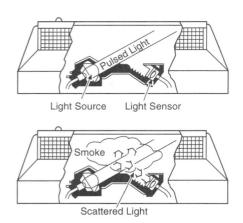

Figure 1.27 *Principle of operation for photoelectric light scattering smoke detector (Top: Clean air; Bottom: With smoke).*

Figure 1.29 Proprietary supervising station. Photograph courtesy of Simplex Time Recorder Co., Gardner, MA.

(a) *Pneumatic Rate-of-Rise Tubing.* A line-type detector comprising small-diameter tubing, usually copper, that is installed on the ceiling or high on the walls throughout the protected area. The tubing is terminated in a detector unit containing diaphragms and associated contacts set to actuate at a predetermined pressure. The system is sealed except for calibrated vents that compensate for normal changes in temperature.

(b) *Spot-Type Pneumatic Rate-of-Rise Detector.* A device consisting of an air chamber, a diaphragm, contacts, and a compensating vent in a single enclosure. The principle of operation is the same as that described for pneumatic rate-of-rise tubing.

(c) *Thermoelectric Effect Detector.* A device whose sensing element comprises a thermocouple or thermopile unit that produces an increase in electric potential in response to an increase in temperature. This potential is monitored by associated control equipment, and an alarm is initiated when the potential increases at an abnormal rate.

(d) *Electrical Conductivity-Type Rate-of-Rise Detector.* A line-type or spot-type sensing element whose resistance changes due to a change in temperature. The rate of change of resistance is monitored by associated control equipment, and an alarm is initiated when the rate of temperature increase exceeds a preset value.

Record Drawings. Drawings (as-built) that document the location of all devices, appliances, wiring sequences, wiring methods, and connections of the components of the fire alarm system as installed.

Record drawings provide information essential to test and maintain the system. These drawings must be prepared for all installations and must be revised as the system changes.

Record of Completion. A document that acknowledges the features of installation, operation (performance), service, and equipment with representation by the property owner, system installer, system supplier, service organization, and the authority having jurisdiction.

Formerly called the Certificate of Completion, the Record of Completion is used to confirm that a system has been installed properly and operates in conformance with this Code. Without a record of completion, there is no way for the authority having jurisdiction to warrant that the

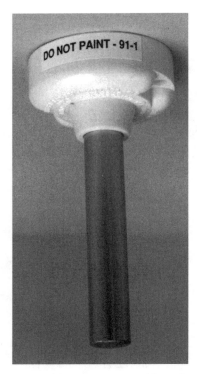

Figure 1.30   *Typical rate-compensated heat detector—*
vertical mounting. Photograph courtesy of Mammoth Fire
Alarms (Thermotech), Lowell, MA.

system was properly installed. Please refer to Figure
1-7.2.1

Relocation. The movement of occupants from a fire
zone to a safe area within the same building.

In hospitals, high-rise buildings or large facilities
where it is impossible to evacuate all occupants, occupants in the fire zone are usually relocated to a safe area
in a separate wing or more protected section of the
building.

Remote Supervising Station Fire Alarm System. A
system installed in accordance with this code to transmit
alarm, supervisory, and trouble signals from one or more protected premises to a remote location at which appropriate
action is taken.

Repeater Facility. Equipment needed to relay signals
between supervisory stations, subsidiary stations, and protected premises.

Repeater Station. The location of the equipment needed
for a repeater facility.

Reset. A control function that attempts to return a system
or device to its normal, nonalarm state.

Restorable Initiating Device. A device whose sensing
element is not ordinarily destroyed in the process of operation. Restoration can be manual or automatic.

Runner. A person other than the required number of
operators on duty at central, supervising, or runner stations
(or otherwise in contact with these stations) available for
prompt dispatching, when necessary, to the protected premises.

It is intended by the Code that the runner be qualified
to perform the required duties at the protected premises,
such as resetting equipment, investigating alarm signals,
and taking corrective action when necessary. This usually requires employing runners with an in-depth knowledge of the systems and equipment within the protected
premises or providing extensive system training to runners without such prior experience.

Runner Service. The service provided by a runner at the
protected premises, including resetting and silencing of all
equipment transmitting fire alarm or supervisory signals to an
off-premises location.

Runner service is generally provided as part of central
station service. A runner is sent to the protected premises
from which the signal was transmitted and takes appropriate action as outlined in the commentary above.

Satellite Trunk. A circuit or path connecting a satellite to
its central or proprietary supervising station.

Scanner. Equipment located at the telephone company
wire center that monitors each local leg and relays status
changes to the alarm center. Processors and associated
equipment might also be included.

Secondary Trunk Facility. That part of a transmission
channel connecting two or more, but fewer than all, leg facilities to a primary trunk facility.

Separate Sleeping Area. An area of the family living unit
in which the bedrooms (or sleeping rooms) are located. Bedrooms (or sleeping rooms) separated by other use areas,
such as kitchens or living rooms (but not bathrooms), are considered as separate sleeping areas.

Shall. Indicates a mandatory requirement.

Shapes of Ceilings. The shapes of ceilings are classified as follows:

Sloping Ceiling. A ceiling having a slope of more than $1^{1}/_{2}$ in./ft (41.7 mm/m). Sloping ceilings are further classified as follows:

Sloping-Peaked Type. A ceiling in which the ceiling slopes in two directions from the highest point. Curved or domed ceilings can be considered peaked with the slope figured as the slope of the chord from highest to lowest point. *See Figure A-5-2.4.4.1.*

Sloping-Shed Type. A ceiling in which the high point is at one side with the slope extending toward the opposite side. *See Figure A-5-2.4.4.2.*

Smooth Ceiling. A ceiling surface uninterrupted by continuous projections, such as solid joists, beams, or ducts, extending more than 4 in. (100 mm) below the ceiling surface.

NOTE: Open truss constructions are not considered to impede the flow of fire products unless the upper member in continuous contact with the ceiling projects below the ceiling more than 4 in. (100 mm).

Should. Indicates a recommendation or that which is advised but not required.

Signal. A status indication communicated by electrical or other means.

Signaling Line Circuit. A circuit or path between any combination of circuit interfaces, control units, or transmitters over which multiple system input signals or output signals, or both, are carried.

Signaling Line Circuit Interface. A system component that connects a signaling line circuit to any combination of initiating devices, initiating device circuits, notification appliances, notification appliance circuits, system control outputs, and other signaling line circuits.

This type of equipment may also be called a *"transponder"* or *"data-gathering panel."* See Figure 1.31.

Signal Transmission Sequence. A DACT that obtains dial tone, dials the number(s) of the DACR, obtains verification that the DACR is ready to receive signals, transmits the signals, and receives acknowledgment that the DACR has accepted that signal before disconnecting (going on-hook).

Single Station Alarm. A detector comprising an assembly incorporating a sensor, control components, and an alarm notification appliance in one unit operated from a power source either located in the unit or obtained at the point of installation.

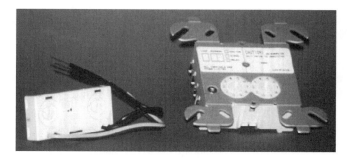

Figure 1.31 *Typical signaling line circuit interfaces. Photograph courtesy R. P. Schifiliti & Associates, Reading, MA.*

A single station alarm may be powered by a battery, an AC power source, or both (AC with battery back-up). See commentary following the definition of Multiple Station Alarm in this section.

Single Station Alarm Device. An assembly incorporating the detector, the control equipment, and the alarm-sounding device in one unit operated from a power supply either in the unit or obtained at the point of installation.

Site-Specific Software. Software that defines the specific operation and configuration of a particular system. Typically, it defines the type and quantity of hardware modules, customized labels, and specific operating features of a system.

Smoke Alarm. A single or multiple station alarm responsive to smoke.

Solid Joist Construction. Refers to ceilings having solid structural or solid nonstructural members projecting down from the ceiling surface for a distance of more than 4 in. (100 mm) and spaced at intervals 3 ft (0.9 m) or less, center to center.

Spacing. A horizontally measured dimension related to the allowable coverage of fire detectors.

Spacing refers to the distance, required by the Code, between each initiating device (according to the device's listing or manufacturer's guidelines) or between each notification appliance.

Spark.* A moving ember.

A-1-4 **Spark.** The overwhelming majority of applications involving the detection of Class A and Class D combustibles with radiant energy-sensing detectors involve the transport of particulate solid materials through pneumatic conveyor ducts or mechanical conveyors. It is common in the industries that include such hazards to refer to a moving piece of burning material as a "spark" and to systems for the detection of such fires as "spark detection systems."

Spark/Ember Detector. A radiant energy fire detector that is designed to detect sparks or embers, or both. These devices are normally intended to operate in dark environments and in the infrared part of the spectrum.

Spark/Ember Detector Sensitivity. The number of watts (or the fraction of a watt) of radiant power from a point source radiator, applied as a unit step signal at the wavelength of maximum detector sensitivity, necessary to produce an alarm signal from the detector within the specified response time.

Spot-Type Detector. A device whose detecting element is concentrated at a particular location. Typical examples are bimetallic detectors, fusible alloy detectors, certain pneumatic rate-of-rise detectors, certain smoke detectors, and thermo-electric detectors.

Story. The portion of a building included between the upper surface of a floor and the upper surface of the floor or roof next above.

Stratification. The phenomenon where the upward movement of smoke and gases ceases due to the loss of buoyancy.

Stratification can prevent smoke and fire gasses from reaching the installed detection device(s). Care must be exercised when installing detection devices in areas subject to this phenomenon such as an atrium. See Section 5-1.4.5. for more details on this phenomenon and detection placement.

Subscriber. The recipient of contractual supervising station signal service(s). In case of multiple, noncontiguous properties having single ownership, the term refers to each protected premises or its local management.

Subsidiary Station. A subsidiary station is a normally unattended location that is remote from the supervising station and linked by a communications channel(s) to the supervising station. Interconnection of signals on one or more transmission channels from protected premises with a communications channel(s) to the supervising station is performed at this location.

Supervising Station. A facility that receives signals and at which personnel are in attendance at all times to respond to these signals.

Supervisory Service. The service required to monitor performance of guard tours and the operative condition of

fixed suppression systems or other systems for the protection of life and property.

Supervisory Signal. A signal indicating the need of action in connection with the supervision of guard tours, the fire suppression systems or equipment, or the maintenance features of related systems.

Supervisory Signal-Initiating Device. An initiating device such as a valve supervisory switch, water level indicator, or low-air pressure switch on a dry-pipe sprinkler system whose change of state signals an off-normal condition and its restoration to normal of a fire protection or life safety system; or a need for action in connection with guard tours, fire suppression systems or equipment, or maintenance features of related systems.

Supplementary. As used in this code, supplementary refers to equipment or operations not required by this code and designated as such by the authority having jurisdiction.

Equipment included as part of the fire alarm system, but not required by the Code, is referred to as supplementary. The authority having jurisdiction, as defined in this section, must designate a system or system device in writing as supplementary. This additional requirement is designed to avoid having the installer or supplier simply label a device or operation as supplementary.

Switched Telephone Network. An assembly of communications facilities and central office equipment operated jointly by authorized service providers that provides the general public with the ability to establish transmission channels via discrete dialing.

System Unit. The active subassemblies at the central station utilized for signal receiving, processing, display, or recording of status change signals; a failure of one of these subassemblies causes the loss of a number of alarm signals by that unit.

Transmission Channel. A circuit or path connecting transmitters to supervising stations or subsidiary stations on which signals are carried.

Transmitter. A system component that provides an interface between signaling line circuits, initiating device circuits, or control units and the transmission channel.

Transponder. A multiplex alarm transmission system functional assembly located at the protected premises.

Trouble Signal. A signal initiated by the fire alarm system or device indicative of a fault in a monitored circuit or component.

WATS (Wide Area Telephone Service). Telephone company service allowing reduced costs for certain telephone call arrangements; it can be in-WATS or 800-number service where calls can be placed from anywhere in the continental U.S. to the called party at no cost to the calling party, or out-WATS, a service whereby, for a flat-rate charge, dependent on the total duration of all such calls, a subscriber can make an unlimited number of calls within a prescribed area from a particular telephone terminal without the registration of individual call charges.

Wavelength.* The distance between the peaks of a sinusoidal wave. All radiant energy can be described as a wave having a wavelength. Wavelength serves as the unit of measure for distinguishing between different parts of the spectrum. Wavelengths are measured in microns (µM), nanometers (nM), or angstroms (Å).

A-1-4 **Wavelength.** The concept of wavelength is extremely important in selecting the proper detector for a particular application. There is a precise interrelation between the wavelength of light being emitted from a flame and the combustion chemistry producing the flame. Specific sub-atomic, atomic, and molecular events yield radiant energy of specific wavelengths. For example, ultraviolet photons are emitted as the result of the complete loss of electrons or very large changes in electron energy levels. During combustion, molecules are violently torn apart by the chemical reactivity of oxygen, and electrons are released in the process, recombining at drastically lower energy levels, thus giving rise to ultraviolet radiation. Visible radiation is generally the result of smaller changes in electron energy levels within the molecules of fuel, flame intermediates, and products of combustion. Infrared radiation comes from the vibration of molecules or parts of molecules when they are in the superheated state associated with combustion. Each chemical compound exhibits a group of wavelengths at which it is resonant. These wavelengths constitute the chemical's infrared spectrum, which is usually unique to that chemical.

This interrelationship between wavelength and combustion chemistry affects the relative performance of various types of detectors with respect to various fires.

Wireless Protection System. A system or a part of a system that can transmit and receive signals without the aid of wire. It may consist of any of the following components:

These new definitions apply to systems covered by Section 3-13.

Wireless Control Panel. A component that transmits/receives and processes wireless signals.

Wireless Repeater. A component used to relay signals between wireless receivers or wireless control panels, or both.

Zone. A defined area within the protected premises. A zone can define an area from which a signal can be received, an area to which a signal can be sent, or an area in which a form of control can be executed.

1-5 Fundamentals

1-5.1 **Common System Fundamentals.** The provisions of this chapter shall apply to Chapters 3 through 7.

The basic requirements for all fire alarm systems, except household fire warning systems, are contained in Chapter 1. Chapter 2, Household Fire Warning Equipment, is a stand-alone chapter with specific requirements for detection in residential family occupancies such as single-family homes, apartments, and condominiums.

1-5.1.1 The provisions of this chapter cover the basic functions of a complete fire alarm system. These systems are primarily intended to provide notification of fire alarm, supervisory, and trouble conditions, alert the occupants, summon appropriate aid, and control fire safety functions.

1-5.1.2 **Equipment.** Equipment constructed and installed in conformity with this code shall be listed for the purpose for which it is used.

It is required that fire alarm products be listed for the fire alarm system applications for which they are used. Because fire alarm systems are used for life safety and property protection, the listing requirements are often more stringent than for those products listed for electrical safety only.

Equipment listings generally contain information pertaining to the permitted use, required ambient conditions in the installed location, mounting orientation, voltage tolerances, etc. Equipment must be installed in conformance with the listing to meet the requirements of the Code.

1-5.1.3 **System Design.** Fire alarm system plans and specifications shall be developed in accordance with this code by persons experienced in the proper design, application, installation, and testing of fire alarm systems.

The Code now requires that fire alarm system designers must be qualified to perform this type of work through training, education and experience.

1-5.2   Power Supplies.

1-5.2.1   Scope. The provisions of this section apply to power supplies used for fire alarm systems.

1-5.2.2   Code Conformance. All power supplies shall be installed in conformity with the requirements of NFPA 70, *National Electrical Code*, for such equipment and with the requirements indicated in this subsection.

1-5.2.3   Power Sources. Fire alarm systems shall be provided with at least two independent and reliable power supplies, one primary and one secondary (standby), each of which shall be of adequate capacity for the application.

Previous editions of NFPA 72 required three sources of power: primary, secondary (standby), and trouble. The requirement for a trouble signal power supply had an exception allowing the secondary power source to be used as the trouble signal power source. Because the majority of fire alarm systems were designed in conformance with this exception, the requirement for a separate trouble power source was eliminated from the Code.

Exception No. 1: Where the primary power is supplied by a dedicated branch circuit of an emergency system in accordance with NFPA 70, National Electrical Code, Article 700, or a legally required standby system in accordance with NFPA 70, National Electrical Code, Article 701, a secondary supply shall not be required.

Exception No. 2: Where the primary power is supplied by a dedicated branch circuit of an optional standby system in accordance with NFPA 70, National Electrical Code, Article 702, which also meets the performance requirements of Article 700 or Article 701, a secondary supply shall not be required.

The exceptions permit a total elimination of the secondary power supply, where these requirements are satisfied.

NOTE to 1-5.2.3, Exceptions No. 1 and No. 2:   A trouble signal is not required where operating power is being supplied by either of the two sources of power indicated in Exceptions No. 1 and No. 2, if they are capable of providing the hours of operation required by 1-5.2.6 and loss of primary power is otherwise indicated (e.g., loss of building lighting).

This is a very important note, since the normal hours of operation for an emergency system meeting the requirements of Article 701 is $1\frac{1}{2}$ hours, and the optional standby system meeting the requirements of Article 702 does not specify an operating time.

Formal Interpretation 85-13
Reference: 1-5.2.3

Background: In reading the 1996 code under NFPA 72, 1-5.2.3, Exception No. 1 stipulates that fire alarm system power supply can be fed from the emergency generator of an emergency system or a legally required standby system.

Question 1: When Exception No. 1 is complied with, are the requirements of 1-5.2.10.4 also required?
Answer: Yes, for generator option.

Question 2: When Exception No. 1 is complied with do the requirements for a 2-hour fuel supply stated in NFPA 70, Section 700-12(b)(2) apply?
Answer: No, 1-5.2.10.4 is more stringent.

Issue Edition: NFPA 72A-1985
Reference: 2-3.2.1, 2-3.4.1
Issue Date: November 1986 ■

Where dc voltages are employed, they shall be limited to no more than 350 volts above earth ground.

1-5.2.4   Primary Supply. The primary supply shall have a high degree of reliability, shall have adequate capacity for the intended service, and shall consist of one of the following:

(a) Light and power service arranged in accordance with 1-5.2.5;

(b) Where a person specifically trained in its operation is on duty at all times, an engine-driven generator or equivalent arranged in accordance with 1-5.2.10.

An engine-driven generator is permitted as a primary power supply because light and power service may not be available at all locations.

1-5.2.5   Light and Power Service.

1-5.2.5.1 A light and power service employed to operate the system under normal conditions shall have a high degree of reliability and capacity for the intended service. This service shall consist of one of the following:

(a) **Two-Wire Supplies.** A two-wire supply circuit shall be permitted to be used for either the primary operating power supply or the trouble signal power supply of the signaling system.

(b) **Three-Wire Supplies.** A three-wire ac or dc supply circuit having a continuous unfused neutral conductor, or a polyphase ac supply circuit having a continuous unfused neutral conductor where interruption of one phase does not prevent operation of the other phase, shall be permitted to be used with one side or phase for the primary operating power supply and the other side or phase for the trouble signal power supply of the fire alarm system.

1-5.2.5.2 Connections to the light and power service shall be on a dedicated branch circuit(s). The circuit(s) and connections shall be mechanically protected. Circuit disconnecting means shall have a red marking, shall be accessible only to authorized personnel, and shall be identified as "FIRE ALARM CIRCUIT CONTROL." The location of the circuit disconnecting means shall be permanently identified at the fire alarm control unit.

These requirements are intended to protect the power supply from tampering, to ensure reliability, to aid in troubleshooting, and to help ensure the safety of those who service the equipment.

1-5.2.5.3 Overcurrent Protection. An overcurrent protective device of suitable current-carrying capacity and capable of interrupting the maximum short-circuit current to which it may be subject shall be provided in each ungrounded conductor. The overcurrent protective device shall be enclosed in a locked or sealed cabinet located immediately adjacent to the point of connection to the light and power conductors.

Connections ahead of the main disconnecting apparatus can be dangerous because, when improperly installed, these connections can allow high fault currents to be carried on the primary supply conductors for the fire alarm system. NFPA 70, *National Electrical Code®*, provides requirements for proper installation. Proper connections must be made using service equipment in accordance with Section 230-82 Exception No. 5. The service equipment to interrupt the fault current must be listed.

Circuit breaker locks are permitted, provided they are listed for use with the circuit breaker. These locks allow the breaker to trip, but do not allow tampering.

1-5.2.5.4 Circuit breakers or engine stops shall not be installed in such a manner as to cut off the power for lighting or for operating elevators.

1-5.2.6 Secondary Supply Capacity and Sources. The secondary supply shall automatically supply the energy to the system within 30 seconds, and without loss of signals, wher-ever the primary supply is incapable of providing the minimum voltage required for proper operation. The secondary (standby) power supply shall supply energy to the system in the event of total failure of the primary (main) power supply or when the primary voltage drops to a level insufficient to maintain functionality of the control equipment and system components. Under maximum normal load, the secondary supply shall have sufficient capacity to operate a protected premises, central station, or proprietary system for 24 hours, or an auxiliary or remote station system for 60 hours; and, at the end of that period, shall be capable of operating all alarm notification appliances used for evacuation or to direct aid to the location of an emergency for 5 minutes. The secondary power supply for emergency voice/alarm communications service shall be capable of operating the system under maximum normal load for 24 hours and then shall be capable of operating the system during a fire or other emergency condition for a period of 2 hours. Fifteen minutes of evacuation alarm operation at maximum connected load shall be considered the equivalent of 2 hours of emergency operation.

The proper amount of battery standby can be calculated. These calculations should include the standby supervisory time as well as the 5-minute alarm time. Where combination systems are used, the secondary supply must be able to power the entire system for the required 24-hour period. Other loads, such as security or building management systems, must be figured into the secondary power calculations.

The emergency voice/alarm communication system must be capable of operating for two (2) hours during the emergency condition because communication to the relocated occupants is necessary.

Central stations and proprietary supervising stations are required to receive trouble signals (produced upon loss of primary power). Runners are sent to investigate and correct the problem, usually within a few hours, depending upon the severity of the problem. Auxiliary and remote supervising stations do not necessarily receive trouble signals. In this case, no runners will be sent to correct the problem. Therefore, it is possible that a protected premises could be left without fire alarm protection for an extended period. For this reason, these systems are required to be powered for a minimum of 60 hours.

The secondary supply shall consist of one of the following:

(a) A storage battery arranged in accordance with 1-5.2.9;

(b) An automatic starting, engine-driven generator arranged in accordance with 1-5.2.10 and storage batteries with 4 hours of capacity arranged in accordance with 1-5.2.9;

(c) Multiple engine-driven generators, one of which is arranged for automatic starting, arranged in accordance with 1-5.2.10, and capable of supplying the energy required herein, with the largest generator out of service. The second generator shall be permitted to be started by pushbutton.

The Code permits pushbutton starting means, but key starting generators are not permitted because keys can be lost.

Operation on secondary power shall not affect the required performance of a fire alarm system. The system shall produce the same alarm, supervisory, and trouble signals and indications (excluding the ac power indicator) when operating from the standby power source as are produced when the unit is operating from the primary power source.

Manufacturers have supplied systems in the past that, in order to save battery power, eliminated annunciation and some supplementary functions when in the standby power mode. The Code prevents this practice by requiring the system to operate with all of the same features as when it is powered by the primary power source.

1-5.2.7   Continuity of Power Supplies.

(a) Where signals could be lost on transfer of power between the primary and secondary sources, rechargeable batteries of sufficient capacity to operate the system under maximum normal load for 15 minutes shall assume the load in such a manner that no signals are lost where either of the following conditions exists:

1. Secondary power is supplied in accordance with 1-5.2.6(a) or 1-5.2.6(b), and the transfer is made manually; or

2. Secondary power is supplied in accordance with 1-5.2.6(c).

(b) Where signals will not be lost due to transfer of power between the primary and secondary sources, one of the following arrangements shall be made:

1. The transfer shall be automatic.

2. Special provisions shall be made to allow manual transfer within 30 seconds of loss of power.

3. The transfer shall be arranged in accordance with 1-5.2.6(a).

(c)* Where a computer system of any kind or size is used to receive or process signals, an uninterruptible power supply (UPS) with sufficient capacity to operate the system for at least 15 minutes, or until the secondary supply is capable of

supplying the UPS input power requirements, shall be required where either of the following conditions apply:

1. The status of signals previously received will be lost upon loss of power.

2. The computer system cannot be restored to full operation within 30 seconds of loss of power.

A-1-5.2.7(c)   An engine-driven generator without standby battery supplement should not be assumed to be capable of a reliable power transfer within 30 seconds of a primary power loss.

(d)* A positive means for disconnecting the input and output of the UPS system while maintaining continuity of power supply to the load shall be provided.

A-1-5.2.7(d)   UPS equipment often contains an internal bypass arrangement to supply the load directly from the line. These internal bypass arrangements are a potential source of failure. UPS equipment also requires periodic maintenance. It is, therefore, necessary to provide a means of promptly and safely bypassing and isolating the UPS equipment from all power sources while maintaining continuity of power supply to the equipment normally supplied by the UPS.

1-5.2.8   Power Supply for Remotely Located Control Equipment.

1-5.2.8.1   Additional power supplies, where provided for control units, circuit interfaces, or other equipment essential to system operation, located remote from the main control unit, shall be comprised of a primary and secondary power supply that shall meet the same requirements as those of 1-5.2.1 through 1-5.2.8 and 1-5.8.6.

1-5.2.8.2   Power supervisory devices shall be arranged so as not to impair the receipt of fire alarm or supervisory signals.

1-5.2.9*   Storage Batteries.

A-1-5.2.9   Rechargeable-(Storage-)Type Batteries.   The following newer types of rechargeable batteries are normally used in protected premises applications:

(a) *Vented Lead-Acid, Gelled, or Starved Electrolyte Battery.*   This rechargeable-type battery is generally used in place of primary batteries in applications having a relatively high current drain or requiring the extended standby capability of much lower currents. The nominal voltage of a single cell is 2 volts, and the battery is available in multiples of 2 volts (e.g., 2, 4, 6, 12). Batteries should be stored according to the manufacturer's recommendations.

(b) *Nickel-Cadmium Battery.*   The sealed-type nickel-cadmium battery generally used in applications where the battery current drain during a power outage is low to moderate (typically up to a few hundred milliamperes) and is fairly constant. Nickel-cadmium batteries are also available in much larger capacities for other applications. The nominal voltage per cell is 1.42 volts, with batteries available in multiples of 1.42 (e.g., 12.78, 25.56). Batteries in storage can be stored in any state of charge for indefinite periods. However, a battery in storage will lose capacity (will self-discharge), depending on storage time and temperature. Typically, batteries stored for more than 1 month require an 8-hour to 14-hour charge period to restore capacity. In service, the battery should receive a continuous constant charging current sufficient to keep it fully charged (typically, the charge rate equals $1/10$ to $1/20$ of the ampere-hour rating of the battery). Because batteries are made up of individual cells connected in series, the possibility exists that, during deep discharge, one or more cells that are low in capacity will reach complete discharge prior to other cells. The cells with remaining life tend to charge the depleted cells, causing a polarity reversal resulting in permanent battery damage. This condition can be determined by measuring the open cell voltage of a fully charged battery (voltage should be a minimum of 1.28 volts per cell multiplied by the number of cells). Voltage depression effect is a minor change in discharge voltage level caused by constant current charging below the system discharge rate.

In some applications of nickel-cadmium batteries (e.g., battery-powered shavers), a memory characteristic also exists. Specifically, where the battery is discharged daily for 1 minute, followed by a recharge, operation for 5 minutes will not result in the rated ampere-hour output. The reason for this is that the battery has developed a 1-minute discharge memory.

(c) *Sealed Lead-Acid Battery.* In a sealed lead-acid battery, the electrolyte is totally absorbed by the separators, and no venting normally occurs. Gas evolved during recharge is internally recombined, resulting in minimal loss of capacity life. A high pressure vent, however, is provided to avoid damage under abnormal conditions.

See Figure 1.32.

1-5.2.9.1 **Location.** Storage batteries shall be so located that the fire alarm equipment, including overcurrent devices, are not adversely affected by battery gases and shall conform to the requirements of NFPA 70, *National Electrical Code*, Article 480. Cells shall be suitably insulated against grounds and crosses and shall be mounted securely in such a manner as not to be subject to mechanical injury. Racks shall be suitably protected against deterioration. Where not located in or adjacent to the fire alarm control panel, the batteries and their charger location shall be permanently identified at the fire alarm control unit.

The new requirement for identification of the location of remotely-located batteries and/or chargers is intended to make system inspections and tests easier.

1-5.2.9.2 **Battery Charging.**

(a) Adequate facilities shall be provided to automatically maintain the battery fully charged under all conditions of normal operation and, in addition, to recharge batteries within 48

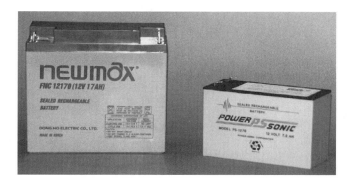

Figure 1.32 *Typical batteries used in a fire alarm system. Photograph courtesy of Mammoth Fire Alarms, Lowell, MA.*

hours after fully charged batteries have been subject to a single discharge cycle as specified in 1-5.2.6. Upon attaining a fully charged condition, the charge rate shall not be so excessive as to result in battery damage.

(b) Supervising stations shall maintain spare parts or units available, which shall be employed to restore failed charging capacity prior to the consumption of $1/2$ of the capacity of the batteries for the supervising station equipment.

(c)* Batteries shall be either trickle- or float-charged.

A-1-5.2.9.2(c) Batteries are trickle-charged where they are off-line and waiting to be put under load in the event of a loss of power.

Float-charged batteries are fully charged and connected across the output of the rectifiers to smooth the output and serve as a standby source of power in the event of a loss of line power.

(d) A rectifier employed as a battery charging supply source shall be of adequate capacity. A rectifier employed as a charging means shall be energized by an isolating transformer.

1-5.2.9.3 **Overcurrent Protection.** The batteries shall be protected against excessive load current by overcurrent devices having a rating not less than 150 percent and not more than 250 percent of the maximum operating load in the alarm condition. The batteries shall be protected from excessive charging current by overcurrent devices or by automatic current-limiting design of the charging source.

1-5.2.9.4 **Metering.** The charging equipment shall provide either integral meters or readily accessible terminal facilities for the connection of portable meters by which the battery voltage and charging current can be determined.

1-5.2.9.5 **Under-Voltage Detection.** An under-voltage detection device shall be provided to detect a failure of the charging source and initiate a trouble signal.

This requirement was part of the metering requirement of the 1989 edition of NFPA 71, *Standard for the Installation, Maintenance, and Use of Signaling Systems for Central Station Service*. The battery charging circuits of all systems are now required to be monitored and to produce a trouble signal upon failure.

1-5.2.10 **Engine-Driven Generator.**

1-5.2.10.1 The installation of engine-driven generators shall conform to the provisions of NFPA 110, *Standard for Emergency and Standby Power Systems*.

Exception: Where restricted by the provisions of this section.

1-5.2.10.2 **Capacity.** The unit shall be of a capacity sufficient to operate the system under the maximum normal load conditions in addition to all other demands placed upon the unit, such as those of emergency lighting.

1-5.2.10.3 **Fuel.** Fuel shall be stored in outside underground tanks wherever possible, and gravity feed shall not be used. Gasoline deteriorates with age. Where gasoline-driven generators are used, fuel shall be supplied from a frequently replenished "working" tank, or other means provided, to ensure that the gasoline is always fresh.

Gravity feed is not permitted because of the potential for fuel leaks, which may pose a fire hazard.

1-5.2.10.4 Sufficient fuel shall be available in storage for 6 months of testing plus the capacity specified in 1-5.2.6. (*For public fire alarm reporting systems, see 4-6.7.3.4.*)

Exception No. 1: Where a reliable source of supply is available at any time on 2-hours' notice, sufficient fuel shall be in storage for 12 hours of operation at full load.

Exception No. 2: Fuel systems using natural or manufactured gas supplied through reliable utility mains shall not be required to have fuel storage tanks unless located in seismic risk zone 3 or greater as defined in ANSI A-58.1, Building Code Requirements for Minimum Design Loads in Buildings and Other Structures.

1-5.2.10.5 A separate storage battery and separate automatic charger shall be provided for starting the engine-driven generator and shall not be used for any other purpose.

1-5.3 **Compatibility.**

1-5.3.1 All initiating devices, notification appliances, and control equipment constructed and installed in conformity with this code shall be listed for the purpose for which they are intended.

The fact that a component is listed is not sufficient for fire alarm application. The component must be tested and listed specifically for use in fire alarm systems.

1-5.3.2 All fire detection devices that receive their power from the initiating device circuit or signaling line circuit of a fire alarm control unit shall be listed for use with the control unit.

A two-wire smoke detector obtains its power from the control unit initiating device circuit. Therefore, it is mandatory that the smoke detector be listed for use with the control unit and its associated initiating device circuit.

The listing organizations have developed specific requirements for this listing process and should be consulted if there is any doubt as to the detector's compatibility with a specific control unit.

1-5.4 **System Functions.**

1-5.4.1 **Protected Premises Fire Safety Functions.**

1-5.4.1.1 Fire safety functions shall be permitted to be performed automatically. The performance of automatic fire safety functions shall not interfere with power for lighting or for operating elevators. This shall not preclude the combination of fire alarm services with other services requiring monitoring of operations.

1-5.4.1.2 The time delay between the activation of an initiating device and the automatic activation of a local fire safety function shall not exceed 90 seconds.

This section, not previously covered in the Code, was added to the 1996 edition because these systems are used for life safety. Critical functions, such as those described in the requirement, must occur immediately.

1-5.4.2 **Alarm Signals.**

1-5.4.2.1* **Coded Alarm Signal.** A coded alarm signal shall consist of not less than three complete rounds of the number transmitted, and each round shall consist of not less than three impulses.

Table A-1-5.4.2.1 Recommended Coded Signal Designations

Location	Coded Signal
4th floor	2-4
3rd floor	2-3
2nd floor	2-2
1st floor	2-1
Basement	3-1
Subasement	3-2

A-1-5.4.2.1 Coded Alarm Signal Designations. The following recommended coded signal designations for buildings having four floors and multiple basements are provided in Table A-1-5.4.2.1.

1-5.4.2.2 Actuation of alarm notification appliances or emergency voice communications shall occur within 90 seconds after the activation of an initiating device.

1-5.4.3 Supervisory Signals.

1-5.4.3.1 Coded Supervisory Signal. A coded supervisory signal shall be permitted to consist of two rounds of the number transmitted to indicate a supervisory off-normal condition, and one round of the number transmitted to indicate the restoration of the supervisory condition to normal.

1-5.4.3.2 Combined Coded Alarm and Supervisory Signal Circuits. Where both coded sprinkler supervisory signals and coded fire or waterflow alarm signals are transmitted over the same signaling line circuit, provision shall be made either to obtain alarm signal precedence or sufficient repetition of the alarm signal to prevent the loss of an alarm signal.

1-5.4.3.3 Visible and audible supervisory signals and visible indication of their restoration to normal shall be indicated within 90 seconds at the following locations:

(a) Control unit (central equipment) for local fire alarm systems;

(b) Building fire command center for emergency voice/ alarm communications systems;

(c) Supervising station location for systems installed in compliance with Chapter 4.

This new section was added in the 1996 edition and provides reporting requirements for supervisory signals. The 90-second requirement corresponds to alarm reporting requirements.

1-5.4.4 Fire alarms, supervisory signals, and trouble signals shall be distinctively and descriptively annunciated.

1-5.4.5* Where status indicators are required to be provided for emergency equipment or fire safety functions, they shall be arranged to reflect the actual status of the associated equipment or function accurately.

A-1-5.4.5 The operability of controlled mechanical equipment (e.g., smoke and fire dampers, elevator recall arrangements, door holders) should be verified by periodic testing. Failure to test and properly maintain controlled mechanical equipment can result in operational failure during an emergency, with potential consequences up to and including loss of life.

1-5.4.6 Trouble Signal.

1-5.4.6.1 General. Trouble signals and their restoration to normal shall be indicated within 200 seconds at the locations identified in 1-5.4.6.2 or 1-5.4.6.3. Trouble signals required to indicate at the protected premises shall be indicated by distinctive audible signals. These audible trouble signals shall be distinctive from alarm signals. Where an intermittent signal is used, it shall sound at least once every 10 seconds, with a minimum duration of $^1/_2$ second. An audible trouble signal shall be permitted to be common to several supervised circuits. The trouble signal(s) shall be located in an area where it is likely to be heard.

1-5.4.6.2 Visible and audible trouble signals and visible indication of their restoration to normal shall be indicated at the following locations:

(a) Control unit (central equipment) for local fire alarm systems;

(b) Building fire command center for emergency voice/ alarm communications systems;

(c) Central station or remote station location for systems installed in compliance with Chapter 4.

1-5.4.6.3 Trouble signals and their restoration to normal shall be visibly and audibly indicated at the proprietary supervising station for systems installed in compliance with Chapter 4.

1-5.4.6.4 Audible Trouble Signal Silencing Switch.

The word *switch* has been replaced by the word *means*, recognizing that it is possible to perform this function with alpha-numeric keypads or touch screens.

1-5.4.6.4.1   A means for silencing the trouble notification appliance(s) shall be permitted only where it is key-operated, located within a locked enclosure, or arranged to provide equivalent protection against unauthorized use. Such a means shall be permitted only where it transfers the trouble indication to a suitably identified lamp or other acceptable visible indicator. The visible indication shall persist until the trouble condition has been corrected. The audible trouble signal shall sound when the silencing means is in its silence position and no trouble exists.

1-5.4.6.4.2   Where an audible trouble notification appliance is also used to indicate a supervisory condition, as permitted in 1-5.4.7(b), a trouble signal silencing switch shall not prevent subsequent sounding of supervisory signals. An audible trouble signal that has been silenced at the protected premises shall automatically re-sound every 24 hours or less until fault conditions are restored to normal.

Trouble signals indicate a fault that may impair system operation. The Code now requires that a silenced trouble signal re-sound at least once every 24 hours until the source of the trouble signal has been identified and corrected by the system operator. This requirement was added to help ensure that trouble signals are not ignored.

1-5.4.7   **Distinctive Signals.**   Audible alarm notification appliances for a fire alarm system shall produce signals that are distinctive from other similar appliances used for other purposes in the same area. The distinction among signals shall be as follows:

(a)   Fire alarm signals shall be distinctive in sound from other signals, and their sound shall not be used for any other purpose. (*See 3-7.2.*)

(b)* Supervisory signals shall be distinctive in sound from other signals. Their sound shall not be used for any other purpose.

A-1-5.4.7(b)   A tamper switch, low pressure switch, or other device intended to cause a supervisory signal when actuated should not be connected in series with the end-of-line supervisory device of initiating device circuits, unless a distinctive signal, different from a trouble signal, is indicated.

Exception: A supervisory signal sound shall be permitted to be used to indicate a trouble condition. Where the same sound is used for both supervisory signals and trouble signals, the distinction between signals shall be by other appropriate means such as visible annunciation.

Formal Interpretation
Reference: 1-5.4.7

Statement: Is it proper, within the meaning of the code, to interconnect the gate valve signal with the trouble signal of the fire alarm system?

Question: Is it proper, within the meaning of the code, to utilize a common audible device to indicate a closed gate valve on a sprinkler system supervisory circuit, as well as to indicate trouble on a separate waterflow alarm circuit, with the understanding that there will be visual means for identifying the specific circuit involved?

Answer: Yes.

Question: Is it proper, within the meaning of the code, to interconnect the gate valve switch(es) on the waterflow alarm circuit so that a closed gate valve will be indicated as a trouble on the waterflow alarm circuit?

Answer: No.

Issue Edition: NFPA 72A-1972
Reference: 3610
Issue Date: June 1974   ■

(c)   Fire alarm, supervisory, and trouble signals shall take precedence, in that respective order of priority, over all other signals.

Exception: Signals from hold-up alarms or other life threatening signals shall be permitted to take precedence over supervisory and trouble signals where acceptable to the authority having jurisdiction.

The Code permits the use of different alarm signals throughout a protected premises. However, every effort should be made to use the same type of signal throughout a protected premises as to avoid confusion among the occupants. Some facilities are better suited for the use of different signals in different areas. For example, hospitals may use coded signals in patient care areas because of concerns for patient safety, and non-coded signals in other areas to notify all occupants of an alarm condition.

1-5.4.8   **Alarm Signal Deactivation.**   A means for turning off the alarm notification appliances shall be permitted only where it is key-operated, located within a locked cabinet, or arranged to provide equivalent protection against unautho-

Formal Interpretation 87-3

Reference: 1-5.4.7

Background: An initiating device circuit has a waterflow device and a valve supervisory device connected to it and by using current limiting techniques provides the distinctive signals (i.e., separate alarm, trouble and supervisory signals) required by NFPA 72, 1-5.4.7.

Question: Does this meet the intent of NFPA 72-1996, 1-5.4.7?

Answer: Yes.

Issue Edition: NFPA 72-1987
Reference: 2-8.5
Issue Date: April 30, 1990 ■

rized use. Such means shall be permitted only if a visible zone alarm indication or the equivalent has been provided as specified in 1-5.7.1 and subsequent alarms on other initiating devices or circuits cause the notification appliances to reactivate. A means that is left in the "off" position when there is no alarm shall operate an audible trouble signal until the means is restored to normal. Where automatically turning off the alarm notification appliances is permitted by the authority having jurisdiction, the alarm shall not be turned off in less than 5 minutes.

Exception: Where otherwise permitted by the authority having jurisdiction, the 5-minute requirement shall not apply.

The intent of this section of the Code is to allow the audible appliances to be silenced to facilitate communication by responders. This section clearly permits deactivation of *all* notification appliances for a time limited to no more than five minutes or to the time limit for deactivation determined by the authority having jurisdiction. In some cases, the alarm system would be designed to deactivate only the audible appliances while allowing the visible notification appliances to remain activated. This design would allow verbal communications among responders, but still provide conveyance of the fire alarm signal.

This section has been revised to include subsequent alarm operation of signaling line circuit (addressable) devices.

1-5.4.9 Supervisory Signal Silencing. A means for silencing a supervisory signal notification appliance(s) shall be permitted only where it is key-operated, located within a locked enclosure, or arranged to provide equivalent protection against unauthorized use. Such a means shall be permitted

only where it transfers the supervisory indication to a lamp or other visible indicator and subsequent supervisory signals in other zones cause the supervisory notification appliance(s) to re-sound. A means that is left in the "silence" position where there is no supervisory off-normal signal shall operate a visible signal silence indicator and cause the trouble signal to sound until the silencing means is restored to normal position.

See Figure 1.33.

1-5.4.10 Presignal Feature. Where permitted by the authority having jurisdiction, systems shall be permitted to have a feature that allows initial fire alarm signals to sound only in department offices, control rooms, fire brigade stations, or other constantly attended central locations and for which human action is subsequently required to activate a general alarm, or a feature that allows the control equipment to delay the general alarm by more than 1 minute after the start of the alarm processing. Where there is a connection to a remote location, it shall activate upon the initial alarm signal.

NOTE: A system provided with an alarm verification feature as permitted by 3-8.2.3 is not considered a presignal system, since the delay in the signal produced is 60 seconds or less and requires no human intervention.

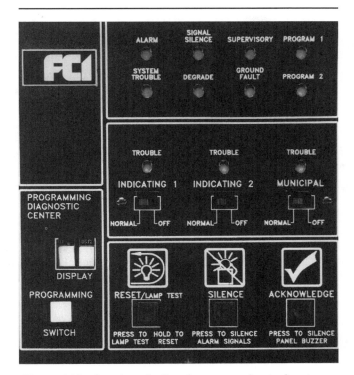

Figure 1.33   Interior of a fire alarm control unit showing alarm, supervisory, and trouble signal indications. Photograph courtesy of Fire Control Instruments, Inc., Newton, MA.

Formal Interpretation 85-9
Reference: 1-5.4.7, 3-8.6.3, 3-8.6.5

Question 1: Is it the intent of the Committee to prohibit the use of a dedicated closed loop circuit to which only normally closed supervisory switches (one or more) are connected where a break in the line or an off-normal supervisory switch produce the same signal at the control unit?
Answer: Yes.

Question 2: Is it the intent of the Committee to permit the use of the same audible signal for both a supervisory signal and a trouble signal?
Answer: Yes.

Question 3: Is it the intent of the Committee to permit silencing an audible supervisory signal?
Answer: Yes.

Question 4: Is it the intent of the Committee to permit an arrangement where silencing an audible trouble signal would prevent the receipt of the first (in case there are several supervisory circuits and the answer to Question 3 is "yes") audible supervisory signal?
Answer: No.

Question 5: Is it the intent of the Committee to prohibit the use of a common trouble signal silencing switch to silence both trouble and supervisory audible signals when operation of the switch to silence the audible signal caused by a trouble condition will prevent the receipt of an audible signal associated with a supervisory signal?
Answer: Yes.

Issue Edition: NFPA 72-1985
Reference: 2-5.5, 3-5.4.2, et al.
Issue Date: October 1985 ■

The remote location referred to in Section 1-5.4.10 is a supervising station or other location where signals are transmitted. Presignal systems rely on human action, which can be unreliable. NFPA 101, *Life Safety Code®*, provides guidance for using this feature. Because this feature delays the general alarm more than one minute, permission of the authority having jurisdiction is required. Caution is recommended when delaying alarm signals.

1-5.4.11 Positive Alarm Sequence.

Section 1-5.4.11 was formerly contained in Chapter 3. Because this material is similar to Presignal Feature in the preceding subsection, it has been relocated to this Chapter.

1-5.4.11.1 Systems having positive alarm features complying with the following shall be permitted where approved by the authority having jurisdiction.

Positive alarm sequence is simply a delayed alarm under specific controlled conditions. Before the advent of alarm verification, this method was used to minimize false alarms. The positive alarm sequence feature can only be used with approval of the authority having jurisdiction.

1-5.4.11.1.1 The signal from an automatic fire detection device selected for positive alarm sequence operation shall be acknowledged at the control unit by trained personnel within 15 seconds of annunciation in order to initiate the alarm investigation phase. If the signal is not acknowledged within 15 seconds, all building and remote signals shall be activated immediately and automatically.

1-5.4.11.1.2 Trained personnel shall have up to 180 seconds during the alarm investigation phase to evaluate the fire condition and reset the system. If the system is not reset during this investigation phase, all building and remote signals shall be activated immediately and automatically.

1-5.4.11.2 If a second automatic fire detector selected for positive alarm sequence is actuated during the alarm investigation phase, all normal building and remote signals shall be activated immediately and automatically.

Formal Interpretation 85-3

Reference: 1-5.4.9

Background: Previous interpretations by this Committee and actions by the membership at the fall of 1984 meeting still leave unclear the Committee's intent on the questions of permitted means to connect supervisory devices to fire alarm control units.

Question 1: If a control unit is arranged to sound the same audible signal for trouble indication as it does for a supervisory signal, is it the intent of the Committee that a supervisory device be permitted to be connected in such a manner that it is not possible to differentiate between an actuated supervisory device or an open circuit trouble condition on the same circuit?

Answer: No.

Question 2: If the answer to Question 1 is "no," would the answer be "yes" if the circuit involved was individually annunciated in some manner?

Answer: No.

Question 3: In a control unit arranged as in Question 1, is it the intent of the Committee to permit an audible trouble signal silencing switch to prevent subsequent sounding of audible supervisory signals?

Answer: No.

Question 4: In a control unit arranged as in Question 1, is it the intent of the Committee to permit an audible supervisory signal silencing switch to prevent subsequent sounding of audible trouble signals or supervisory signals?

Answer: Yes.

Issue Edition: NFPA 72A-1985
Reference: 2-5.5, 3-5.4.2, et al.
Issue Date: June 1985 ∎

1-5.4.11.3 If any other initiating device is actuated, all building and remote signals shall be activated immediately and automatically.

Subsections 1-5.4.11.1 through 1-5.4.11.3 help to eliminate the human unreliability factor from the use of the positive alarm sequence feature by requiring time limitations and automatic activation when additional initiating devices are activated.

1-5.4.11.4* The system shall provide means to bypass the positive alarm sequence.

A-1-5.4.11.4 The bypass means is intended to enable automatic or manual day/night/weekend operation.

1-5.5 Performance and Limitations.

1-5.5.1 Voltage, Temperature, and Humidity Variation. Equipment shall be designed so that it is capable of performing its intended functions under the following conditions:

(a)* At 85 percent and at 110 percent of the nameplate primary (main) and secondary (standby) input voltage(s);

A-1-5.5.1(a) This requirement does not preclude transfer to secondary supply at less than 85 percent of nominal primary voltage, provided the requirements of 1-5.2.6 are met.

(b) At ambient temperatures of 32°F (0°C) and 120°F (49°C);

(c) At a relative humidity of 85 percent and an ambient temperature of 86°F (30°C).

1-5.5.2 Installation and Design.

1-5.5.2.1* All systems shall be installed in accordance with the specifications and standards approved by the authority having jurisdiction.

A-1-5.5.2.1 Specifications. Fire alarm specifications can include some or all of the following:

(a) The address of the protected premises;

(b) The owner of the protected premises;

(c) The authority having jurisdiction;

(d) The applicable codes, standards, and other design criteria to which the system is required to comply;

(e) The type of building construction and occupancy;

(f) The fire department response point(s) and annunciator location(s);

(g) The type of fire alarm system to be provided;

(h) Calculations (e.g., secondary supply and voltage drop calculations);

(i) The type(s) of fire alarm-initiating devices, supervisory alarm-initiating devices, and evacuation notification appliances to be provided;

(j) The intended area(s) of coverage;

(k) A complete list of detection, evacuation signaling, and annunciator zones;

(l) A complete list of fire safety control functions;

(m) A complete sequence of operations detailing all inputs and outputs.

1-5.5.2.2 Devices and appliances shall be so located and mounted that accidental operation or failure is not caused by vibration or jarring.

1-5.5.2.3 All apparatus requiring rewinding or resetting to maintain normal operation shall be restored to normal as promptly as possible after each alarm and kept in normal condition for operation.

1-5.5.2.4 Equipment shall be installed in locations where conditions do not exceed the voltage, temperature, and humidity limits specified in 1-5.5.1.

Exception: Equipment specifically listed for use in locations where conditions can exceed the upper and lower limits specified in 1-5.5.1.

1-5.5.3 To reduce the possibility of damage by induced transients, circuits and equipment shall be properly protected in accordance with the requirements set forth in NFPA 70, *National Electrical Code*, Article 800.

1-5.5.4* **Wiring.**   The installation of all wiring, cable, and equipment shall be in accordance with NFPA 70, *National Electrical Code*, and specifically with Article 760, Article 770, and Article 800, where applicable. Optical fiber cables shall be protected against mechanical injury in accordance with Article 760.

A-1-5.5.4 **Wiring and Equipment.**   The installation of all fire alarm system wiring should take into account the fire alarm system manufacturer's published installation instructions and the limitations of the applicable product listings or approvals.

1-5.5.5 **Grounding.** All systems shall test free of grounds.

Exception: Parts of circuits or equipment that are intentionally and permanently grounded to provide ground-fault detection, noise suppression, emergency ground signaling, and circuit protection grounding.

1-5.5.6 **Initiating Devices.**

1-5.5.6.1 Initiating devices of the manual or automatic type shall be selected and installed so as to minimize nuisance alarms.

1-5.5.6.2 Fire alarm boxes of the manually operated type shall comply with 3-8.1.

1-5.6 **Protection of Control Equipment.**   In areas that are not continuously occupied, automatic smoke detection shall be provided at each control unit(s) location to provide notification of fire at that location.

Exception: Where ambient conditions prohibit installation of automatic smoke detection, automatic heat detection shall be permitted.

This requirement applies even in areas protected by automatic sprinklers. The exception allows the use of heat detectors where conditions are not suitable for smoke detectors. The same problem may arise when placing control equipment in similar areas. Therefore, it may be necessary to condition the area to allow placement of control equipment if relocating the equipment elsewhere is not an option or where this is not possible or desirable. Always check the listing of the control equipment. See commentary following definition of control equipment in section 1-4 of the Code.

1-5.7 **Zoning and Annunciation.**

1-5.7.1 **Visible Zone Alarm Indication.**   Where required, the location of an operated initiating device shall be visibly indicated by building, floor, fire zone, or other approved subdivision by annunciation, printout, or other approved means. The visible indication shall not be canceled by the operation of an audible alarm silencing means.

Local building codes may require each floor of a building to be zoned separately for smoke detectors, waterflow switches, manual pull stations and other approved automatic fire detection devices. Addressable systems typically provide individual device status on the Fire Alarm Control Unit (FACU), which may satisfy those requirements. See Figures 1.34 and 1.35.

1-5.7.1.1 The primary purpose of fire alarm system annunciation is to enable responding personnel to identify the location of a fire quickly and accurately and to indicate the status of emergency equipment or fire safety functions that might affect the safety of occupants in a fire situation. All

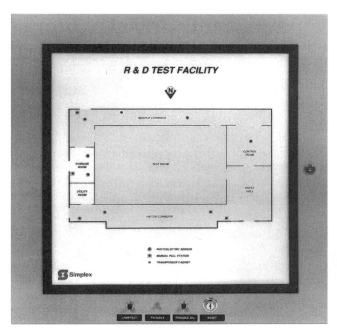

Figure 1.34 Typical remote fire alarm annunciator. Photograph courtesy of Simplex Time Recorder Co., Gardner, MA.

required annunciation means shall be readily accessible to responding personnel and shall be located as required by the authority having jurisdiction to facilitate an efficient response to the fire situation.

The authority having jurisdiction determines the type and location of any required annunciation means in specific building types. Common locations for annunciation are lobbies, guard's desks and Fire Command Centers.

Figure 1.35 Typical back-lit, labeled annunciators. Photograph courtesy of ESL Sentrol, Inc., Tualitin, OR.

1-5.7.1.2* Zone of Origin. Fire alarm systems serving two or more zones shall identify the zone of origin of the alarm initiation by annunciation or coded signal.

A-1-5.7.1.2 System Zoning. Fire alarm system annunciation should, as a minimum, be sufficiently specific to identify the origin of a fire alarm signal in accordance with the following:

(a) Where a floor exceeds 20,000 ft² (1860 m²) in area, the floor should be subdivided into detection zones of 20,000 ft² (1860 m²) or less, consistent with the existing smoke and fire barriers on the floor.

(b) Where a floor exceeds 20,000 ft² (1860 m²) in area and is undivided by smoke or fire barriers, detection zoning should be determined on a case-by-case basis in consultation with the authority having jurisdiction.

(c) Waterflow switches on sprinkler systems that serve multiple floors, areas exceeding 20,000 ft² (1860 m²), or areas inconsistent with the established detection system zoning should be annunciated individually.

(d) In-duct smoke detectors on air-handling systems that serve multiple floors, areas exceeding 20,000 ft² (1860 m²), or areas inconsistent with the established detection system zoning should be annunciated individually.

(e) Where a floor area exceeds 20,000 ft² (1860 m²), additional zoning should be provided. The length of any zone should not exceed 300 ft (91 m) in any direction. Where the building is provided with automatic sprinklers throughout, the area of the alarm zone should be permitted to coincide with the allowable area of the sprinkler zone.

1-5.7.1.3 Visual annunciators shall be capable of displaying all zones in alarm. Where all zones in alarm are not displayed simultaneously, there shall be visual indication that other zones are in alarm.

This new requirement was added to allow the responding firefighters to locate the zone source of the alarm, even in areas where the visual annunciator may not display the information for each zone covered by the system.

1-5.7.2 Alarm annunciation at the fire command center shall be by means of audible and visible indicators.

1-5.7.3 For the purpose of alarm annunciation, each floor of the building shall be considered as a separate zone. Where a floor is subdivided by fire or smoke barriers and the fire plan for the protected premises allows relocation of occupants from the zone of origin to another zone on the same floor, each zone on the floor shall be annunciated separately for purposes of alarm location.

The second requirement was added to the 1996 edition of the Code to address horizontal relocation of occupants on the same floor of a protected premises. Where horizontal relocation of occupants within a single floor is planned, whole floor annunciation would not be useful.

1-5.7.6 Where the system serves more than one building, each building shall be indicated separately.

1-5.8 Monitoring Integrity of Installation Conductors and Other Signaling Channels.

1-5.8.1 All means of interconnecting equipment, devices, and appliances and wiring connections shall be monitored for the integrity of the interconnecting conductors or equivalent path so that the occurrence of a single open or a single ground-fault condition in the installation conductors or other signaling channels and their restoration to normal shall be automatically indicated within 200 seconds.

Connections to devices and appliances must be made so that the opening of any installer's connection to the device or appliance will cause a trouble signal. Many installers "loop" the conductor around the terminal without cutting the conductor and making the necessary two connections. If the wire is disconnected from the terminal, there may be no indication of trouble. This practice is in violation of the Code. Where a listed device is furnished with pigtail connections, the installer must use separate "in/out" wires for each circuit passing into or through the device in order to prevent pigtail connections in the installation wiring.

However, addressable devices on signaling line circuits typically use an interrogation routine to monitor for integrity. This type of circuit may be wired without duplicate terminals and are often "T-tapped." The control unit interrogates each device on a regular basis, and "knows" when a device has become disconnected. Therefore, this is an acceptable practice for Class B signaling line circuits, when the designer allows it.

See Figures 1.36 and 1.37.

NOTE: The provision of a double loop or other multiple path conductor or circuit to avoid electrical monitoring is not acceptable.

Exception No. 1: Styles of initiating device circuits, signaling line circuits, and notification appliance circuits tabulated in Tables 3-5, 3-6, and 3-7.1 that do not have an "X" under "Trouble" for the abnormal condition indicated.

Exception No. 2: Shorts between conductors, other than as required by 1-5.8.3, 1-5.8.4, and 1-5.8.5.2 and Tables 3-5, 3-6, and 3-7.1, shall not be subject to this requirement.

Exception No. 3: A noninterfering shunt circuit, provided that a fault circuit condition on the shunt circuit wiring results only in the loss of the noninterfering feature of operation.

Exception No. 4: Connections to and between supplementary system components, provided that single open, ground, or short circuit conditions of the supplementary equipment or interconnecting means, or both, do not affect the required operation of the fire alarm system.

Exception No. 5: The circuit of an alarm notification appliance installed in the same room with the central control equipment, provided that the notification appliance circuit conductors are installed in conduit or equivalently protected against mechanical injury.

Exception No. 6: A trouble signal circuit.

Exception No. 7: Interconnection between equipment within a common enclosure.

NOTE: This code does not have jurisdiction over the monitoring integrity of conductors within equipment, devices, or appliances.

The requirement for monitoring applies only to installation conductors. The wiring within equipment,

Figure 1.36   Initiating device base showing duplicate terminals. Photograph courtesy of Radionics, Salinas, CA.

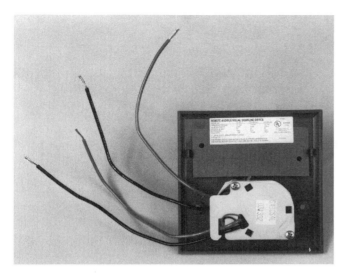

Figure 1.37 Notification appliance showing duplicate leads. Photograph courtesy of R.P. Shifilitti & Associates (Gentex), Reading, MA.

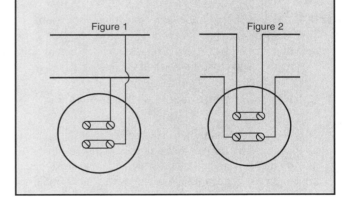

Formal Interpretation 75-5

Reference: 1-5.8.1

Question: Is it the intent of 1-5.8.1, with reference to initiating device circuits, that all wires installed by the installer be supervised?

Answer: Yes. a) In Figure 1 the two field installed wires to the screw terminals of the normally open device are not supervised and therefore unacceptable. b) In Figure 2 all four of the installed wires connected to the normally open device are supervised and, therefore, acceptable.

Issue Edition: NFPA 72A-1975
Reference: 2411
Issue Date: August 1977 ■

devices, or appliances is not required to be monitored for integrity.

Exception No. 8: Interconnection between enclosures containing control equipment located within 20 ft (6 m) where the conductors are installed in conduit or equivalently protected against mechanical injury.

Exception No. 9: Conductors for ground detection where a single ground does not prevent the required normal operation of the system.

Exception No. 10: Central station circuits serving notification appliances within a central station.

Exception No. 11: Pneumatic rate-of-rise systems of the continuous line type in which the wiring terminals of such devices are connected in multiple across electrically supervised circuits.

1-5.8.2 Interconnection means shall be arranged so that a single break or single ground fault does not cause an alarm signal.

1-5.8.3 An open, ground, or short circuit fault on the installation conductors of one alarm notification appliance circuit shall not affect the operation of any other alarm notification circuit.

1-5.8.4 The occurrence of a wire-to-wire short circuit fault on any alarm notification appliance circuit shall result in a trouble signal at the protected premises.

Exception No. 1: A circuit employed to produce a supplementary local alarm signal, provided that the occurrence of a short circuit on the circuit in no way affects the required operation of the fire alarm system.

Exception No. 2: The circuit of an alarm notification appliance installed in the same room with the central control equipment, provided that the notification appliance circuit conductors are installed in conduit or equivalently protected against mechanical injury.

Exception No. 3: Central station circuits serving notification appliances within a central station.

1-5.8.5 Monitoring Integrity of Emergency Voice/Alarm Communications Systems.

1-5.8.5.1* Monitoring Integrity of Speaker Amplifier and Tone-Generating Equipment. Where speakers are used to produce audible fire alarm signals, the following shall apply:

(a) Failure of any audio amplifier shall result in an audible trouble signal.

(b) Failure of any tone-generating equipment shall result in an audible trouble signal.

Exception: Tone-generating and amplifying equipment enclosed as integral parts and serving only a single, listed loudspeaker shall not be required to be monitored.

A-1-5.8.5.1   Backup amplifying and evacuation signal-generating equipment is recommended with automatic transfer upon primary equipment failure to ensure prompt restoration of service in the event of equipment failure.

1-5.8.5.2   Where a two-way telephone communications circuit is provided, its installation wires shall be monitored for a short circuit fault that would cause the telephone communications circuit to become inoperative.

1-5.8.6   Monitoring Integrity of Power Supplies.

1-5.8.6.1   All primary and secondary power supplies shall be monitored for the presence of voltage at the point of connection to the system. Failure of either supply shall result in a trouble signal in accordance with 1-5.4.6. The trouble signal also shall be visually and audibly indicated at the protected premises. Where the DACT is powered from a protected premises fire alarm system control unit, power failure indication shall be in accordance with this paragraph.

This section has been revised to clearly indicate that a trouble signal is required for failure of either the primary or secondary power supplies.

Exception No. 1:   A power supply for supplementary equipment.

Exception No. 2:   The neutral of a three-, four-, or five-wire ac or dc supply source.

Exception No. 3:   In a central station, the main power supply, provided the fault condition is otherwise indicated so as to be obvious to the operator on duty.

Exception No. 4:   The output of an engine-driven generator that is part of the secondary power supply, provided the generator is tested weekly in accordance with Chapter 7.

1-5.8.6.2   Power supply sources and electrical supervision for digital alarm communications systems shall be in accordance with 1-5.2 and 1-5.8.1.

NOTE: Since digital alarm communicator systems establish communications channels between the protected premises and the central station via the public switched telephone network, the requirement to supervise circuits between the protected premises and the central station (*see 1-5.8.1*) is considered to be met where the communications channel is periodically tested in accordance with 4-5.3.2.1.10.

1-5.8.6.3   The primary power failure trouble signal for the DACT shall not be transmitted until the actual battery capacity is depleted by at least 25 percent, but by not more than 50 percent.

Since different batteries discharge at different rates, the DACTs will transmit at different times. This requirement prevents jamming of telephone lines at the supervising station during the first moments of a widespread power outage.

1-6   System Interfaces.   The requirements by which fire alarm systems interface with other fire protective systems and fire safety functions can be found in Chapter 3.

1-7   Documentation

1-7.1   Approval and Acceptance.

1-7.1.1   The authority having jurisdiction shall be notified prior to installation or alteration of equipment or wiring. At its request, complete information regarding the system or system alterations, including specifications, wiring diagrams, battery calculation, and floor plans shall be submitted for approval.

1-7.1.2   Before requesting final approval of the installation, where required by the authority having jurisdiction, the installing contractor shall furnish a written statement to the effect that the system has been installed in accordance with approved plans and tested in accordance with the manufacturer's specifications and the appropriate NFPA requirements.

1-7.2   Completion Documents.

1-7.2.1*   A record of completion (*see Figure 1-7.2.1*) shall be prepared for each system. Parts 1, 2, and 4 through 10 shall be completed after the system is installed and the installation wiring has been checked. Part 3 shall be completed after the operational acceptance tests have been completed. A preliminary copy of the record of completion shall be given to the system owner and, where requested, to other authorities having jurisdiction after completion of the installation wiring tests, and a final copy shall be provided after completion of the operational acceptance tests.

The Record of Completion was previously referred to as the Certificate of Completion. This document is very important to the owner, maintainer and the authority having jurisdiction because it provides information necessary to maintain, repair and modify the system. A record of completion is required for all installed fire Alarm Systems.

Record of Completion

Name of Protected Property: _____

Address: _____

Rep. of Protected Prop. (name/phone): _____

Authority Having Jurisdiction: _____

Address/Phone Number: _____

1.   Type(s) of System or Service

_____NFPA 72, Chapter 3 — Local

If alarm is transmitted to location(s) off premises, list where received:

_____

_____NFPA 72, Chapter 3 — Emergency Voice/Alarm Service

Quantity of voice/alarm channels: _____   Single: _____   Multiple: _____

Quantity of speakers installed: _____   Quantity of speaker zones: _____

Quantity of telephones or telephone jacks included in system: _____

_____NFPA 72, Chapter 4 — Auxiliary

Indicate type of connection:

Local energy: _____   Shunt: _____   Parallel telephone: _____

Location and telephone number for receipt of signals:

_____

_____NFPA 72, Chapter 4 — Remote Station

Alarm: _____

Supervisory:

_____NFPA 72, Chapter 4 — Proprietary

If alarms are retransmitted to public fire service communications center or others, indicate location and telephone number of the organization receiving alarm:

_____

Indicate how alarm is retransmitted:

_____

_____NFPA 72, Chapter 4 — Central Station

The Prime Contractor:

Central Station Location:

Figure 1-7.2.1   *Record of Completion.*

Means of transmission of signals from the protected premises to the central station:

_____ McCulloh   _____ Multiplex   _____ One-Way Radio

_____ Digital Alarm Communicator   _____ Two-Way Radio   _____ Others _____

Means of transmission of alarms to the public fire service communications center:

(a) _____

(b) _____

System Location: _____

	Organization Name/Phone	Representative Name/Phone
Installer	_____	_____
Supplier	_____	_____

Service Organization_____

Location of Record (As-Built) Drawings:

_____

Location of Owners Manuals:

_____

Location of Test Reports:

_____

A contract, dated_____ , for test and inspection in accordance with NFPA standard(s) No(s).
dated _____ , is in effect.

2.   Record of System Installation

(Fill out after installation is complete and wiring checked for opens, shorts, ground faults, and improper branching, but prior to conducting operational acceptance tests.)

This system has been installed in accordance with the NFPA standards as shown below, was inspected by _____
on _____ , includes the devices shown below, and has been in service
since _____ .

_____   NFPA 72, Chapters   1   3   4   5   6   7   (circle all that apply)

_____   NFPA 70, *National Electrical Code*, Article 760

_____   Manufacturer's Instructions

_____   Other (specify):_____

Signed: _____   Date: _____

Organization: _____

3.   Record of System Operation

All operational features and functions of this system were tested by _____ on _____ , and
found to be operating properly in accordance with the requirements of:

_____   NFPA 72, Chapters   1   3   4   5   6   7   (circle all that apply)

_____   NFPA 70, *National Electrical Code*, Article 760

_____   Manufacturer's Instructions

_____   Other (specify): _____

Signed: _____   Date: _____

Organization: _____

Figure 1-7.2.1   Continued

4. Alarm-Initiating Devices and Circuits (use blanks to indicate quantity of devices)

    MANUAL

    (a) _____ Manual Stations _____ Noncoded, Activating _____ Transmitters _____ Coded

    (b) _____ Combination Manual Fire Alarm and Guard's Tour Coded Stations

    AUTOMATIC

    Coverage: Complete: _____   Partial: _____

    (a) _____ Smoke Detectors   _____ Ion   _____ Photo

    (b) _____ Duct Detectors   _____ Ion   _____ Photo

    (c) _____ Heat Detectors   _____ FT   _____ RR   _____ FT/RR   _____ RC

    (d) _____ Sprinkler Waterflow Switches: _____ Transmitters   _____ Noncoded, Activating   _____ Coded

    (e) _____ Other (list): _____

5. Supervisory Signal-Initiating Devices and Circuits (use blanks to indicate quantity of devices)

    GUARD'S TOUR

    (a) _____ Coded Stations

    (b) _____ Noncoded Stations, Activating _____ Transmitters

    (c) _____ Compulsory Guard Tour System Comprised of _____ Transmitter Stations and _____ Intermediate Stations

    NOTE: Combination devices recorded under 4(b) and 5(a).

    SPRINKLER SYSTEM

    (a) _____ Coded Valve Supervisory Signaling Attachments

                          Valve Supervisory Switches, Activating _____ Transmitters

    (b) _____ Building Temperature Points

    (c) _____ Site Water Temperature Points

    (d) _____ Site Water Supply Level Points

    Electric Fire Pump:

    (e) _____ Fire Pump Power

    (f) _____ Fire Pump Running

    (g) _____ Phase Reversal

    Engine-Driven Fire Pump:

    (h) _____ Selector in Auto Position

    (i) _____ Engine or Control Panel Trouble

    (j) _____ Fire Pump Running

    Engine-Driven Generator:

    (k) _____ Selector in Auto Position

    (l) _____ Control Panel Trouble

    (m) _____ Transfer Switches

    (n) _____ Engine Running

    Other Supervisory Function(s) (specify): _____

_____

_____

_____

Figure 1-7.2.1 Continued

6.   Alarm Notification Appliances and Circuits

Quantity of indicating appliance circuits connected to the system: _____

Types and quantities of alarm indicating appliances installed:

(a)   _____ Bells   _____ Inch

(b)   _____ Speakers

(c)   _____ Horns

(d)   _____ Chimes

(e)   _____ Other: _____

(f)   _____ Visual Signals   Type: _____

_____ with audible   _____ w/o audible

(g)   _____ Local Annunciator

7.   Signaling Line Circuits

Quantity and Style (see NFPA 72, Table 3-6) of signaling line circuits connected to system:

Quantity: _____   Style: _____

8.   System Power Supplies

(a)   Primary (Main):Nominal Voltage: _____Current Rating: _____

Overcurrent Protection: Type: _____Current Rating: _____

Location:

(b)   Secondary (Standby):

_____ Storage Battery: Amp-Hour Rating

Calculated capacity to drive system, in hours: _____ 24   _____ 60

Engine-driven generator dedicated to fire alarm system:

Location of fuel storage:

(c)   Emergency or Standby System used as backup to Primary Power Supply, instead of using a Secondary Power Supply:

Emergency System described in NFPA 70, Article 700

Legally Required Standby System described in NFPA 70, Article 701

_____ Optional Standby System described in NFPA 70, Article 702, which also meets the performance requirements of Article 700 or 701

9.   System Software

(a)   Operating System Software Revision Level(s): _____

(b)   Application Software Revision Level(s): _____

(c)   Revision Completed by: _____
(name)   (firm)

10.   Comments:

_____
(signed) for Central Station or Alarm Service Company   (title)   (date)

Figure 1-7.2.1 *Continued*

Frequency of routine tests and inspections, if other than in accordance with the referenced NFPA standards(s): _____

_____

_____

System deviations from the referenced NFPA standard(s) are: _____

_____

_____

_____

(signed) for Central Station or Alarm Service Company   (title)   (date)

Upon completion of the system(s) satisfactory test(s) witnessed (if required by the authority having jurisdiction):

_____

(signed) for Central Station or Alarm Service Company   (title)   (date)

Figure 1-7.2.1

A-1-7.2.1 The requirements of Chapter 7 should be used to perform the installation wiring and operational acceptance tests required when completing the record of completion.

1-7.2.2 Every system shall include the following documentation, which shall be delivered to the owner or the owner's representative upon final acceptance of the system.

(a)* An owner's manual and installation instructions covering all system equipment; and

(b) Record drawings.

A-1-7.2.2(a) The owner's manual should include:

(a) A detailed narrative description of the system inputs, evacuation signaling, ancillary functions, annunciation, intended sequence of operations, expansion capability, application considerations, and limitations.

(b) Operator instructions for basic system operations, including alarm acknowledgment, system reset, interpretation of system output (LEDs, CRT display, and printout), operation of manual evacuation signaling and ancillary function controls, and change of printer paper.

(c) A detailed description of routine maintenance and testing as required and recommended and as would be provided under a maintenance contract, including testing and maintenance instructions for each type of device installed. This information should include the following:

1. A listing of the individual system components that require periodic testing and maintenance;

2. Step-by-step instructions detailing the requisite testing and maintenance procedures, and the intervals at which these procedures shall be performed, for each type of device installed;

3. A schedule that correlates the testing and maintenance procedures recommended by A-1-7.2.2(c)2 with the listing recommended by A-1-7.2.2(c)1.

(d) Detailed troubleshooting instructions for each trouble condition generated from the monitored field wiring, including opens, grounds, and loop failures. These instructions should include a list of all trouble signals annunciated by the system, a description of the condition(s) that causes such trouble signals, and step-by-step instructions describing how to isolate such problems and correct them (or how to call for service, as appropriate).

(e) A service directory, including a list of names and telephone numbers of those who provide service for the system.

1-7.2.3 Central Station Fire Alarm Systems. It shall be conspicuously indicated by the prime contractor (*see* Chapter 4) that the fire alarm system providing service at a protected premises complies with all applicable requirements of this code by providing a means of verification as specified in either 1-7.2.3.1 or 1-7.2.3.2.

1-7.2.3.1 The installation shall be certificated.

1-7.2.3.1.1 Central station fire alarm systems providing service that complies with all requirements of this code shall be certificated by the organization that has listed the prime

trol unit or, where no control unit exists, on or near a fire alarm system component, and shall identify the central station and, where applicable, the prime contractor by name and telephone number.

1-7.3   Records.   A complete unalterable record of the tests and operations of each system shall be kept until the next test and for 1 year thereafter. The record shall be available for examination and, where required, reported to the authority having jurisdiction. Archiving of records by any means shall be permitted if hard copies of the records can be provided promptly when requested.

Exception: Where off-premises monitoring is provided, records of all signals, tests, and operations recorded at the supervising station shall be maintained for not less than 1 year

contractor, and a document attesting to this certification shall be located on or near the fire alarm system control unit or, where no control unit exists, on or near a fire alarm system component.

1-7.2.3.1.2   A central repository of issued certification documents, accessible to the authority having jurisdiction, shall be maintained by the organization that has listed the central station.

1-7.2.3.2   The installation shall be placarded.

1-7.2.3.2.1   Central station fire alarm systems providing service that complies with all requirements of this code shall be conspicuously marked by the prime contractor to indicate compliance. The marking shall be by means of one or more securely affixed placards.

1-7.2.3.2.2   The placard(s) shall be 20 in.^{2} (130 cm^2) or larger, shall be located on or near the fire alarm system con-

2

Household Fire Warning Equipment

Contents, Chapter 2

This chapter was previously a stand-alone document, NFPA 74-1989, entitled *Standard for the Installation, Maintenance, and Use of Household Fire Warning Equipment.*

A-2   Household Fire Warning Protection.

(a) *Fire Danger in the Home.*   Fire is the third leading cause of accidental death. Residential occupancies account for most fire fatalities, and most of these deaths occur at night during the sleeping hours.

Most fire injuries also occur in the home. It is estimated that 1.5 million Americans are injured by fire each year. Many never resume normal lives.

It is estimated that each household will experience three (usually unreported) fires per decade and two fires serious enough to report to a fire department per lifetime.

(b) *Fire Safety in the Home.*   This code is intended to provide reasonable fire safety for persons in family living units. Reasonable fire safety can be produced through a three-point program:

1. Minimizing fire hazards;
2. Providing a fire warning system;
3. Having and practicing an escape plan.

(c) *Minimizing Life Safety Hazards.*   This code cannot protect all persons at all times. For instance, the application of this code might not provide protection against the three traditional fatal fire scenarios:

1. Smoking in bed;
2. Leaving children home alone;
3. Cleaning with flammable liquids such as gasoline.

However, Chapter 2 can lead to reasonable safety from fire where the three points under A-2(b) are observed.

(d) *Fire Warning System.*   There are two types of fire to which household fire warning equipment needs to respond. One is a rapidly developing, high heat fire. The other is a slow, smoldering fire. Either can produce smoke and toxic gases.

Household fires are especially dangerous at night when the occupants are asleep. Fires produce smoke and deadly gases that can overcome occupants while they are asleep. Furthermore, dense smoke reduces visibility. Most fire casualties are victims of smoke and gas inhalation rather than burns. To warn against a fire, Chapter 2 requires smoke detectors in accordance with 2-2.1.1.1 and recommends heat or smoke detectors in all other major areas. (*See* 2-2.1.1.1.)

(e) *Family Escape Plan.*   There often is very little time between the detection of a fire and the time it becomes deadly. This interval can be as little as 1 minute or 2 minutes. Thus, this code requires detection means to give a family

some advance warning of the development of conditions that become dangerous to life within a short period of time. Such warning, however, could be wasted unless the family has planned in advance for rapid exit from their residence. Therefore, in addition to the fire warning system, this code requires exit plan information to be furnished.

Planning and practicing for fire conditions with a focus on rapid exit from the residence are important. Drills should be held so that all family members know the action to be taken. Each person should plan for the possibility that exit out of a bedroom window could be necessary. An exit out of the residence without the need to open a bedroom door is essential.

(f) *Special Provisions for the Disabled.*   For special circumstances where life safety of an occupant(s) depends upon prompt rescue by others, the fire warning system should include means of prompt, automatic notification to those who are to be depended upon for rescue.

2-1   Introduction.

2-1.1*   Scope.   This chapter contains minimum requirements for the selection, installation, operation, and maintenance of fire warning equipment for use within family living units. The requirements of the other chapters shall not apply.

Exception: Where specifically indicated.

Chapter 2 deals strictly with household fire detection and warning equipment and fire alarm systems used in family living units. (See Section 1-4, definition of *family living unit*).

Although this is a stand-alone chapter, all definitions for this chapter, and other chapters of the Code are located in Chapter 1.

A-2-1.1   Chapter 2 does not attempt to cover all equipment, methods, and requirements that might be necessary or advantageous for the protection of lives and property from fire.

NFPA 72 is a "minimum code," and it provides a number of requirements related to household fire warning equipment that are deemed to be the practical and necessary minimum for average conditions at the present state of the art.

2-1.2   Purpose.

2-1.2.1   Household fire warning systems shall be designed and installed to provide sufficient warning of a fire to enable occupants to escape. It is recognized that household fire warning systems might not be of material assistance to all occupants, such as persons intimate with the ignition of a fire.

2-1.2.2 This chapter is primarily concerned with life safety, not with protection of property. It presumes that a family has an exit plan.

Family living unit fires lead all other types of occupancies as the site of fire-related deaths in the United States. Smoke detectors required by the Code, and other means of detection recommended in this Chapter provide warning but do not extinguish the fire. After the detector does its job, it is the responsibility of the occupants to follow their emergency exit plan. Home fire statistics, recommendations for fire safety and life safety, fire warning system capabilities, a more detailed explanation of an escape plan, and special provisions for the disabled are listed in Section A-2. of this Code.

2-1.3 General.

2-1.3.1 A control and associated equipment, a multiple or single station alarm(s), or any combination thereof shall be permitted to be used as a household fire warning system, provided the requirements of 2-1.3.7 are met.

The minimum requirement is for a smoke detector on each level of the home, outside each separate sleeping area and in each bedroom (new construction only) of a home. This requirement can be met using multiple-station alarm devices. See Figure 2.1.

The Code also allows the use of a complete household fire alarm system that contains system-type (i.e., connected to a control panel) smoke detectors along with

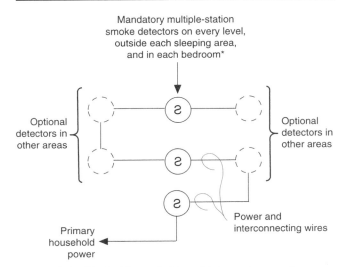

Figure 2.1 *Typical household multiple-station smoke detector system (*new construction only).*

Mandatory smoke detector on every level, outside each sleeping area, and in each bedroom*

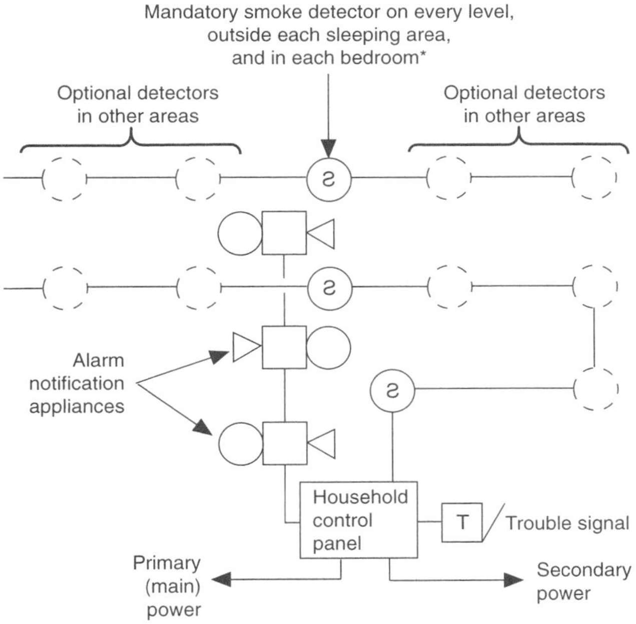

Figure 2.2 *Typical household fire alarm system with separate control panel (*new construction only).*

other devices, such as heat detectors and alarm notification appliances. See Figure 2.2. Obviously, the complete system would be used in place of, and not in addition to, the interconnected 120-VAC multiple-station smoke alarm devices. See 2-2.2.1.

2-1.3.2 Detection and alarm systems for use within the protected household are covered by this chapter.

This chapter covers the family living unit only, and does not provide requirements for common or tenantless areas such as apartment building lobbies or hallways. Chapter 3 describes the requirements for protected premises systems in common areas. Equipment signals to an off-premises location are covered in Chapter 4.

2-1.3.3 Supplementary functions, including the extension of an alarm beyond the household, shall be permitted and shall not interfere with the performance requirements of this chapter.

These supplementary functions include the connection to a remote station, a central station, or to another remote monitoring location. See Chapter 4 for information regarding supervising station connection requirements.

2-1.3.4 Where the authority having jurisdiction requires a household fire warning system to comply with the requirements of Chapter 4 or any other chapters of this code, the requirements of Section 2-2 shall still apply.

The primary objective of a household fire warning system is life safety. Although some jurisdictions require a household fire alarm system (utilizing a control panel) to be connected to a supervising station, the basic requirements of the occupant warning features in this chapter must be followed regardless of any supervising station connection.

2-1.3.5 The definitions of Section 1-4 shall apply.

2-1.3.6 This chapter does not exclude the use of fire alarm systems complying with other chapters of this code in household applications, provided all of the requirements of this chapter are met or exceeded.

For example, a listed commercial fire alarm system, installed in accordance with Chapter 3 of this Code, that meets the basic detection and warning requirements of this chapter would be an acceptable alternative to the more typical systems listed for household use described in this chapter.

2-1.3.7 All devices, combinations of devices, and equipment to be installed in conformity with this chapter shall be approved or listed for the purposes for which they are intended.

The approval or listing of fire alarm system devices is typically conducted by a testing laboratory. It is the responsibility of the authority having jurisdiction to either accept or reject the laboratory's approval or listing. The authority having jurisdiction may require a product to be listed or labeled, but the listing or label alone does not constitute approval. The authority having jurisdiction has the right to approve products or systems that are not labeled or listed.

2-1.3.8 A device or system of devices having materials or forms that differ from those detailed in this chapter shall be permitted to be examined and tested according to the intent of the chapter and, if found equivalent, shall be permitted to be approved.

The authority having jurisdiction is responsible for approval or disapproval of devices and equipment used in, and the placement of, fire alarm systems based on the information presented for the equipment and systems proposed. The authority having jurisdiction may also request tests of that equipment to determine equivalency.

2-1.3.9 Equivalency. Nothing in this code is intended to prevent the use of systems, methods, or devices of equivalent or superior quality, strength, fire resistance, effectiveness, durability, and safety over those prescribed by this code, provided technical documentation is submitted to the authority having jurisdiction to demonstrate equivalency and the system, method, or device is approved for the intended purpose.

The authority having jurisdiction has the sole responsibility for accepting a proposed system, method, or device as equivalent.

2-2   Basic Requirements.

2-2.1   Required Protection.

2-2.1.1* This code requires the following detectors within the family living unit.

A-2-2.1.1 Experience has shown that all hostile fires in family living units generate smoke to some degree. This is also true with respect to heat buildup from fires. However, the results of full-scale experiments conducted over the past several years in the U.S., using typical fires in family living units, indicate that detectable quantities of smoke precede detectable levels of heat in nearly all cases. In addition, slowly developing, smoldering fires can produce smoke and toxic gases without a significant increase in the room's temperature. Again, the results of experiments indicate that detectable quantities of smoke precede the development of hazardous atmospheres in nearly all cases.

For the above reasons, the required protection in this code utilizes smoke detectors as the primary life safety equipment for providing a reasonable level of protection against fire.

Of course, it is possible to install fewer detectors than required in this code. It could be argued that the installation of only one fire detector, whether a smoke or heat detector, offers some life-saving potential. While this is true, it is the opinion of the committee that developed Chapter 2 that the smoke detector requirements as stated in 2-2.1.1 are the minimum that should be considered.

The installation of additional detectors of either the smoke or heat type should result in a higher degree of protection. Adding detectors to rooms that are normally closed off from the required detectors increases the escape time because the fire does not need to build to the higher level necessary to force smoke out of the closed room to the required detector. As a consequence, it is recommended that the householder consider the installation of additional fire protection devices. However, it should be understood that Chapter 2 does not require additional detectors over and above those called for in 2-2.1.1.

2-2.1.1.1 Smoke detectors shall be installed outside of each separate sleeping area in the immediate vicinity of the bedrooms and on each additional story of the family living unit, including basements and excluding crawl spaces and unfinished attics. In new construction, a smoke detector also shall be installed in each sleeping room.

In order to gain the full benefit of smoke detection, it is imperative that detectors be located as outlined in this chapter. However, repeated false alarms from smoke detectors, triggered by common household conditions, undermine the response of family members to the alarm. These nuisance alarms are most often experienced when smoke detectors are improperly placed in kitchens, bathrooms, garages, and basements with dirt floors or moisture problems. In locations with conditions likely to cause frequent nuisance alarms, heat detectors or other fire detection devices less likely to be affected by false triggering conditions may offer additional protection. In any case, the requirements of the Code regarding placement of smoke detectors near sleeping areas (or inside the bedrooms in new construction) should be met and these other devices installed only as additional means of protection.

2-2.1.1.2* For family living units with one or more split levels (i.e., adjacent levels with less than one full story separation between levels), a smoke detector required by 2-2.1.1.1 shall be permitted for an adjacent lower level, including basements. *See* Figure A-2-2.1.1.2.

Exception: Where there is an intervening door between one level and the adjacent lower level, a smoke detector shall be installed on the lower level.

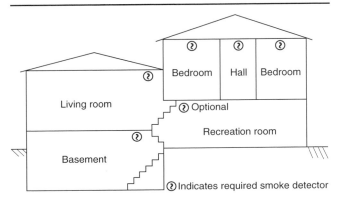

Figure A-2-2.1.1.2 *Split level arrangement. Smoke detectors are required where shown. Smoke detectors are optional where a door is not provided between living room and recreation room.*

2-2.1.1.3 Automatic sprinkler systems provided in accordance with NFPA 13D, *Standard for the Installation of Sprinkler Systems in One- and Two-Family Dwellings and Manufactured Homes*, or NFPA 13R, *Standard for the Installation of Sprinkler Systems in Residential Occupancies up to and Including Four Stories in Height*, shall be interconnected to sound alarm notification appliances throughout the dwelling where a fire warning system is provided.

This interconnection should be approved by the authority having jurisdiction. There are presently no listed multiple-station smoke alarms that have the necessary connections to accept the attachment of additional devices, such as waterflow switches. A listed fire alarm control unit or combination burglar/fire alarm control unit already serving the the residence would have to be used to accept interconnection of the water flow switch. Additionally, a waterflow switch could be connected to a separate fire alarm system control unit to sound notification appliances throughout the family living unit.

2-2.2*   Alarm Notification Appliances. Each automatic alarm-initiating device shall cause the operation of an alarm that shall be clearly audible in all bedrooms over background noise levels with all intervening doors closed. The tests of audibility level shall be conducted with all household equipment that might be in operation at night in full operation.

Examples of such equipment are window air conditioners and room humidifiers. (*See* A-2-2.2 for additional information.)

Because many residential fire deaths occur as a result of smoke inhalation rather than burns, the National Fire Protection Association recommends sleeping with bedroom doors closed to limit the spread of smoke in the event of a fire at night. To ensure that smoke detector alarms can be heard by residents inside closed sleeping areas, detectors must be tested with bedroom doors shut and in conditions as similar as possible to those normally occurring in the home at night. Thus, if air conditioning units or humidifiers are in routine use in a home, those appliances should be operating when the detectors are tested.

A-2-2.2 At times, depending upon conditions, the audibility of detection devices could be seriously impaired where occupants are within the bedroom area. For instance, there might be a noisy window air conditioner or room humidifier generating an ambient noise level of 55 dBA or higher. The detection devices' alarms need to penetrate through the closed doors and be heard over the bedroom's noise levels with sufficient intensity to awaken sleeping occupants therein. Test data indicate that detection devices having sound pressure ratings of 85 dBA of 10 ft (3 m) and installed outside the bedrooms can produce about 15 dBA over ambient noise levels of 55 dBA in the bedrooms. This is likely to be sufficient to awaken the average sleeping person.

The measurements required here can be made using a sound pressure level meter described in Chapter 6.

Detectors located remote from the bedroom area might not be loud enough to awaken the average person. In such cases, it is recommended that detectors be interconnected in such a way that the operation of the remote detector causes an alarm of sufficient intensity to penetrate the bedrooms. The interconnection can be accomplished by the installation of a fire detection system, by the wiring together of multiple station alarm devices, or by the use of line carrier or radio frequency transmitters/receivers.

Additional notification appliances connected to, and powered through, the dry contacts of a single- or multiple-station smoke detector relay may also be used to comply with this requirement. See Figures 2.3 and 2.4.

2-2.2.1 In new construction, where more than one smoke detector is required by 2-2.1, detectors shall be arranged so that operation of any smoke detector causes the alarm in all smoke detectors within the dwelling to sound.

Exception: Configurations that provide equivalent distribution of the alarm signal.

The requirement is to use listed multiple-station smoke detectors, as a minimum, in all new construction. Equivalent distribution of the alarm signal can also be achieved by notification appliances connected to a household control panel. Wireless transmitters also may be used to activate notification appliances distributed throughout the home.

2-2.2.2* **Standard Signal.** Newly installed alarm notification appliances used with a household fire warning system and single and multiple station smoke alarms shall produce the audible emergency evacuation signal described in ANSI S3.41, *Audible Emergency Evacuation Signal*. Signals from different notification appliances shall not be required to be synchronized.

A-2-2.2.2 The use of the distinctive three-pulse temporal pattern fire alarm evacuation signal required by 3-7.2(a) had previously been recommended for this purpose by this code since 1979. It has since been adopted as both an American National Standard (ANSI S3.41, *Audible Emergency Evacuation Signal*) and an International Standard (ISO 8201, *Audible Emergency Evacuation Signal*).

Figure 2.3 *Typical notification appliances that could be used in a household system. Top: Bell. Photograph courtesy of Wheelock, Long Branch, NJ. Bottom: Mini-horn. Photograph courtesy of Gentex Corp., Zeeland, MI.*

Copies of both of these standards are available from the Standards Secretariat, Acoustical Society of America, 335 East 45th Street, New York, NY 10017-3483. Telephone 212-661-9404, ext. 562.

The standard fire alarm evacuation signal is a three-pulse temporal pattern using any appropriate sound. The pattern

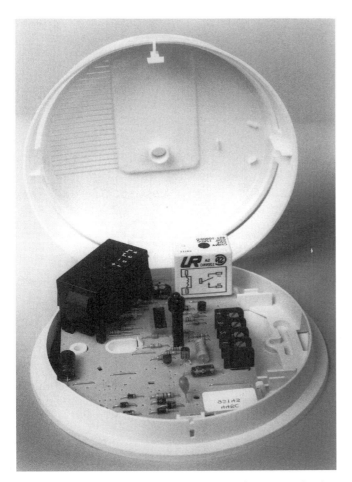

Figure 2.4 *Single/multiple-station smoke detector with relay contacts for remote notification appliance. Photograph courtesy of Mammoth Fire Alarms (ESL/Sentrol Inc.), Lowell, MA.*

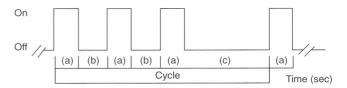

Key:
Phase (a) signal is "on" for 0.5 sec ± 10%
Phase (b) signal is "off" for 0.5 sec ± 10%
Phase (c) signal is "off" for 1.5 sec ± 10% [(c) = (a) + 2(b)]
Total cycle lasts for 4 sec ± 10%

Figure A-2-2.2.2(a) *Temporal pattern parameters.*

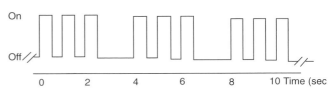

Figure A-2-2.2.2(b) *Temporal pattern imposed on signaling appliances that emit a continuous signal while energized.*

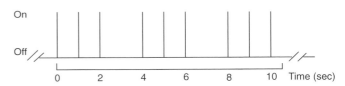

Figure A-2-2.2.2(c) *Temporal pattern imposed on a single-stroke bell or chime.*

consists of an "on" phase (a) lasting 0.5 second ± 10 percent followed by an "off" phase (b) lasting 0.5 second ± 10 percent, for three successive "on" periods, which are followed by an "off" phase (c) lasting 1.5 seconds ± 10 percent [*see* Figures A-2-2.2.2(a) and (b)]. The signal should be repeated for a period appropriate for the purposes of evacuation of the building, but for not less than 180 seconds. A single-stroke bell or chime sounded at "on" intervals lasting 1 second ± 10 percent, with a 2-second ± 10 percent "off" interval after each third "on" stroke, may be permitted [*see* Figure A-2-2.2.2(c)].

The minimum repetition time may be permitted to be manually interrupted.

2-2.3 **Alarm Notification Appliances for the Hearing Impaired.** In a household occupied by one or more hearing impaired persons, each initiating device shall cause the operation of a visible alarm signal(s) in accordance with 2-4.4.2.

Since hearing deficits are often not apparent, the responsibility for advising the appropriate persons shall be that of the hearing impaired party. The responsibility for compliance shall be that of the occupants of the family living unit.

Exception: A listed tactile signal shall be permitted to be employed.

This section of the Code addresses the needs of the hearing impaired who cannot respond to auditory alarms alone. The exception allows the use of bed shakers and personal vibrating alarm pagers. These tactile notification devices (defined in Section 1-4 of the Code) are used primarily at night. At other times, the Code permits the use of a visible signal of lower intensity. See Section 2-4.4.2. Chapter 6 describes how Underwriters Laboratories investigates the effectiveness of strobe lights for the notification of the hearing impaired.

2-3   Power Supplies.

2-3.1 General.

2-3.1.1 All power supplies shall have sufficient capacity to operate the alarm signal(s) for at least 4 continuous minutes.

This requirement applies to battery-operated smoke detectors as well as to smoke detectors powered by 120 VAC or by a control panel power supply. Four minutes is considered sufficient time to warn occupants that a fire condition exists.

2-3.1.2 There shall be a primary (main) and a secondary (standby) power source. For electrically powered household fire warning equipment, the primary (main) power source shall be ac; the secondary (standby) power source shall be a battery.

This requirement addresses the increased need to have a functioning smoke detector during a power outage. Periodic testing of batteries is vital to ensure that this secondary power supply to the detector will function. During a power outage, occupants may greatly increase the risk of fire by using candles, lanterns, space heaters and other equipment not usually in operation in the home environment. Therefore, the smoke detector plays an even more important role in warning residents of a fire in these conditions. As stated in Exception 5, this requirement does not apply to AC powered detectors in existing construction, which implies that the main requirement for secondary power applies only to new construction.

Exception No. 1:   Where the primary (main) power source is an emergency circuit or a legally required standby circuit capable of operating the system for at least 24 hours in the normal condition, followed by not less than 4 minutes of alarm, a secondary (standby) source shall not be required.

Exception No. 2:   Where the primary (main) power source is a circuit of an optional standby system capable of operating the system for at least 24 hours, followed by not less than 4 minutes of alarm, that meets the requirements for either an emergency system or a legally required standby system as defined in NFPA 70, National Electrical Code, Articles 700 and 701, respectively, a secondary (standby) supply shall not be required.

Exception No. 3:   Detectors and alarms powered from a monitored dc circuit of a control unit where power for the control unit meets the requirements of Section 2-3 and the circuit remains operable upon loss of primary (main) ac power.

Exception No. 4:   A detector and a wireless transmitter that serves only that detector shall be permitted to be powered from a monitored battery primary (main) source where part of a listed, monitored low power radio (wireless) system. A secondary (standby) source shall not be required.

Exception No. 5:   In existing construction, either an ac primary power source, as described in 2-3.2, or a monitored battery primary (main) power source, as described in 2-3.3, shall be permitted. A secondary (standby) source shall not be required.

Exception No. 6:   Visible notification appliances required by 2-4.4.2.

Exception No. 7:   Where the primary (main) power source is nonelectrical, a secondary (standby) source shall not be required. The requirements of 2-3.5 shall apply.

This requirement applies specifically to spring-wound heat detector alarm devices.

2-3.2 Primary Power Supply—AC.

2-3.2.1 An ac primary (main) power source shall be a dependable commercial light and power supply source. A visible "power on" indicator shall be provided.

The "power-on" indicator is required on both smoke detectors and control panels used in household systems powered by an AC power source.

Formal Interpretation 78-2
Reference: 2-3.2.1

Background: Paragraph 2-3.2.1 states that "A visible 'power-on' indicator shall be provided." It is our understanding that this "power-on" indicator requirement is to alert the homeowner that the unit was being powered and no unintentional interruption of the power to the detector has taken place.

The problem at hand involves a 120-VAC residential smoke detector also equipped with an alternative power supply in the form of a battery. The battery is fully monitored, in accordance with 2-3.3 with or without ac power being applied.

Question: With the above background and under the above conditions, is it the Committee's intent that such a smoke detector as the one described above be required to have a "power-on" indicator?

Answer: Yes.

Issue Edition: NFPA 74-1978
Reference: 2-1.2.1
Issue Date: January 1980
Reissued: January 1986   ∎

2-3.2.2 All electrical systems designed to be installed by other than a qualified electrician shall be powered from a source not in excess of 30 volts that meets the requirements for power limited fire alarm circuits as defined in NFPA 70, *National Electrical Code*, Article 760.

The voltage limit regulation is included in the Code to reduce the shock and fire hazards associated with the 120 VAC wiring. Most jurisdictions require licensed electricians to install all 120 VAC outlets and connections. Some jurisdictions permit a licensed fire alarm technician to install any 120-VAC connection that is associated with the fire alarm system. It is imperative that the authority having jurisdiction be consulted as to the installation requirements of the fire alarm system is being installed.

2-3.2.3 A restraining means shall be used at the plug-in of any cord connected installation.

Plug-in type detectors can only be effective when the power source supply is not interrupted. Accidental bumping of the plug is likely in most household situations. This requirement was added to reduce the risk of unplugging the equipment and applies to plug-in type detectors as well as panel powered household systems with plug-in type connections to AC power (e.g., a plug-in-type transformer).

2-3.2.4 AC primary (main) power shall be supplied either from a dedicated branch circuit or the unswitched portion of a branch circuit also used for power and lighting. Operation of a switch (other than a circuit breaker) or a ground-fault circuit-interrupter shall not cause loss of primary (main) power.

Exception No. 1: Single or multiple station alarms with a supervised rechargeable standby battery that provides at least 4 months of operation with a fully charged battery.

Exception No. 2: Where a ground-fault circuit-interrupter serves all electrical circuits within the household.

When installing single- or multiple-station smoke detectors, it is good practice to connect the detector's power to a branch circuit serving electrical outlets in a habitable area such as the living room or family room. This is done to ensure that if for any reason the circuit breaker is in the "off" position, the condition will be noticed more quickly because lights and other appliances used frequently in the home will not operate. The power connection to a household control panel can be connected in the same way. When connecting to a power circuit that serves other

appliances, the installer must ensure that the circuit is not overloaded, causing the circuit breaker to frequently trip. Some state and local codes may require this branch circuit connection to be made to a dedicated circuit breaker. Consult with the authority having jurisdiction to determine if local codes or regulations differ from Code requirements in this section.

2-3.2.5 Neither loss nor restoration of primary (main) power shall cause an alarm signal.

Exception: An alarm signal shall be permitted within the household but shall not exceed 2 seconds.

Generally, a loss or restoration of power will not cause any alarm signal. However, the new exception permits a 120-VAC single- and multiple-station smoke detector to sound briefly (2 seconds or less) to alert the household that there has been a power interruption. The new Exception was added to the 1996 edition of the Code to address this issue.

2-3.2.6 Where a secondary (standby) battery is provided, the primary (main) power supply shall be of sufficient capacity to operate the system under all conditions of loading with any secondary (standby) battery disconnected or fully discharged.

2-3.3 **Primary Power Supply—Monitored Battery.** Household fire warning equipment shall be permitted to be powered by a battery, provided that the battery is monitored to ensure that the following conditions are met:

(a) All power requirements are met for at least 1 year of battery life, including monthly testing.

(b) A distinctive audible trouble signal sounds before the battery is incapable of operating (from causes such as aging or terminal corrosion) the device(s) for alarm purposes.

Battery powered smoke detectors, (see Figure 2.5) once installed, are often ignored by household occupants. NFPA studies indicate that nearly twenty percent of installed detectors do not operate, primarily because of dead or missing batteries. These requirements address the maintenance steps necessary to ensure that the smoke detector will function reliably. The trouble signal requirement was added to allow occupants to be alerted to a imminent battery failure. However, many homeowners or tenants do not recognize the trouble signal and may think it is a nuisance alarm. Establishing a routine battery replacement program is important to keep smoke detectors functioning. A popular program is the "change your clocks, change your smoke detector batteries" plan. This is effective only in areas where daylight savings time is adopted. In other areas, increasing public awareness of regular battery testing is a greater challenge.

(c) For a unit employing a lock-in alarm feature, automatic transfer is provided from alarm to a trouble condition.

(d) The unit is capable of producing an alarm signal for at least 4 minutes at the battery voltage at which a trouble signal is normally obtained, followed by not less than 7 days of trouble signal operation.

(e) The audible trouble signal is produced at least once every minute for 7 consecutive days.

(f) Acceptable replacement batteries are clearly identified by the manufacturer's name and model number on the unit near the battery compartment.

(g) A readily noticeable, visible indication is displayed when a primary battery is removed from the unit.

(h) Any unit that uses a nonrechargeable battery as a primary power supply that is capable of a 10-year or greater service life, including testing, and meets the requirements of 2-3.3(b) through (e) shall not be required to have a replaceable battery.

Formal Interpretation 78-1

Reference: 2-3.3

Question 1: Is it the intent of 2-3.3 (a) for all power requirements to be met for at least one year's life at 90°F including weekly testing?

Answer: No.

Question 2: Is it the intent of 2-3.3 that inferior batteries (batteries with poor shelf life or incapable of meeting the power requirements over the normal environmental conditions of a household) can be used provided they meet the requirements of 2-3.3 (a) for at least six months?

Answer: No.

Question 3: Is it the intent of 2-3.3 that the requirements of 2-3.3 (a) do not have to be met provided the requirements of 2-3.3 (d) and 2-3.3 (e) are met?

Answer: No.

Issue Edition: NFPA 74-1978

Reference: 2-1.3.1, 2-1.3.1(a)

Issue Date: November 1978

Reissued: January 1986 ▪

Figure 2.5   Battery operated smoke detector. Photograph courtesy of Mammoth Fire Alarms (BRK First Alert), Lowell, MA.

2-3.4 **Secondary (Standby) Power Supply.**

2-3.4.1 Removal or disconnection of a battery used as a secondary (standby) power source shall cause a distinctive audible or visible trouble signal.

2-3.4.2 Acceptable replacement batteries shall be clearly identified by manufacturer's name and model number on the unit near the battery compartment.

2-3.4.3 Where required by law for disposal reasons, rechargeable batteries shall be removable.

2-3.4.4 **Automatic Recharging.**

2-3.4.4.1 Automatic recharging shall be provided where a rechargeable battery is used as the secondary (standby) supply. The supply shall be capable of operating the system for at least 24 hours in the normal condition, followed by not less than 4 minutes of alarm. Loss of the secondary (standby) source shall sound an audible trouble signal at least once every minute.

The authority having jurisdiction should require the submission of battery calculations for household fire alarm systems. These calculations are used to used to determine the capacity required for standby and alarm time requirements for the household fire alarm system when the standby power is supplied by rechargeable batteries.

2-3.4.4.2 The battery shall be recharged within 4 hours where power is provided from a circuit that can be switched on or off by means other than a circuit breaker, or within 48 hours where power is provided from a circuit that cannot be switched on or off by means other than a circuit breaker.

2-3.4.5 Where automatic recharging is not provided, the battery shall be monitored to ensure that the following conditions are met:

(a) All power requirements are met for at least 1 year of battery life.

(b) A distinctive audible trouble signal sounds before the battery capacity has been depleted below the level required to produce an alarm signal for 4 minutes.

2-3.5 **Primary Power — Nonelectrical.** A suitable spring-wound mechanism shall provide power for the nonelectrical portion of a listed single station alarm. A visible indication shall be provided to show that sufficient operating power is not available.

2-4 Equipment Performance.

2-4.1 **General.** The failure of any nonreliable or short-life component that renders the detector inoperable shall be readily apparent to the occupant of the living unit without the need for test.

This subsection requires the supervision of the circuitry in the smoke detector and some form of audible or visible indication of detector component failure. One acceptable means of indication could be a fail-safe feature, where the detector fails in the alarm mode. Other acceptable methods include use of visible or audible signals, such as indicator lights or a distinctive audible signal.

2-4.2 **Smoke Detectors.** Each smoke detector shall detect abnormal quantities of smoke that can occur in a dwelling, shall properly operate in the normal environmental conditions of a household, and shall be in compliance with ANSI/UL 268, *Standard for Safety Smoke Detectors for Fire Protective Signaling Systems*, or ANSI/UL 217, *Standard for Safety Single and Multiple Station Smoke Detectors*.

ANSI/UL 268 is the standard for "system" smoke detectors. These detectors are connected to and powered by a control panel and may also have integral notification appliances, depending on the model and manufacturer. All AC-powered, battery-powered, or combination AC- or battery-powered single- and multiple-station smoke

alarms must comply with ANSI/UL 217 in order to be listed. Smoke alarms are not typically connected to a control panel unless they have been specifically listed for that purpose. Both the ionization type and the photoelectric type smoke detectors are available under ANSI/UL 268 or ANSI/UL 217. See Chapter 5 for more details on these detectors.

2-4.3* **Heat Detectors.**

A-2-4.3 The linear space rating is the maximum allowable distance between heat detectors. The linear space rating is also a measure of detector response time to a standard test fire when tested at the same distance. The higher the rating, the faster the response time. This code recognizes only those heat detectors with ratings of 50 ft (15 m) or more.

The heat detector types that are allowed by this requirement are either fixed-temperature or rate-of-rise detectors. See Figures 2.6 and 2.7.

2-4.3.1 Each heat detector, including a heat detector integrally mounted on a smoke detector, shall detect abnormally high temperature or rate-of-temperature rise, and all such detectors shall be listed for not less than 50-ft (15-m) spacing.

For similar reasons, in applying rate-of-rise heat detectors, which respond to rapid temperature increases, one should consider the environment in which these detectors are to be installed. Areas near dishwashers, hot air vents, and ovens are examples of areas to be avoided. See Figure 2.7.

Figure 2.6 *Combination smoke detector and heat detector. Photograph courtesy of Mammoth Fire Alarms (ESL/Sentrol Inc.), Lowell, MA.*

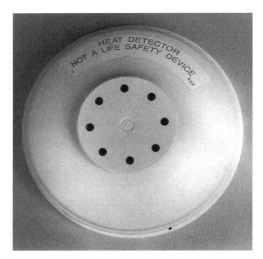

Figure 2.7   *Low profile rate-of-rise heat detector.*
Photograph courtesy of Mammoth Fire Alarms (Edwards
Systems Technology), Lowell, MA.

Formal Interpretation 89-1

Reference: 2-4.2

Question: Is it the intent of the Committee that only devices which comply with the definition of a smoke detector in 1-4 meet the intent of 2-4.2?

Answer: No. Paragraph 2-1.3.8 allows for any device or system that can demonstrate equivalent performance to be "approved." Data currently exists that could be used by any interested party to demonstrate such equivalency for a device sensing carbon monoxide. If an analysis of this data shows an equivalent performance, carbon monoxide sensing devices could be listed or approved as meeting the requirements of NFPA 72.

Issue Edition: NFPA 74-1989
Reference: 1-4, 4-2.1

Issue Date: January 2, 1990   ∎

2-4.3.2*   Fixed temperature detectors shall have a temperature rating at least 25°F (14°C) above the normal ambient temperature and shall not be rated 50°F (28°C) higher than the maximum anticipated ambient temperature in the room or space where installed.

A-2-4.3.2   A heat detector with a temperature rating somewhat in excess of the highest normally expected ambient

temperature is specified in order to avoid the possibility of premature response of the heat detector to nonfire conditions.

Some areas or rooms of the family living unit can experience ambient temperatures considerably higher than those in the normally occupied living spaces. Examples are unfinished attics, the space near hot air registers, and some furnace rooms. This fact should be considered in the selection of the appropriate temperature rating for fixed-temperature heat detectors to be installed in these areas or rooms.

2-4.4   **Alarm Signaling Intensity.**

2-4.4.1   All alarm-sounding appliances shall have a minimum rating of 85 dBA at 10 ft (3 m).

Exception: An additional sounding appliance intended for use in the same room as the user, such as a bedroom, may have a sound pressure level as low as 75 dBA at 10 ft (3 m).

The requirement of 85 dBA at 10 ft (3 m) is measured under specific conditions and marked on the notification appliance. The minimum rating of a notification appliance is not measured in the field. The sound level needed to awaken someone varies; however, the recommended sound level for household occupancies is 15 dBA above ambient conditions.

Where a protected premises fire alarm system is intended to provide warning to occupants of individual family living units or households (e.g., apartments, dormitory rooms, hotel rooms, etc.) contained within a commercial occupancy, the minimum sound pressure level required at the pillow in a bedroom, by Chapter 6, is the greater of 15 dBA above the average ambient, 5 dBA above the maximum sound pressure level lasting 60 seconds or longer, or 70 dBA. This may require individual notification appliances in each family living unit or household bedroom. For more information regarding notification appliances, refer to Chapter 6.

2-4.4.2   Visible notification appliances used in rooms where a hearing impaired person(s) sleeps shall have a minimum rating of 177 candela for a maximum room size of 14 ft × 16 ft (4.27 m × 4.88 m). For larger rooms, the visible notification appliance shall be located within 16 ft (4.88 m) of the pillow. Visible notification appliances in other areas shall have a minimum rating of 15 candela.

Exception: Where a visible notification appliance in a sleeping room is mounted more than 24 in. (610 mm) below the ceiling, a minimum rating of 110 candela shall be permitted.

There are federal laws and regulations that affect the placement and intensity of visible signals. The intensity levels required by this subsection were established through testing conducted by Underwriters Laboratories. These levels have been accepted by the Technical Committee on Household Fire Warning Equipment and the engineering community as sufficient for use in household fire alarm systems.

A 110 candela visible notification appliance, installed at least 24 inches below the ceiling, will generally provide sufficient light intensity to awaken a sleeping person. A 177 candela appliance is required when mounted on the ceiling because the light signal may be attenuated by the smoke layer. See Figures 2.8 and 2.9.

Figure 2.8 *Smoke detector with integral notification appliance for the hearing impaired. Photograph courtesy of BRK/First Alert, Aurora, IL.*

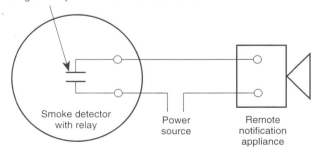

Figure 2.9 *Single-station smoke detector with remote notification appliance. Drawing courtesy of Fire Protection Alliance, Stockbridge, GA.*

2-4.5 Control Equipment.

2-4.5.1 The control equipment shall be automatically restoring upon restoration of electrical power.

2-4.5.2 The control equipment shall be of a type that "locks in" on an alarm condition. Smoke detection circuits shall not be required to lock in.

The control panel must have the "lock-in" feature. However, the circuits powering single- or multiple-station smoke detectors are not required to lock-in. System-connected smoke detectors utilizing a control panel are required by ANSI/UL 268 to provide a lamp or equivalent on a spot-type detector head or base to identify it as the unit from which the alarm was initiated. ANSI/UL 268 requires that "the means incorporated to identify the initiation of an alarm shall remain activated after the smoke has dissipated from within the detector." The lock-in feature is, therefore, required on all spot-type smoke detectors connected to a control panel.

2-4.5.3 If a reset switch is provided, it shall be of a self-restoring type.

2-4.5.4 An alarm-silencing switch or an audible trouble-silencing switch shall not be required to be provided.

Exception: Where the switch's silenced position is indicated by a readily apparent signal.

2-4.5.5 Each electrical fire warning system and each single station smoke detector shall have an integral test means to allow the householder to check the system and the sensitivity of the detector(s).

This sensitivity test assesses either the outside limits of detector operation, as allowed by the testing laboratories, or the detector's sensitivity "window" of operation. If a more exact sensitivity reading is necessary, see Chapters 5 and 7. See Figures 2.10 and 2.11.

2-4.6 Monitoring Integrity of Installation Conductors. All means of interconnecting initiating devices or notification appliances shall be monitored for the integrity of the interconnecting pathways up to the connections to the device or appliance so that the occurrence of a single open or single ground fault, which prevents normal operation of the system, is indicated by a distinctive trouble signal.

Figure 2.10   Smoke detector with integral sensitivity testing means. Photograph courtesy of Gentex, Inc., Zeeland, MI

Figure 2.11   Close-up of sensitivity testing means. Photograph courtesy of Gentex, Inc., Zeeland, MI.

Exception No. 1: Conductors connecting multiple station alarms, provided a single fault on the wiring cannot prevent single station operation of any of the interconnected detectors.

Exception No. 2: Circuits extending from single or multiple station alarms to required remote notification appliances, provided operation of the test feature on any detector causes all connected appliances to activate.

The monitoring of installation conductors only applies to the devices and appliances connected to a control panel installed in accordance with the requirements of Chapter 3.

2-4.7   Combination System.

2-4.7.1 Where common wiring is employed for a combination system, the equipment for other than the fire warning signaling system shall be connected to the common wiring of the system so that short circuits, open circuits, grounds, or any fault in this equipment or interconnection between this equipment and the fire warning system wiring does not interfere with the supervision of the fire warning system or prevent alarm or trouble signal operation.

In order to comply with this requirement, it may be necessary to provide additional detection beyond the basic requirements of 2-2.1.1.1. With some combination fire/burglar alarm control panels, a fault on the burglar alarm system wiring may affect the operation of the entire control panel. If this situation might occur, protection of the wiring and control panel accessories should be considered. This protection could take the form of additional detectors (heat or smoke detectors, depending on the location) or physical protection of the wiring using one or more metal raceways. See Figure 2.12.

2-4.7.2 In a fire/burglar system, the operation shall be as follows:

(a) A fire alarm signal shall take precedence or be clearly recognizable over any other signal even when the nonfire alarm signal is initiated first.

(b) Distinctive alarm signals shall be used so that fire alarms can be distinguished from other functions such as burglar alarms. The use of a common sounding appliance for fire and burglar alarms shall be permitted where distinctive signals are used. (*See* 2-2.2.2.)

The distinctive alarm signals called for in 2-4.7.2(b) can be coded signals, where a steady tone for the burglar alarm and a three-pulse temporal pattern for the fire alarm can be used, with one appliance supplying both tones. See Subsection 2-2.2.2.

2-4.8   Low Power Wireless Systems. Household fire warning systems utilizing low power wireless transmission of signals within the protected household shall comply with the requirements of Section 3-13.

The requirements for supervision of placement do not apply in this case because it is assumed that homeowners or occupants of a family living unit will not remove or tamper with their own equipment.

Exception: Paragraph 3-13.4.5 shall not apply.

2-4.9 Supervising Station Systems.

2-4.9.1 Any communications method described in Section 4-5 shall be permitted for transmission of signals from household fire warning equipment to a supervising station. All of the provisions of Section 4-5 shall apply, as appropriate.

Exception No. 1: Only one telephone line shall be required for one- and two-family residences.

Exception No. 2: Each DACT shall be required to be programmed to call a single DACR number only.

Exception No. 3: Each DACT serving a one- or two-family residence shall transmit a test signal to its associated receiver at least monthly.

The new exceptions allow the homeowner to utilize a Digital Alarm Communicator Transmitter (DACT), defined in Section 1-4 of the Code, without the additional restrictions imposed on protected premises (commercial) systems. Obviously, should the homeowner desire the additional protection requirements for commercial systems, the Code would allow it to be used in a home, provided the family living unit Code requirements that apply are also met.

2-4.9.2* On receipt of an alarm signal from household fire warning equipment, the supervising station shall immediately (within 90 seconds) retransmit the alarm to the public fire communications center.

Exception: The supervising station shall be permitted to contact the residence for verification of an alarm condition and, where acceptable assurance is provided within 90 seconds that the fire service is not needed, retransmission of an alarm to the public service fire communications center shall not be required.

This new section was added to the 1996 edition. The exception permits supervising station personnel to place a verification call before retransmitting the alarm signal. Verification should be conducted by means of a password.

Figure 2.12 *Combination fire/burglar alarm control unit. Photograph courtesy of Radionics, Inc., Salinas, CA.*

A-2-4.9.2   Where the exception to 2-4.9.2, which provides for screening alarm signals to minimize response to false alarms, is to be implemented, the following should be considered:

(a) Was the verification call answered at the protected premises?

(b) Did the respondent provide proper identification?

(c) Is it necessary for the respondent to identify the cause of the alarm signal?

(d) Should the public service fire communications center be notified and advised that an alarm signal was received, including the response to the verification call, when an authorized respondent states that fire service response is not desired?

(e) Should the public service fire communications center be notified and advised that an alarm signal was received, including the response to the verification call, for all other situations, including both a hostile fire and no answer to the verification call?

(f) What other actions should be required by a standard operating procedure?

2-5   Installation.

2-5.1   General.

2-5.1.1   General Provisions.

2-5.1.1.1*   All equipment shall be installed in a workmanlike manner.

"Workmanlike manner" means that the installation is neat, safe, easily maintained, and complies with all appropriate codes and standards.

A-2-5.1.1.1   Where homeowner inspection, testing, and maintenance are required or assumed, the equipment should be installed in an accessible manner.

2-5.1.1.2   All devices shall be so located and mounted that accidental operation is not caused by jarring or vibration.

2-5.1.1.3   All installed household fire warning equipment shall be mounted so as to be supported independently of its attachment to wires.

2-5.1.1.4   All equipment shall be restored to normal as promptly as possible after each alarm or test.

2-5.1.1.5   The supplier or installing contractor shall provide the owner with:

(a) An instruction booklet illustrating typical installation layouts.

(b) Instruction charts describing the operation, method and frequency of testing, and proper maintenance of household fire warning equipment.

(c) Printed information for establishing a household emergency evacuation plan.

(d) Printed information to inform owners where they can obtain repair or replacement service, and where and how parts requiring regular replacement (such as batteries or bulbs) can be obtained within 2 weeks.

2-5.1.2   **Interconnection of Detectors or Multiple Station Alarms.**

(a) Where the interconnected wiring is unsupervised, no more than 18 multiple station alarms shall be interconnected in a multiple station configuration.

(b) Where the interconnecting wiring is supervised, the number of interconnected detectors shall be limited to 64.

The intent of this requirement is to recognize that systems designed for use in large residences differ from those intended for use in smaller residences. It is important to remember that the configuration allowed in 2-5.1.2(a) incorporates heat detectors that may be connected to multiple-station smoke detectors or to their unsupervised circuits. The number of multiple-station smoke detectors used to protect a residence is limited to 12 (see 2-5.1.2.2). The cautions regarding misapplication of devices mentioned earlier in the chapter should be reviewed. Refer to Section 2-2.1.1.1 and commentary.

2-5.1.2.1*   Interconnection that causes other alarms to sound shall be limited to an individual family living unit. Remote annunciation from single or multiple station alarms shall be permitted.

A-2-5.1.2.1   One of the common problems associated with residential smoke detectors is the nuisance alarms that are usually triggered by products of combustion from cooking, smoking, or other household particulates. While an alarm for such a condition is anticipated and tolerated by the occupant of a family living unit through routine living experience, the alarm is not permitted where it also sounds alarms in other family living units or in common use spaces. Nuisance alarms caused by cooking are a very common occurrence, and inspection authorities should be aware of the possible ramifications where the coverage is extended beyond the limits of the family living unit.

2-5.1.2.2   No more than 12 smoke alarms shall be interconnected in a multiple station connection. The remainder of the alarms shall be permitted to be of other types.

Multiple-station applications where smoke alarms and heat detectors are connected together on the same circuit must utilize equipment that is listed as compatible.

2-5.2* Detector Location and Spacing.

A-2-5.2 One of the most critical factors of any fire alarm system is the location of the fire detecting devices. This appendix is not a technical study. It is an attempt to provide some fundamentals on detector location. For simplicity, only those types of detectors recognized by Chapter 2 (i.e., smoke and heat detectors) are discussed. In addition, special problems requiring engineering judgment, such as locations in attics and in rooms with high ceilings, are not covered.

For more information on special problems, see Chapter 5.

2-5.2.1* Smoke Detectors.

A-2-5.2.1 Smoke Detection.

(a) *Where to Locate the Required Smoke Detectors in Existing Construction.* The major threat from fire in a family living unit occurs at night when everyone is asleep. The principal threat to persons in sleeping areas comes from fires in the remainder of the unit; therefore, a smoke detector(s) is best located between the bedroom areas and the rest of the unit. In units with only one bedroom area on one floor, the smoke detector(s) should be located as shown in Figure A-2-5.2.1(a).

In family living units with more than one bedroom area or with bedrooms on more than one floor, more than one smoke detector is required, as shown in Figure A-2.5.2.1(b).

In addition to smoke detectors outside of the sleeping areas, Chapter 2 requires the installation of a smoke detector on each additional story of the family living unit, including the basement. These installations are shown in Figure A-2-5.2.1(c). The living area smoke detector should be installed in the living room or near the stairway to the upper level, or in both locations. The basement smoke detector should be installed in close proximity to the stairway leading to the floor above. Where installed on an open-joisted ceiling, the detector should be placed on the bottom of the joists. The detector should be positioned relative to the stairway so as to intercept smoke coming from a fire in the basement before the smoke enters the stairway.

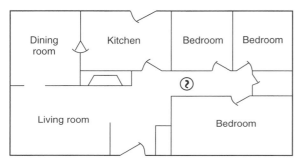

Figure A-2-5.2.1(a) *A smoke detector should be located between the sleeping area and the rest of the family living unit.*

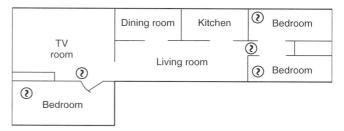

Figure A-2-5.2.1(b) *In family living units with more than one sleeping area, a smoke detector should be provided to protect each sleeping area in addition to detectors required in bedrooms.*

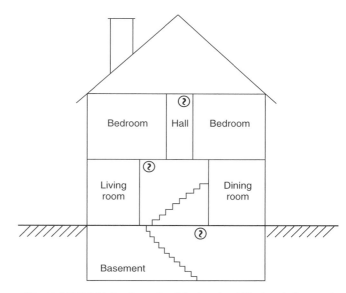

Figure A-2-5.2.1(c) *A smoke detector should be located on each story.*

(b) *Where to Locate the Required Smoke Detectors in New Construction.* All of the smoke detectors specified in A-2-5.2.1(a) for existing construction are required, and, in addition, a smoke detector is required in each bedroom.

(c) *Are More Smoke Detectors Desirable?* The required number of smoke detectors might not provide reliable early warning protection for those areas separated by a door from the areas protected by the required smoke detectors. For this reason, it is recommended that the householder consider the use of additional smoke detectors for those areas for increased protection. The additional areas include the basement, bedrooms, dining room, furnace room, utility room, and hallways not protected by the required smoke detectors. The installation of smoke detectors in kitchens, attics (finished or unfinished), or garages is not normally recommended, as these locations occasionally experience conditions that can result in improper operation.

2-5.2.1.1 Smoke detectors in rooms with ceiling slopes greater than 1 ft in 8 ft (1 m in 8 m) horizontally shall be located at the high side of the room.

2-5.2.1.2 A smoke detector installed in a stairwell shall be so located as to ensure that smoke rising in the stairwell cannot be prevented from reaching the detector by an intervening door or obstruction.

Doors at the tops of stairwells will prevent smoke flow in the upward direction. The stairwell acts as a dead air space and will trap smoke below. This prevents smoke from reaching a detector located in the stairwell.

2-5.2.1.3 A smoke detector installed to detect a fire in the basement shall be located in close proximity to the stairway leading to the floor above.

Doors at the top of stairways prevent smoke from flowing upward. The stairwell acts as a dead air space, trapping smoke below and preventing smoke from reaching a detector located in the stairwell. Detectors should always be mounted on the bottom of floor joists in unfinished construction. Placing the detector in the pockets between joists may prevent or delay detector response.

2-5.2.1.4 The smoke detector installed to comply with 2-2.1.1.1 on a story without a separate sleeping area shall be located in close proximity to the stairway leading to the floor above.

This practice will allow the detector to sense smoke rising up the stairway to occupant sleeping areas.

2-5.2.1.5* Smoke detectors shall be mounted on the ceiling at least 4 in. (102 mm) from a wall or on a wall with the top of the detector not less than 4 in. (102 mm) nor more than 12 in. (305 mm) below the ceiling.

Exception: Where the mounting surface might become considerably warmer or cooler than the room, such as a poorly insulated ceiling below an unfinished attic or an exterior wall, the detectors shall be mounted on an inside wall.

A-2-5.2.1.5 **Smoke Detector Mounting—Dead Air Space.** The smoke from a fire generally rises to the ceiling, spreads out across the ceiling surface, and begins to bank down from the ceiling. The corner where the ceiling and wall meet is an air space into which the smoke could have difficulty penetrating. In most fires, this dead air space measures about 4 in. (0.1 m) along the ceiling from the corner and about 4 in. (0.1 m) down the wall, as shown in Figure A-2-5.2.2(b). Detectors should not be placed in this dead air space.

Smoke and heat detectors should be installed in those locations recommended by the manufacturer, except in those cases where the space above the ceiling is open to the outside and little or no insulation is present over the ceiling. Such cases result in the ceiling being excessively cold in the winter or excessively hot in the summer. Where the ceiling is significantly different in temperature from the air space below, smoke and heat have difficulty reaching the ceiling and a detector that is located on that ceiling. In this situation, placement of the detector on a sidewall, with the top 4 in. to 12 in. (0.1 m to 0.3 m) from the ceiling, is recommended.

The situation described above for uninsulated or poorly insulated ceilings can also exist, to a lesser extent, in the case of outside walls. The recommendation is to place the smoke detector on a sidewall. However, where the sidewall is an exterior wall with little or no insulation, an interior wall should be selected. It should be recognized that the condition of inadequately insulated ceilings and walls can exist in multi-family housing (apartments), single-family housing, and mobile homes.

In those family living units employing radiant heating in the ceiling, the wall location is the recommended location. Radiant heating in the ceiling can create a hot-air, boundary layer along the ceiling surface, which can seriously restrict the movement of smoke and heat to a ceiling-mounted detector.

2-5.2.1.6 Smoke detectors shall not be located within kitchens or garages, or in other spaces where temperatures can fall below 40°F (4°C) or exceed 100°F (38°C). Smoke detectors shall not be located closer than 3 ft (0.9 m) horizontally from:

(a) The door to a kitchen.

(b) The door to a bathroom containing a tub or shower.

(c) The supply registers of a forced air heating or cooling system, and outside of the airflow from those registers.

Exception: Detectors specifically listed for the application.

Detectors producing nuisance alarms should be relocated away from areas that produce excessive moisture, cooking vapors or exhaust. Additional relief may be provided by using photoelectric type smoke detection. These devices are more resistant to nuisance alarms caused by moisture in bathrooms as well as moisture and cooking vapors in kitchen areas

2-5.2.2* **Heat Detectors.**

Heat detectors are not considered life safety devices. They should be used only to provide additional detection in areas not suitable for smoke detectors.

A-2-5.2.2 **Heat Detection.**

(a) *General.* While Chapter 2 does not require heat detectors as part of the basic protection scheme, it is recommended that the householder consider the use of additional heat detectors for the same reasons presented under A-2-5.2.1(c). The additional areas lending themselves to protection with heat detectors are the kitchen, dining room, attic (finished or unfinished), furnace room, utility room, basement, and integral or attached garage. For bedrooms, the installation of a smoke detector is recommended over the installation of a heat detector for protection of the occupants from fires in their bedrooms.

(b) *Heat Detector Mounting — Dead Air Space.* Heat from a fire rises to the ceiling, spreads out across the ceiling surface, and begins to bank down from the ceiling. The corner where the ceiling and the wall meet is an air space into which heat has difficulty penetrating. In most fires, this dead air space measures about 4 in. (0.1 m) along the ceiling from the corner and 4 in. (0.1 m) down the wall as shown in Figure A-2-5.2.2(b). Heat detectors should not be placed in this dead air space.

The placement of the detector is critical where maximum speed of fire detection is desired. Thus, a logical location for a detector is the center of the ceiling. At this location, the detector is closest to all areas of the room.

If the detector cannot be located in the center of the ceiling, an off-center location on the ceiling may be permitted to be used.

The next logical location for mounting detectors is on the sidewall. Any detector mounted on the sidewall should be located as near as possible to the ceiling. A detector mounted on the sidewall should have the top of the detector between 4 in. and 12 in. (0.1 m and 0.3 m) from the ceiling.

(c) *The Spacing of Detectors.* Where a room is too large for protection by a single detector, several detectors should be used. It is important that they be properly located so all parts of the room are covered. (*For further information on the spacing of detectors, see* Chapter 5.)

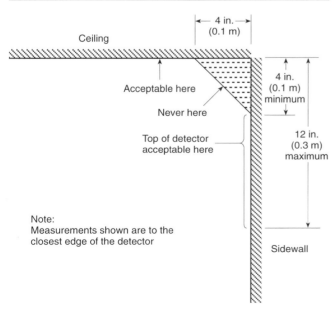

Figure A-2-5.2.2(b)   *Example of proper mounting for detectors.*

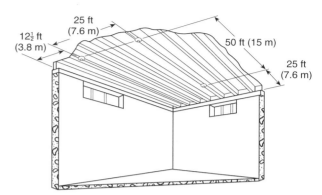

Figure A-2-5.2.2(d)   *Open joists, attics, and extra high ceilings are some of the areas that require special knowledge for installation.*

(d) *Where the Distance Between Detectors Should Be Further Reduced.* The distance between detectors is based on data obtained from the spread of heat across a smooth ceiling. Where the ceiling is not smooth, the placement of the detector should be tailored to the situation.

For instance, with open wood joists, heat travels freely down the joist channels so that the maximum distance between detectors [50 ft (15 m)] may be permitted to be used. However, heat has trouble spreading across the joists, so the distance in this direction should be $\frac{1}{2}$ the distance allowed between detectors, as shown in Figure A-2-5.2.2(d), and the distance to the wall is reduced to $12\frac{1}{2}$ ft (3.8 m). Since $\frac{1}{2} \times 50$ ft (15 m) is 25 ft (7.6 m), the distance between detectors across open wood joists should not exceed 25 ft (7.6 m), as shown in Figure A-2-5.2.2(d), and the distance to the wall is reduced [$\frac{1}{2} \times 25$ ft (7.6 m)] to 12.5 ft (3.8 m). Paragraph 2-5.2.2.4 requires that detectors be mounted on the bottom of the joists and not up in joist channels.

Walls, partitions, doorways, ceiling beams, and open joists interrupt the normal flow of heat, thus creating new areas to be protected.

2-5.2.2.1 On smooth ceilings, heat detectors shall be installed within the strict limitations of their listed spacing.

2-5.2.2.2 For sloped ceilings having a rise greater than 1 ft in 8 ft (1 m in 8 m) horizontally, the detector shall be located on or near the ceiling at or within 3 ft (0.9 m) of the peak. The spacing of additional detectors, if any, shall be based on a horizontal distance measurement, not on a measurement along the slope of the ceiling.

2-5.2.2.3* Heat detectors shall be mounted on the ceiling at least 4 in. (102 mm) from a wall or on a wall with the top of the detector not less than 4 in. (102 mm) nor more than 12 in. (305 mm) below the ceiling.

Exception: Where the mounting surface might become considerably warmer or cooler than the room, such as a poorly insulated ceiling below an unfinished attic or an exterior wall, the detectors shall be mounted on an inside wall.

A-2-5.2.2.3   See A-2-5.2.1.5.

2-5.2.2.4   In rooms with open joists or beams, all ceiling-mounted detectors shall be located on the bottom of such joists or beams.

2-5.2.2.5*   Detectors installed on an open-joisted ceiling shall have their smooth ceiling spacing reduced where this spacing is measured at right angles to solid joists; in the case of heat detectors, this spacing shall not exceed $1/2$ of the listed spacing.

A-2-5.2.2.5   In addition to the special requirements for heat detectors installed on ceilings with exposed joists, reduced spacing also might be required due to other structural characteristics of the protected area, possible drafts, or other conditions that could affect detector operation.

2-5.3   **Wiring and Equipment.**   The installation of wiring and equipment shall be in accordance with the requirements of NFPA 70, *National Electrical Code*, Article 760.

The installation of all fire alarm system wiring should take into account the fire alarm system manufacturer's published installation instructions and the limitations of the applicable product listings or approvals.

2-6   Maintenance and Tests.

2-6.1*   **Maintenance.**   Where batteries are used as a source of energy, they shall be replaced in accordance with the recommendations of the alarm equipment manufacturer.

Exception: Batteries described in 2-3.3(h).

A-2-6.1   Good fire protection requires that the equipment be periodically maintained. If the householder is unable to perform the required maintenance, a maintenance agreement should be considered.

2-6.2*   **Tests.**

A-2-6.2   It is a good practice to establish a specific schedule for these tests.

2-6.2.1   **Single and Multiple Station Smoke Alarms.**   Homeowners shall inspect and test smoke alarms and all connected appliances in accordance with the manufacturer's instructions at least monthly.

2-6.2.2   **Fire Alarm Systems.**   Homeowners shall test systems in accordance with the manufacturer's instructions and shall have every household fire alarm system having a control panel tested by a qualified service technician at least every 3 years. This test shall be conducted according to the methods of Chapter 7.

Homeowners who are unable to perform monthly maintenance should consider a maintenance contract or agreement with a qualified service company.

2-7   Markings and Instructions.

All household fire warning equipment or systems shall be plainly marked with the following information on the unit:

(a)   Manufacturer's or listee's name, address, and model number;

(b)   A mark or certification that the unit has been approved or listed by a testing laboratory;

(c)   Electrical rating (where applicable);

(d)   Temperature rating (where applicable);

(e)   Spacing rating (where applicable);

(f)   Operating instructions;

(g)   Test instructions;

(h)   Maintenance instructions;

(I)   Replacement and service instructions.

Exception: Where space limitations prohibit inclusion of 2-7(g), (h), and (i), a label or plaque suitable for permanent attachment within the living unit, or a manufacturer's manual, shall be provided with the equipment and referenced on the equipment. In the case of a household fire warning system, the required information shall be prominently displayed at the control panel.

3

Protected Premises
Fire Alarm Systems

Contents, Chapter 3

3-1   Scope.

This chapter provides requirements for the application, installation, and performance of fire alarm systems, including fire alarm and supervisory signals, within protected premises.

3-2   General.

The systems covered in this chapter are intended to be used for the protection of life by automatically indicating the necessity for evacuation of the building or fire area, and for the protection of property through the automatic notification of responsible persons and for the automatic activation of fire safety functions. The requirements of the other chapters shall also apply.

The wording in the first sentence of Section 3-2 has changed from "primarily for the protection of life and secondarily for the protection of property," which appeared in earlier editions of NFPA 72. This change is to emphasize that the protection of both life and property are given equal and full consideration.

Exception No. 1: Where the requirements of other chapters conflict with the requirements of this chapter.

This exception is to ensure that all fire alarm systems installed within the protected premises shall first comply with Chapter 3 and then comply with the requirements of other chapters. The other chapters may add requirements to the protected premises system installation, but must not replace, or conflict with, the requirements of Chapter 3.

Exception No. 2: For household fire warning equipment protecting a single living unit, see Chapter 2.

Fire alarm systems as discussed in Chapter 3 can also be used in single living units or as household fire warning systems, provided the requirements of Chapter 2 are satisfied. However, the fire alarm systems covered under Chapter 2 are usually more economical and easier to operate for household applications.

Exception No. 3: For the performance, installation, and operation requirements of continuously attended fire alarm system supervising stations and subsidiary stations and the transmission and communications channels used to convey signals between the protected premises and the supervising station, see Chapter 4.

This exception clearly delineates responsibility of all off-premises signaling requirements to Chapter 4.

3-2.1 Systems requiring transmission of signals to continuously attended locations providing supervising station service (e.g., central station, proprietary, remote station) shall also comply with the applicable requirements of Chapter 4.

Reference to Chapter 4 applies to signals transmitted to off-premises locations, utilizing central supervising stations, proprietary supervising stations, remote supervising station systems, and auxiliary systems. The Chapter 4 requirements apply to the transmitter located at the protected premises and the transmission channel between the protected premises and the remotely located supervising station.

There are essentially two choices that can be made where a single multi-building contiguous property has its proprietary supervising station in one of the on-site buildings. Each building can have its own protected premises system and be connected to its on-site supervisory station through a transmitter and transmission channel that meet the requirements of Section 4-5. Or, the individual building systems can be directly connected to the supervising station using signaling line circuits (SLCs). These systems must comply with the requirements of Section 3-11 for interconnected fire alarm control units. For a single-building property, the initiating devices and notification appliances are either directly connected or connected through zone or floor sub-control units to the supervising station using SLC circuits. Where sub-control units are used, the requirements of Section 3-11 apply. Regardless of the method of connection, for both single- and multiple-building properties, the supervising station facilities must comply with the requirements of Section 4-3.

3-2.2 All protected premises fire alarm systems shall be maintained and tested in accordance with Chapter 7.

Testing of fire alarm systems is of paramount importance if the system reliability is to be ensured. 3-2.2 applies to all installed fire alarm systems. Because the testing of fire alarm systems is critical, many jurisdictions feel the need to develop and enforce their own fire alarm system testing requirements. The purpose of 3-2.2 is to provide the authorities having jurisdiction with a mandatory requirement that can be enforced, to test all fire alarm systems in accordance with this Code.

3-2.3 Fire alarm systems provided for evacuation of occupants shall have one or more notification appliances listed for the purpose on each floor of the building and so located that they have the characteristics for public mode described in Chapter 6.

When the protected premises fire alarm system's purpose is to notify and evacuate the occupants of the protected premises, Chapter 6 requirements for public mode signaling must be followed.

The phrase "listed for the purpose" appears throughout the code. Many devices and appliances are listed, but they may be listed for other non-fire alarm applications. For example, there are listed notification appliances that do not meet all the requirements for fire alarm use and are intended for background music or other applications. Only those notification appliances specifically listed for fire alarm use are acceptable to meet the "listed for the purpose" requirement of 3-2.3 (See 1-5.1.2).

3-2.4* The system shall be so designed and installed that attack by fire:

(a) In an evacuation zone, causing loss of communications to this evacuation zone, shall not result in loss of communications to any other evacuation zone.

"Evacuation zone" is not defined in the code, however "notification zone" is defined and when that definition is combined with the definition of "zone," the term "evacuation zone" makes sense. An evacuation zone can be an area of a floor, an entire floor, or several floors that are always intended to be evacuated simultaneously. See also 3-4.2 and 3-12.7. See Figure 3.1.

(b) Causing failure of equipment or a fault on one or more installation wiring conductors of one communications path shall not result in total loss of communications to any evacuation zone.

This paragraph essentially requires the use of separate communications paths, circuits, or some other arrangement to avoid the total loss of communications when the fire attacks the installation wiring conductors of one communications path. For example, if a fire attacked a communications path riser that connected all of the notification appliances on multiple floors, it is possible that there would be a total loss of communications to all the floors connected to that riser. However, if the notification appliances on each floor were served by a separate riser, then a fire attack on one communications path riser would not result in a total loss of communications to all the connected floors. See Exception No. 6 to 3-2.4 (b) and A-3-2.4. See Figure 3.1.

Exception No. 1 to (a) and (b): Systems that, on alarm, automatically sound evacuation signals throughout the protected premises.

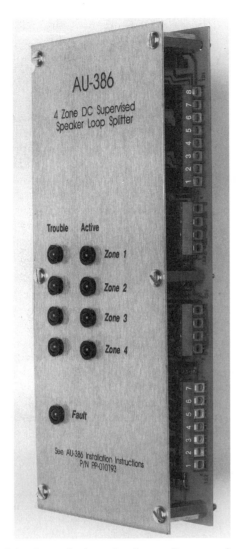

Figure 3.1   Supervised speaker loop splitters can be used to separate speaker circuits to evacuation zones. Photograph courtesy of Audiosone Corp., Stratford, CT

This entire section applies only to fire alarm systems that provide selective evacuation of a protected premises.

Exception No. 2 to (a) and (b):   Where there is a separate means acceptable to the authority having jurisdiction for voice communications to each floor or evacuation zone.

An example of a "separate means" of communication that may be acceptable to an authority having jurisdiction would be a separate paging system that is either a stand alone system or integral to a background music

system, with the controls for the paging system located in the Fire Command Center (see 3-12.6.5).

Exception No. 3 to (b):   The fire command center and the central control equipment.

Exception No. 4 to (b):   Where the installation wiring is enclosed in a 2-hour rated cable assembly or enclosed in a 2-hour rated enclosure, other than a stairwell.

This exception has been changed to allow the use of 2-hour rated cable assemblies.

Exception No. 5 to (b):   Where the installation wiring is enclosed within a 2-hour rated stairwell in a fully sprinklered building in accordance with NFPA 13, Standard for the Installation of Sprinkler Systems.

This exception applies only to stairwells that are not classified as exit stairwells by other codes.

Exception No. 6 to (b):   When the evacuation zone is directly attacked by fire within the zone.

A-3-2.4   This requirement is intended to limit damage to a fire alarm system, resulting from a fire, to the area in which the fire occurs. The concern is maintaining the operability of the system in areas beyond, but threatened by, the fire.
Conformance to this requirement could entail that:

(a) Where common risers or trunk circuits are used:

1. Separately routed, redundant risers or trunk circuits be provided, arranged so that one or more circuit faults on one riser or trunk circuit cause the system to switch over automatically to its associated, alternate circuit without loss of function. This capability should allow full system operation with a damaged or severed riser or trunk circuit.

2. Primary and alternate conductors for redundant circuits be separated by 2-hour fire-resistive construction.

(b) Where multiple individual circuits are routed in a common riser, conduit, raceway, cable, bundle of conductors, or other arrangement resulting in close physical proximity and resultant susceptibility to common misfortune, such circuits be Class A, capable of full operation over a single open or single ground fault.

(c) Where Class A circuits are required, they be installed so that the supply and return conductors are routed separately. Supply and return risers should be separated by at least 2-hour rated fire construction.

The survivability requirements previously applied only to emergency voice/alarm communication systems. They now have been extended to all types of systems installed in buildings where the fire response plan permits either selective evacuation or relocation of the building occupants to a "safe" area during a fire emergency. The survivability requirements have been relocated to Section 3-2 because the intent now is to have them apply to non-voice systems used to evacuate the building occupants by floor or zone. If occupants are allowed or required to remain in the building, it is essential that the fire alarm system remain operational so that additional floors or zones can be evacuated as needed.

3-2.5 Software and Firmware Control.

3-2.5.1 All software and firmware provided with a fire alarm system shall be listed for use with the fire alarm control unit.

3-2.5.2 A record of installed software and firmware version numbers shall be maintained at the location of the fire alarm control unit.

3-2.5.3 All software and firmware shall be protected from unauthorized changes through the use of "access levels."

3-2.5.4 All changes shall be tested in accordance with 7-1.6.2.

This new section of the Code has been inserted to acknowledge the special requirements that must be followed when using the computerized, microprocessor-based fire alarm systems. The term "firmware" is contained in the definition of "Operating System Software" and is synonymous with the latter term. The term "software" is not specifically defined in the Code, but it means the programming specific to the fire alarm system within the protected premises. Because a single programming change could effect the entire operation of the fire alarm system, the software and firmware must be protected from unauthorized use and the revision or "REV" number of the installed software and firmware must be recorded. If, upon subsequent inspections, the authority having jurisdiction finds a different "REV" number of software or firmware installed than was installed at the time of the acceptance test, then the additional tests performed in accordance with 7-1.6.2 should be reviewed.

3-3 Applications.

Protected premises fire alarm systems include one or more of the following features:

(a) Manual alarm signal initiation;

(b) Automatic alarm signal initiation;

(c) Monitoring of abnormal conditions in fire suppression systems;

(d) Activation of fire suppression systems;

(e) Activation of fire safety functions;

(f) Activation of alarm notification appliances;

(g) Emergency voice/alarm communications;

(h) Guard's tour supervisory service;

(i) Process monitoring supervisory systems;

(j) Activation of off-premises signals;

(k) Combination systems;

(l) Integrated systems.

3-4 System Performance and Integrity.

The title for this section has changed and speaks to design issues concerned with the circuit performance and operation of the protected premises fire alarm system.

3-4.1 The purpose of this section is to provide information to be used in the design and installation of protected premises fire alarm systems for the protection of life and property.

3-4.2 Notification zones shall be consistent with the emergency response or evacuation plan for the protected premises. The boundaries of notification zones shall be coincident with building outer walls, building fire or smoke compartment boundaries, floor separations, or other fire safety subdivisions.

As stated in section 3-2.4, "notification zone" has been defined. The boundaries described in this section further define the term "evacuation zone" as used in section 3-2.4. It is possible to have multiple notification appliance circuits in an evacuation zone. In such cases, the affected notification appliances must all sound within that evacuation zone. Also, if the evacuation plan is for selective evacuation, then the requirements of 3-2.4 must be followed with regards to survivability.

3-4.3* **Circuit Designations.** Initiating device, notification appliance, and signaling line circuits shall be designated by class or style, or both, depending on the circuits' capability to continue to operate during specified fault conditions.

This section requires that the class or style of circuits must be designated based on the designer's (or owner's) requirements. Therefore, unless another code, the authority having jurisdiction or the project's specifications has designated a class or style of circuit to be used, it is the designer's responsibility to designate the circuit classifications. Although 3-4.3 states that "circuits shall be designated by class or style, or both," if a circuit performance is specified as a certain style, it would be redundant to classify the circuit using both class and style.

A-3-4.3
Class A circuits are considered to be more reliable than Class B circuits because they remain fully operational during the occurrence of a single open or a single ground fault, while Class B circuits remain operational only up to the location of an open fault. However, neither Class A nor Class B circuits remain operational during a wire-to-wire short.

For both Class A and Class B initiating device circuits, a wire-to-wire short is permitted to cause an alarm on the system based on the rationale that a wire-to-wire short is the result of a double fault (e.g., both circuit conductors have become grounded), while the code only considers the consequences of single faults. For many applications, an alarm caused by a wire-to-wire short is not permitted, and limitation to a simple Class A designation is not adequate. Introducing the style designation has made it possible to specify the exact performance required during a variety of possible fault conditions.

Limitation to Class A and Class B circuits only poses a more serious problem for signaling line circuits. Though a Class A signaling line circuit remains fully operational during the occurrence of a single open or single ground fault, a wire-to-wire short disables the entire circuit. The risk of such a catastrophic failure is unacceptable to many system designers, users, and authorities having jurisdiction. Once again, using the style designation makes it possible to specify either full system operation during a wire-to-wire short (Style 7) or a level of performance in between that of a Style 7 and a minimum function Class A circuit (Style 2).

A specifier can specify a circuit as either Class A or Class B where system performance during wire-to-wire shorts is of no concern, or it can specify, by the appropriate style designation, where the system performance during a wire-to-wire short and other multiple fault conditions *is* of concern.

The type of circuit to be used by the designer often depends on the number of devices on the circuit, the amount of detection that would be lost during a fault condition and the impact on life safety or property protection that the loss of detection would have. The Code does not require a specific class or style of circuit to be used, but simply defines the operation of each and leaves the decision to the designer or owner. Reliable operation under the environmental and physical conditions present in the protected premises under consideration is one of the more important factors to be considered for circuit class or style selection.

3-4.3.1 Class. Initiating device, notification appliance, and signaling line circuits shall be permitted to be designated as either Class A or Class B, depending on the capability of the circuit to transmit alarm and trouble signals during nonsimultaneous single circuit fault conditions as specified by the following:

The performance of either class of circuit is dependent on two factors: how it is physically wired and how the fire alarm control unit operates in the specified fault condition.

(a) Circuits capable of transmitting an alarm signal during a single open or a nonsimultaneous single ground fault on a circuit conductor shall be designated as Class A.

(b) Circuits not capable of transmitting an alarm beyond the location of the fault conditions specified in 3-4.3.1(a) shall be designated as Class B.

Faults on both Class A and Class B circuits shall result in a trouble condition on the system in accordance with the requirements of 1-5.8.

3-4.3.2 Style. Initiating device, notification appliance, and signaling line circuits shall be permitted to be designated by style also, depending on the capability of the circuit to transmit alarm and trouble signals during specified simultaneous multiple circuit fault conditions in addition to the single circuit fault conditions considered in the designation of the circuits by class.

The style of circuit to be used by the designer again depends on the number of devices on the circuit, the amount of detection that would be lost during the specified fault conditions and the impact on life safety or property protection that the loss of detection would have. The Code does not require a specific style of circuit to be used, but simply defines the operation of each and leaves the decision to the designer or owner. Reliable operation under the environmental and physical conditions present in the protected premises under consideration is one of the more important factors to be considered for circuit style selection.

The performance of any of the circuit styles is dependent on two factors: how it is physically wired and how the fire alarm control unit operates in the specified fault conditions.

(a) An initiating device circuit shall be permitted to be designated as either Style A, B, C, D, or E, depending on its ability to meet the alarm and trouble performance requirements shown in Table 3-5, during a single open, single ground, wire-to-wire short, and loss of carrier fault condition.

(b) A notification appliance circuit shall be permitted to be designated as either Style W, X, Y, or Z, depending on its ability to meet the alarm and trouble performance requirements shown in Table 3-7.1, during a single open, single ground, and wire-to-wire short fault condition.

(c) A signaling line circuit shall be permitted to be designated as either Style 0.5, 1, 2, 3, 3.5, 4, 4.5, 5, 6, or 7, depending on its ability to meet the alarm and trouble performance requirements shown in Table 3-6, during a single open, single ground, wire-to-wire short, simultaneous wire-to-wire short and open, simultaneous wire-to-wire short and ground, simultaneous open and ground, and loss of carrier fault conditions.

3-4.4*
All styles of Class A circuits using physical conductors (e.g., metallic, optical fiber) shall be installed such that the outgoing and return conductors, exiting from and returning to the control unit, respectively, are routed separately. The outgoing and return (redundant) circuit conductors shall not be run in the same cable assembly (i.e., multiconductor cable), enclosure, or raceway.

This requirement ensures that the Class A circuit's operation is not defeated by the cutting of the cable or the failure of a section of cable due to attack by fire in a single location. See A-3-4.4.

Exception No. 1:   For a distance not to exceed 10 ft (3 m) where the outgoing and return conductors enter or exit the initiating device, notification appliance, or control unit enclosures; or

Exception No. 2:   Where the vertically run conductors are contained in a 2-hour rated cable assembly or enclosed (installed) in a 2-hour rated enclosure other than a stairwell; or

It is now acceptable to use cable assemblies that have been listed with a fire rating of 2 hours.

Exception No. 3:   Where permitted and where the vertically run conductors are enclosed (installed) in a 2-hour rated stairwell in a building fully sprinklered in accordance with NFPA 13, Standard for the Installation of Sprinkler Systems.

This exception applies only to stairwells that are not classified as exit stairwells by other codes.

Exception No. 4:   Where looped conduit/raceway systems are provided, single conduit/raceway drops to individual devices or appliances shall be permitted.

Exception No. 5:   Where looped conduit/raceway systems are provided, single conduit/raceway drops to multiple devices or appliances installed within a single room not exceeding 1000 ft² (92.9 m²) in area shall be permitted.

A-3-4.4
A goal of 3-4.4 is to provide adequate separation between the outgoing and return cables. This separation is required to help ensure protection of the cables from physical damage. The recommended minimum separation to prevent physical damage is 1 ft (305 mm) where the cable is installed vertically and 4 ft (1.22 m) where the cable is installed horizontally.

The amount of separation distance recommendation was added to the Appendix because each installation will present different physical conditions and the amount of separation distance depends on those conditions. When Class A circuits are required by the designer, reliability of operation under the specified fault conditions is obviously a design consideration. The separation of the outgoing and return conductors is an additional requirement to help ensure reliable operation.

3-4.5   Signaling Paths.

3-4.5.1
The class or style of signaling paths (circuits) shall be determined from an evaluation based on the path performance detailed in this code and on engineering judgment.

It is important to evaluate the type of circuit (Initiating Device Circuit or Signaling Line Circuit) that is to be used based on the planned installation technique. If "T" tapping is used, the performance of the type of circuit originally chosen may not be valid with the "T" taps in the circuit. Engineering judgment would be used in conjunction with the requirements of 3-4.5.2 to determine the overall reliability desired from the class or style circuit needed to meet the fire protection goals of the owner.

3-4.5.2
Where determining the integrity and reliability of the interconnecting signaling paths (circuits) installed within the protected premises, the following influences shall be considered:

(a) The transmission media utilized;

(b) The length of the circuit conductors;

(c) The total building area covered by and the quantity of initiating devices and notification appliances connected to a single circuit;

(d) The nature of the hazard present within the protected premises;

(e) The functional requirements of the system necessary to provide the level of protection required for the system;

(f) The size and nature of the population of the protected premises.

3-5*   Performance of Initiating Device Circuits (IDC).

The assignment of class designations or style designations, or both, to initiating device circuits shall be based on their performance capabilities under abnormal (fault) conditions in accordance with the requirements of Table 3-5.

A-3-5   Tables 3-5 and 3-6 should be used as follows:

(a) It should be determined whether the initiating devices are:

1. Directly connected to the initiating device circuit.

2. Directly connected to a signaling line circuit interface on a signaling line circuit.

3. Directly connected to an initiating device circuit, which in turn is connected to a signaling line circuit interface on a signaling line circuit.

(b) The style of signaling performance required should be determined. The columns marked A through Eα in Table 3-5,

and 0.5 through 7α in Table 3-6 are arranged in ascending order of performance and capacities.

(c) Upon determining the style of the system, the charts, singularly or together, specify the maximum number of devices, equipment, premises, and buildings permitted to be incorporated into an actual protected premises installation.

(d) In contrast, where the number of devices, equipment, premises, and buildings (in addition to signaling ability) in an installation is known, a required system style can be determined.

(e) The prime purpose of the tables is to enable identification of minimum performance for styles of initiating device circuits and signaling line circuits. It is not the intention that the styles be construed as grades. That is, a Style 3 system is not superior to a Style 2 system, or vice versa. In fact, a particular style might better provide adequate and reliable signaling for an installation than a more complex style. The quantities tabulated under each style do, unfortunately, tend to imply that a given style is superior to the style to its left. The increased quantities for the higher style numbers are based on the ability to signal an alarm during an abnormal condition in addition to signaling the same abnormal condition.

(f) The tables allow users, designers, manufacturers, and the authority having jurisdiction to identify minimum performance of present and future systems by determining the trou-

Table 3-5   Performance of Initiating Device Circuits (IDC)

	Class	B			B			B			A			A		
	Style	A			B			C			D			Eα		
R = Required capability X = Indication required at protected premises and as required by Chapter 4 α = Style exceeds minimum requirements for Class A		Alarm	Trouble	Alarm receipt capability during abnormal condition	Alarm	Trouble	Alarm receipt capability during abnormal condition	Alarm	Trouble	Alarm receipt capability during abnormal condition	Alarm	Trouble	Alarm receipt capability during abnormal condition	Alarm	Trouble	Alarm receipt capability during abnormal condition
Abnormal Condition		1	2	3	4	5	6	7	8	9	10	11	12	13	14	15
A. Single open			X			X			X			X	X			X	X
B. Single ground			R			X	R		X	R		X	R			X	R
C. Wire-to-wire short		X			X				X		X					X	
D. Loss of carrier (if used)/channel interface									X							X	

Table 3-6 Performance of Signaling Line Circuits (SLC)

Class	B			B			A			B			B			B			B			A			A			A		
Style	0.5			1			2α			3			3.5			4			4.5			5α			6α			7α		
M = May be capable of alarm with wire-to-wire short; R = Required capability; X = Indication required at protected premises and as required by Chapter 4; α = Style exceeds minimum requirements for Class A	Alarm	Trouble	Alarm receipt capability during abnormal condition	Alarm	Trouble	Alarm receipt capability during abnormal condition	Alarm	Trouble	Alarm receipt capability during abnormal condition	Alarm	Trouble	Alarm receipt capability during abnormal condition	Alarm	Trouble	Alarm receipt capability during abnormal condition	Alarm	Trouble	Alarm receipt capability during abnormal condition	Alarm	Trouble	Alarm receipt capability during abnormal condition	Alarm	Trouble	Alarm receipt capability during abnormal condition	Alarm	Trouble	Alarm receipt capability during abnormal condition	Alarm	Trouble	Alarm receipt capability during abnormal condition
Abnormal Condition	1	2	3	4	5	6	7	8	9	10	11	12	13	14	15	16	17	18	19	20	21	22	23	24	25	26	27	28	29	30
A. Single open		X			X			X	R		X			X			X			X	R		X	R		X	R		X	R
B. Single ground		X			X	R		X	R		X	R		X			X	R		X			X	R		X	R		X	R
C. Wire-to-wire short								M			X			X			X			X			X			X			X	R
D. Wire-to-wire short & open								M			X			X			X			X			X			X			X	
E. Wire-to-wire short & ground							X	M			X			X			X			X			X			X			X	
F. Open and ground							X	R			X			X			X			X			X		X	X			X	R
G. Loss of carrier (if used)/channel interface														X			X			X			X			X			X	

ble and alarm signals received at the control unit for the specified abnormal conditions.

(g) The overall system reliability is considered to be equal from style to style where the capacities are at the maximum allowed.

(h) Upon determining the style of the system, the tables indicate specifics such as the maximum number of devices, equipment, and protected buildings permitted to be incorporated into an actual installation for a protected premises fire alarm system.

(i) The number of automatic fire detectors connected to an initiating device circuit is limited by good engineering practice. Where a large number of detectors are connected to one initiating device circuit covering a widespread area, pinpointing the source of alarm becomes difficult and time consuming.

On certain types of detectors, a trouble signal results from faults in the detector. When this occurs where there are large numbers of detectors on an initiating device circuit, locating the faulty detector also becomes difficult and time consuming.

3-6* Performance of Signaling Line Circuits (SLC).

The assignment of class designations or style designations, or both, to signaling line circuits shall be based on their performance capabilities under abnormal (fault) conditions in accordance with the requirements of Table 3-6.

A-3-6 See A-3-5.

The capacity limitations of initiating device and signaling line circuits have not been defined. Many manufacturers publish the capacity limitations of their systems, but if there is a question as to the number of devices allowed, common sense should dictate. The capacities allowed on a single circuit should be determined based on the amount of detection that would be lost during a fault condition

Formal Interpretation 79-8

Reference: 3-5, 3-6

Background: This is a request for a formal interpretation in regards to the use of addressable initiating devices. Since the control unit or the central supervising station is in two-way communication with these devices, then:

Question 1: Is it the intent to categorize the description of performance of the circuit these devices are on as a "Signaling Line Circuit," rather than an "Initiating Device Circuit"?

Answer: Yes.

Question 2: If the style of this addressable communication circuit has a different performance (style number) than the remaining portions of the multiplex pathways, would these different circuit performance levels have to be individually specified in order to adequately describe the system?

Answer: Yes.

Issue Edition: NFPA 72D-1979

Reference: 3-9, 3-10

Issue Date: June 1985   ∎

Formal Interpretation 87-1

Reference: Tables 3-5, 3-6

Question 1: Is it the intent of the Committee to categorize the description of performance of the circuit these devices are on as a "Signaling Line Circuit" illustrated in Table 3-6 rather than an Initiating Device Circuit as described in Table 3-5?

Answer: Yes.

Question 2: If the style of this communication circuit has a different performance (style number as illustrated in Table 3-6) than the remaining portions of the multiplex pathways, would these different circuit performance levels have to be individually specified in order to adequately describe the system?

Answer: Yes.

Issue Edition: NFPA 72D-1987

Reference: Tables 2-12.1, 2-13.1

Issue Date: June 1987

Reissued: January 1989   ∎

Formal Interpretation 86-1

Reference: Table 3-6

Question: For systems where the main (primary) and standby (secondary) power are transmitted over the same circuit separate from the signaling circuit, is it the intent of 72 to require redundancy (i.e., two such circuits) of the power circuit, when a Style 7 signaling line circuit is required?

Answer: Yes.

Issue Edition: NFPA 72D-1986

Reference: Entire Standard

Issue Date: August 1986

Replaces F.I. 86-1 issued February 1986   ∎

and the impact on life safety or property protection that the loss of detection would have. Reliable operation under the environmental and physical conditions present in the protected premises under consideration is one of the more important factors to be considered when determining the number of devices to be installed on a single circuit. The original Table 3-5 and Table 3-6 included device capacity limitations based on the class or style of the circuit used. The capacity sections only applied to proprietary supervising station systems and was an attempt to make the circuits "equal."

3-7   Notification Appliance Circuits (NAC).

3-7.1 **Performance.** The assignment of class designations or style designations, or both, to notification appliance circuits shall be based on their performance capabilities under abnormal (fault) conditions in accordance with the requirements of Table 3-7.1.

3-7.2 **Distinctive Evacuation Signal.**

(a)* Paragraph 1-5.4.7 requires that fire alarm signals be distinctive in sound from other signals and that this sound not be used for any other purpose. To meet this requirement, the fire alarm signal used to notify building occupants of the need to evacuate (leave the building) shall be in accordance with ANSI S3.41, *Audible Emergency Evacuation Signal.*

A-3-7.2(a) The use of the distinctive three-pulse temporal pattern fire alarm evacuation signal required by 3-7.2(a) became effective July 1, 1996, for new systems installed after that date. It had previously been recommended for this

Table 3-7.1 Notification Appliance Circuits (NAC)

	Class	B	B	B	A
X = Indication required at protected premises	Style	W	X	Y	Z
		Trouble indication at protected premises / Alarm capability during abnormal conditions	Trouble indication at protected premises / Alarm capability during abnormal conditions	Trouble indication at protected premises / Alarm capability during abnormal conditions	Trouble indication at protected premises / Alarm capability during abnormal condition
Abnormal condition		1 2	3 4	5 6	7 8
Single open		X	X X	X	X X
Single ground		X	X	X X	X X
Wire-to-wire short		X	X	X	X

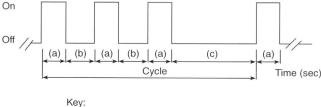

Key:
Phase (a) signal is "on" for 0.5 sec ± 10%
Phase (b) signal is "off" for 0.5 sec ± 10%
Phase (c) signal is "off" for 1.5 sec ± 10% [(c) = (a) + 2(b)]
Total cycle lasts for 4 sec ± 10%

Figure A-3-7.2(a)(1) *Temporal pattern parameters.*

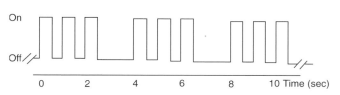

Figure A-3-7.2(a)(2) *Temporal pattern imposed on signaling appliances that emit a continuous signal while energized.*

purpose by this code since 1979. It has since been adopted as both an American National Standard (ANSI S3.41, *Audible Emergency Evacuation Signal*) and an International Standard (ISO 8201, *Audible Emergency Evacuation Signal*).

Copies of both of these standards are available from the Standards Secretariat, Acoustical Society of America, 335 East 45th Street, New York, NY 10017-3483. Telephone 212-661-9404, ext. 562.

The standard fire alarm evacuation signal is a three-pulse temporal pattern using any appropriate sound. The pattern consists of an "on" phase (a) lasting 0.5 second ± 10 percent followed by an "off" phase (b) lasting 0.5 second ± 10 percent, for three successive "on" periods, which are then followed by an "off" phase (c) lasting 1.5 seconds ± 10 percent [*see* Figures A-3-7.2(a)(1) and (2)]. The signal should be repeated for a period appropriate for the purposes of evacuation of the building, but for not less than 180 seconds. A single-stroke bell or chime sounded at "on" intervals lasting 1 second ± 10 percent, with a 2-second ± 10 percent "off" interval after each third "on" stroke, may be permitted [*see* Figure A-3-7.2(a)(3)].

The minimum repetition time shall be permitted to be manually interrupted.

It is important to note that the American National Standard Audible Emergency Evacuation Signal is required only where immediate evacuation of a building or a zone of a building is desired. The goal is to have anyone hearing the signal in any building in the United States (or in other countries that adopt ISO standards requiring the same evacuation signal) immediately recognize the signal as a fire alarm evacuation signal.

(b) The use of the American National Standard Audible Emergency Evacuation Signal shall be restricted to situations where it is desired that all occupants hearing the signal evacuate the building immediately. It shall not be used where, with the approval of the authority having jurisdiction, the planned action during a fire emergency is not evacuation, but relocation, of the occupants from the affected area to a safe area within the building, or their protection in place (e.g., high-rise buildings, health care facilities, penal institutions).

Figure A-3-7.2(a)(3) *Temporal pattern imposed on a single-stroke bell or chime.*

3-8   System Requirements.

See also Section 5-8.

3-8.1   Manual Fire Alarm Signal Initiation.

3-8.1.1   Fire alarm boxes shall be listed for the intended application, installed in accordance with Chapter 5, and tested in accordance with Chapter 7.

3-8.1.2   For fire alarm systems employing automatic fire detectors or waterflow detection devices, at least one fire alarm box shall be provided to initiate a fire alarm signal. This fire alarm box shall be located where required by the authority having jurisdiction.

One reason that at least one fire alarm box is required is to allow an alarm to be transmitted if the automatic fire detectors or sprinkler system are out of service during repairs or during a test (this presumes there is a contingency plan to address a fire emergency during the out of service time or during a test and that the supervising station is aware of both the plan and acknowledges receipt of the alarm). Typical locations for these manual fire alarm boxes could be at the automatic sprinkler control valve area or at the telephone operator's station. Another reason the fire alarm box is required is to permit a building occupant to initiate an alarm signal prior to the actuation of an automatic initiating device, providing earlier warning of a fire emergency. Where there is more than one fire alarm box on a zone circuit, and where it is practical, placing one at the end of the circuit allows for both testing of the fire alarm box and ensuring that the entire zone circuit is intact.

Exception: Fire alarm systems dedicated to elevator recall control and supervisory service as permitted in 3-8.14.1.

3-8.1.3   Where signals from fire alarm boxes and other fire alarm initiating devices within a building are transmitted over the same signaling line circuit, there shall be no interference with fire alarm box signals when both types of initiating devices are operated at or near the same time. Provision of the shunt noninterfering method of operation shall be permitted for this performance.

Note that this requirement applies only to systems that have devices reporting to the control panel using a signaling line circuit arrangement. This does not apply to systems utilizing initiating device circuits, since these circuits do not distinguish which initiating device initiated the alarm. The requirement was first introduced to apply to spring-wound coded devices that could only transmit a fixed number of rounds of code. If the first device was interfered with by the simultaneous alarm transmission of another device on the same circuit, the first transmission could be lost or garbled.

3-8.2   Automatic Fire Alarm Signal Initiation.

3-8.2.1   Automatic alarm-initiating devices shall be listed for the intended application and installed in accordance with Chapter 5.

The phrases "listed for the purpose" and "listed for the intended application" have similar meanings. "Listed for the intended application" means, for example, that if a device is to be used in a low-temperature environment or a wet location, the device should be listed for that environment. Also, if a device is to be used for releasing service, it should be listed for that application as well. "Listed for the purpose," as used in this Code, means listed for fire alarm system use.

3-8.2.2   Automatic alarm-initiating devices having integral trouble contacts shall be wired on the initiating device circuit so that a trouble condition within a device does not impair the alarm transmission from any other initiating device.

It is possible to disable part or all of an initiating device circuit beyond the initiating device in trouble if the integral trouble contacts are connected improperly. Most automatic alarm initiating devices do not have integral trouble contacts; therefore, this requirement does not apply.

At one time, photoelectric smoke detectors used a tungsten filament lamp as a light source, and the integrity of the filament was required to be monitored. A common method used for reporting an open filament was to open a normally closed trouble contact within the smoke detector, which was wired in series with the initiating circuit.

To comply with this requirement, the trouble contacts must be wired in series, with the end-of-line resistor at the end of the circuit after the last initiating device. See Figures 3.2 and 3.3.

NOTE: Although a trouble signal is required when a plug-in initiating device is removed from its base, it is not considered as a trouble condition within the device and the requirement of 3-8.2.2 does not apply.

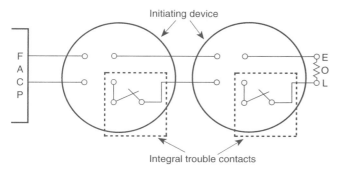

Figure 3.2 <u>Incorrect</u> *method of connection of integral trouble contacts. Drawing courtesy of the Fire Protection Alliance, Inc., Stockbridge, GA.*

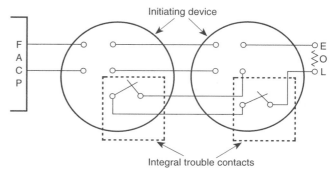

Figure 3.3 <u>Correct</u> *method of connection of integral trouble contacts. Drawing courtesy of the Fire Protection Alliance, Inc., Stockbridge, GA.*

The note following 3-8.2.2 exempts initiating devices that plug-in to a base from the requirement. However, it is important to remember that the removal of a plug-in initiating device from its base will interrupt the initiating device circuit (IDC) and could affect the alarm receipt capability at the fire alarm control unit.

3-8.2.3* Systems equipped with alarm verification features shall be permitted, provided:

(a) A smoke detector continuously subjected to a smoke concentration above alarm threshold magnitude initiates a system alarm within 1 minute.

(b) Actuation of an alarm-initiating device other than a smoke detector causes a system alarm signal within 15 seconds.

A-3-8.2.3 The alarm verification feature should not be used as a substitute for proper detector location/applications or regular system maintenance. Alarm verification features are intended to reduce the frequency of false alarms caused by transient conditions. They are not intended to compensate for design errors or lack of maintenance.

Alarm verification can be very useful in reducing false alarms from transient conditions, but not from conditions that remain relatively constant, such as a wet environment or areas subject to insect infestation. The alarm verification feature can reduce both malicious and accidental false alarms caused by such acts as the spraying of aerosols into a smoke detector or a gust of wind blowing dust or contaminants into the detector. The feature should not be installed/programmed in a system until a thorough investigation of the causes of false alarms has been accomplished. Alarm verification refers to specific timing sequences of detector/system operation. See Figure 3.4.

3-8.2.4 Where individual alarm-initiating devices are used to control the operation of equipment as permitted by 1-5.4.1.1, this control capability shall remain operable even when all of the initiating devices connected to the same circuit are in an alarm state.

This requirement does not allow the use of multiple, two-wire, circuit-powered smoke detectors with an auxiliary relay to be installed on the same initiating device circuit. In addition, this type of smoke detector (with an auxiliary relay) should not be installed on the same initiating device circuit with other devices, such as manual fire alarm boxes, waterflow switches or heat detectors. The reason for this requirement is that with a device such as a heat detector in alarm on the initiating device circuit, power is removed from the remaining smoke detector(s) with the relay. Without power, the smoke detectors with the control relay cannot operate. This problem does not exist with detectors installed on signaling line circuits because each device reports separately and is not affected by the operation of other devices on the circuit.

3-8.2.5* Systems that require the operation of two automatic detection devices to initiate the alarm response shall be permitted, provided:

A-3-8.2.5 See A-3-8.2.5(c).

(a) They are not prohibited by the authority having jurisdiction.

(b) There are at least two automatic detection devices in each protected space.

(c)* The area protected by an automatic detection device is no more than $1/2$ the maximum area for the detector as determined by the application of Chapter 5.

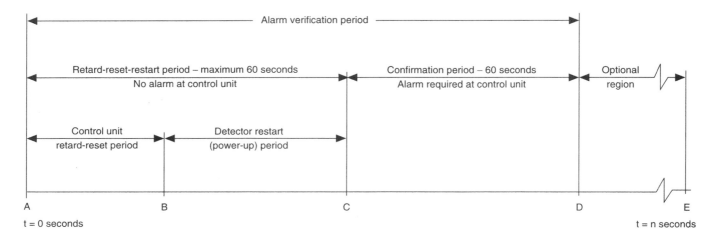

A – Smoke detector goes into alarm.

AB – Retard-reset period (control unit) – Control unit senses detector in alarm and retards (delays) alarm signal, usually by de-energizing power to the detector. Length of time varies with design.

BC – Restart period (detector power-up time) – Power to the detector is reapplied and time is allowed for detector to become operational for alarm. Time varies with detector design.

AC – Retard-reset-restart period – No alarm obtained from control unit. Maximum permissible time is 60 seconds.

CD – Confirmation period – Detector is operational for alarm at point C. If detector is still in alarm at point C, control unit will alarm. If detector is not in alarm, system returns to standby. If the detector realarms at any time during the confirmation period, the control unit will alarm.

DE – Optional region – Either an alarm can occur at control unit or restart of the alarm verification cycle can occur.

AD – Alarm verification period – Consists of the retard-reset-restart and confirmation periods.

Figure 3.4   Alarm verification timing diagram. Drawing courtesy of Underwriters Laboratories, Inc., Northbrook, IL.

A-3-8.2.5(c)   Alarm response, as identified, is intended to be the activation of the building evacuation (notification) alarms.

Example:
Given: Area to be protected: 30 ft × 60 ft (9.14 m × 18.28 m). No air movement.
Detector listed to cover 900 ft^2 (83.6 m^2).
Solution: Per 3-8.2.5(c), area is required to be reduced by $^1/_2$, or 900 ft^2 ÷ 2 = 450 ft^2 (83.6 m^2 ÷ 2 = 41.8 m^2) per detector. Number of devices required per Chapter 5.
30 ft × 60 ft = 1800 ft^2 (9.14 m × 18.28 m = 167.2 m^2).
1800 ft^2 ÷ 450 ft^2 (167.2 m^2 ÷ 41.8 m^2) per detector = 4 detectors.
Therefore, four detectors are required.
Where the system requires the operation of two detection devices to initiate an automatic suppression system and the first automatic detection device causes the activation of notification appliances within the protected space, the maximum area per detector is not intended to be reduced by $^1/_2$.
Example (for fire extinguishing and suppression systems):
Given: Area to be protected: 30 ft × 60 ft (9.14 m × 18.28 m). No air movement.
Detectors listed to cover 900 ft^2 (83.6 m^2). Number of detectors required per Chapter 5.
Solution: 30 ft × 60 ft = 1800 ft^2 (9.14 m × 18.28 m = 167.2 m^2).
1800 ft^2 ÷ 900 ft^2 (167.2 m^2 ÷ 83.6 m^2) per detector = 2 detectors.
Therefore, two detectors are required.

The example is not clear when compared to the requirements and guidance provided in Chapter 5. The intent of the requirement is to have more detectors than normally required by Chapter 5, but less than the amount of detectors required if half-spacing requirements were followed.

(d)   The alarm verification feature is not used.

The common names used for this type of configuration are "cross zoning" and "priority matrix zoning." The most common use of this configuration is with special hazard extinguishing systems. In this configuration, false alarms are minimized because more than one detector in alarm is required to activate the special hazard system. Another similar design option is requiring any second detector in alarm on the same zone before sounding the general alarm.

3-8.3*   **Concealed Detectors.**   Where a remote alarm indicator is provided of an automatic fire detector in a concealed location, the location of the detector and the area protected by the detector shall be prominently indicated at the remote alarm indicator by a permanently attached placard or by other approved means.

In a conventional fire alarm system (non-addressable) a remote alarm indicator, usually a red Light Emitting Diode (LED) mounted on a single gang plate, is the most common method of indicating an alarm from a concealed detector. It is important to locate and mark the remote alarm indicator so that the detector in alarm can be found easily. Engraved phenolic plates permanently attached to the remote alarm indicator are generally considered the most appropriate way of complying with this requirement. See Figures 3.5 and 3.6.

In an addressable fire alarm system, the Liquid Crystal Display (LCD), located at the fire alarm control unit or remote annunciator, will provide detailed detector location information and would be an acceptable alternative to compliance with this requirement.

A-3-8.3   Embossed plastic tape, pencil, ink, or crayon should not be considered to be a permanently attached placard.

3-8.4 Automatic Drift Compensation.   Where automatic drift compensation of sensitivity for a fire detector is provided, the control unit shall identify the affected detector when the limit of compensation is reached.

Automatic drift compensation of sensitivity is used to eliminate false alarms from smoke detectors that experience dust or dirt build-up within the detection chamber or that respond to minor changes in the environment.

Figure 3.6   *Remote indicator used for concealed detectors. Photograph (System Sensor) courtesy of Mammoth Fire Alarms, Inc., Lowell, MA.*

The feature allows the detector to maintain its original sensitivity by compensating for effects caused by outside sources. If the compensated value places the detector's sensitivity outside its listed window of sensitivity, the control unit indicates that maintenance is needed.

3-8.5 **Waterflow Alarm Signal Initiation.**

3-8.5.1   The provisions of 3-8.5 shall apply to sprinkler system signaling attachments that initiate an alarm indicating a flow of water in the system. Waterflow initiating devices shall be listed for the intended application and installed in accordance with Chapter 5.

Waterflow devices are also required by NFPA 13, *Standard for the Installation of Sprinkler Systems*, and additional information regarding their placement and use may be found in NFPA 13. See Figures 3.7 and 3.8.

3-8.5.2   A dry-pipe or preaction sprinkler system that is supplied with water by a connection beyond the alarm initiating device of a wet-pipe system shall be equipped with a separate waterflow alarm initiating pressure switch or other approved means to initiate a waterflow alarm.

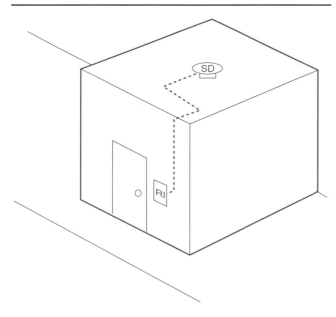

Figure 3.5   *Concealed smoke detector (SD) in locked room with remote indicator (RI). Drawing courtesy of FIREPRO Incorporated, Burlington, MA*

Figure 3.7   Vane-type waterflow switch. Photograph courtesy of Potter Electric Signal Company, St. Louis, MO.

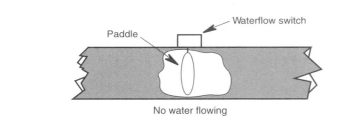

No water flowing

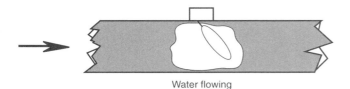

Water flowing

Figure 3.8   Typical waterflow switch operation. Drawing courtesy of FIREPRO Incorporated, Burlington, MA.

3-8.5.3 The number of waterflow switches permitted to be connected to a single initiating device circuit shall not exceed five.

The maximum number of waterflow switches the Code allows on a single IDC is five. However, design considerations for a given installation or requirements by the authority having jurisdiction might limit the number further. The reasoning here is to avoid the potential loss of all waterflow signals in a large sprinklered facility and to avoid confusion as to the location of the activated sprinkler area. In a large protected area, consideration should be given to separately zoning each main waterflow switch.

The maximum floor area an automatic sprinkler system supplied by any one sprinkler riser can protect ranges from 40,000 to 52,000 square feet. (See NFPA 13, 4-2.1) The size of the building may warrant a smaller area breakdown by zone to respond to the fire department needs for locating the source of the fire quickly.

3-8.6 **Supervisory Signal Initiation.**

3-8.6.1 **General.** The provisions of 3-8.6 shall apply to the monitoring of sprinkler systems, other fire suppression systems, and other systems for the protection of life and property for the initiation of a supervisory signal indicating an off-normal condition that could adversely affect the performance of the system.

3-8.6.1.1 Supervisory devices shall be listed for the intended application and installed in accordance with Chapter 5.

3-8.6.1.2 The number of supervisory devices permitted to be connected to a single initiating device circuit shall not exceed 20.

The Code allows up to 20 supervisory devices on a single initiating device circuit. However, design considerations for a given installation might limit the number even further. Locating a supervisory device that is in an off-normal position may prove to be difficult and time consuming when 20 devices are on one circuit. The Code does not differentiate between types of supervisory devices (e.g., gate valve versus special hazard control unit), and this also may add to the difficulty of locating the device off-normal. Care should be taken when determining the number and type of supervisory devices installed on an initiating device circuit.

3-8.6.2* Provisions shall be made for supervising the conditions that are essential for the proper operation of sprinkler and other fire suppression systems.

Exception: Those conditions related to water mains, tanks, cisterns, reservoirs, and other water supplies controlled by a municipality or a public utility.

A-3-8.6.2 Supervisory systems are not intended to provide indication of design, installation, or functional defects in the supervised systems or system components and are not a substitute for regular testing of those systems in accordance with the applicable standard.

Supervised conditions should include, but should not be limited to:

    (a) Control valves 1½ in. (38.1 mm) or larger.

See Figures 3.9 and 3.10.

    (b) Pressure:
        Dry-pipe system air
        Pressure tank air
        Pre-action system supervisory air
        Steam for flooding systems
        Public water.

See Figure 3.11.

    (c) Water tanks:
        Level
        Temperature.

    (d) Building temperature (including areas such as valve closet and fire pump house).

    (e) Fire pumps:
        Electric:
            Running (alarm or supervisory)
            Power failure
            Phase reversal.
        Engine-driven:
            Running (alarm or supervisory)
            Failure to start
            Controller off "automatic"
            Trouble (e.g., low oil, high temperature, overspeed).
        Steam turbine:
            Running (alarm or supervisory)
            Steam pressure
            Steam control valves.

    (f) Fire suppression systems appropriate to the system employed.

3-8.6.3 Signals shall distinctively indicate the particular function (e.g., valve position, temperature, or pressure) of the system that is off-normal and also indicate its restoration to normal.

NOTE: Cancellation of the off-normal signal may be permitted as a restoration signal, unless separate recording of all changes of state is a specific requirement. *See* Chapter 4.

3-8.6.4 A dry-pipe sprinkler system equipped for waterflow alarm signaling shall be supervised for off-normal system air pressure.

The intent of 3-8.6.4 is to require the monitoring of the air pressure of the dry-pipe sprinkler system. This will result in the receipt of a supervisory signal prior to loss of total air pressure and the subsequent filling of the piping network with water.

3-8.6.5 A control valve shall be supervised to initiate a distinctive signal indicating movement of the valve from its normal position. The off-normal signal shall remain until the valve is restored to its normal position. The off-normal signal shall be obtained during the first two revolutions of the hand wheel or during ⅕ of the travel distance of the valve control apparatus from its normal position.

This subsection provides the requirements in cases where the valves on connections to water supplies are supervised. There are four methods of supervision allowed by NFPA 13. (See NFPA 13, subsection 4-6.1.1.3.) The method described in 3-8.6.5 meets the requirements of NFPA 13, 4-6.1.1.3, (a) and (b). Some property insurers may require that the valves on connections to water supplies be locked in the open position using either frangible shackle locks or hardened shackle locks. Electrical supervision does provide more information to management regarding the position of these valves, but as is the case with all of these requirements, the primary goal is to increase the reliability of water availability to the fire suppression system. See Figures 3.9 and 3.10.

3-8.6.6 An initiating device for supervising the position of a control valve shall not interfere with the operation of the valve, obstruct the view of its indicator, or prevent access for valve maintenance.

3-8.6.7 **Pressure Supervision.** Pressure sources shall be supervised to obtain two separate and distinct signals, one indicating that the required pressure has been increased or decreased, and the other indicating restoration of the pressure to its required value.

    (a) A pressure supervisory signal-initiating device for a pressure tank shall indicate both high and low pressure conditions. A signal shall be obtained where the required pressure is increased or decreased 10 psi (70 kPa) from the required pressure value.

    (b) A pressure supervisory signal-initiating device for a dry-pipe sprinkler system shall indicate both high and low pressure conditions. A signal shall be obtained when the required pressure is increased or decreased 10 psi (70 kPa) from the required pressure value.

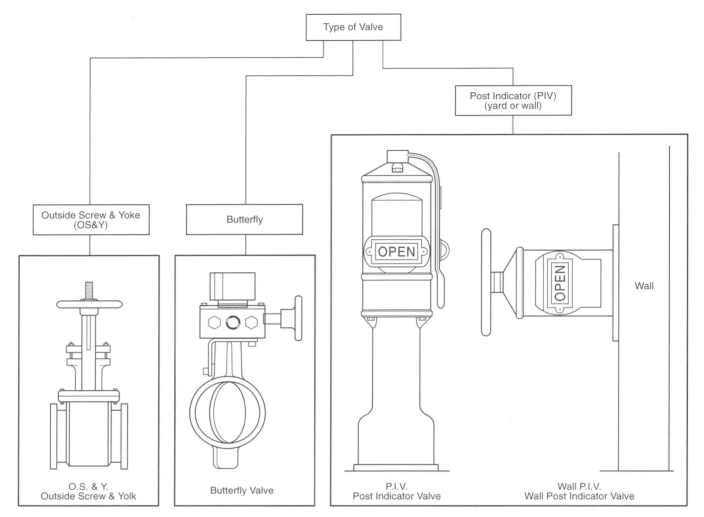

Figure 3.9   *Types of control valves that require a supervisory switch. Drawing courtesy of Potter Electric Signal Company, St. Louis, MO.*

(c) A steam pressure supervisory initiating device shall indicate a low pressure condition. A signal shall be obtained where the pressure is reduced to a value that is 110 percent of the minimum operating pressure of the steam operated equipment supplied.

(d) An initiating device for supervising the pressure of sources other than those specified in 3-8.6.7(a) through (c) shall be provided as required by the authority having jurisdiction.

See Figure 3.11.

3-8.6.8 **Water Temperature Supervision.** Exposed water storage containers shall be supervised to obtain two separate and distinct signals, one indicating that the temperature of the water has been lowered to 40°F (4.4°C), and the other indicating restoration to a temperature above 40°F (4.4°C).

The water intended for fire suppression system use stored in the outside containers must not be allowed to freeze; therefore, the temperature is monitored.

3-8.7 **Signal Annunciation.** Protected premises fire alarm systems shall be arranged to annunciate alarm, supervisory, and trouble signals in accordance with 1-5.7.

The requirement for three separate and distinct signals has been included in previous editions of the Code. All protected premises systems must indicate all three signals in a distinct fashion.

The system trouble LED and audible notification appliance cannot be used to indicate a supervisory condition (see 1-5.4.7). However, the system trouble audible notification appliance can be used in conjunction with a separate supervisory LED indication to comply with 3-8.7.

Figure 3.10 *Control valve supervisory switch. Photograph courtesy of Potter Electric Signal Company, St. Louis, MO.*

3-8.8 Signal Initiation from Automatic Fire Suppression System Other than Waterflow.

3-8.8.1 The operation of an automatic fire suppression system installed within the protected premises shall cause an alarm signal at the protected premises fire alarm control unit.

This subsection requires that an alarm condition at the automatic suppression system will cause an alarm indication at the protected premises fire alarm system. The automatic suppression system should be connected to the protected premises control unit as a separate zone or point. When an alarm condition occurs on the automatic suppression system, the protected premises system would operate as intended and activate all the notification appliances, etc.

3-8.8.2 A supervisory signal shall indicate the off-normal condition and its restoration to normal as appropriate to the system employed.

In addition to indicating alarm conditions to the protected premises fire alarm system, the off-normal condition of the automatic suppression system must also

be indicated. This supervisory signal must be able to be differentiated from a trouble condition such as a broken wire in the interconnection wiring between the protected premises fire alarm system and the automatic suppression system. See 3-8.8.3.

3-8.8.3 The integrity of each fire suppression system actuating device and its circuit shall be supervised in accordance with 1-5.8.1 and with other applicable NFPA standards.

The word "supervised" as used here means the same as "monitored for integrity" as used elsewhere in this Code.

Because fire suppression and alarm systems are not exercised on a daily basis, the integrity of the systems is electrically monitored for proper operation regardless of the type of system (protected premises fire alarm system or special hazard suppression system).

3-8.9 Pump Supervision. Automatic fire pumps and special service pumps shall be supervised in accordance with NFPA 20, *Standard for the Installation of Centrifugal Fire Pumps,* and the authority having jurisdiction.

3-8.9.1 Supervision of electric power supplying the pump shall be made on the line side of the motor starter. All phases and phase reversal shall be supervised.

The word "supervision" is used in 3-8.9 and 3-8.9.1 for the same reason indicated in the commentary following 3-8.8.3.

Figure 3.11 *High/low pressure switch. Photograph courtesy of Potter Electric Signal Company, St. Louis, MO.*

3-8.9.2 Where both sprinkler supervisory signals and pump running signals are transmitted over the same signaling circuits, provisions shall be made to obtain pump running signal preference.

Exception: Where the circuit is so arranged that no signals can be lost.

3-8.10 Tampering.

3-8.10.1 Automatic fire suppression system alarm-initiating devices and supervisory signal-initiating devices and their circuits shall be so designed and installed that they cannot be readily subject to tampering, opening, or removal without initiating a signal. This provision specifically includes junction boxes installed outside of buildings to facilitate access to the initiating device circuit.

Junction boxes installed outside of buildings must be either tamper-proof or provided with a device to initiate a signal when the box is opened. Generally this is a supervisory signal, not an alarm signal; however, the authority having jurisdiction should be consulted prior to using this type of connection to determine the signal to be transmitted.

3-8.10.2* Where a valve is installed in the connection between a signal attachment and the fire suppression system to which it is attached, such a valve shall be supervised in accordance with the requirements of Chapter 5.

Supervision of the valve means monitoring the status of the valve and providing an indication of an off-normal condition of the valve being supervised, at the protected premises fire alarm system. As previously noted, "supervised" as used in this Code, has the same meaning as "monitored for integrity."

A-3-8.10.2 Sealing or locking such a valve in the open position or removing the handle from the valve does not meet the intent of this requirement.

3-8.11 Guard's Tour Supervisory Service.

Guard's tour supervisory service is used to provide fire protection surveillance during the hours when occupants are not in a building; to facilitate and control the movement of persons into, out of, and within a building; and to carry out procedures for the orderly conduct of specific operations in a building or on the surrounding property.

Guard's tour supervisory services designed to continually report the performance of a guard are often found in connection with protected premises fire alarm systems utilizing off-premises reporting through central or proprietary supervising stations.

3-8.11.1 Guard's tour reporting stations shall be listed for the application.

3-8.11.2 The number of guard's tour reporting stations, their locations, and the route to be followed by the guard for operating the stations shall be approved for the particular installation in accordance with NFPA 601, *Standard for Security Services in Fire Loss Prevention.*

3-8.11.3 A permanent record indicating every time each signal-transmitting station is operated shall be made at the main control unit. Where intermediate stations that do not transmit a signal are employed in conjunction with signal-transmitting stations, distinctive signals shall be transmitted at the beginning and end of each tour of a guard, and a signal-transmitting station shall be provided at intervals not exceeding 10 stations. Intermediate stations that do not transmit a signal shall be capable of operation only in a fixed sequence.

3-8.12 Suppressed (Exception Reporting) Signal System.

This tour arrangement is somewhat less flexible than supervised tours but has the advantages of the absence of interconnected wires between the preliminary stations and the reduction of signal traffic. The usual arrangement is to have the guard transmit only start and finish signals that must be received at the central point at programmed reception times.

3-8.12.1 The system shall comply with the provisions of 3-8.11.2.

3-8.12.2 The system shall transmit a start signal to the signal-receiving location and shall be initiated by the guard at the start of continuous tour rounds.

3-8.12.3 The system shall automatically transmit a delinquency signal within 15 minutes after the predetermined actuation time if the guard fails to actuate a tour station as scheduled.

3-8.12.4 A finish signal shall be transmitted within a predetermined interval after the guard's completion of each tour of the premises.

3-8.12.5 For periods of over 24 hours during which tours are continuously conducted, a start signal shall be transmitted at least every 24 hours.

3-8.12.6 The start, delinquency, and finish signals shall be recorded at the signal-receiving location.

3-8.13 Combination Systems.

3-8.13.1* Fire alarm systems shall be permitted to share components, equipment, circuitry, and installation wiring with nonfire alarm systems.

A-3-8.13.1 The provisions of this paragraph apply to the types of equipment used in common for fire alarm systems (such as fire alarm, sprinkler supervisory, or guard's tour service) and for other systems (such as burglar alarm or coded paging systems) and to methods of circuit wiring common to both types of systems.

3-8.13.2 Where common wiring is employed for combination systems, the equipment for other than fire alarm systems shall be permitted to be connected to the common wiring of the system. Short circuits, open circuits, or grounds in this equipment or between this equipment and the fire alarm system wiring shall not interfere with the monitoring for integrity of the fire alarm system or prevent alarm or supervisory signal transmissions.

Common wiring between fire alarm and non-fire alarm systems means receiving signal inputs or providing outputs for either system on the same wires. This common wiring could be device power, initiating device, or notification appliance circuits. In any case, a short, a ground, or an open circuit in the common wiring caused by the non-fire equipment should not prevent the receipt of an alarm, a trouble, a supervisory signal, or prevent the notification appliances from operating.

3-8.13.3 To maintain the integrity of fire alarm system functions, the removal, replacement, failure, or maintenance procedure on any hardware, software, or circuit not required to perform any of the fire alarm system functions shall not cause loss of any of these functions.

Exception: Where the hardware, software, and circuits are listed for fire alarm use.

This subsection addresses the problem of field maintenance or equipment failure of non-fire alarm system components. Equipment not required for the operation of the fire alarm system that is modified, removed, or is malfunctioning in any way must not impair the opera-

tion of the fire alarm system. Subsection 3-8.13 permits fire alarm systems to be installed using some components not specifically listed for fire alarm use. Most applications of this permitted use involve interconnection of the listed fire alarm system with paging systems, burglar alarm systems, HVAC control systems and process monitoring systems. Users have also connected fire alarm systems to supplementary equipment such as business computers and monitors not listed for fire alarm use, in order to display system conditions to operators in more detail than may be available from the fire alarm systems alone.

However, authorities having jurisdiction have found that some installations do not satisfy the intent of the Code sections covering combination systems. In some installations, program changes or other repairs to the non-fire alarm equipment delayed fire alarm signals or prevented their display altogether. In general, incorrectly applied and interconnected systems may prevent one or more of the fire alarm system functions from operating as intended. To guard against such failures, it is usually necessary to retest the complete fire alarm system or extensive portions of the system after all changes and repairs have been made to the non-fire alarm components of the system.

Where a non-fire alarm system component is listed for fire alarm use, the listing agency has investigated the timing aspects of the signals, as well as temperature characteristics and other extensive fire alarm safety factors, and found the product suitable for the purpose. Consequently, the exception to 3-8.13.3 exempts this specifically listed equipment from the requirements of 3-8.13.3.

3-8.13.4 Speakers used as alarm notification appliances on fire alarm systems shall not be used for nonemergency purposes.

Exception: Where the fire command center is constantly attended by a trained operator, selective paging shall be permitted.

The fire alarm notification appliances cannot be used for general paging functions unless the exception applies. Conversely, standard background music or paging speakers cannot be used as fire alarm notification appliances. The speakers must be listed for fire alarm use. The term "constantly attended" means that personnel are in the fire command center 24 hours a day, seven days a week, not only during the operation of the facility.

3-8.13.5 In combination systems, fire alarm signals shall be distinctive, clearly recognizable, and take precedence over any other signal even when a nonfire alarm signal is initiated first.

This requirement does not mean that two separate notification appliances must be used. A single appliance may be used if it can supply two different, distinctive signals; the fire alarm signal always takes precedence. However, the notification appliance circuit must comply with 3-8.13.2.

3-8.13.6 Where the authority having jurisdiction determines that the information being displayed or annunciated on a combination system is excessive and could cause confusion and delay response to a fire emergency, the authority having jurisdiction shall be permitted to require that the display or annunciation of information for the fire alarm system be separate from and have priority over information for the nonfire alarm systems.

3-8.14 Elevator Recall for Fire Fighters' Service.

3-8.14.1* System-type smoke detectors located in elevator lobbies, elevator hoistways, and elevator machine rooms used to initiate fire fighters' service recall shall be connected to the building fire alarm system. In facilities without a building fire alarm system, these smoke detectors shall be connected to a dedicated fire alarm system control unit that shall be designated as "elevator recall control and supervisory panel" on the record drawings. Unless otherwise required by the authority having jurisdiction, only the elevator lobby, elevator hoistway, and the elevator machine room smoke detectors shall be used to recall elevators for fire fighters' service.

Elevator lobby, elevator hoistway and elevator machine room smoke detectors are the only smoke detectors required to initiate elevator recall in accordance with ANSI/ASME A17.1-1995, Safety Code for Elevators and Escalators (See A17.1b-1995, Section 211.3b., Figure 3.12), which requires recall of elevators to the designated level when these detectors are activated.

The new type of control unit, designated the "Elevator Recall Control and Supervisory Panel," will now be required for use in buildings where smoke detectors for elevator recall are installed but are not required to have (and do not have) a fire alarm system.

A-3-8.14.1 Dedicated fire alarm system control units are required for elevator recall by 3-8.14.1 in order that the elevator recall systems be monitored for integrity and have primary and secondary power meeting the requirements of this Code.

The control unit used for this purpose should be located in an area that is normally occupied and should have audible and visible indicators to annunciate supervisory (elevator recall) and trouble conditions; however, no form of general occupant notification or evacuation signal is required or intended by 3-8.14.1.

The new elevator control unit should be placed in an area that is constantly attended for monitoring, especially when it is installed as a stand-alone control. See Figure 3.13.

3-8.14.2 Each elevator lobby, elevator hoistway, and elevator machine room smoke detector shall be capable of initiating elevator recall when all other devices on the same initiating device circuit have been manually or automatically placed in the alarm condition.

This requirement simply restates 3-8.2.4. Generally, unless the required smoke detectors are installed on individual fire alarm initiating device circuits without any other fire alarm devices installed on those circuits, the smoke detectors should be powered separately from the control unit. Smoke detectors installed on signaling line circuits will not be affected.

3-8.14.3 Smoke detectors shall not be installed in elevator hoistways.

Exception: Where the top of the elevator hoistway is protected by automatic sprinklers.

Smoke detectors installed in elevator hoistways require continuous maintenance and are a source of numerous false or nuisance alarms. For this reason, the Code has specifically deleted them from the elevator hoistway unless the top of the hoistway is protected by an automatic sprinkler system. If sprinklers are installed at the top of the hoistway, then the smoke detector is needed to provide the recall feature before the heat detector or waterflow switch on the hoistway sprinkler system activates. See Figure 3.12, Rule 211.3b.

3-8.14.4 Where ambient conditions prohibit installation of automatic smoke detection, other appropriate automatic fire detection shall be permitted.

The intent of this subsection is to prevent nuisance alarms from smoke detectors installed in areas that are inappropriate for their use. Where the designer, the authority having jurisdiction or another code requires a detector in areas where the ambient conditions are unsuitable for a smoke detector, this subsection allows

211.3b Smoke Detectors. System type smoke detectors conforming to the requirements of UL 268 shall be installed in each elevator lobby and associated machine room in accordance with NFPA 72, Chapter 3. Smoke detectors are not required in elevator lobbies at unenclosed landings. System type smoke detectors may be installed in any hoistway and shall be installed in hoistways which are sprinklered (see Rule 102.2).

(1) The activation of a smoke detector in any elevator lobby, other than at the designated level, shall cause all cars that serve that lobby to return nonstop to the designated level. The activation of a smoke detector in any elevator hoistway shall cause all elevators having any equipment located in the hoistway, and any associated elevators of a group automatic operation, to return nonstop to the designated level, except that smoke detectors in hoistways installed at or below the lowest landing of recall, when activated, shall cause the car to be sent to the upper level of recall. The operation shall conform to the requirements of Rule 211.3a.

(2) If the smoke detector at the designated level is activated, the operation shall conform to the requirements of Rule 211.3a, except that the cars shall return to an alternate level approved by the enforcing authority, unless the designated-level three-position Phase I switch (Rule 211.3a) is in the "ON" position.

(3) The activation of a smoke detector in any elevator machine room, except a machine room at the designated level, shall cause all elevators having any equipment located in that machine room, and any associated elevators of a group automatic operation, to return nonstop to the designated level. The activation of a smoke detector in any elevator machine room at the designated level shall cause all elevators having any equipment located in that machine room, and any associated elevators of a group automatic operation, to return nonstop to the alternate level, or the appointed level when approved by the authority having jurisdiction.

(4) Elevators shall only react to the first smoke detector zone which is activated for that group.

(5) Phase I operation, when initiated by a smoke detector, shall be maintained until canceled by moving the Phase I Switch to the "BYPASS" position [see also Rule 211.3a(10)]. Smoke detectors and/or smoke detector systems shall not be self-resetting.

NOTE (Rule 211.3b): In buildings with fire alarm systems, see NFPA 72 for additional information.

Figure 3.12 Excerpts from ANSI/ASME A17.1-1995. Courtesy of the American Society of Mechanical Engineers. Reprinted with Permission.

the use of any other type of detector that would be stable and still provide necessary detection.

3-8.14.5 When actuated, each elevator lobby, elevator hoistway, and elevator machine room smoke detector shall initiate an alarm condition on the building fire alarm system and shall visibly indicate, at the control unit and required remote annunciators, the alarm initiation circuit or zone from which the alarm originated. Actuation from elevator hoistway and elevator machine room smoke detectors shall cause separate and distinct visible annunciation at the control unit and required annunciators to alert fire fighters and other emergency personnel that the elevators are no longer safe to use. Actuation of these detectors shall not be required to sound the building evacuation alarm where the alarm condition is indicated at a constantly attended location.

The intent of this subsection is to ensure that the area, or zone of alarm (floor, room, etc.,) be indicated on the fire alarm control unit and the remote annunciator. The Code now requires that the elevator hoistway smoke detector (if one is present) and the elevator machine

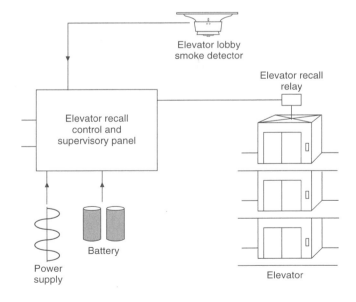

Figure 3.13 Elevator recall system. Drawing courtesy of FIREPRO Incorporated, Burlington, MA.

room smoke detector(s) be connected to the fire alarm control unit and remote annunciator as a separate zone or point of alarm indication. The detectors located in the elevator hoistway and elevator machine room actuate recall, but are not required to sound the building evacuation alarm. Their annunciation, however, must conform to the requirements as stated.

Exception: Where approved by the authority having jurisdiction, the elevator hoistway and machine room detectors shall be permitted to initiate a supervisory signal.

This option is provided to minimize the nuisance alarms from smoke detectors in these areas. The elevator recall system would still operate, but the fire alarm signal would not sound. The option should be used only where trained personnel are constantly in attendance and can immediately respond to the supervisory signal and investigate the cause of the signal. Means should be provided for initiating the fire alarm signal if the investigation of the supervisory signal cause indicates building evacuation is necessary. In addition to having trained personnel constantly in attendance, it is recommended that if the supervisory signal is not acknowledged within a given period of time (3-10 minutes), the fire alarm system will automatically and immediately initiate an alarm.

3-8.14.6*   For each group of elevators within a building, three separate elevator control circuits shall be terminated at the designated elevator controller within the group's elevator machine room(s). The operation of the elevators shall be in accordance with ANSI/ASME A17.1, *Safety Code for Elevators and Escalators*, Rules 211.3 through 211.8. The smoke detectors shall actuate the three elevator control circuits as follows:

(a) The smoke detector located in the designated elevator recall lobby shall actuate the first elevator control circuit. In addition, where the elevator is equipped with front and rear doors, the smoke detectors in both lobbies at the designated level shall actuate the first elevator control circuit.

(b) The smoke detectors in the remaining elevator lobbies shall actuate the second elevator control circuit.

(c) The smoke detectors in elevator hoistways and the elevator machine room(s) shall actuate the third elevator control circuit. In addition, where the elevator machine room is located at the designated level, that elevator machine room smoke detector shall also actuate the first elevator control circuit.

Three elevator control circuits are needed for proper operation of the recall sequence and for safe use by the

fire department. The first circuit is needed to prevent recalling the elevators and discharging passengers to the designated floor when the designated floor is the fire location, and to provide for an alternate recall location (determined by the authority having jurisdiction) when the designated floor is reporting a fire condition. The second circuit configuration (part (b)) provides for standard recall to the designated floor when any other elevator lobby, machine room or elevator hoistway smoke detector is in alarm. The operation of the third circuit, (part (c)), although not defined in the Code, is intended for the safety of firefighters who may be using the elevators to bring equipment to staging areas in a high rise building. This third circuit's feature could override the firefighter's key operation and bring the elevators to a safe level of discharge prior to equipment shutdown due to fire in the hoistway or machine room. The third circuit could also sound a warning in the elevator cab to notify the fire department personnel using the elevators during the fire to immediately move to a safe floor and exit the elevator. See Figure 3.14.

A-3-8.14.6   It is recommended that the installation be in accordance with Figures A-3-8.14.6(a) and (b). Figure A-3-8.14.6(a) should be used where the elevator is installed at the same time as the building fire alarm system. Figure A-3-8.14.6(b) should be used where the elevator is installed after the building fire alarm system.

3-8.15   **Elevator Shutdown.**

This subsection is the result of additional sprinkler protection requirements of other codes for elevator

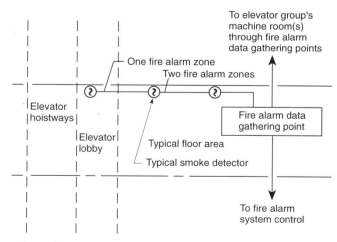

Figure A-3-8.14.6(a)   *Elevator zone—elevator and fire alarm system installed at same time.*

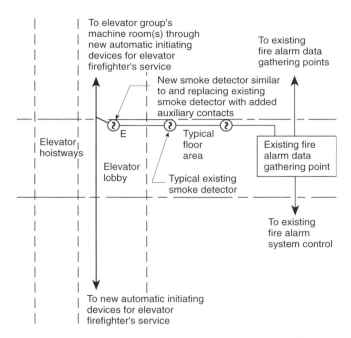

Figure A-3-8.14.6(b) *Elevator zone—elevator installed after fire alarm system.*

machine rooms and hoistways. (See ANSI/ASME A17.1b-1995, Section 102. Rule 102.2, (c) (3) & (4). See Figure 3.15. The purpose of elevator shutdown prior to sprinkler operation is to avoid the hazard of a wet elevator braking system. If the elevator brakes are wet, there is the danger of the elevator rising uncontrollably to the top of the shaft or, depending on the load in the car, descending uncontrollably to the bottom of the shaft.

3-8.15.1* Where heat detectors are used to shut down elevator power prior to sprinkler operation, the detector shall have both a lower temperature rating and a higher sensitivity [often characterized by a lower response time index (RTI)] as compared to the sprinkler.

A-3-8.15.1 A lower response time index is intended to provide detector response prior to the sprinkler response, since a lower temperature rating alone might not provide earlier response. The listed spacing rating of the heat detector should be 25 ft (7.6 m) or greater.

This requirement is extremely important. Often, it is not understood that a 135°F heat detector may not respond prior to a 165°F sprinkler head despite the

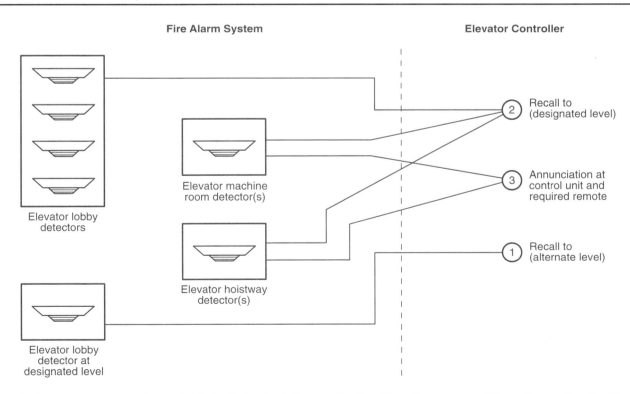

Figure 3.14 *Optional connection to third circuit for fire fighter notification. Drawing courtesy of Bruce Fraser, Simplex Time Recorder Company, Gardner, MA.*

Rule 102.2

(c) (3) Means shall be provided to automatically disconnect the main line power supply to the affected elevator upon or prior to the application of water from sprinklers located in the machine room or in the hoistway. This means shall be independent of the elevator control and shall not be self-resetting. The activation of sprinklers outside of the hoistway or machine room shall not disconnect the main line power supply.

  (4) Smoke detectors shall not be used to activate sprinklers in these spaces or to disconnect the main line power supply.

Figure 3.15   Excerpts from ANSI/ASME A17.1-1995, Courtesy of the American Society of Mechanical Engineers. Reprinted with Permission.

---

obvious differences in temperature sensitivity. The response time is based on the RTI of both devices and must be known prior to design and installation of heat detectors for elevator shutdown. Because the RTI for heat detectors is not readily available, one must use a sensitive heat detector (40 foot or higher listed spacing) in the locations where heat detectors are required.

3-8.15.2 Where heat detectors are used for elevator power shutdown prior to sprinkler operation, they shall be placed within 2 ft (610 mm) of each sprinkler head and be installed in accordance with the requirements of Chapter 5. Alternatively, engineering methods (such as specified in Appendix B) shall be permitted to be used to select and place heat detectors to ensure response prior to any sprinkler head operation under a variety of fire growth rate scenarios.

3-8.15.3* Where pressure or waterflow switches are used to shut down elevator power immediately upon or prior to the discharge of water from sprinklers, the use of devices with time delays shall not be permitted.

The intent is to shut down elevator power as soon as water flows in the automatic sprinkler system that is protecting the elevator hoistway and machine room. These waterflow devices should be installed in the cross main or branch line serving the automatic sprinkler system in those areas.

A-3-8.15.3 Care should be taken to ensure that elevator power cannot be interrupted due to water pressure surges in the sprinkler system.

3-8.16 **Trouble Signals to Supervising Station.**

There have been fire alarm systems that have used normally de-energized relays to transmit this information off-premises and it was found that certain trouble conditions involving loss of operating power prevented transmission of a trouble signal. For protected premises that are normally not occupied, this could result in a system being out of service for a unacceptable amount of time. In addition, where addressable relays or modules are used to provide this information off-premises, faults such as short circuit faults on the signaling line circuit, could prevent activation of the addressable relay.

3-8.16.1 Relays or modules providing transmission of trouble signals to a supervising station shall be arranged to provide fail-safe operation.

3-8.16.2 Means provided to transmit trouble signals to supervising stations shall be so arranged to transmit a trouble signal to the supervising station for any trouble condition received at the protected premises control unit, including loss of primary or secondary power.

3-9   Fire Safety Control Functions.

The control of preprogrammed fire safety control functions is automatically initiated by the fire alarm system in response to fire alarm signals. Depending on the system configuration, either all of the fire safety functions are initiated by any alarm signal, or selected fire safety functions may be initiated in response to a specific initiating device or zone in alarm (e.g., doors released or fans shut down on the floor of fire origin only). In some cases, as in subsection 3-8.14.5, a supervisory signal from the fire alarm control unit will initiate the fire safety control function.

3-9.1 **Scope.** The provisions of this section cover the minimum requirements for the interconnection of fire safety control functions (e.g., fan control, door control) to the fire alarm system. These fire safety functions are not intended to provide notification of alarm, supervisory, or trouble conditions; to alert or control occupants; or to summon aid.

3-9.2 **General.**

3-9.2.1 An auxiliary relay connected to the fire alarm system used to initiate control of fire safety functions shall be

---

located within 3 ft (1 m) of the controlled circuit or device. The auxiliary relay shall function within the voltage and current limitations of the control unit. The installation wiring between the fire alarm system control unit and the auxiliary relay shall be monitored for integrity.

For example, the fan control for a fan located on the fourth floor may be located in the basement. Subsection 3-9.2.1 requires the monitoring for integrity of the wiring from the fire alarm system control unit to the fire alarm system fan control relay, which actuates the fan control in the basement, not the fan located on the fourth floor. The distance between the fire alarm system fan control relay and the fan control should not exceed 3 ft (1 m).

Exception: Control devices that operate on loss of power or on loss of power to the auxiliary relay shall be considered self-monitoring for integrity.

This subsection and its exception are based on Chapter 7 (7-6.5.5) of NFPA 101, *Life Safety Code®*. This addition has been made to ensure that when auxiliary relays are used to cause the operation of smoke dampers, fire dampers, fan controls, smoke doors, and fire doors, and when they are connected to the building fire alarm system, the requirements of monitoring the interconnecting wiring for integrity apply. The exception would be devices that are installed in a fail-safe manner such as magnetic door hold-open devices. If a wire connection providing power to one of these devices is broken, the door closes. This is the expected operation during a fire condition, therefore the device fails "safe," and the wiring does not need to be monitored for integrity.

3-9.2.2 Fire safety functions shall not interfere with other operations of the fire alarm control system.

This requirement is similar to the requirements for combination systems. One way to ensure that the fire safety functions do not interfere with any operations of the fire alarm system is to use auxiliary relays, listed for use with the fire alarm control unit, to isolate the fire safety function from the control unit. Fire safety functions connected directly to a notification appliance circuit, for instance, may disable the notification appliance circuit if a fault condition occurs in the fire safety function control equipment.

3-9.2.3 Transfer of data over listed serial communications ports shall be permitted to be a means of interfacing between the fire alarm control unit and fire safety function control devices.

3-9.2.4 The fire safety function control devices shall be listed as compatible with the fire alarm control unit so as not to interfere with the control unit's operation.

Generally, the fire safety control function devices are auxiliary relays. These relays must be listed specifically to operate with the fire alarm control unit and not be "off the shelf" items from an electronics supply store.

3-9.2.5 The interfaced systems shall be acceptance tested together in the presence of the authority having jurisdiction to ensure proper operation of the fire alarm system and the interfaced system(s).

The testing of the interfaced systems is extremely important. The proper interface and subsequent proper operation of the fire safety function is often dependent on more than one contractor and design engineer. See Figure 3.16.

To ensure that the individuals involved communicate and work together to interface the fire safety function properly, all parties involved should be present when the authority having jurisdiction is present for the acceptance test.

3-9.2.6 Where manual controls for emergency control functions are required to be provided, they shall provide visible indication of the status of the associated control circuits.

As used here, "emergency control functions" are the same as "fire safety control functions" mentioned previously in this Code. The required visible status indication can be by a labeled LED annunciator (or equivalent means) or by the labeled position of a toggle or rotary switch.

3-9.3 Heating, Ventilation, and Air Conditioning (HVAC) Systems.

3-9.3.1 The provisions of 3-9.3 apply to the basic method by which a fire alarm system interfaces with the HVAC systems.

3-9.3.2 All detection devices used to cause the operation of smoke dampers, fire dampers, fan control, smoke doors, and fire doors shall be monitored for integrity in accordance with 1-5.8 where connected to the fire alarm system serving the protected premises.

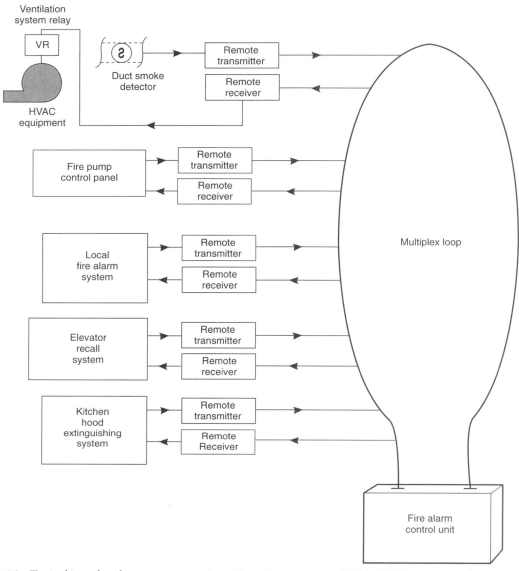

***Figure* 3.16**   *Typical interfaced systems connections. Drawing courtesy of FIREPRO Incorporated, Burlington, MA.*

This requirement precludes the use of stand-alone detection devices (such as duct-type smoke detectors) unless there is no fire alarm system in the building. In most instances, the detection devices referred to in 3-9.3.2 will be the automatic initiating devices used to detect and report a fire emergency. The wiring to these devices is required to be monitored for integrity by subsection 1-5.8. See Figure 3.16.

3-9.3.3 Connections between fire alarm systems and the HVAC system for the purpose of monitoring and control shall operate and be monitored in accordance with applicable NFPA standards.

3-9.3.4 Where the fire alarm control unit activates the HVAC system for the purpose of smoke control, the automatic alarm-initiating zones shall be coordinated with the smoke-control zones they actuate.

This requirement demands coordination between the fire alarm system designer and installer and the smoke control system designer and installer. Because the smoke zones are often smaller and more numerous than fire zones, this requirement will most often lead to additional zone requirements for the fire alarm system.

3-9.4   Door Release Service.

3-9.4.1   The provisions of 3-9.4 apply to the methods of connection of door hold release devices and to integral door hold release, closer, and smoke detection devices.

3-9.4.2   All detection devices used for door hold release service, whether integral or stand alone, shall be monitored for integrity in accordance with 1-5.8 where connected to the fire alarm system serving the protected premises.

3-9.4.3   All door hold release and integral door release and closure devices used for release service shall be monitored for integrity in accordance with 3-9.2.

Generally, door hold release magnetic devices are installed in a "fail-safe" manner in that they release on loss of power. If these types of devices are connected in such a fashion to the fire alarm system, the wiring to the control relay or circuit does not need to be monitored for integrity (See the exception to 3-9.2.1.).

3-9.4.4   Magnetic door holders that allow doors to close upon loss of operating power shall not be required to have a secondary power source.

The purpose of magnetic door release devices are to keep doors open under normal conditions and allow the doors to close in smoke and fire conditions. If the designer or the authority having jurisdiction wishes for the doors to remain open even under a primary power failure, then the magnetic door holders must be placed on a circuit with secondary power. The Code recognizes that this is optional and, therefore, does not require secondary power for these devices. See Figure 3.17.

3-9.5   Door Unlocking Devices.

3-9.5.1   Any device or system intended to effect the locking/unlocking of emergency exits shall be connected to the fire alarm system serving the protected premises.

3-9.5.2   All emergency exits connected in accordance with 3-9.5.1 shall unlock upon receipt of any fire alarm signal by means of the fire alarm system serving the protected premises.

Figure 3.17   Typical magnetic door hold release device. Photograph courtesy of Edwards Systems Technology/EST, Cheshire, CT.

3-9.5.3   All emergency exits connected in accordance with 3-9.5.1 shall unlock upon loss of the primary power to the fire alarm system serving the protected premises. The secondary power supply shall not be utilized to maintain these doors in the locked condition.

This subsection confirms by the description of the required operation that the wiring from the interface to the unlocking device need not be monitored for integrity because it will operate in a fail safe manner upon loss of power. This requirement also infers that the power to the door unlocking devices is connected to the same circuit that is supplying primary power to the fire alarm control unit.

3-10   Suppression System Actuation.

3-10.1   Fire alarm systems listed for releasing service shall be permitted to provide automatic or manual actuation of fire suppression systems.

3-10.2   The integrity of each releasing device (e.g., solenoid, relay) shall be supervised in accordance with applicable NFPA standards.

Figure 3.18 Wet chemical kitchen hood and duct cylinder with control head. Photograph courtesy of Kidde-Fenwal Protection Systems, Ashland, MA.

In 3-10.2, the word "supervised" means the same as "monitored for integrity" as used elsewhere in this Code. See commentary following 3-8.8.3. See Figure 3.18.

3-10.3 The integrity of the installation wiring shall be monitored in accordance with the requirements of Chapter 1.

3-10.4 Fire alarm systems used for fire suppression releasing service shall be provided with a disconnect switch to allow system testing without activating the fire suppression systems. Operation of the disconnect switch shall cause a trouble signal at the fire alarm control unit.

This feature is extremely important. Very often the contractor who will be testing the fire alarm system is not proficient in the operation of fire suppression systems. The supervised disconnect switch allows the fire alarm system contractor to perform maintenance or tests on the fire alarm system without inadvertently activating the suppression system.

3-10.5 Sequence of operation shall be consistent with the applicable suppression system standards.

3-10.6* Each space protected by an automatic fire suppression system actuated by the fire alarm system shall contain one or more automatic fire detectors installed in accordance with Chapter 5.

A-3-10.6 Automatic fire suppression systems referred to in 3-10.6 include, but are not limited to, preaction and deluge sprinkler systems, carbon dioxide systems, halon systems, and dry chemical systems.

3-10.7 Suppression systems or groups of systems shall be controlled by a single control unit that monitors the associated initiating devices, actuates the associated releasing device(s), and controls the associated agent release notification appliances. Where the releasing panel is located in a protected premises having a separate fire alarm system, it shall be monitored by, but shall not be dependent on or affected by, the operation or failure of the protected premises fire alarm system.

There have been several instances where multi-tier releasing arrangements have resulted in inadvertent system discharges. These inadvertent discharges have occurred during normal system testing and maintenance, and occasionally because of system wiring faults unrelated to the required operation of the releasing system.

Another issue addressed by this new subsection is control units not listed for releasing service being used to provide initiating inputs to listed releasing control units. In these systems, the initiating devices are connected to a control unit which is not listed for releasing service, the releasing devices (solenoids or squibs, for example) are activated by a second control unit which is listed for releasing service, and the outputs from the first control unit provides the inputs to the second (releasing) control unit. These arrangements are particularly prone to inadvertent suppression system release and border on misapplication of equipment.

Exception: Where the configuration of multiple control units is listed for releasing device service, and where a trouble condition or manual disconnect on either control unit causes a trouble or supervisory signal, the initiating devices on one control unit shall be permitted to actuate releasing devices on another control unit.

3-11*   Interconnected Fire Alarm Control Units. Fire alarm systems shall be permitted to be either integrated systems combining all detection, notification, and auxiliary functions in a single system or a combination of component subsystems. Fire alarm system components shall be permitted to share control equipment or shall be able to operate as stand alone subsystems, but, in any case, they shall be arranged to function as a single system. All component subsystems shall be capable of simultaneous, full load operation without degradation of the required, overall system performance.

This section applies both where the interconnected control units are the products of a single manufacturer and where they are the products of two or more manufacturers, regardless of when installed. The section covers the requirements for the interconnection, monitoring, and compatibility of the control units.

A-3-11 This code contemplates field installations interconnecting two or more listed control units, possibly from different manufacturers, that together fulfill the requirements of this code.

Such an arrangement should preserve the reliability, adequacy, and integrity of all alarm, supervisory, and trouble signals and interconnecting circuits intended to be in accordance with the provisions of this code.

Where interconnected control units are in separate buildings, consideration should be given to protecting the interconnecting wiring from electrical and radio frequency interference.

There are many reasons for interconnecting control units. In some cases an existing building may need additional power supplies and notification appliances to conform to a new law, such as the Americans with Disabilities Act (ADA), and the original fire alarm system may not be able to accommodate the necessary changes. It also may not be feasible to attempt to modify or expand the existing fire alarm control unit due to lack of parts, economics, or the system configuration. Some newer systems can consist of two or more subsystems (control units) connected to a single- or- multiple "master" control unit(s). In other cases, a new wing is added to a building, and the contractor plans to use a new, separate, and different manufacturer's system for the new wing. The problem is that the new wing's fire alarm system must operate as if it were part of the original system installed many years before.

Until this section appeared in the Code, there had been no guidance regarding the correct procedures to follow when interconnecting the control panels or fire alarm systems as described in these examples. The goal of 3-11 and A-3-11 is to advise the fire alarm system designer, the installer, and the authority having jurisdiction that consideration must be given regarding the appropriate, most reliable, and acceptable method of interconnection.

It is not the intent of this section to allow two small fire alarm control units to be interconnected to avoid the installation of a single larger fire alarm control unit in a newly constructed building. It is the intent of the Code

that the system designer use the appropriately sized equipment designed to meet the fire alarm system needs of the owner of a newly constructed or renovated building.

3-11.1 The method of interconnection of control units shall meet the monitoring requirements of 1-5.8 and NFPA 70, *National Electrical Code*, Article 760, and shall be achieved by the following recognized means:

    (a) Properly rated electrical contacts;

    (b) Compatible digital data interfaces;

    (c) Other listed methods.

3-11.2 Where approved by the authority having jurisdiction, interconnected control units providing localized detection, evacuation signaling, and auxiliary functions shall be permitted to be monitored by a fire alarm system as initiating devices.

Section 3-11.2 covers such installations where, for example, multiple buildings on a contiguous property, under single ownership, and with individual protected premises fire alarm systems, may be monitored by a single fire alarm control unit. This control unit may also serve the building it is located in as part of the protected premises fire alarm system. The purpose of this arrangement may be to have only one off-premises connection. In any case, the building with the "master" control unit that is providing the monitoring of the other systems should be identified at the building, at the remote annunciator, and at the supervising station receiving the fire alarm signal.

3-11.2.1 Each interconnected control unit shall be separately monitored for alarm, trouble, and supervisory conditions.

This subsection clarifies the intent of the whole section, but seems to conflict with 3-11.2. Each interconnected fire alarm control unit must be monitored for alarm first. Then each interconnected fire alarm control unit must be monitored for supervisory conditions. This requirement means that if the satellite FACU interconnected to the master FACU experiences a trouble condition for any reason, that trouble condition will report to the master FACU as a supervisory condition, indicating the interconnected fire alarm control unit is off-normal. The interconnection between the fire alarm control units is monitored for integrity and if that circuit experiences a fault condition, a trouble condition for that circuit (zone or point) is indicated at the master fire alarm control unit.

3-11.2.2 Interconnected control unit alarm signals shall be permitted to be monitored by zone or combined as common signals, as appropriate.

3-11.3 Protected premises fire alarm control units shall be capable of being reset or silenced only from the control unit at the protected premises.

Exception: Where otherwise specifically permitted by the authority having jurisdiction.

This subsection has been added to require the on-site restoration to normal of fire alarm systems. It is expected that if a fire alarm system is in alarm or trouble, a technician should be required to investigate the causes of that alarm or trouble, and then reset the fire alarm control unit.

3-12 Emergency Voice/Alarm Communications.

3-12.1 **Emergency Voice/Alarm Communications Service.** Emergency voice/alarm communications service shall be provided by a system with automatic or manual voice capability that is installed to provide voice instructions to the building occupants where it is intended that there be only partial or selective evacuation or directed relocation of building occupants in the event of a fire.

Exception: Where emergency voice/alarm communications are used to automatically and simultaneously notify all occupants to evacuate the protected premises during a fire emergency, manual or selective paging shall not be required but, where provided, shall meet the requirements of Section 3-12.

3-12.2 **Application.** This section describes the requirements for emergency voice/alarm communications. The primary purpose is to provide dedicated manual and automatic facilities for the origination, control, and transmission of information and instructions pertaining to a fire alarm emergency to the occupants (including fire department personnel) of the building. It is the intent of this section to establish the minimum requirements for emergency voice/alarm communications.

3-12.3 Monitoring of the integrity of speaker amplifiers, tone-generating equipment, and two-way telephone communications circuits shall be in accordance with 1-5.8.5.

3-12.4 **Survivability.**

3-12.4.1 A fire command center shall be provided in accordance with 3-12.6.5.

Exception: Where emergency voice/alarm communications are used to automatically and simultaneously notify all occupants to evacuate the protected premises during a fire emergency, a fire command center shall not be required, but, where provided, shall meet the requirements of Section 3-12.

Many designers are using voice/alarm communications in buildings other than high-rise or large buildings. In most of these smaller building designs, the voice/alarm communications system sounds the alarm throughout the building for total evacuation rather than relocation. In those instances, the Code now allows these systems to be used without the requirement for a Fire Command Center. If a designer chooses to use a Fire Command Center, the requirements of 3-12.6.5 apply.

3-12.4.2 The fire command center and the central control unit shall be located within a minimum 1-hour rated fire-resistive area and shall have a minimum 3-ft (1-m) clearance about the face of the fire command center control equipment.

Exception: Where approved by the authority having jurisdiction, the fire command center control equipment shall be permitted to be located in a lobby or other approved space.

3-12.4.3 Where the fire command center control equipment is remote from the central control equipment:

(a) The interconnecting wiring shall be provided with mechanical protection by installing the wiring in metal conduit or metal raceway.

(b) The interconnecting wiring shall be provided with resistance to attack from a fire by routing the wiring through areas whose characteristics are at least equal to the limited combustible characteristics defined in NFPA 90A, *Standard for the Installation of Air Conditioning and Ventilating Systems.*

(c) Where the interconnecting wiring exceeds 100 ft (30 m), additional resistance to attack from a fire shall be provided by either:

1. Installing the wiring in metal conduit or metal raceway in a 2-hour fire rated enclosure; or

2. Enclosing the wiring in a 2-hour fire rated cable assembly and installing the cable in metal conduit or metal raceway.

Emergency voice/alarm communications service is generally used where relocation or partial evacuation (by floor or zone) of the building occupants is part of the fire response plan. If occupants are allowed to remain in the building, system survivability must be considered. It is essential that the fire alarm system remain operational during the fire emergency so that additional floors or zones can be evacuated as needed. See 3-2.4 for additional survivability requirements.

3-12.5 Power Supplies.

3-12.5.1 The wiring between the central control equipment and the primary power supply also shall be routed through areas whose characteristics are at least equal to the limited-combustible characteristics as defined in NFPA 90A, *Standard for the Installation of Air Conditioning and Ventilating Systems.*

3-12.5.2 The secondary (standby) power supply shall be provided in accordance with 1-5.2.6.

Exception: Where emergency voice/alarm communications are used to notify all occupants automatically and simultaneously to evacuate the protected premises during a fire emergency in meeting the requirements of 1-5.2.6, the secondary supply shall be required to be capable of operating the system during a fire or other emergency condition for a period of 5 minutes rather than 2 hours.

As in subsection 3-12.4.1, when the voice/alarm communications system is not being designed to relocate the building occupants, but is providing an evacuation signal, using prerecorded voice instructions for example, the battery standby requirements match the fire alarm system type used in the building (24 or 60 hours) and the alarm must sound for 5 minutes.

3-12.6 Voice/Alarm Signaling Service.

3-12.6.1* General. The purpose of the voice/alarm signaling service is to provide an automatic response to the receipt of a signal indicative of a fire emergency. Subsequent manual control capability of the transmission and audible reproduction of evacuation tone signals, alert tone signals, and voice directions on a selective and all-call basis, as determined by the authority having jurisdiction, is also required from the fire command center.

Exception No. 1: Where the fire command center or remote monitoring location is constantly attended by trained operators, and operator acknowledgment of receipt of a fire alarm signal is received within 30 seconds, automatic response shall not be required.

Exception No. 2: Where emergency voice/alarm communications are used to notify all occupants automatically and simultaneously to evacuate the protected premises during a fire emergency, the ability to give voice directions on a selec-

tive basis shall not be required but, where provided, shall meet the requirements of Section 3-12.

A-3-12.6.1 It is not the intention that emergency voice/alarm communications service be limited to English-speaking populations. Emergency messages should be provided in the language of the predominant building population. Where there is a possibility of isolated groups that do not speak the predominant language, multilingual messages should be provided. It is expected that small groups of transients unfamiliar with the predominant language will be picked up in the traffic flow in the event of emergency and are not likely to be in an isolated situation.

3-12.6.2 Multichannel Capability. Where required by the authority having jurisdiction, the system shall allow the application of an evacuation signal to one or more zones and, at the same time, shall allow voice paging to the other zones selectively or in any combination.

3-12.6.3 Functional Sequence.

3-12.6.3.1 In response to an initiating signal indicative of a fire emergency, the system shall automatically transmit, either immediately or after a delay acceptable to the authority having jurisdiction, the following:

(a) An alert tone of 3 seconds' to 10 seconds' duration followed by a message (or messages where multichannel capability is provided) shall be repeated at least three times to direct the occupants of the alarm signal initiation zone and other zones in accordance with the building's fire evacuation plan; or

(b) An evacuation signal to the alarm signal initiation zone and other zones in accordance with the building's fire evacuation plan.

Exception: Where emergency voice/alarm communications are used to notify all occupants automatically and simultaneously to evacuate the protected premises during a fire emergency, and the functional sequence described in 3-12.6.3.1(a) is provided, the capability to notify portions of the protected premises selectively shall not be required but, where provided, shall meet the requirements of Section 3-12.

3-12.6.3.2 Failure of the message described by 3-12.6.3.1(a), where used, shall sound the evacuation signal automatically. Provisions for manual initiation of voice instructions or evacuation signal generation shall be provided.

Exception No. 1: Other functional sequences shall be permitted where approved by the authority having jurisdiction.

Exception No. 2: Where emergency voice/alarm communications are used to notify all occupants automatically and simultaneously to evacuate the protected premises during a fire emergency, provision for manual initiation of voice instructions shall not be required but, where provided, shall meet the requirements of Section 3-12.

3-12.6.3.3 Live voice instructions shall override all previously initiated signals on that channel and shall have priority over any subsequent automatically initiated signals on that channel. Where multichannel application is required, subsequent alarms shall be activated in accordance with 3-12.6.2.

Exception: Where emergency voice/alarm communications are used to notify all occupants automatically and simultaneously to evacuate the protected premises during a fire emergency, the ability to give live voice instructions shall not be required but, where provided, shall meet the requirements of Section 3-12.

3-12.6.3.4 Where provided, manual controls for emergency voice/alarm communications shall be arranged to provide visible indication of the on/off status for their associated evacuation zones.

3-12.6.4 Voice and Tone Devices. The alert tone preceding any message shall be permitted to be a part of the voice message or to be transmitted automatically from a separate tone generator.

3-12.6.5 Fire Command Center.

3-12.6.5.1* A fire command center shall be provided near a building entrance or other location approved by the authority having jurisdiction. The fire command center shall provide a communications center for the arriving fire department and shall provide for control and display of the status of detection, alarm, and communications systems. The fire command center shall be permitted to be physically combined with other building operations and security centers as permitted by the authority having jurisdiction. Operating controls for use by the fire department shall be clearly marked.

A-3-12.6.5.1 The choice of the location(s) for the fire command center should also consider the ability of the fire alarm system to operate and function during any probable single event.

3-12.6.5.2 The fire command center shall control the emergency voice/alarm communications signaling service and, where provided, the two-way telephone communications service. All controls for manual initiation of voice instructions and evacuation signals shall be located or secured to restrict access to trained and authorized personnel.

All of the requirements in 3-12.6.5 are in addition to the obvious coordination necessary with the responding fire department. Voice communication systems can be very effective both by calming occupants in areas remote from the fire and by directing others toward safety. All of the guidance and requirements in this section must be complemented with adequate training of the fire service personnel who will be responsible for using the equipment.

3-12.6.5.3 Where there are multiple fire command centers, the center in control shall be identified by a visible indication at that center.

3-12.6.6 Loudspeakers.

3-12.6.6.1 Loudspeakers and their enclosures shall be listed for voice/alarm signaling service and installed in accordance with Chapter 6.

Loudspeakers used for background music, etc., are not acceptable unless they have been specifically listed for fire alarm system use.

3-12.6.6.2* There shall be at least two loudspeakers in each paging zone of the building, so located that signals can be clearly heard regardless of the maximum noise level produced by machinery or other equipment under normal conditions of occupancy. *See Section 6-3.*

"Paging zone" is not defined in the Code. A common definition of a paging zone is a zone where all speakers are selected by the same switch or switching circuit. (See the definition of Notification Zone.) It can be inferred from 3-12.6.6.2 that a paging zone can be different than an evacuation zone. While this may be true in many cases, some emergency voice/alarm communications systems paging and evacuation zones are served by the same speakers and control equipment. The two-loudspeaker requirement is in consideration for survivability as discussed in subsection 3-2.4. See Figure 3.19.

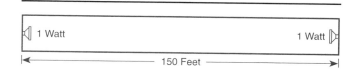

Figure 3.19   Paging zone showing typical speaker location in a corridor. Drawing courtesy of Simplex Time Recorder Company, Gardner, MA.

Figure 3.20 A *"typical" evacuation signal zone - one floor of office occupancy with ceiling mounted speakers. Drawing courtesy of Edwards Systems Technology/EST, Cheshire, CT.*

A-3-12.6.6.2 The design and layout of the loudspeaker audible notification appliances should be arranged such that they do not interfere with the operations of the emergency response personnel. Speakers located in the vicinity of the fire command center should be arranged so they do not cause audio feedback when the system microphone is used. Speakers installed in the area of two-way telephone stations should be arranged so that the sound pressure level emitted does not preclude the effective use of the two-way telephone system. Circuits for paging zones and telephone zones should be separated, shielded, or otherwise arranged to prevent audio cross talk between circuits.

This problem is often overlooked. Loudspeakers located too near a microphone at a command center or a fire fighters' telephone can interfere with critical communications during a fire emergency.

3-12.6.6.3 Each elevator car shall be equipped with a single loudspeaker connected to the paging zone serving the elevator group in which the elevator car is located.

3-12.6.6.4 Each enclosed stairway shall be equipped with loudspeakers connected to a separate paging zone.

Exception: Stairwells not exceeding two stories in height.

3-12.7 Evacuation Signal Zoning.

3-12.7.1 Where two or more evacuation signaling zones are provided, such zones shall be arranged consistent with the fire or smoke barriers within the protected premises. Undivided fire or smoke areas shall not be divided into multiple evacuation signaling zones.

NOTE: This section does not prohibit provision of multiple notification appliance circuits within a single evacuation signaling zone (i.e., separate circuits for audible and visible signals, redundant circuits provided to enhance survivability, or multiple circuits necessary to provide sufficient power/capacity).

"Evacuation signaling zones" are not defined in this Code. (See the definition of Notification Zone.). As stated in 3-2.4, an evacuation zone can be an area of a floor, an entire floor, or several floors that are always intended to be evacuated simultaneously. Subsection 3-12.7.1 requires that the evacuation signaling zones (also called "evacuation zones" elsewhere in the code) be consistent with the fire or smoke barriers within the protected premises. See Figure 3.20.

3-12.7.2 Where multiple notification appliance circuits are provided within an single evacuation signaling zone, all of the notification appliances within the zone shall be arranged to activate simultaneously, either automatically or by actuation of a common, manual control.

Exception: Where the different notification appliance circuits within an evacuation signaling zone perform separate functions (e.g., presignal and general alarm signals, predischarge and discharge signals).

3-12.8 Two-Way Telephone Communications Service.

3-12.8.1 Two-way telephone communications equipment shall be listed for two-way telephone communications service and installed in accordance with 3-12.8.

3-12.8.2 Two-way telephone communications service, where provided, shall be available for use by the fire service. Additional uses, where specifically permitted by the authority having jurisdiction, shall be permitted to include signaling and communications for a building fire warden organization, signaling and communications for reporting a fire and other emergencies (e.g., voice call box service, signaling, and communications for guard's tour service), and other uses. Variation of equipment and system operation provided to facilitate additional use of the two-way telephone communications service shall not adversely affect performance where used by the fire service.

Two-way telephone communications service is provided normally because fire department hand-held radios may be ineffective in buildings with a great deal of structural steel. The authority having jurisdiction can waive this requirement if the hand-held radios used by the fire department personnel in the jurisdiction work effectively in the specific building in question. See Figure 3.21.

3-12.8.3* Two-way telephone communications service shall be capable of permitting the simultaneous operation of any five telephone stations in a common talk mode.

A-3-12.8.3 Consideration should be given to the type of fire fighters' telephone handset used in areas where high ambient noise levels exist or areas where high noise levels could exist during a fire condition. Push-to-talk handsets, handsets containing directional microphones, or handsets containing other suitable noise-canceling features can be used.

3-12.8.4 A notification signal at the fire command center, distinctive from any other alarm or trouble signal, shall indicate the off-hook condition of a calling telephone circuit. Where a selective talk telephone communications service is

Figure 3.21   Two-way telephone communications service in use. Photograph courtesy of Simplex Time Recorder Company, Gardner, MA.

supplied, a distinctive visible indicator shall be furnished for each selectable circuit so that all circuits with telephones off-hook are continuously and visibly indicated.

Exception: Where emergency voice/alarm communications are used to notify all occupants automatically and simultaneously to evacuate the protected premises during a fire emergency, signals from the two-way telephone system shall be required to indicate only at a location approved by the authority having jurisdiction.

3-12.8.5 A switch for silencing the audible call-in signal sounding appliance shall be permitted, provided it is key-operated, in a locked cabinet, or provided with equivalent protection from use by unauthorized persons. Such a switch shall be permitted, provided that it operates a visible indicator and sounds a trouble signal whenever the switch is in the silence position where there are no telephone circuits in an off-hook condition. Where a selective talk telephone system is used, such a switch shall be permitted, provided subsequent telephone circuits going off-hook operate the distinctive off-hook audible signal sounding appliance.

3-12.8.6 Minimum Systems. As a minimum (for fire service use only), two-way telephone systems shall be common talk (i.e., a conference or party line circuit) providing at least one telephone station or jack per floor and at least one telephone station or jack per exit stairway. In buildings equipped with a fire pump(s), a telephone station or jack shall be provided in each fire pump room.

3-12.8.7 Fire Warden Use. Where the two-way telephone system is intended to be used by fire wardens in addition to the fire service, the minimum requirement shall be a selective talk system (where phones are selected from the fire command center). Systems intended for fire warden use shall provide telephone stations or jacks as required for fire service use and additional telephone stations or jacks as necessary to provide at least one telephone station or jack in each voice paging zone. Telephone circuits shall be selectable from the fire command center either individually or, where approved by the authority having jurisdiction, by floor or stairwell.

The paging zone is normally the same as the evacuation zone. See comment for subsection 3-12.6.6.2. In most designs, a fire fighters' telephone station or jack is located in each stairwell and elevator lobby.

3-12.8.8 Where the control equipment provided does not indicate the location of the caller (common talk systems), each telephone station or phone jack shall be clearly and permanently labeled to allow the caller to readily identify his/her location to the fire command center by voice.

3-12.8.9 Where telephone jacks are provided, a sufficient quantity of portable handsets, as determined by the authority having jurisdiction, shall be stored at the fire command center for distribution during an incident to responding personnel.

The required quantity of portable handsets is often a percentage of the telephone jacks in the building. The authority having jurisdiction should be consulted for the acceptable quantity of portable handsets.

3-13* Special Requirements for Low Power Radio (Wireless) Systems.

A-3-13 Special Requirements for Low Power Radio (Wireless) Systems.

(a) The term "wireless" has been replaced with "low power radio" to eliminate potential confusion with other transmission media such as optical fiber cables.

(b) Low power radio devices are required to comply with the applicable low power requirements of Title 47, *Code of Federal Regulations*, Part 15.

Listed wireless fire alarm systems have numerous applications. Historical buildings where wire or cable

installation will damage the building or impact the historical significance of the property have used wireless fire alarm systems successfully. Industrial buildings that are remote from the main facility is another example of a wireless fire alarm system application. See Figures 3.22 and 3.23.

3-13.1 Compliance with this section shall require the use of low power radio equipment specifically listed for the purpose.

NOTE: Equipment listed solely for household use does not comply with this requirement.

3-13.2 Power Supplies. A primary battery (dry cell) shall be permitted to be used as the sole power source of a low power radio transmitter where all of the following conditions are met:

(a) Each transmitter shall serve only one device and shall be individually identified at the receiver/control unit.

(b) The battery shall be capable of operating the low power radio transmitter for not less than 1 year before the battery depletion threshold is reached.

(c) A battery depletion signal shall be transmitted before the battery has depleted to a level insufficient to support alarm transmission after 7 additional days of normal operation. This

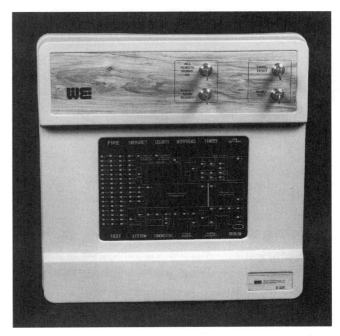

Figure 3.22   Low power wireless combination system control unit. Photograph courtesy of World Electronics, Coral Gables, FL.

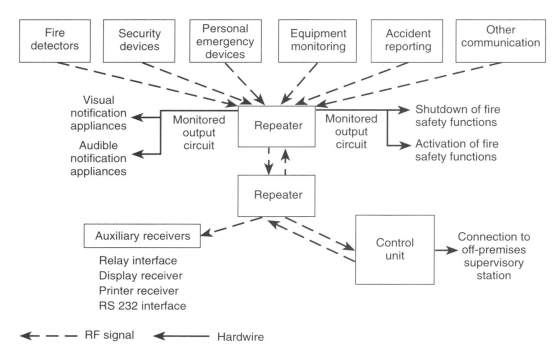

Figure 3.23   *Combination system including a low power radio (wireless) fire alarm system with other non-fire equipment. Drawing courtesy of World Electronics, Inc., Coral Springs, FL.*

signal shall be distinctive from alarm, supervisory, tamper, and trouble signals; shall visibly identify the affected low power radio transmitter; and, when silenced, shall automatically re-sound at least once every 4 hours.

(d) Catastrophic (open or short) battery failure shall cause a trouble signal identifying the affected low power radio transmitter at its receiver/control unit. When silenced, the trouble signal shall automatically re-sound at least once every 4 hours.

(e) Any mode of failure of a primary battery in a low power radio transmitter shall not affect any other low power radio transmitter.

3-13.3 Alarm Signals.

3-13.3.1 When actuated, each low power radio transmitter shall automatically transmit an alarm signal.

NOTE: This requirement is not intended to preclude verification and local test intervals prior to alarm transmission.

3-13.3.2 Each low power radio transmitter shall automatically repeat alarm transmission at intervals not exceeding 60 seconds until the initiating device is returned to its normal condition.

3-13.3.3 Fire alarm signals shall have priority over all other signals.

3-13.3.4 The maximum allowable response delay from activation of an initiating device to receipt and display by the receiver/control unit shall be 90 seconds.

3-13.3.5 An alarm signal from a low power radio transmitter shall latch at its receiver/control unit until manually reset and shall identify the particular initiating device in alarm.

3-13.4 Supervision.

3-13.4.1 The low power radio transmitter shall be specifically listed as using a transmission method that shall be highly resistant to misinterpretation of simultaneous transmissions and to interference (e.g., impulse noise and adjacent channel interference).

3-13.4.2 The occurrence of any single fault that disables transmission between any low power radio transmitter and the receiver/control unit shall cause a latching trouble signal within 200 seconds.

Exception: Where Federal Communications Commission (FCC) regulations prevent meeting the 200-second requirement, the time period for a low power radio transmitter with only a single, connected alarm-initiating device shall be permitted to be increased to four times the minimum time interval permitted for a 1-second transmission up to:

(a) Four hours maximum for a transmitter serving a single initiating device.

(b) Four hours maximum for a retransmission device (repeater) where disabling of the repeater or its transmission does not prevent the receipt of signals at the receiver/control unit from any initiating device transmitter.

3-13.4.3 A single fault on the signaling channel shall not cause an alarm signal.

3-13.4.4 The normal periodic transmission from a low power radio transmitter shall ensure successful alarm transmission capability.

3-13.4.5 Removal of a low power radio transmitter from its installed location shall cause immediate transmission of a distinctive supervisory signal that indicates its removal and individually identifies the affected device.

Exception: This requirement shall not apply to household fire warning systems.

3-13.4.6 Reception of any unwanted (interfering) transmission by a retransmission device (repeater) or by the main receiver/control unit, for a continuous period of 20 seconds or more, shall cause an audible and visible trouble indication at the main receiver/control unit. This indication shall identify the specific trouble condition as an interfering signal.

4

Supervising Station Fire Alarm Systems

Contents, Chapter 4

This chapter presents the requirements for three of the supervising station services: central station, proprietary, and remote station. It also presents the requirements for the various transmission technologies, and the requirements for public fire reporting and auxiliary systems.

4-1 Scope.

This chapter covers the requirements for the proper performance, installation, and operation of fire alarm systems at a continuously attended supervising station and between the protected premises and the continuously attended supervising station.

This chapter of the code covers the requirements for a protected premises fire alarm system connected to, and monitored by, a continuously attended supervising station. This station may be either a central station, proprietary supervising station, remote supervising station, or, in the case of an auxiliary connection to a public fire reporting system, a public fire service communication center. See Figure 4.1.

4-2 Fire Alarm Systems for Central Station Service.

NOTE: The requirements of Chapters 1 and 7 and Section 4-5 shall apply to central station fire alarm systems, unless they conflict with the requirements of this section.

4-2.1 **Scope.** This section describes the general requirements and use of fire alarm systems to provide central station service as defined in Section 1-4.

4-2.2 **General.**

4-2.2.1 These systems include the central station physical plant, exterior communications channels, subsidiary stations, and signaling equipment located at the protected premises.

See Figure 4.2.

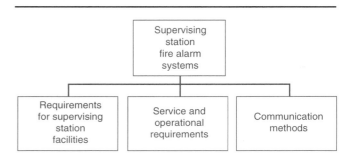

Figure 4.1 *Supervising station fire alarm systems. Drawing courtesy of R.P. Schifiliti Associates, Inc., Reading, MA.*

Figure 4.2   Typical Central Station Facility. Photograph courtesy of AFA Protective Systems, Inc., Syosset, NY.

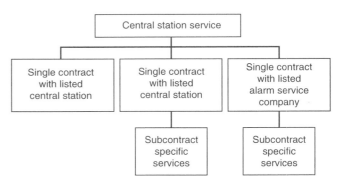

Figure 4.3   Subscriber contracts for central station services. Drawing courtesy of R.P. Schifiliti Associates, Inc., Reading, MA.

4-2.2.2*   This section applies to central station service, which consists of the following elements: installation of fire alarm transmitters; alarm, guard, supervisory, and trouble signal monitoring; retransmission; associated record keeping and reporting; testing and maintenance; and runner service. These services shall be provided under contract to a subscriber by one of the following:

(a) A listed central station that provides all of the elements of central station service with its own facilities and personnel.

(b) A listed central station that provides, as a minimum, the signal monitoring, retransmission, and associated record keeping and reporting with its own facilities and personnel and that might subcontract all or any part of the installation, testing and maintenance, and runner service.

(c) A listed fire alarm service-local company that provides the installation and testing and maintenance with its own facilities and personnel and that subcontracts the monitoring, retransmission, and associated record keeping and reporting to a listed central station. The required runner service shall be provided by the listed fire alarm service-local company with its own personnel or the listed central station with its own personnel.

See Figure 4.3.

A-4-2.2.2   There are related types of contract service that often are provided from, or controlled by, a central station but that are neither anticipated by, nor consistent with, the provisions of 4-2.2.2. Although 4-2.2.2 does not preclude such arrangements, a central station company is expected to recognize, provide for, and preserve the reliability, adequacy, and integrity of those supervisory and alarm services intended to be in accordance with the provisions of 4-2.2.2.

Central station service consists of eight distinct elements. Installation, testing, maintenance and runner service are based at the protected premises. Management or operation of the system, monitoring of signals from the protected premises, retransmission of signals and record keeping are based at the supervising station. This code chapter provides the guidelines for three ways of accomplishing central station service: a central station may provide all eight elements; a central station may provide the four elements at the supervising station and subcontract one or more of the four elements at the protected premises; a listed fire alarm service (local company) may provide the four elements at the protected premises and subcontract the supervising station duties to a central station. In this latter arrangement, if the listed fire alarm service (local company) does not provide the runner service, then runner service must be provided by the central station.

Typically, central station service provides the four elements at the supervising station and subcontracts one or more of the elements at the protected premises. Most commonly, the central station subcontracts part of the installation—notably the installation of waterflow alarm and sprinkler supervisory devices, which are subcontracted to a sprinkler system contractor—while providing all other elements at the protected premises.

Many fire alarm system installers connect protected premises fire alarm systems to a location remote from the protected premises that monitors signals. Relatively few such arrangements meet the requirements of Section 4-2 and should not be called "central station service." Only service that incorporates all eight elements of central station service provided by listed alarm service providers who design, specify, install, test, maintain, and use the system in accordance with the requirements of Section 4-2 should be called "central station service."

4-2.2.3 The prime contractor shall conspicuously indicate that the fire alarm system providing service at a protected premises complies with all the requirements of this code by providing a means of third party verification, as specified in 4-2.2.3.1 or 4-2.2.3.2.

To ensure the inherent higher level of protection that central station service provides, the code requires the prime contractor to verify that the entire fire alarm system meets the requirements of the Code. The prime contractor may supply a certificate issued by the organization that has listed the central station. Or, the prime contractor may post a placard that indicates compliance. By intent, the code does not provide details of the process by which the organization that has listed the prime contractor issues the required certificate. Rather, the code leaves these details up to the procedures and practices of the listing organization. Unless an authority having jurisdiction specifies one of these two methods, the prime contractor—either the central station or the listed fire alarm service (local company) may choose the method of verification.

4-2.2.3.1 The installation shall be certificated.

4-2.2.3.1.1 Fire alarm systems providing service that complies with all requirements of this code shall be certified by the organization that has listed the central station, and a document attesting to this certification shall be located on or near the fire alarm system control unit or, if no control unit exists, on or near a fire alarm system component.

4-2.2.3.1.2 A central repository of issued certification documents, accessible to the authority having jurisdiction, shall be maintained by the organization that has listed the central station.

The organization that has listed the central station determines the detailed procedures that result in the issuing of a certificate. This organization produces a document, or certificate, for a specific protected premises. This certificate indicates that the central station fire alarm system installed by the prime contractor complies with the requirements of the code.

As mentioned previously, the code does not provide details of the process by which the organization that has listed the prime contractor issues the required certificate. Rather, the code leaves these details up to the procedures and practices of the listing organization. For example, Underwriters Laboratories, Inc. (UL) has a central station fire alarm certificate program that meets the intent of this requirement of the code.

In UL's program, a UL listed prime contractor submits an application form to UL asking UL to issue a certificate for the specific installation. UL reviews the details supplied on the application form, and if UL judges that the installation described on the application form meets the requirements of the code, and UL's own requirements, it will issue the certificate. To help maintain the integrity of the certification process, UL annually inspects a statistically significant sampling of certified installations for each listed prime contractor. To comply with the requirements of 4-2.2.3.1.2, UL maintains a computer-operated database of certified installations at UL's headquarters in Northbrook, IL. Authorities having jurisdiction may access this database by means of UL's certificate verification service (ULCVS).

Any other organization that lists prime contractors (see the definition of "listed" in 1-4) could also develop a central station fire alarm certificate program.

4-2.2.3.2 The installation shall be placarded.

> **Formal Interpretation 89-1**
> Reference: 4-2.2.3.2
>
> *Question:* Is placarding (which is not defined as is certification) intended to be an independent method of verification by the central station with no third party agency being involved as assurance?
> *Answer:* No.
>
> *Issue Edition:* NFPA 71-1989
> *Reference:* 1-2.3.1
> *Issue Date:* June 22, 1992 ∎

The review process of the authority having jurisdiction will provide such third party involvement.

4-2.2.3.2.1 Fire alarm systems providing service that complies with all requirements of this code shall be conspicuously marked by the central station to indicate compliance. The marking shall be by one or more securely affixed placards that meet the requirements of the organization that has listed the central station and requires the placard.

4-2.2.3.2.2 The placard(s) shall be 20 in.^{2} (130 cm^2) or larger, shall be located on or near the fire alarm system control unit or, if no control unit exists, on or near a fire alarm system component, and shall identify the central station by name and telephone number.

Formal Interpretation 96-1
Reference: 4-2.2.3 through 4-2.2.5

Question 1: Do paragraphs 4-3.2.3 [4-2.2.3 in 1996 edition] through 4-3.2.5 [4-2.2.5 in 1996 edition] require that compliance of central station fire alarm systems with the requirements of NFPA 72 be verified by a third party inspection
Answer: No.

Question 2: Does the following sentence express the intent of the Technical Committee?
"Although inspection may be a part of a listing organization's verification procedure, the Code requires only that such third party verification be accomplished by certification or by placarding."
Answer: Yes.

Issue Edition: NFPA 72-1993
Reference: 4-3.2.3 through 4-3.2.5.
Issue Date: December 20, 1996 ■

An authority having jurisdiction may choose the posting of a placard by the central station as the method for the prime contractor to verify compliance with the code. Or, if the authority having jurisdiction does not have a preference, the prime contractor may choose this method of verifying compliance with the code. In either case, the prime contractor posts the placard on or near the fire alarm system control unit or, if no control unit exists, on or near a significant component of the fire alarm system. During the initial acceptance test of a placarded system, the authority having jurisdiction will be given the opportunity to inspect the system to assure the accuracy of the placard.

4-2.2.4* Fire alarm system service not complying with all requirements of Section 4-2 shall not be designated as central station service.

A-4-2.2.4 It is the responsibility of the prime contractor to remove all compliance markings (certification markings or placards) when a service contract goes into effect that conflicts in any way with the requirements of 4-2.2.4.

If a change is made that invalidates the designation "central station service," the prime contractor must remove the certificate, placard, or other designation stating that the system is providing central station service. Enforcement of this requirement is held by the authority having jurisdiction.

4-2.2.5* For the purpose of Section 4-2, the subscriber shall notify the prime contractor in writing of the identity of the authority(ies) having jurisdiction.

A-4-2.2.5 The prime contractor should be aware of statutes, public agency regulations, or certifications regarding fire alarm systems that might be binding on the subscriber. The prime contractor should identify for the subscriber which agencies could be an authority having jurisdiction and, where possible, advise the subscriber of any requirements or approvals being mandated by these agencies.

The subscriber has the responsibility for notifying the prime contractor of those private organizations that are being designated as an authority having jurisdiction. The subscriber also has the responsibility to notify the prime contractor of changes in the authority having jurisdiction, such as where there is a change in insurance companies. Although the responsibility is primarily the subscriber's, the prime contractor should also take the responsibility to seek out these private authorities having jurisdiction through the subscriber. The prime contractor has the responsibility for maintaining current records on the authority(ies) having jurisdiction for each protected premises.

The most prevalent public agency involved as an authority having jurisdiction with regard to fire alarm systems is the local fire department or fire prevention bureau. These are normally city or county agencies with statutory authority, and their approval of fire alarm system installations might be required. At the state level, the fire marshal's office is most likely to serve as the public regulatory agency.

The most prevalent private organizations involved as authorities having jurisdiction are insurance companies. Others include insurance rating bureaus, insurance brokers and agents, and private consultants. It is important to note that these organizations have no statutory authority and become authorities having jurisdiction only when designated by the subscriber.

With both public and private concerns to satisfy, it is not uncommon to find multiple authorities having jurisdiction involved with a particular protected premises. It is necessary to identify all authorities having jurisdiction in order to obtain all the necessary approvals for a central station fire alarm system installation.

It is essential that the subscriber and the prime contractor identify all of the authorities having jurisdiction involved at a particular protected premises. Although this responsibility rests primarily with the subscriber, the subscriber would normally only know the private authorities having jurisdiction. The prime contractor would most often know any additional public authorities having jurisdiction. Thus, a joint effort will most effectively resolve this important requirement.

4-2.3 Facilities.

4-2.3.1 The central station building or that portion of a building occupied by a central station shall conform to the construction, fire protection, restricted access, emergency lighting, and power facilities requirements of the latest edition of ANSI/UL 827, *Standard for Safety Central-Station for Watchman, Fire-Alarm and Supervisory Services*.

The ANSI/UL standard provides detailed protection features that help to maintain the integrity and continuity of the physical central station. In 4-2.2.2, the code requires that a nationally-recognized testing laboratory acceptable to the authority having jurisdiction lists the central station and examines the protection features required by this ANSI/UL standard for compliance.

4-2.3.2 Subsidiary station buildings or those portions of buildings occupied by subsidiary stations shall conform to the construction, fire protection, restricted access, emergency lighting, and power facilities requirements of the latest edition of ANSI/UL 827, *Standard for Safety Central-Station for Watchman, Fire-Alarm and Supervisory Services*.

This requirement, and those that follow in subsections 4-2.3.2.1 through 4-2.3.2.7, reflect the fact that under normal operating conditions, no one staffs a subsidiary station. Usually, a subsidiary station serves a particular geographic area. It concentrates signals from many protected premises and transmits those concentrated signals to a supervising station. A malfunction at a subsidiary station can substantially impair the successful transmission of signals from the properties it serves. Thus, these requirements help ensure the overall operational reliability of the subsidiary station. They also help ensure the integrity of the transmission path between the subsidiary station and the supervising station.

4-2.3.2.1 All intrusion, fire, power, and environmental control systems for subsidiary station buildings shall be monitored by the central station in accordance with 4-2.3.

The central station staff manages the subsidiary station by monitoring critical building systems. This helps to ensure the operational integrity of the subsidiary station.

4-2.3.2.2 The subsidiary facility shall be inspected at least monthly by central station personnel for the purpose of verifying the operation of all supervised equipment, all telephones, all battery conditions, and all fluid levels of batteries and generators.

The central station staff also manages the integrity of the subsidiary station by inspecting it monthly. Not simply a "stop by" visit, this inspection verifies the continuity of the systems installed at the subsidiary station.

4-2.3.2.3 In the event of the failure of equipment at the subsidiary station or the communications channel to the central station, a backup shall be operational within 90 seconds. Restoration of a failed unit shall be accomplished within 5 days.

The subsidiary station must have backup equipment and, in the event of a failure, the central station must be able to place the backup equipment into operation within 90 seconds. A technician must repair or replace any defective equipment within 5 days.

4-2.3.2.4 There shall be continuous supervision of each communications channel between the subsidiary station and the central station.

The equipment connected to each communications channel between the subsidiary station and the central station must continuously monitor for channel integrity. In most cases, the equipment uses some form of continuous multiplex transmission technology. Equipment using "T1 network technology" has become quite common for this purpose.

4-2.3.2.5 When the communications channel between the subsidiary station and the supervising station fails, the communications shall be switched to an alternate path. Public switched telephone network facilities shall be used only as an alternate path.

In addition to a fully redundant communication channel, the central station may use public telephone utility provided "dial up, make good" service. The central station initiates the "dial up, make good" to re-establish the communication channel between the subsidiary station and the central station.

4-2.3.2.6 In the subsidiary station, there shall be either a cellular telephone or an equivalent communications path that is independent of the telephone cable between the subsidiary station and the serving wire center.

This requirement ensures that service personnel can establish communication with the central station upon arrival at a totally impaired subsidiary station.

4-2.3.2.7 A plan of action to provide for restoration of services specified by this code shall exist for each subsidiary station.

4-2.3.2.7.1 This plan shall provide for restoration of services within 4 hours of any impairment causing loss of signals from the subsidiary station to the central station.

The central station must formulate a written plan for restoring service from a subsidiary station. This plan must encompass all services not already covered by 4-2.3.2.3. Such restoration must occur within 4 hours. Commonly the organization that lists a central station utilizing a subsidiary station will review such a plan.

As with all emergency plans, the central station must test this plan's accuracy and validity. By performing this annual exercise, the implementing personnel have an opportunity to become thoroughly familiar with the procedure. Such an exercise also helps keep the plan up-to-date, and disclose changes at either the subsidiary station or the central station that may affect the integrity of the plan.

4-2.3.2.7.2 There shall be an exercise to demonstrate the adequacy of the plan at least annually.

4-2.4 Equipment.

4-2.4.1 The central station and all subsidiary stations shall be so equipped to receive and record all signals in accordance with 4-5.4. Circuit-adjusting means for emergency operation shall be permitted to be automatic or to be provided through manual operation upon receipt of a trouble signal. Computer aided alarm and supervisory signal processing hardware and software shall be listed for the specific application.

This subsection permits these specially-trained central station operators (see Section 4-2.5 for training requirements) to manually operate circuit-adjusting means. Equipment may also automatically operate circuit-adjusting means. The organization listing the central station must also specifically list any computer aided alarm and supervisory signal processing hardware and software for central station service.

4-2.4.2 Power supplies shall comply with the requirements of Chapter 1.

4-2.4.3 Transmission means shall comply with the requirements of Section 4-5.

4-2.4.4* Two independent means shall be provided to retransmit a fire alarm signal to the appropriate public fire service communications center.

NOTE: The use of a universal emergency number (e.g., 911 public safety answering point) does not meet the intent of this code for the principal means of retransmission.

A-4-2.4.4 Two telephone lines (numbers) at the central station connected to the public switched telephone network, each having its own telephone instrument connected, and two telephone lines (numbers) available at the public fire service communications center to which a central station operator can retransmit an alarm meet the intent of this requirement.

In subsection 4-2.2.2, the code declares that retransmitting fire alarm signals to the public fire service communication center is one of the eight elements of central station service. The central station must have a reasonably secure means to retransmit.

In most cases the central station will use the public switched telephone network to dial the seven-digit fire reporting number (additional digits are needed if the central station must dial an area code). The code states that the central station may not use 911 as the primary means of retransmission because: a) the central station may not be located in the same community as the protected premises, b) a public service answering point (PSAP) staffed with personnel who are not a part of the public fire department answer most 911 and enhanced 911 emergency telephone calls, and, c) the vast majority of 911 calls concern non-fire emergencies. Following this requirement helps to avoid the bottleneck that sometimes occurs at the PSAP.

4-2.4.4.1 Where the principal means of retransmission is not equipped to allow the center to acknowledge receipt of each fire alarm report, both means shall be used to retransmit.

4-2.4.4.2* Where required by the authority having jurisdiction, one of the means of retransmission shall be supervised so that interruption of retransmission circuit (channel) communications integrity results in a trouble signal at the central station.

A-4-2.4.4.2 The following methods have been used successfully for supervising retransmission circuits (channels):

(a) An electrically supervised circuit (channel) provided with suitable code sending and automatic recording equipment.

(b) A supervised circuit (channel) providing suitable voice transmitting, receiving, and automatic recording equipment. The circuit may be permitted to be a telephone circuit that:

1. Cannot be used for any other purpose;

2. Is provided with a two-way ring down feature for supervision between the fire department communications center and the central station;

3. Is provided with terminal equipment located on the premises at each end; and

4. Is provided with 24-hour standby power.

NOTE: Local on-premises circuits are not required to be supervised.

(c) Radio facilities using transmissions over a supervised channel with supervised transmitting and receiving equipment. Circuit continuity ensured by any means at intervals not exceeding 8 hours is satisfactory.

In certain cases, the authority having jurisdiction may require the central station to have a retransmission channel—monitored for integrity—for each public fire service communication center that serves its protected premises. This is important because the public switched telephone network may be unreliable, or, an extreme hazard to property or life may exist at one or more protected premises.

4-2.4.4.3 The retransmission means shall be tested in accordance with Chapter 7.

4-2.4.4.4 The retransmission signal and the time and date of retransmission shall be recorded at the central station.

This requirement does not mandate the central station to record the time and date automatically. However, when computer-based automation systems manage the receipt and retransmission of signals, the central station can and should automatically record the time and date.

Further, based on a somewhat broad interpretation of subsection 4-2.4.1, most central stations record the actual telephone call to the public fire service communication center. This tape recording, along with records of signals received at the central station, often helps investigators reconstruct the sequence of events that occurred during a major fire.

4-2.5 Personnel.

4-2.5.1 The central station shall have sufficient personnel (a minimum of two persons) on duty at the central station at all times to ensure attention to signals received.

The central station must have two operators on duty at all times at the central station. By mandating that two operators be present, the Code maximizes the likelihood that at least one of the operators will be fully alert to incoming signals.

4-2.5.2 Operation and supervision shall be the primary functions of the operators, and no other interest or activity shall take precedence over the protective service.

The Code requires that the operators have no other duties that would distract them from the prompt, effective handling of signals.

4-2.6 Operations.

4-2.6.1 Disposition of Signals.

4-2.6.1.1 Alarm signals initiated by manual fire alarm boxes, automatic fire detectors, waterflow from the automatic sprinkler system, or actuation of other fire suppression system(s) or equipment shall be treated as fire alarms.

The central station shall:

(a)* Immediately retransmit the alarm to the public fire service communications center.

A-4-2.6.1.1(a) The term "immediately" in this context is intended to mean "without unreasonable delay." Routine handling should take a maximum of 90 seconds from receipt of an alarm signal by the central station until the initiation of retransmission to the public fire service communications center.

A central station must give highest priority to the prompt handling and retransmission of fire alarm signals. Under the most adverse circumstances, some transmission technologies may have already taken up to 15 minutes to complete the transmission of a signal from the protected premises to the central station. (See subsections 4-5.3.2.1.4 and 4-5.3.2.1.5.)

(b) Dispatch a runner or technician to the protected premises to arrive within 1 hour after receipt of a signal where equipment needs to be manually reset by the prime contractor.

(c) Notify the subscriber by the quickest available method.

(d) Provide notice to the subscriber or authority having jurisdiction, or both, where required.

Exception: Where the alarm signal results from a prearranged test, the actions specified by 4-2.6.1.1(a) and (c) shall not be required.

Where permitted by the authority having jurisdiction, the runner or technician need respond only when the prime contractor must manually reset equipment at the protected premises. If permitted by the authority having jurisdiction, the runner or technician need not respond when the prime contractor can reset the equipment automatically or remotely, or can make provision for the subscriber or some other trained individual to reset the equipment. See Section 3-11.3.

Notifying the subscriber by telephone is generally the quickest available method. The "notice" mentioned in subsection 4-2.6.1.1(d) is usually a written notice.

4-2.6.1.2 Guard's Signal.

4-2.6.1.2.1 Upon failure to receive a guard's regular signal within a 15-minute maximum grace period, the central station shall:

(a) Communicate without unreasonable delay with personnel at the protected premises.

(b) Dispatch a runner to the protected premises to arrive within 30 minutes of the delinquency where communications cannot be established.

(c) Report all delinquencies to the subscriber or authority having jurisdiction, or both, where required.

If the central station cannot promptly contact personnel at the protected premises, then it should dispatch a runner to investigate why the guard missed a signal. Once dispatched, the runner must arrive at the protected premises within 30 minutes. This means the runner may actually arrive 45 minutes after the guard missed the signal. Even so, in actual cases a responding runner has found the guard injured or ill and, by summoning medical assistance, has saved the guard's life.

4-2.6.1.2.2 Failure of the guard to follow a prescribed route in transmitting signals shall be handled as a delinquency.

Guard's tour supervision by a central station mandates a compulsory tour arrangement. The central station can monitor every reporting station along a route. Or, the central station can monitor only a few of the stations along a route of stations. But in either case, the guard must follow a prescribed route, proceeding from station to station in a fixed sequence. To leave the prescribed route is a delinquency.

4-2.6.1.3* Upon receipt of a supervisory signal from a sprinkler system, other fire suppression system(s), or other equipment, the central station shall:

A-4-2.6.1.3 It is anticipated that the central station will first attempt to notify designated personnel at the protected premises. When such notification cannot be made, it might be appropriate to notify law enforcement or the fire department, or both. For example, if a valve supervisory signal is received where protected premises are not occupied, it is appropriate to notify the police.

The central station must handle supervisory signals promptly and accurately. These signals may indicate that something has impaired a vital protective system.

The runner or technician only responds when the central station operator cannot resolve the restoration of the supervisory signal to normal by contacting designated personnel.

(a)* Communicate immediately with the person(s) designated by the subscriber.

A-4-2.6.1.3(a) The term "immediately" in this context is intended to mean "without unreasonable delay." Routine handling should take a maximum of 4 minutes from receipt of a supervisory signal by the central station until the initiation of communications with a person(s) designated by the subscriber.

(b) Dispatch a runner or maintenance person (arrival time not to exceed 1 hour) to investigate.

Exception: Where an abnormal condition is restored to normal in accordance with a scheduled procedure determined by 4-2.6.1.3(a).

(c) Notify the fire department or law enforcement agency, or both, where required.

(d) Notify the authority having jurisdiction when sprinkler systems or other fire suppression system(s) or equipment has been wholly or partially out of service for 8 hours.

(e) When service has been restored, provide notice, where required, to the subscriber or the authority having jurisdiction, or both, as to the nature of the signal, time of occurrence, and restoration of service when equipment has been out of service for 8 hours or more.

Exception: Where the supervisory signal results from a prearranged test, the actions specified by 4-2.6.1.3(a), (c), and (e) shall not be required.

4-2.6.1.4 Upon receipt of trouble signals or other signals pertaining solely to matters of equipment maintenance of the fire alarm systems, the central station shall:

(a)* Communicate immediately with persons designated by the subscriber.

A-4-2.6.1.4(a) The term "immediately" in this context is intended to mean "without unreasonable delay." Routine handling should take a maximum of 4 minutes from receipt of a trouble signal by the central station until initiation of the investigation by telephone.

The central station must handle trouble signals promptly and accurately. These signals indicate that the fire alarm system itself is wholly or partly out of service.

The personnel, dispatched to arrive within 4 hours, must initiate repairs. This generally means that a technician, rather than a runner, must respond.

(b) Dispatch personnel to arrive within 4 hours to initiate maintenance, if necessary.

(c) Provide notice, where required, to the subscriber or the authority having jurisdiction, or both, as to the nature of the interruption, time of occurrence, and restoration of service, when the interruption is more than 8 hours.

4-2.6.1.5 All test signals received shall be recorded to indicate date, time, and type.

(a) Test signals initiated by the subscriber, including those for the benefit of an authority having jurisdiction, shall be acknowledged by central station personnel whenever the subscriber or authority inquires.

(b)* Any test signal not received by the central station shall be investigated immediately and appropriate action taken to reestablish system integrity.

A-4-2.6.1.5(b) The term "immediately" in this context is intended to mean "without unreasonable delay." Routine handling should take a maximum of 4 minutes from receipt of a trouble signal by the central station until initiation of the investigation by telephone.

The central station must handle test signals immediately. The Code defines this as within four minutes. These signals help to ensure that the fire alarm system functions properly. Likewise, the central station must cooperate with any authority having jurisdiction who inquires regarding test signals.

(c) The central station shall dispatch personnel to arrive within 1 hour where protected premises equipment needs to be manually reset after testing.

The runner or technician need respond only when he or she must manually reset equipment at the protected premises. The runner or technician need not respond when the prime contractor can reset the equipment automatically or remotely, or can make provision for the subscriber or some other trained individual to reset the equipment.

4-2.6.2 **Record Keeping and Reporting.**

4-2.6.2.1 Complete records of all signals received shall be retained for at least 1 year.

4-2.6.2.2 The central station shall make arrangements to furnish reports of signals received to the authority having jurisdiction in a form it finds acceptable.

When an authority having jurisdiction requests reports from a central station, the central station must provide the reports in a useful and usable form.

4-2.7 **Testing and Maintenance.**

4-2.7.1 Testing and maintenance for central station service shall be performed in accordance with Chapter 7.

4-2.7.2 The prime contractor shall provide each of its representatives and each alarm system user with a unique personal identification code.

4-2.7.3 In order to authorize the placing of an alarm system into test status, a representative of the prime contractor or an alarm system user shall first provide the central station with his or her personal identification code.

The prime contractor issues each of its representatives and each alarm system user a unique personal identification code (see 4-2.7.2) and requires its use (see 4-2.7.3) in order to carefully control those who may place the system into a test mode. This new requirement helps to maintain the security and operational integrity of the system. Without this precaution, the central station has no way of verifying that the person placing the fire alarm system into test status has authorization to do so.

4-3 Proprietary Supervising Station Systems.

NOTE: The requirements of Chapters 1 and 7 and Section 4-5 apply to proprietary fire alarm systems, unless they conflict with the requirements of this section.

4-3.1 **Scope.** This section describes the operational procedures for the supervising facilities of proprietary fire alarm systems. It provides the minimum requirements for the facilities, equipment, personnel, operation, and testing and maintenance of the proprietary supervising station.

4-3.2 **General.**

4-3.2.1 Proprietary supervising stations shall be operated by trained, competent personnel in constant attendance who are responsible to the owner of the protected property. *(See 4-3.5.3.)*

Management of a facility protected by a proprietary fire alarm system, uses that system to oversee the built-in fire protection systems. In this role of "management tool," the proprietary fire alarm system can help assure that all other fire protection systems remain in-service.

4-3.2.2   The protected property shall be either a contiguous property or noncontiguous properties under one ownership.

From a single proprietary supervising station, an owner may oversee the protection features at one or more properties. These properties may contiguously occupy a single piece of land or non-contiguous portions of land.

4-3.2.3   Where a protected premises master control unit is integral to or collocated with the supervising station equipment, the requirements of Section 4-5 shall not apply.

This subsection recognizes that in some cases the proprietary fire alarm system may have a master fire alarm control unit, as defined in 1-4, located in the proprietary supervising station. In these cases, the transmission technology requirements described in 4-5 would not apply. Rather, the system would use initiating device circuits, signaling line circuits, and notification appliance circuits described in Chapter 3. Section 4-3 would provide the requirements for all other aspects of such a proprietary fire alarm system.

4-3.2.4*   The systems of this section shall be permitted to be interconnected to other systems intended to make the premises safer in the event of fire or other emergencies indicative of hazards to life or property.

In addition to the information contained in section 4-3, refer to subsection 3-8.14, Elevator Recall for Fire Fighters' Service, to section 3-9, Fire Safety Control Functions, and to section 3-10, Suppression System Actuation.

A-4-3.2.4   The following functions are included in Appendix A to provide guidelines for utilizing building systems and equipment in addition to proprietary fire alarm equipment in order to provide life safety and property protection.

Building functions that should be initiated or controlled during a fire alarm condition include, but should not be limited to, the following:

(a) Elevator operation consistent with ANSI A17.1, *Safety Code for Elevators and Escalators.*

(b) Unlocking of stairwell and exit doors. (*See* NFPA 80, *Standard for Fire Doors and Fire Windows*, and NFPA *101, Life Safety Code.*)

(c) Release of fire and smoke dampers. (*See* NFPA 90A, *Standard for the Installation of Air Conditioning and Ventilat-*

ing Systems, and NFPA 90B, *Standard for the Installation of Warm Air Heating and Air Conditioning Systems.*)

(d) Monitoring and initiating of self-contained automatic fire extinguishing system(s) or suppression system(s) and equipment. (*See* NFPA 11, *Standard for Low-Expansion Foam*; NFPA 11A, *Standard for Medium- and High-Expansion Foam Systems*; NFPA 12, *Standard on Carbon Dioxide Extinguishing Systems*; NFPA 12A, *Standard on Halon 1301 Fire Extinguishing Systems*; NFPA 13, *Standard for the Installation of Sprinkler Systems*; NFPA 14, *Standard for the Installation of Standpipe and Hose Systems*; NFPA 15, *Standard for Water Spray Fixed Systems for Fire Protection*; and NFPA 17, *Standard for Dry Chemical Extinguishing Systems.*)

(e) Lighting control necessary to provide essential illumination during fire alarm conditions. (*See* NFPA 70, National Electrical Code, and NFPA 101, *Life Safety Code.*)

(f) Emergency shutoff of hazardous gas.

(g) Control of building environmental heating, ventilating, and air conditioning equipment to provide smoke control. (*See* NFPA 90A, *Standard for the Installation of Air Conditioning and Ventilating Systems.*)

(h) Control of process, data processing, and similar equipment as necessary during fire alarm conditions.

4-3.3   Facilities.

4-3.3.1   The proprietary supervising station shall be located in a fire-resistive, detached building or in a suitable cut-off room and shall not be near or exposed to the hazardous parts of the premises protected.

The requirements of subsection 4-3.3.1 help to maintain a high degree of physical integrity for the proprietary supervising station.

4-3.3.2   Access to the proprietary supervising station shall be restricted to those persons directly concerned with the implementation and direction of emergency action and procedure.

The proprietary supervising station must not become a congregating place for guards, fire fighters, or other personnel. The presence of such persons may interfere with the operators and distract them from giving proper attention to signal traffic. If management locates the proprietary supervising station within a guard house where guards admit vehicles and personnel to the premises, management should provide some segregation. This will help to ensure that operators may effectively and efficiently handle the signal traffic without distraction.

4-3.3.3   The proprietary supervising station, as well as remotely located power rooms for batteries or engine-driven generators, shall be provided with portable fire extinguishers that comply with the requirements of NFPA 10, *Standard for Portable Fire Extinguishers.*

Personnel in a proprietary supervising station must have the means to handle a small fire in the supervising station or in the power rooms for batteries or engine-driven generators. Management should refer to the requirements of NFPA 600, *Standard on Industrial Fire Brigades*. These requirements ensure that management has properly organized and trained personnel to safely use the fire extinguishers provided.

4-3.3.4 Emergency Lighting System.

4-3.3.4.1 The proprietary supervising station shall be provided with an automatic emergency lighting system. The emergency source shall be independent of the primary lighting source.

4-3.3.4.2 The emergency lighting shall be sufficient to carry on operation of the supervising station for a 26-hour period and shall be tested monthly.

4-3.3.5 Where 25 or more protected buildings or premises are connected to a subsidiary station, both of the following shall be provided at the subsidiary station:

(a) Automatic means for receiving and recording signals under emergency-staffing conditions;

(b) A telephone.

A proprietary supervising station may receive signals from many buildings on a very large premises or from several non-contiguous premises through a subsidiary station. When 25 or more protected buildings or premises transmit through a subsidiary station, management must equip that subsidiary station so that it can be staffed by operators in an emergency. For example, if the signaling path between the subsidiary station and the proprietary supervising station fails, operators will travel to the subsidiary station, staff it, and operate it independently from the proprietary supervising station. See Figure 4.4.

4-3.4 Equipment.

4-3.4.1 This section shall apply to signal-receiving equipment in a proprietary supervising station.

4-3.4.2 Provision shall be made to designate the building in which a signal originates. The floor, section, or other subdivision of the building shall be designated at the proprietary supervising station or at the building protected.

Exception: Where the area, height, or special conditions of occupancy make detailed designation unessential as

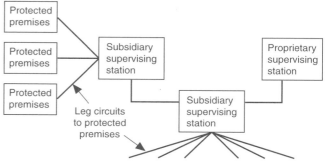

Figure 4.4 Proprietary supervising station facilities— subsidiary stations. Drawing courtesy of R.P. Schifiliti Associates, Inc., Reading, MA.

approved by the authority having jurisdiction. This detailed designation shall utilize indicating appliances acceptable to the authority having jurisdiction.

In order to effectively manage the built-in fire protection features of the protected premises, the Code requires that the information received by the proprietary fire alarm system must have sufficient detail, subject to possible waiver by the authority having jurisdiction.

4-3.4.3 The proprietary supervising station shall have, in addition to a recording device, two different means for alerting the operator when each signal is received indicating a change of state of any connected initiating device circuit. One of these means shall be an audible signal and shall persist until manually acknowledged. This shall include the receipt of alarm signals, supervisory signals, and trouble signals, including signals indicating restoration to normal.

Subsection 4-3.4.3 acknowledges that the operators in the supervising station may attend to other duties. To alert the operators to incoming signals, the Code requires two means of notification.

4-3.4.4 Where suitable means is provided in the proprietary supervising station to identify readily the type of signal received, a common audible indicating appliance shall be permitted to be used for alarm, supervisory, and trouble indication.

Subsection 1-5.4.7 requires distinctive signals. Subsection 4-3.4.4 modifies this requirement to permit a common audible notification appliance in the supervising station, as long as other means readily identify the type of signal.

4-3.4.5   At a proprietary supervising station, an audible trouble signal shall be permitted to be silenced, provided the act of silencing does not prevent the signal from operating immediately upon receipt of a subsequent trouble signal.

4-3.4.6   All signals required to be received by the proprietary supervising station that show a change in status shall be automatically and permanently recorded, including time and date of occurrence. This record shall be in a form that expedites operator interpretation in accordance with any one of the following:

(a) In the event that a visual display is used that automatically provides change of status information for each required signal, including type and location of occurrence, any form of automatic permanent visual record shall be permitted. The recorded information shall include the content described above. The visual display shall show status information content at all times and shall be distinctly different after the operator has manually acknowledged each signal. Acknowledgment shall produce recorded information indicating the time and date of acknowledgment.

Subsection 4-3.4.6(a) describes an annunciator that continuously shows the status of every point in the system that generates a signal. At a glance, the operator can see the status of every point. With this type of visual display, the proprietary fire alarm system may use any type of permanent visual record. Such systems most often use a logging-type printer.

(b) In the event that a visual display is not provided, required signal content information shall be automatically recorded on duplicate permanent visual recording instruments.

One recording instrument shall be used for recording all incoming signals, while the other shall be used for required fire, supervisory, and trouble signals only. Failure to acknowledge a signal shall not prevent subsequent signals from recording. Restoration of the signal to its prior or normal condition shall be recorded.

When the proprietary fire alarm system does not provide a visual display, it may use two printers. One printer will record all signals. The other printer will record only fire alarm, supervisory, and trouble signals. The printer must format the output to allow the operator to easily read, interpret, and act on the information provided.

(c) In the event that a system combines the use of a sequential visual display and recorded permanent visual presentation, the required signal content information shall be displayed and recorded. The visual information component shall be either retained on the display until manually acknowl-

edged or periodically repeated at intervals not greater than 5 seconds, for durations of 2 seconds each, until manually acknowledged. Each new displayed status change shall be accompanied by an audible indication that shall persist until manual acknowledgment of the signal is performed.

There shall be a means provided for the operator to redisplay the status of required signal initiating inputs that have been acknowledged but not yet restored to a normal condition. Where the system retains the signal on the visual display until manually acknowledged, subsequent recorded presentations shall not be inhibited upon failure to acknowledge. Fire alarm signals shall be segregated on a separate visual display in this configuration.

The visual display unit described in subsection 4-3.4.6(c) presents one or more lines of information, but does not simultaneously display the status of all points in the proprietary fire alarm system. The operator must scroll through the display once he or she has acknowledged each signal. To help the operator give proper precedence to fire alarm signals, the signals must either appear on a separate display, or the system must give them priority status on a common display. The system must still provide a permanent visual record, but the Code does not specify the type of printer.

Exception: Fire alarm signals shall not be required to be segregated on a separate display where given priority status on the common visual display.

4-3.4.7   The maximum elapsed time from sensing a fire alarm at an initiating device or initiating device circuit until it is recorded or displayed at the proprietary supervising station shall not exceed 90 seconds.

Some authorities having jurisdiction have interpreted subsection 4-3.4.7 to severely limit the type of transmission technology used for a proprietary fire alarm system. They believe this subsection requires a much higher level of performance than do the requirements of subsections 4-5.3.2.1.5, 4-5.3.2.3.3, or 4-5.3.5.2(a), (b), and (c). They also believe this subsection further reinforces the requirements in subsections 4-5.3.1.2.3(a) and 4-5.3.4.1(a), and assert that whatever transmission technology the proprietary fire alarm system uses, that technology must guarantee delivery of a fire alarm signal within 90 seconds of actuation of the initiating device, in all circumstances. However, the Code states such requirements assuming a normal transmission pathway. Such a requirement would not apply to an impaired transmission pathway. As long as the transmission pathway meets the requirements under normal circumstances, the pathway complies with the requirement of subsection 4-3.4.7.

4-3.4.8 To facilitate the prompt receipt of fire alarm signals from systems handling other types of signals that can produce multiple simultaneous status changes, the requirements of either of the following shall be met:

(a) In addition to the maximum processing time for a single alarm, the system shall record simultaneous status changes at a rate not slower than either a quantity of 50, or 10 percent of the total number of initiating device circuits connected, within 90 seconds, whichever number is smaller, without loss of any signal.

(b) In addition to the maximum processing time, the system shall either display or record fire alarm signals at a rate not slower than one every 10 seconds, regardless of the rate or number of status changes occurring, without loss of any signals.

Exception: Where fire alarm, waterflow alarm, and sprinkler supervisory signals and their associated trouble signals are the only signals processed by the system, the rate of recording shall not be slower than one signal every 30 seconds.

The requirements of this subsection helps to ensure the prompt receipt of fire alarm signals when a proprietary fire alarm system receives other types of signals from initiating devices that may produce many status changes at the same time. This subsection reinforces the requirement for active multiplex transmission systems in subsection 4-5.3.1.2.3(c). It also applies that requirement to other transmission technologies.

However, when a proprietary fire alarm system handles only fire alarm signals, waterflow alarm signals, sprinkler supervisory signals, and their associated trouble signals, the Exception relieves such a system from meeting the requirements of this subsection.

4-3.4.9 Trouble signals required in 1-5.8 and their restoration to normal shall be automatically indicated and recorded at the proprietary supervising station within 200 seconds.

This subsection also limits the transmission technologies used for a proprietary fire alarm system. This subsection modifies, in part, the requirements in subsection 4-5.3.1.2.3(b) for active multiplex transmission systems. Subsection 4-5.3.1.2.3(b) requires Type 1 active multiplex systems to transmit all signals, including trouble signals, within 90 seconds.

The requirements of subsection 4-3.4.9 also impose the 200 second limit on the receipt of trouble signals on the other transmission technologies.

4-3.4.10 The recorded information for the occurrence of any trouble condition of signaling line circuit, leg facility, or trunk facility that prevents receipt of alarm signals at the proprietary supervising station shall be such that the operator is able to determine the presence of the trouble condition. Trouble conditions in a leg facility shall not affect or delay receipt of signals at the proprietary supervising station from other leg facilities on the same trunk facility.

The last sentence of this subsection effectively mandates that the transmission technology preserve the signals from other leg facilities when one leg facility experiences trouble. Management would have to analyze each transmission technology chosen to determine if it could meet the requirements of this subsection. For example, if management chooses to use an active multiplex transmission technology, this requirement dictates that the system meet the requirements of a Type 1 or 2 active multiplex system.

4-3.5 Personnel.

4-3.5.1 At least two operators, one of whom shall be permitted to be a runner, shall be on duty at all times.

Exception: Where the means for transmitting alarms to the fire department is automatic, at least one operator shall be on duty at all times.

The Exception implies that where management provides automatic retransmission of fire alarm signals, the sole operator could respond as a runner.

4-3.5.2 When the runner is not in attendance at the proprietary supervising station, the runner shall establish two-way communications with the station at intervals not exceeding 15 minutes.

4-3.5.3 The primary duties of the operator(s) shall be to monitor signals, operate the system, and take such action as shall be required by the authority having jurisdiction. The operator(s) shall not be assigned any additional duties that would take precedence over the primary duties.

The Code expects the operators to have no other duties that would distract from the prompt, effective handling of signals.

4-3.6 Operations.

4-3.6.1 Communications and Transmission Channels.

4-3.6.1.1 All communications and transmission channels between the proprietary supervising station and the protected premises master control unit (panel) shall be operated manually or automatically once every 24 hours to verify operation.

4-3.6.1.2 When a communications or transmission channel fails to operate, the operator shall immediately notify the person(s) identified by the owner or authority having jurisdiction.

4-3.6.2 All operator controls at the proprietary supervising station(s) designated by the authority having jurisdiction shall be operated at each change of shift.

4-3.6.3 If operator controls fail, the operator shall immediately notify the person(s) identified by the owner or authority having jurisdiction.

The requirements of subsections 4-3.6.1, 4-3.6.1.1, 4-3.6.2, and 4-3.6.3 help the proprietary supervising station to continue operating. Exercising the communication channels and operating controls allows the operator to more quickly identify potential failures. Also, by promptly contacting designated persons when a failure occurs, the operators will help to ensure that repairs begin as soon as possible.

4-3.6.4 Indication of a fire shall be promptly retransmitted to the public fire service communications center or other locations acceptable to the authority having jurisdiction, indicating the building or group of buildings from which the alarm has been received.

4-3.6.5* The means of retransmission shall be acceptable to the authority having jurisdiction and shall be in accordance with Section 4-2, 4-4, 4-6, or 4-7.

Exception: Secondary power supply capacity shall be as required in Chapter 1.

A-4-3.6.5 It is the intent of this code that the operator within the proprietary supervising station should have a secure means of immediately retransmitting any signal indicative of a fire to the public fire department communications center. Automatic retransmission using an approved method installed in accordance with Sections 4-2, 4-3, 4-4, 4-6, and 4-7 is the best method for proper retransmission. However, a manual means may be permitted to be used, consisting of either a manual connection following the requirements of Sections 4-2, 4-4, and 4-7, or, for proprietary supervising stations serving only contiguous properties, a means in the

form of a municipal fire alarm box installed within 50 ft (15 m) of the proprietary supervising station in accordance with Section 4-6 may be permitted.

This subsection requires that the proprietary supervising station retransmit signals to the public fire service communication center by means of a central station fire alarm system, a remote station fire alarm system, or an auxiliary fire alarm system. It also permits a proprietary supervising station covering a contiguous property to retransmit signals to the public fire service communication center by means of a municipal fire alarm box.

The authority having jurisdiction would have to make an exception to the requirements of this subsection if management desired to retransmit alarms by ordinary telephone.

4-3.6.6* Retransmission by coded signals shall be confirmed by two-way voice communications indicating the nature of the alarm.

A-4-3.6.6 Regardless of the type of retransmission facility used, telephone communications between the proprietary supervising station and the fire department should be available at all times and should not depend on a switchboard operator.

Management should provide the proprietary supervising station with a connection to the public switched telephone network that does not require private branch exchange (PBX) switchboard manual intervention to obtain access to the network. In fact, management should provide a direct connection to the network that completely bypasses the PBX. This will allow operators in the proprietary supervising station to make a telephone call, even when the PBX fails.

4-3.6.7 Dispositions of Signals.

4-3.6.7.1 Alarms. Upon receipt of a fire alarm signal, the proprietary supervising station operator shall initiate action to:

(a) Immediately notify the fire department, the plant fire brigade, and such other parties as the authority having jurisdiction requires.

(b) Promptly dispatch a runner to the alarm location (travel time shall not exceed 1 hour).

(c) Restore the system to its normal operating condition as soon as possible after disposition of the cause of the alarm signal.

4-3.6.7.2 Guard's Tour Delinquency. Where a regular signal is not received from a guard within a 15-minute maximum grace period, or where a guard fails to follow a prescribed route in transmitting the signals (where a prescribed route has been established), it shall be treated as a delinquency signal. When a guard's tour delinquency occurs, the proprietary supervising station operator shall initiate action to:

(a) Communicate at once with the protected areas or premises by telephone, radio, calling back over the system circuit, or other means acceptable to the authority having jurisdiction.

(b) Dispatch a runner to investigate the delinquency, where communications with the guard cannot be promptly established (travel time shall not exceed 1/2 hour).

4-3.6.7.3   Supervisory Signals.

Upon receipt of sprinkler system and other supervisory signals, the proprietary supervising station operator shall initiate action to:

(a) Where required, communicate immediately with the designated person(s) to ascertain the reason for the signal.

(b) Where required, dispatch a runner or maintenance person (travel time not to exceed 1 hour) to investigate, unless supervisory conditions are promptly restored to normal.

(c) Where required, notify the fire department.

(d) Where required, notify the authority having jurisdiction when sprinkler systems are wholly or partially out of service for 8 hours or more.

(e) Where required, provide written notice to the authority having jurisdiction as to the nature of the signal, time of occurrence, and restoration of service, when equipment has been out of service for 8 hours or more.

4-3.6.7.4   Trouble Signals.

Upon receipt of trouble signals or other signals pertaining solely to matters of equipment maintenance of the fire alarm system, the proprietary supervising station operator shall initiate action to:

(a) Where required, communicate immediately with the designated person(s) to ascertain reason for the signal.

(b) Where required, dispatch a runner or maintenance person (travel time not to exceed 1 hour) to investigate.

(c) Where required, notify the fire department.

(d) Where required, notify the authority having jurisdiction when interruption of normal service will exist for 4 hours or more.

(e) Where required, provide written notice to the authority having jurisdiction as to the nature of the signal, time of occurrence, and restoration of service, when equipment has been out of service for 8 hours or more.

The requirements in subsection 4-3.6.7 almost match those in subsection 4-2.6.1 for central station fire alarm systems with the following notable differences:

- Upon receipt of an alarm, the proprietary supervising station must always dispatch a runner.
- The authority having jurisdiction may require the proprietary supervising station to notify the fire department upon receipt of a supervisory signal or a trouble signal. This will alert the fire department to impaired protection.
- The runner must respond to a trouble signal within one hour.
- The proprietary supervising station must notify the authority having jurisdiction if an interruption to service that produces a trouble signal persists for 4 hours.

4-3.6.8   Record Keeping and Reporting.

4-3.6.8.1

Complete records of all signals received shall be retained for at least 1 year.

4-3.6.8.2

The proprietary supervising station shall make arrangements to furnish reports of signals received to the authority having jurisdiction in a form it finds acceptable.

When an authority having jurisdiction requests reports from a proprietary supervising station, the supervising station must provide the reports in a useful and usable form.

4-3.7   Testing and Maintenance.

Testing and maintenance of proprietary fire alarm systems shall be performed in accordance with Chapter 7.

4-4   Remote Supervising Station Fire Alarm Systems.

NOTE: The requirements of Chapters 1 and 7 and Section 4-5 shall apply to remote supervising station fire alarm systems, unless they conflict with the requirements of this section.

4-4.1   Scope.

This section is intended to apply where central station service is neither required nor elected. It describes the installation, maintenance, testing, and use of a remote supervising station fire alarm system that serves properties under various ownership from a remote supervising station where trained, competent personnel are in constant attendance. It covers the minimum requirements for the remote supervising station physical facilities, equipment, operating personnel, response, retransmission, signals, reports, and testing.

An authority having jurisdiction may require a remote station fire alarm system when that authority does not need the level of protection offered by a central station fire alarm system. Or, the management of a facility may choose to provide a remote station fire alarm system when management does not believe it needs the level of protection offered by a central station fire alarm system or a proprietary fire alarm system.

When it first appeared in the Code in 1961, the requirements for a remote station fire alarm system provided a means of transmitting fire alarm, supervisory, and trouble signals from a protected premises to the public fire service communication center. In most of these cases, the municipality did not have a public fire reporting system, and no one could provide central station service to that locale.

Sometimes the municipality did not have a constantly-attended public fire service communication center. Rather, officials relied on a multiple-location fire telephone system. When an individual placed a telephone call to the fire reporting number, telephones in several locations throughout the community rang. These locations included local businesses, as well as the homes of the fire chief and other fire officers. A switch at each telephone could activate sirens throughout the community that would summon the volunteer fire fighters.

In these communities, officials would have to find an alternate location to receive signals from remote station fire alarm systems. Often, officials chose a local 24-hour telephone answering service used by doctors, dentists, or tradesmen to receive the remote station fire alarm signals. In some cases, the officials chose a gasoline service station or local restaurant that remained open around the clock, to receive the remote station fire alarm signals.

In later years, some alarm system installers who chose not to provide listed central station service set up monitoring centers to receive remote station fire alarm signals. In turn, some listed central station operating companies began to provide equipment that would meet the requirements for remote station fire alarm systems to receive such signals.

4-4.2   General.

4-4.2.1   Remote supervising station fire alarm systems shall provide an automatic audible and visible indication of alarm and, where required, of supervisory and trouble conditions at

a location remote from the protected premises and a manual or automatic permanent record of these conditions.

Unlike central station and proprietary fire alarm systems that keep records automatically, remote station fire alarm systems may record signals in a manually-written log.

4-4.2.2   This section does not require the use of audible signal notification appliances other than those required at the remote supervising station. If it is desired to provide fire alarm evacuation signals in the protected premises, the alarm signals, circuits, and controls shall comply with the provisions of Chapter 3 and Chapter 6 in addition to the provisions of this section.

A remote station fire alarm system only needs to provide audible and visible notification at the remote supervising station. If either an authority having jurisdiction or management desires to provide audible and visible notification appliances throughout a protected premises, they should refer to the requirements in Chapters 3 and 6 of the Code.

4-4.2.3   The loading capacities of the remote supervising station equipment for any approved method of transmission shall be as designated in Section 4-5.

Remote station fire alarm systems have the full range of transmission technologies available, if they meet the requirements of Section 4-5 of this Code.

4-4.3*   Facilities.

A-4-4.3   As a minimum, the room or rooms containing the remote supervising station equipment should have a 1-hour fire rating, and the entire structure should be protected by an alarm system complying with Chapter 3.

The Code recommends, but does not require, that the remote station have a one-hour fire rated cut-off from the rest of the facility, and that it have the protection of a fire alarm system.

4-4.3.1   Where a remote supervising station connection is used to transmit an alarm signal, the signal shall be received at the public fire service communications center, at a fire station, or at the similar governmental agency that has a public responsibility for taking prescribed action to ensure response upon receipt of a fire alarm signal.

Exception: Where such an agency is unwilling to receive alarm signals or will permit the acceptance of another location by the authority having jurisdiction, such alternate location shall have personnel on duty at all times who are trained to receive the alarm signal and immediately retransmit it to the fire department.

When conditions meet the requirements of the Exception, the authority having jurisdiction may accept any suitable location to act as the remote supervising station. For example, an authority having jurisdiction could permit a listed central station to receive these signals. This would constitute remote station service, but not central station service.

4-4.3.2 Supervisory and trouble signals shall be handled at a constantly attended location having personnel on duty who are trained to recognize the type of signal received and to take prescribed action. This shall be permitted to be a location other than that at which alarm signals are received.

In some installations, a remote station fire alarm system transmits fire alarm signals to the public fire service communication center, and transmits supervisory signals and trouble signals to another location acceptable to the authority having jurisdiction.

4-4.3.3 Where locations other than the public fire service communications center are used for the receipt of signals, access to receiving equipment shall be restricted in accordance with requirements of the authority having jurisdiction.

This requirement helps to ensure the security and operational integrity of the remote station receiving equipment.

4-4.4 Equipment.

4-4.4.1 Signal-receiving equipment shall indicate receipt of each signal both audibly and visibly.

4-4.4.1.1 Audible signals shall meet the requirements of Chapter 6 for the private operating mode.

See subsection 6-3.3.

4-4.4.1.2 Means for silencing alarm, supervisory, and trouble signals shall be provided and shall be so arranged that subsequent signals shall re-sound.

Silencing one signal must not prevent a subsequent signal from causing the audible notification appliance to sound.

4-4.4.1.3 A trouble signal shall be received when the system or any portion of the system at the protected premises is placed in a bypass or test mode.

The requirements in subsection 4-4.4.1.3 prevent any type of so-called silent disconnect switch at the protected premises. Operation of any disconnect switch must produce a trouble signal at the remote supervising station.

4-4.4.1.4 An audible and visible indication shall be provided upon restoration from any off-normal condition.

It is not sufficient to indicate restoration to normal by merely extinguishing a lamp. The audible notification appliance at the remote supervising station must also sound.

4-4.4.1.5 Where suitable visible means are provided in the remote supervising station to identify readily the type of signal received, a common audible notification appliance shall be permitted to be used.

4-4.4.2 Power supplies shall comply with the requirements of Chapter 1.

Exception: In a remote supervising station fire alarm system where the alarm and supervisory signals are transmitted over a listed supervised one-way radio system, 24 hours of secondary (standby) power shall be permitted in lieu of 60 hours, as required in 1-5.2.6, at the radio alarm repeater station receivers (RARSR), provided that personnel are dispatched to arrive within 4 hours after detection of failure to initiate maintenance.

Many of the one-way radio alarm system sites that house an RARSR have a minimal footprint in equipment rooms. For example, at the top of a high-rise building, alarm service providers lease such space at a very high cost. This small footprint does not give adequate room for 60 hours of standby batteries. The Exception permits the use of 24 hours of standby power as long as a technician will arrive within 4 hours of the receipt of a trouble signal from the RARSR.

4-4.4.3 Transmission means shall comply with the requirements of Section 4-5.

Remote station fire alarm systems have the full range of transmission technologies available from section 4-5, as long as they meet any requirements of Section 4-4.

4-4.4.4 Retransmission of an alarm signal, where required, shall be by one of the following methods, which appear in descending order of preference as follows:

(a) A dedicated circuit that is independent of any switched telephone network. This circuit shall be permitted to be used for voice or data communications.

(b) A one-way (outgoing only) telephone at the remote supervising station that utilizes the public switched telephone network. This telephone shall be used primarily for voice transmission of alarms to a telephone at the public fire service communications center that cannot be used for outgoing calls.

(c) A private radio system using the fire department frequency, where permitted by the fire department.

(d) Other methods acceptable to the authority having jurisdiction.

The vast majority of remote supervising stations will use a retransmission method that complies with the requirements of subsection 4-4.4.4(b). The telephone utility switch blocks incoming telephone calls to this number, allowing operators at the remote supervising station to make outgoing calls on this line, but not to receive incoming calls.

4-4.5 **Personnel.** Sufficient personnel shall be available at all times to receive alarm signals at the remote supervising station and to take immediate appropriate action. Duties pertaining to other than operation of the remote supervising station receiving and retransmitting equipment shall be permitted subject to the approval of the authority having jurisdiction.

Although operators at the remote supervising station may have other duties requiring approval of the authority having jurisdiction, the code limits the extent to which these other duties may interfere with the proper handling of signals.

4-4.6 **Operations.**

4-4.6.1 Where the remote supervising station is at a location other than the public fire service communications center, alarm signals shall be immediately retransmitted to the public fire service communications center.

4-4.6.2 Upon receipt of an alarm, supervisory, or trouble signal by the remote supervising station other than the public fire service communications center, it shall be the responsibility of the operator on duty to notify the owner or the owner's designated representative immediately.

Promptly contacting designated persons when a failure occurs, helps to ensure that repairs begin as soon as possible.

4-4.6.3 A permanent record of the time, date, and location of all signals and restorations received; the action taken; and the results of all tests shall be maintained for at least 1 year and made available to the authority having jurisdiction. These records shall be permitted to be created by manual means.

Most often, a manually-maintained log book will contain these required records.

4-4.6.4 All operator controls at the remote supervising station shall be operated at the beginning of each shift or change in personnel and the status of all off-normal conditions noted and recorded.

The requirements of subsection 4-4.6.4 help to ensure the operational continuity of the supervising station. By exercising operating controls, operators will more quickly identify potential harmful failures.

4-4.7 **Testing and Maintenance.** Testing and maintenance for remote supervising stations shall be performed in accordance with Chapter 7.

4-5   Communications Methods for Supervising Station Fire Alarm Systems.

NOTE: The requirements of Chapters 1 and 7 apply to continuously attended supervising station fire alarm systems, unless they conflict with the requirements of this section.

4-5.1 **Scope.** This section describes the requirements for the methods of communications between the protected premises and the supervising station. These include the transmitter located at the protected premises, the transmission channel between the protected premises and the supervising station or subsidiary station, and, where used, any subsidiary station and its communications channel, and the signal receiving, processing, display, and recording equipment at the supervising station. *See Figure 4-5.1.*

The Code makes a full range of transmission technologies available to all of the supervising station services. This gives designers maximum flexibility in choosing the transmission technology most appropri-

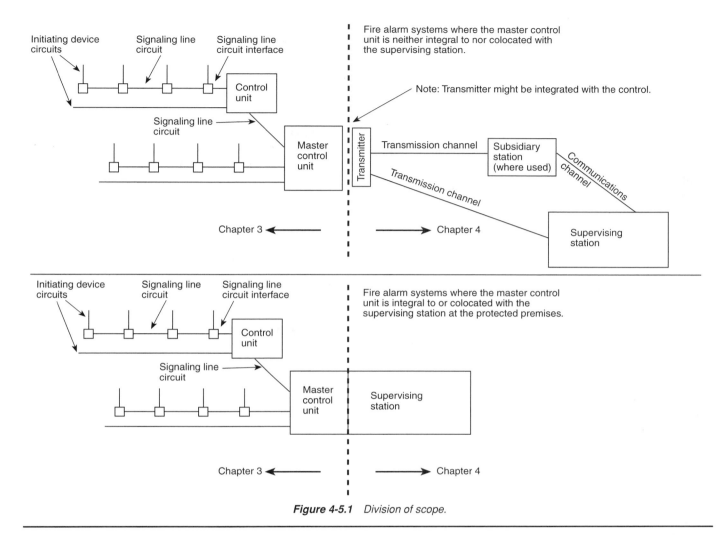

Figure 4-5.1   Division of scope.

ate for the particular application. Available technologies include active multiplex, including derived local channel; digital alarm communicator systems; digital alarm radio systems; McCulloh systems; two-way radio frequency multiplex systems; one-way radio alarm systems; directly-connected noncoded systems; and private microwave radio systems. However, specific requirements of each section of Chapter 4 may limit the use of a particular transmission technology. Of the available supervising station services, the Code most limits the public fire reporting system. In fact, the requirements so limit the transmission technologies for this service that Section 4-6 fully defines them. See Figure 4.5.

Exception: Transmission channels owned by and under the control of the protected premises owner that are not facilities leased from a supplier of communications service capabilities

such as video cable, telephone, and similar services that are also offered to other customers.

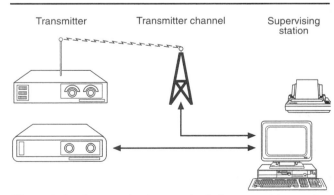

Figure 4. 5   Communications methods for supervising station fire alarm systems. Drawing courtesy of R.P. Schifiliti Associates, Inc., Reading, MA.

4-5.2 General.

4-5.2.1 Applicable Requirements.

Over the years, as the code encompassed new transmission technologies, it most often compared them to the performance capabilities of the McCulloh system—the first transmission technology used by supervising station systems. This intermixed certain operational requirements among the various technologies. Therefore, this subsection becomes a caveat to make certain that when applying a particular transmission technology, authorities having jurisdiction, as well as system designers, installers, and users, do not ignore critical operational requirements that appear in only one section.

4-5.2.1.1 Where the protected premises master control unit is neither integral to nor collocated with the supervising station, the communications methods of Section 4-5 shall be used to connect the protected premises to either a subsidiary station (where used) or a supervising station providing central station service in accordance with Section 4-2, proprietary service in accordance with Section 4-3, or remote station service in accordance with Section 4-4. These communications methods shall be permitted to include active multiplex circuits that are part of a supervising station, including systems utilizing derived channels; digital alarm communicator systems, including digital alarm radio systems; McCulloh systems; two-way radio frequency (RF) multiplex systems; one-way radio alarm systems; or directly-connected noncoded systems.

4-5.2.1.2* Nothing in this chapter shall be interpreted as prohibiting the use of listed equipment using alternate communications methods that provide a level of reliability and supervision consistent with the requirements of Chapter 1 and the intended level of protection.

A-4-5.2.1.2 It is not the intent of the requirements of Section 4-5 to limit the use of listed equipment using alternate communications methods, provided these methods demonstrate performance characteristics that are equal to or superior to those technologies described in Section 4-5. Such demonstration of equivalency is to be evidenced by the equipment using the alternate communications methods meeting all the requirements of Chapter 1, including those that deal with such factors as reliability, monitoring for integrity, and listing. It is further expected that suitable proposals stating the requirements for such technology will be submitted for inclusion in subsequent editions of this code.

4-5.2.2 Equipment.

4-5.2.2.1 Fire alarm system equipment and installations shall comply with Federal Communication Commission (FCC)

rules and regulations, as applicable, concerning electromagnetic radiation; use of radio frequencies; and connection to the public switched telephone network of telephone equipment, systems, and protection apparatus.

This subsection recognizes that the Federal Communications Commission (FCC) has jurisdiction over the installation requirements for certain communication equipment used to transmit signals from a protected premises to a supervising station.

4-5.2.2.2 Radio receiving equipment shall be installed in compliance with NFPA 70, *National Electrical Code®,* Article 810.

When a particular transmission technology uses television or radio equipment, that equipment must be installed in compliance with the appropriate article of NFPA 70, *National Electrical Code.*®

4-5.2.2.3 The external antennas of all radio transmitting and receiving equipment shall be protected in order to minimize the possibility of damage by static discharge or lightning.

4-5.2.3 Adverse Conditions.

4-5.2.3.1 For active and two-way RF multiplex systems, the occurrence of an adverse condition on the transmission channel between a protected premises and the supervising station that prevents the transmission of any status change signal shall be automatically indicated and recorded at the supervising station. This indication and record shall identify the affected portions of the system so that the supervising station operator can determine the location of the adverse condition by trunk or leg facility, or both.

Interrogation and response transmission back and forth along the communication path monitors the integrity of active multiplex transmission technology. The satisfactory exchange of data ensures that all trunks and legs remain operational. If the system does not successfully complete an interrogation and response sequence, this indicates the possible failure of a trunk or a leg. In such a case, this subsection provides for the detailed notification of the supervising station.

4-5.2.3.2 For a one-way radio alarm system, the system shall be supervised to ensure that at least two independent radio alarm repeater station receivers (RARSR) are receiving signals for each radio alarm transmitter (RAT) during each 24-hour period. The occurrence of a failure to receive a signal by either RARSR shall be automatically indicated and recorded at the supervising station. The indication shall identify which RARSR has failed to receive such supervisory signals. It is not necessary for properly received test signals to be indicated at the supervising station.

The satisfactory receipt of at least one transmission every 24 hours by at least two RARSRs monitors the integrity of one-way radio transmission technology. If receivers do not receive such a signal, then this section provides for the detailed notification of the supervising station.

4-5.2.3.3 For active and two-way RF multiplex systems that are part of a central station fire alarm system, restoration of normal service to the affected portions of the system shall be automatically recorded. When normal service is restored, the first status change of any initiating device circuit, or any initiating device directly connected to a signaling line circuit, or any combination thereof that occurred at any of the affected premises during the service interruption, also shall be recorded.

A central station must automatically record restoration of interrupted service and report the first status change on any initiating device circuit. This means that for each initiating device circuit, the equipment at the protected premises must retain and later report the first status change that occurs during the transmission interruption.

4-5.2.4 Dual Control.

4-5.2.4.1 Dual control, where required, shall provide for redundancy in the form of a standby circuit or a similar alternate means of transmitting signals over the primary trunk portion of a transmission channel. The same method of signal transmission shall be permitted to be used over separate routes, or alternate methods of signal transmission shall be permitted to be utilized. Public switched telephone network facilities shall be used only as an alternate method of transmitting signals.

Although dual control does not provide full redundancy for every trunk and leg, it does offer an option that an authority having jurisdiction or system designer can choose to help assure receipt of signals during interruptions to the primary trunk. Most often, technology called DataPhone Select-A-Station (DSAS) and offered by the public telephone utility provides this redundancy. When the primary trunk fails, this technology allows the supervising station to either automatically or manually dial into the public switched telephone network and

establish an alternate path for the signals that would normally transmit over the primary trunk. Telephone technicians sometimes refer to this arrangement as "dial up, make good."

4-5.2.4.2 Where utilizing facilities leased from a telephone company, that portion of the primary trunk facility between the supervising station and its serving wire center shall not be required to comply with the separate routing requirement of the primary trunk facility. Dual control, where used, shall require supervision as follows:

(a) Dedicated facilities that are available on a full-time basis, and whose use is limited to signaling purposes as defined in this code, shall be exercised at least once every hour.

(b) Public switched telephone network facilities shall be exercised at least once every 24 hours.

To ensure that the dual control system can use the alternate path, it must be exercised by operators. If the alternate path is dedicated solely for this purpose, it must be exercised hourly. If it is part of the public switched telephone network, which is typical, it must be exercised daily.

4-5.3 Communications Methods.

4-5.3.1 Active Multiplex Transmission Systems.

4-5.3.1.1 The multiplex transmission channel shall terminate in a transmitter at the protected premises and in a system unit at the supervising station. The derived channel shall terminate in a transmitter at the protected premises and in derived channel equipment at a subsidiary station location or a telephone company wire center. The derived channel equipment at the subsidiary station location or a telephone company wire center shall select or establish the communications with the supervising station.

The protected facility may own its own equipment or it may lease some or all of the equipment from an alarm service provider. In the case of "derived local channel" the protected facility or the alarm service provider may lease some of the equipment from the public telephone utility.

The protected facility may own some or all of the transmission facilities or may lease some or all of the transmission facilities from an alarm service provider. The protected facility or the alarm service provider may lease the transmission facilities from the public telephone utility. See Figure 4.6.

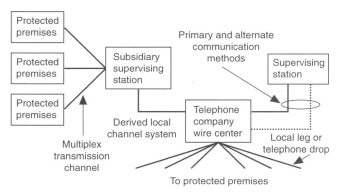

Figure 4.6   Communications methods—active multiplex transmission systems. Drawing courtesy of R.P. Schifiliti Associates, Inc., Reading, MA.

4-5.3.1.2* Operation of the transmission channel shall conform to the requirements of this code whether channels are private facilities, such as microwave, or leased facilities furnished by a communications utility company. Where private signal transmission facilities are utilized, the equipment necessary to transmit signals shall also comply with the requirements for duplicate equipment or replacement of critical components, as described in 4-5.4.2. The trunk transmission channels shall be dedicated facilities for the main channel. For Type 1 multiplex systems, the public switched telephone network facilities shall be permitted to be used for the alternate channel.

Exception: Derived channel scanners with no more than 32 legs shall be permitted to use the public switched telephone network for the main channel.

One manufacturer of derived local channel equipment offers an option of connecting the scanner in the telephone utility wire center to the supervising station by means of a dial-up modem that essentially performs as if it were a DACT. The exception to this subsection limits the loading of such a scanner to no more than 32 legs (protected premises).

A-4-5.3.1.2 Where derived channels are used, normal operating conditions of the telephone equipment are not to inhibit or impair the successful transmission of signals. These normal conditions include, but are not limited to:

(a) Intraoffice calls with a transponder on the originating end.

(b) Intraoffice calls with a transponder on the terminating end.

(c) Intraoffice calls with transponders on both ends.

(d) Receipt and origination of long distance calls.

(e) Calls to announcement circuits.

(f) Permanent signal receiver off-hook tone.

(g) Ringing with no answer, with transponder on either the originating or the receiving end.

(h) Calls to tone circuits (i.e., service tone, test tone, busy, or reorder).

(i) Simultaneous with voice source.

(j) Simultaneous with data source.

(k) Tip and ring reversal.

(l) Cable identification equipment.

4-5.3.1.2.1 Derived channel signals shall be permitted to be transmitted over the leg facility, which shall be permitted to be shared by the telephone equipment under all normal on-hook and off-hook operating conditions.

This subsection embodies the whole concept of derived local channel; it shares the leg facility used by the "plain, old telephone service" (POTS) for a particular protected premises. The derived local channel system permits the transmission of alarm, supervisory, and trouble signals, even when normal telephone communication uses the leg. An interrogation and response sequence, similar to that used by other active multiplex systems, monitors the integrity of the leg.

4-5.3.1.2.2 Where derived channel equipment uses the public switched telephone network to communicate with a supervising station, such equipment shall meet the requirements of 4-5.3.2.

4-5.3.1.2.3 The maximum end-to-end operating time parameters allowed for an active multiplex system are as follows:

(a) The maximum allowable time lapse from the initiation of a single fire alarm signal until it is recorded at the supervising station shall not exceed 90 seconds. When any number of subsequent fire alarm signals occur at any rate, they shall be recorded at a rate no slower than one every 10 additional seconds.

This subsection was designed to ensure that the system will complete an interrogation and response sequence at least every 90 seconds. As an alternative, the system may provide some other means to ensure alarm receipt in that time frame. For example, a designer could devise equipment that transmits alarm signals from the protected premises at a time other than during the normal interrogation and response sequence. However, most systems complete the interrogation and response sequence within 90-seconds.

(b)* The maximum allowable time lapse from the occurrence of an adverse condition in any transmission channel until recording of the adverse condition is started shall not exceed 90 seconds for Type 1 and Type 2 systems, and 200 seconds for Type 3 systems. (*See 4-5.3.1.3.*)

The reporting of an adverse condition on a Type 3 system within 200 seconds allows for a system, described in subsection 4-5.3.1.2.3, in which the equipment at the protected premises transmits alarm signals at a time other than during the normal interrogation and response sequence. This requirement ensures that the interrogation and response sequence will occur at least every 200 seconds.

A-4-5.3.1.2.3(b) Derived channel systems comprise Type 1 and Type 2 systems only.

(c) In addition to the maximum operating time allowed for fire alarm signals, the requirements of one of the following paragraphs shall be met:

1. A system unit having more than 500 initiating device circuits shall be able to record not less than 50 simultaneous status changes in 90 seconds.

2. A system unit having fewer than 500 initiating device circuits shall be able to record not less than 10 percent of the total number of simultaneous status changes within 90 seconds.

These requirements ensure that the portion of the multiplex system that processes and records status changes can do so with sufficient speed to handle a reasonable volume of signal traffic, based on the systems signal capacity.

Exception: Proprietary supervising station systems.

NOTE: Operating time requirements for proprietary supervising station systems are specified in 4-3.4.7 through 4-3.4.9.

4-5.3.1.3 **System Classification.** Active multiplex systems are divided into three categories based upon their ability to perform under adverse conditions of their transmission channels. The system classifications are as follows:

(a) A Type 1 system shall have dual control as described in 4-5.2.4. An adverse condition on a trunk or leg facility shall not prevent the transmission of signals from any other trunk or leg facility, except those signals normally dependent on the portion of the transmission channel in which the adverse condition has occurred. An adverse condition limited to a leg facility shall not interrupt normal service on any trunk or other

leg facility. The requirements of 4-5.2.1, 4-5.2.2, and 4-5.2.3 shall be met by Type 1 systems.

To meet the requirements for Type 1, an active multiplex system must isolate each system leg and trunk from other legs and trunks on the system. To do this, the equipment uses a device called a "closed window bridge" wherever two or more trunks or two or more legs converge. Coupling circuitry within the closed window bridge allows signals to pass, but keeps a fault on one leg or trunk from interfering with signals from another leg or trunk. The public telephone utility may supply the bridge and locate it in a public telephone wire center or at a protected premises. The protected property or the alarm service provider may own or lease the bridge and determine its location. For example, equipment could multiplex the fire alarm systems for the individual stores in a shopping mall and for the mall common areas through a bridge located at an equipment room in the mall.

Type 1 systems also employ dual control, as described in subsection 4-5.2.4. This gives the system an alternate transmission path should the primary trunk fail.

(b) A Type 2 system shall have the same requirements as a Type 1 system.

A Type 2 system also uses a closed window bridge to provide isolation between trunks and legs. Type 2 systems need not provide dual control. Thus, a Type 2 system has no alternate transmission path should the primary trunk fail.

Exception: Dual control of the primary trunk facility shall not be required.

(c) A Type 3 system shall automatically indicate and record at the supervising station the occurrence of an adverse condition on the transmission channel between a protected premises and the supervising station. The requirements of 4-5.2 shall be met.

Exception: The requirements of 4-5.2.4 shall not apply.

Type 3 systems have no requirement for isolation between legs and trunks. They commonly employ an "open window bridge." While this device allows the coupling of signals wherever two or more legs or two or more trunks converge, an adverse condition on one leg or trunk may affect the operation of other legs or trunks.

Table 4-5.3.1.4   Loading Capacities for Active Multiplex Systems

	System Type		
	Type 1	Type 2	Type 3
A. Trunks			
Maximum number of fire alarm service initiating device circuits per primary trunk facility	5120	1280	256
Maximum number of leg facilities for fire alarm service per primary trunk facility	512	128	64
Maximum number of leg facilities for all types of fire alarm service per secondary trunk facility[1]	128	128	128
Maximum number of all types of initiating device circuits per primary trunk facility in any combination[1]	10,240	2560	512
Maximum number of leg facilities for all types of fire alarm service per primary trunk facility in any combination[1]	1024	256	128
B. System Units at the Supervising Station			
Maximum number of all types of initiating device circuits per system unit[1]	10,240[2]	10,240[2]	10,240[2]
Maximum number of fire protecting buildings and premises per system unit	512[2]	512[2]	512[2]
Maximum number of fire fire alarm service initiating device circuits per system unit	5120[2]	5120[2]	5120[2]
C. Systems Emitting from Subsidiary Station	Same as B	Same as B	Same as B

[1]Includes every initiating device circuit (e.g., waterflow, fire alarm, supervisory, guard, burglary, hold-up).
[2]Paragraph 4-5.3.1.5 applies.

4-5.3.1.4 **System Loading Capacities.** The capacities of active multiplex systems are based on the overall reliability of the signal receiving, processing, display, and recording equipment at the supervising and subsidiary stations, and the capability to transmit signals during adverse conditions of the signal transmission facilities. Table 4-5.3.1.4 establishes the allowable capacities.

The loading of trunks depends on the capability of the type of system. Since a Type 1 system has a redundant primary trunk (dual control) and isolation between legs and trunks, it has the greatest permitted trunk capacity. A Type 2 system does not have dual control, but does have isolation between legs and trunks. It has less trunk capacity than that a Type 1 system, but more than a Type 3 system. A Type 3 system has neither dual control nor isolation between legs and trunks. It has the least trunk capacity of the three types.

4-5.3.1.5 **Exceptions to Loading Capacities Listed in Table 4-5.3.1.4.** Where the signal receiving, processing, display, and recording equipment are duplicated at the supervising station and a switchover can be accomplished in not more than 30 seconds with no loss of signals during this period, the capacity of a system unit shall be unlimited.

This subsection modifies Part B of Table 4-5.3.1.4. However, to meet this requirement an active multiplex system would have to employ complete redundancy of all critical components, and complete a switch-over in

30 seconds with no loss of signals. Those systems that meet this requirement generally process all incoming signals in tandem. That is, the standby unit functions fully at all times and simply continues to function normally when the main unit fails. Operators would actually "change-over" only those incidental peripheral devices that have no required redundancy.

4-5.3.2 **Digital Alarm Communicator Systems.**

4-5.3.2.1 **Digital Alarm Communicator Transmitter (DACT).**

4-5.3.2.1.1 A DACT shall be connected to the public switched telephone network upstream of any private telephone system at the protected premises. In addition, the connections to the public switched telephone network shall be under the control of the subscriber for whom service is being provided by the supervising station fire alarm system, and special attention shall be required to ensure that this connection is made only to a loop start telephone circuit and not to a ground start telephone circuit.

Exception: Where public cellular telephone service is utilized as a secondary means of transmission, the requirements of this paragraph shall not apply to the cellular telephone service.

The DACT connects to the public switched telephone network so that it may seize the line to which it is connected. This seizure disconnects any private telephone equipment beyond the DACT's point of connection. This arrangement gives the DACT control over the line at all times.

On a loop-start telephone line, the public telephone utility continuously supplies voltage from the first telephone utility wire center. The vast majority of residential telephone connections use loop-start lines. In contrast, almost all business telephone connections, particularly those employing private branch exchange (PBX) connections, use ground-start lines. In order to obtain dial tone and operating power on a ground-start line, the user equipment momentarily connects one side of the line to earth ground. Since the public telephone utility does not supply voltage to an idle ground-start line, the DACT cannot use the presence of voltage to monitor the integrity of the ground-start line as it can with a loop-start line.

Functionally, a DACT can signal over a ground-start line and frequently does so when used as part of a burglar alarm system. However, the DACT can only monitor a loop-start line for integrity.

The exception is necessary since public cellular telephone systems do not use telephone lines. Thus, when the public cellular telephone system is used as a secondary means of signal transmission, the requirements of this subsection do not apply to the cellular portion of the system.

4-5.3.2.1.2 All information exchanged between the DACT at the protected premises and the digital alarm communicator receiver (DACR) at the supervising or subsidiary station shall be by digital code or equivalent. Signal repetition, digital parity check, or some equivalent means of signal verification shall be used.

The functional requirements of this subsection rule out the use of an analog or digital voice tape dialer to transmit fire alarm signals. Such a device dials a predetermined telephone number and then plays a voice message, such as, "There is a fire at 402 Spruce Street." Over the years, officials have reported many cases where a voice tape dialer malfunctions and endlessly repeats its message, tying up a vital emergency telephone line in a public fire service communication center. The code strictly forbids the use of analog or digital voice tape dialers.

4-5.3.2.1.3* A DACT shall be configured so that when it is required to transmit a signal to the supervising station, it shall seize the telephone line (going off-hook) at the protected premises, disconnect an outgoing or incoming telephone call,

and prevent its use for outgoing telephone calls until signal transmission has been completed. A DACT shall not be connected to a party line telephone facility.

A-4-5.3.2.1.3 In order to give the DACT the ability to disconnect an incoming call to the protected premises, telephone service should be of the type that provides for timed-release disconnect. In some telephone systems (step-by-step offices), timed-release disconnect is not provided.

In order to ensure reliability for transmission of fire alarm, supervisory, and trouble signals, this requirement and recommendation give the DACT exclusive control over the telephone service to which it is connected.

4-5.3.2.1.4 A DACT shall have the means to satisfactorily obtain an available dial tone, dial the number(s) of the DACR, obtain verification that the DACR is ready to receive signals, transmit the signal, and receive acknowledgment that the DACR has accepted that signal. In no event shall the time from going off-hook to on-hook exceed 90 seconds per attempt.

This subsection describes the normal sequence of operation for a DACT. Upon initiation of an alarm, a supervisory, or a trouble signal, the DACT seizes the line, obtains dial tone, dials the number of the DACR, receives a "handshake" signal from the DACR, transmits its data, receives an acknowledgment signal—sometimes called the "kiss-off signal"—from the DACR, and hangs up. Each attempt of this calling and verification sequence must take no longer than 90 seconds to complete.

4-5.3.2.1.5* A DACT shall have suitable means to reset and retry where the first attempt to complete a signal transmission sequence is unsuccessful. A failure to complete connection shall not prevent subsequent attempts to transmit an alarm where such alarm is generated from any other initiating device circuit or signaling line circuit, or both. Additional attempts shall be made, until the signal transmission sequence has been completed, up to a minimum of five and a maximum of 10 attempts.

Where the maximum number of attempts to complete the sequence is reached, an indication of the failure shall be made at the premises.

A-4-5.3.2.1.5 A DACT can be programmed to originate calls to the DACR telephone lines (numbers) in any alternating sequence. The sequence can consist of single or multiple calls to one DACR telephone line (number), followed by single or multiple calls to a second DACR telephone line (number), or any combination thereof that is consistent with the minimum/maximum attempt requirements in 4-5.3.2.1.5.

The DACT, as described in subsection 4-5.3.2.1.4, must make at least five attempts to complete the sequence. However, it must not make more than ten attempts in order to prevent a malfunctioning DACT from tying up the DACR.

Under the most adverse circumstances, where the DACT finally completes a transmission on the last, or tenth, attempt, at a maximum of 90 seconds per attempt (see subsection 4-5.3.2.1.4) nearly 900 seconds or fifteen minutes could have elapsed. Based on this potential delay, some authorities having jurisdiction may not accept digital alarm communicator systems (DACS) for proprietary fire alarm systems (see subsections 4-3.4.7 and 4-3.4.9).

4-5.3.2.1.6 DACT Transmission Channels.

4-5.3.2.1.6.1 A DACT shall employ one of the following combinations of transmission channels:

(a) Two telephone lines (numbers) See Figure 4-7;

(b) One telephone line (number) and one cellular telephone connection;

(c) One telephone line (number) and a one-way radio system;

(d) One telephone line (number) equipped with a derived local channel;

Some public telephone companies offer so-called "cut line" detection to supervising station alarm system providers. This service uses derived local channel equipment to detect adverse conditions on a telephone line. When a supervising station uses cut line supervision, it may operate its DACS with a single telephone line connected to each DACT.

(e) One telephone line (number) and a one-way private radio alarm system;

(f) One telephone line (number) and a private microwave radio system;

(g) One telephone line (number) and a two-way RF multiplex system.

4-5.3.2.1.6.2 The following requirements shall apply to all combinations in 4-5.3.2.1.6.1:

(a) Both channels shall be supervised in a manner appropriate for the means of transmission employed.

(b) Both channels shall be tested at intervals not exceeding 24 hours.

Note the additional requirements in subsection 4-5.3.2.1.10.

Exception No. 1:   For public cellular telephone service, a verification (test) signal shall be transmitted at least monthly.

Exception No. 2:   Where two telephone lines (numbers) are used, it shall be permitted, until June 1, 1998, to test the primary telephone line (number) at 24-hour intervals without testing the secondary line (number). After June 1, 1998, where two telephone lines (numbers) are used, it shall be permitted to test each telephone line (number) at alternating 24-hour intervals.

(c) The failure of either channel shall send a trouble signal on the other channel within 4 minutes.

As important as it is to monitor the integrity of the transmission means, it is equally important to avoid nuisance trouble signals. The permissible 4-minute delay in transmitting a trouble signal allows for momentary, or even somewhat longer, interruptions in the transmission path, such as might occur during a storm.

(d) When one transmission channel has failed, all status change signals shall be sent over the other channel.

Exception: Where used in combination with a DACT, a derived local channel shall not be required to send status change signals other than those indicating that adverse conditions exist on the telephone line (number).

(e) The primary channel shall be capable of delivering an indication to the DACT that the message has been received by the supervising station.

A one-way radio alarm system could not meet this requirement. Thus, it could not serve as the primary transmission means.

(f) The first attempt to send a status change signal shall utilize the primary channel.

Exception: Where the primary channel is known to have failed.

(g) Simultaneous transmission over both channels shall be permitted.

(h) Failure of telephone lines (numbers) or cellular service shall be annunciated locally.

4-5.3.2.1.7 DACT Transmission Means.

4-5.3.2.1.7.1 A DACT shall be connected to two separate means of transmission at the protected premises. The DACT shall be capable of selecting the operable means of transmission in the event of failure of the other. The primary means of transmission shall be a telephone line (number) connected to the public switched network.

If the DACT detects that one of the two transmission means has failed (loss of voltage on a wire line, loss of one-way radio alarm service, or loss of cellular telephone service), it must switch to the other operable means. See Figure 4.7.

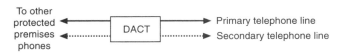

Figure 4.7 *Connections to a DACT. Drawing courtesy of R.P. Schifiliti Associates, Inc., Reading, MA.*

Formal Interpretation 87-1
Reference: 4-5.3.2.1.6

Question: NFPA 72, 4-5.3.2.1.6, states: "A DACT shall be connected to two separate means of transmission at the protected premises." To comply with this section, would it be required that the utility company install two (2) separate incoming phone lines, each one entering the building at a separate location?

Answer: No.

Issue Edition: NFPA 71-1987
Reference: 5-2.6
Issue Date: April 1988 ■

4-5.3.2.1.7.2 The first transmission attempt shall utilize the primary means of transmission.

4-5.3.2.1.8 Each DACT shall be programmed to call a second DACR line (number) where the signal transmission sequence to the first called line (number) is unsuccessful.

To help avoid a possible disarrangement of the transmission path on the receiving end of the digital alarm communicator system, this requirement specifies that the DACT must call a second number if calling the first

number does not result in completion of the transmission.

4-5.3.2.1.9* Where long distance telephone service (including WATS) is used, the second telephone number shall be provided by a different long distance service provider, where available.

A-4-5.3.2.1.9 The requirement for use of two different long distance providers is to prevent a lost signal due to a fault in one long distance provider's network. This requirement is not meant to apply in local situations where signal traffic is strictly within the area covered by one local telephone company.

Since it is never certain whether a subscriber has changed long distance providers, it is recommended that, where direct dialer service is used, a telephone call should be forced onto a specific long distance provider's network by using the dialing prefix carrier identification code (CIC) specific to each long distance provider.

4-5.3.2.1.10 Each DACT shall automatically initiate and complete a test signal transmission sequence to its associated DACR at least once every 24 hours. A successful signal transmission sequence of any other type within the same 24-hour period shall be considered sufficient to fulfill the requirement to verify the integrity of the reporting system, provided signal processing is automated so that 24-hour delinquencies are individually acknowledged by supervising station personnel.

At least once every 24 hours each DACT must initiate a signal to verify the end-to-end integrity of the digital alarm communicator system. If the receiving or processing equipment at the supervising station has sufficient intelligence to automatically keep track of signal traffic, any incoming signal from a particular DACT may serve to satisfy this requirement, as long as the receiver receives one signal during every 24-hour period.

Note the additional requirements in subsection 4-5.3.2.1.6.2(b) and Exceptions 1 and 2. Particularly note that after June 1, 1998, when the DACT connects to two telephone lines, the daily test must alternate between telephone lines.

4-5.3.2.1.11* Where a DACT is programmed to call a telephone line (number) that is call forwarded to the line (number) of the DACR, a means shall be implemented to verify the integrity of the call forwarding feature every 4 hours.

A-4-5.3.2.1.11 Since call forwarding requires equipment at a telephone company central office that might occasionally interrupt the call forwarding feature, a signal should be initiated whereby the integrity of the forwarded telephone line (number) that is being called by DACTs is verified every 4 hours. This can be accomplished by a single DACT, either in service or used solely for verification, that automatically initiates and completes a transmission sequence to its associated DACR every 4 hours. A successful signal transmission sequence of any other type within the same 4-hour period should be considered sufficient to fulfill this requirement.

Call forwarding should not be confused with WATS or 800 service. The latter, differentiated from the former by dialing the 800 prefix, is a dedicated service used mainly for its toll-free feature; all calls are preprogrammed to terminate at a fixed telephone line (number) or to a dedicated line.

Occasionally, a supervising station will maintain one or more telephone numbers in a local calling area that the telephone equipment will call forward to another number connected to the DACR. When the supervising station employs this practice, it must verify the integrity of the call forward instruction every 4 hours. Most often it does this by having the technicians program the automatic test signal of six of the DACTs in the service area that use the call forwarded number so they each initiate their test signals at 4 hour intervals during a 24-hour period.

4-5.3.2.2 Digital Alarm Communicator Receiver (DACR).

4-5.3.2.2.1 Equipment.

4-5.3.2.2.1.1 Spare DACRs shall be provided in the supervising or subsidiary station and shall be able to be switched into the place of a failed unit within 30 seconds after detection of failure.

NOTE: One spare DACR may be permitted to serve as a backup for up to five DACRs in use.

The presence of a spare DACR does not by itself satisfy the requirements of this subsection. In order to meet the switching requirement in 4-5.3.2.2.1.1, someone must provide adequate written instructions and train the personnel on duty in the supervising station to accomplish the switch-over. Preferably, the connections to the unit should end in a fashion that permits rapid, error-free reconnection to the spare unit. For example, multiple telephone line connections could terminate in a plug and jack assembly that would permit rapid disconnection and rapid reconnection to the second unit.

4-5.3.2.2.1.2 The number of incoming telephone lines to a DACR shall be limited to eight lines.

Exception: Where the signal receiving, processing, display, and recording equipment at the supervising or subsidiary station is duplicated and a switchover can be accomplished in less than 30 seconds with no loss of signal during this period, the number of incoming lines to the unit shall be permitted to be unlimited.

The code allows a maximum of eight incoming lines to be connected to a single DACR. This helps to prevent overloading a DACR's ability to receive and process signals promptly. However, some fully-automated supervising station facilities may exist to take advantage of the exception.

4-5.3.2.2.2 Transmission Channel.

4-5.3.2.2.2.1* The DACR equipment at the supervising or subsidiary station shall be connected to a minimum of two separate incoming telephone lines (numbers). If the lines (numbers) are in a single hunt group, they shall be individually accessible; otherwise, separate hunt groups shall be required. These lines (numbers) shall be used for no other purpose than receiving signals from a DACT. These lines (numbers) shall be unlisted.

A-4-5.3.2.2.2.1 The timed-release disconnect considerations as outlined in A-4-5.3.2.1.3 apply to the telephone lines (numbers) connected to a DACR at the supervising station.

It might be necessary to consult with appropriate telephone service personnel to ensure that numbers assigned to the DACR can be individually accessed even where they are connected in rotary (a hunt group).

Hunt groups provided by some older public telephone utility central office equipment may have the potential for locking on to a defective line. This disables all lines in the hunt group. The requirements in subsection 4-5.3.2.2.2.1 help to ensure that the design of the digital alarm communicator system receiving network has a high degree of reliability.

4-5.3.2.2.2.2 The failure of any telephone line (number) connected to a DACR due to loss of line voltage shall be annunciated visually and audibly in the supervising station.

The DACR must connect to loop-start telephone lines with voltage normally present. The DACR will monitor this voltage to assure an operable line up to the first public telephone utility wire center.

4-5.3.2.2.2.3* The loading capacity for a hunt group shall be in accordance with Table 4-5.3.2.2.2.3 or be capable

Table 4-5.3.2.2.2.3 Loading Capacities for Hunt Groups

System Loading at the Supervising Station	Number of Lines in Hunt Group				
	1	2	3	4	5 to 8
With DACR lines processed in parallel					
Number of initiating circuits	N/A	5000	10,000	20,000	20,000
Number of DACTs[1]	N/A	500	1500	3000	3000
With DACR lines processed serially (put on hold, then answered one at a time)					
Number of initiating circuits	N/A	3000	5000	6000	6000
Number of DACTs[1]	N/A	300	800	1000	1000

N/A: Not Acceptable.

[1]Table 4-5.3.2.2.2.3 is based on an average distribution of calls and an average connected time of 30 seconds for a message. The loading figures in the table presume that the lines are in a hunt group (i.e., DACT can access any available line). Note that a single-line DACR is NOT ACCEPTABLE (N/A) for any of the configurations shown.

of demonstrating a 90 percent probability of immediately answering an incoming call.

(a) Each supervised burglar alarm (open/close) or each suppressed guard tour transmitter shall reduce the allowable DACTs as follows:

1. Up to a four-line hunt group, by 10;

2. Up to a five-line hunt group, by seven;

3. Up to a six-line hunt group, by six;

4. Up to a seven-line hunt group, by five;

5. Up to an eight-line hunt group, by four.

(b) Each guard tour transmitter shall reduce the allowable DACTs as follows:

1. Up to a four-line hunt group, by 30;

2. Up to a five-line hunt group, by 21;

3. Up to a six-line hunt group, by 18;

4. Up to a seven-line hunt group, by 15;

5. Up to an eight-line hunt group, by 12.

A-4-5.3.2.2.2.3 In determining system loading, Table 4-5.3.2.2.2.3 can be used, or it should be demonstrated that there is a 90 percent probability of incoming line availability. Table 4-5.3.2.2.2.3 is based on an average distribution of calls and an average connected time of 30 seconds per message. Therefore, where it is proposed to use Table 4-5.3.2.2.2.3 to determine system loading, if any factors are disclosed that could extend DACR connect time so as to increase the average connect time, the alternate method of determining system loading should be used. Higher (or possibly lower) loadings might be appropriate in some applications.

(a) Some factors that could increase (or decrease) the capacity of a hunt group follow:

1. Shorter (or longer) average message transmission time can influence hunt group capacity.

2. The use of audio monitoring (listen-in) slow scan video or other similar equipment can significantly increase the connected time for a signal and reduce effective hunt group capacity.

3. The clustering of active burglar alarm signals can generate high peak loads at certain hours.

4. Inappropriate scheduling of 24-hour test signals can generate excessive peak loads.

(b) Demonstration of a 90 percent probability of incoming line availability can be accomplished by the following in-service monitoring of line activity:

1. Incoming lines are assigned to telephone hunt groups. When a DACT calls the main number of a hunt group, it can connect to any currently available line in that hunt group.

2. The receiver continuously monitors the "available" status of each line. A line is available when it is waiting for an incoming call. A line is unavailable for any of the following reasons:

a. Currently processing a call

b. Line in trouble

c. Audio monitoring (listen-in) in progress

d. Any other condition that makes the line input unable to accept calls.

3. The receiver monitors the "available" status of the hunt group. A hunt group is available when any line in it is available.

4. A message is emitted by the receiver when a hunt group is unavailable for more than 1 minute out of 10 minutes. This message references the hunt group and the degree of overload.

The loading of a DACR helps to determine the overall reliability of a digital alarm communicator system. System designers may use one of two options to determine loading capacity. They may use Table 4-5.3.2.2.2.3. Or, they may ensure 90 percent availability. Larger-capacity supervising stations that employ a computer-based automation system to oversee the handling of signals normally use the second option. Such a system can monitor traffic to maintain the necessary reliability as loading increases due to the addition of new customers.

4-5.3.2.2.2.4* A signal shall be received on each individual incoming DACR line at least once every 24 hours.

A-4-5.3.2.2.2.4 The verification of the 24-hour DACR line test should be done early enough in the day to allow repairs to be made by the telephone company.

Depending on the number of lines involved and the design and complexity of the particular hunt group arrangements, the supervising station automation system may perform these tests automatically. Or, the supervising station operators may initiate the test signals manually while sequentially creating a busy signal on each line in a hunt group.

4-5.3.2.2.2.5 The failure to receive a test signal from the protected premises shall be treated as a trouble signal. *(See 4-2.6.1.4.)*

The daily test signal serves to verify the end-to-end functioning of the system. It monitors the integrity of the system and guards against the loss of both telephone lines connected to the DACT. It may also detect the malfunctioning of an entire hunt group at the DACR. In larger supervising stations, the computer-based automation system often oversees the test signals. Smaller supervising stations may use a manual logging system to keep track of the test signals.

4-5.3.2.3 Digital Alarm Radio System (DARS).

See subsections 4-5.3.2.1.6.1(c) and (e) and subsection 4-5.3.2.1.6.2(e).

4-5.3.2.3.1 In the event that any DACT signal transmission is unsuccessful, the information shall be transmitted by means of the digital alarm radio transmitter (DART). The DACT shall continue its normal transmission sequence as required by 4-5.3.2.1.5.

When a digital alarm radio system (DARS) is provided as the secondary transmission path for a DACT, the DACT must still continue to attempt to complete the call to the DACR. Subsection 4-5.3.2.1.6.2(g) permits simultaneous transmission by both the DACT and DART.

4-5.3.2.3.2 The DARS shall be capable of demonstrating a minimum of 90 percent probability of successfully completing each transmission sequence.

To fulfill this requirement, engineers must complete radio propagation studies that would satisfy the specified 90 percent reliability factor. The Code requires similar studies for a one-way radio alarm system. See subsections 4-5.3.5.2 and A-4-5.3.5.2.

4-5.3.2.3.3 Transmission sequences shall be repeated a minimum of five times. The digital alarm radio transmitter (DART) transmission shall be permitted to be terminated in less than five sequences where the DACT successfully communicates to the DACR.

In order to ensure overall system reliability, the system must make a sufficient number of attempts to complete the signal transmission.

4-5.3.2.3.4 Each DART shall automatically initiate and complete a test signal transmission sequence to its associated digital alarm radio receiver (DARR) at least once every 24 hours. A successful DART signal transmission sequence of any other type within the same 24-hour period shall be considered sufficient to fulfill the requirement to test the integrity of the reporting system, provided signal processing is automated so that 24-hour delinquencies are individually acknowledged by supervising station personnel.

The requirement of this subsection dovetails with subsection 4-5.3.2.1.6.2 (b). When a DACT connects to a single telephone line as the primary transmission path, and through a digital alarm radio system (DARS) as the secondary transmission path, the system must conduct a test at least once every 24 hours for each transmission path.

4-5.3.2.4 Digital Alarm Radio Transmitter (DART). A DART shall transmit a digital code or equivalent by use of radio transmission to its associated digital alarm radio receiver (DARR). Signal repetition, digital parity check, or an equivalent means of signal verification shall be used. The DART shall comply with applicable FCC rules consistent with its operating frequency.

This requirement ensures that a DART uses digital information or other coded signal radio transmission to communicate the status of the fire alarm system at the protected premises. This requirement precludes the use of voice information transmission.

4-5.3.2.5 Digital Alarm Radio Receiver (DARR).

4-5.3.2.5.1 Equipment.

4-5.3.2.5.1.1 A spare DARR shall be provided in the supervising station and shall be able to be switched into the place of a failed unit within 30 seconds after detection of failure.

4-5.3.2.5.1.2 Facilities shall be provided at the supervising station for supervisory and control functions of subsidiary and repeater station radio receiving equipment. This shall be accomplished via a supervised circuit where the radio equipment is remotely located from the supervising or subsidiary station. The following conditions shall be supervised at the supervising station:

(a) Failure of ac power supplying the radio equipment;

(b) Receiver malfunction;

(c) Antenna and interconnecting cable malfunction;

(d) Indication of automatic switchover of the DARR;

(e) Data transmission line between the DARR and the supervising or subsidiary station.

Monitoring the integrity of these functions helps to ensure overall system reliability. A large supervising station equipped with a computer-based automation system will use that system to perform most or all of these functions.

4-5.3.3 McCulloh Systems.

McCulloh systems are the oldest form of transmission between a protected premises and a supervising station. Coded transmitters at a protected premises connect in series with transmitters at other protected premises and with receiving equipment at the supervising station. The interconnected circuits must maintain continuous metallic circuit continuity. This allows the dc current to flow from the power supply at the supervising station, out over the series circuit, and through the coded contacts of the transmitters at the protected premises. Where the public telephone utility does not wish to maintain circuits that will offer continuous metallic circuit continuity, an alternative exists (see subsection 4-5.3.2.6). This alternative system converts the McCulloh-type

system into a multiplex system between public telephone utility company wire centers. Then, it reconverts the multiplex system to a McCulloh system in order to deliver the signals to the supervising station.

Initiation of an alarm, supervisory, or trouble signal at the protected premises actuates the associated transmitter. As the code wheel of the actuated transmitter turns, it alternately breaks the circuit and connects the circuit to earth ground.

Under normal circumstances, the breaking of the circuit operates receiving equipment that records the coded pulses at the supervising station. Operators, either manually or by a computer-based automation system at the supervising station, convert these pulses to information that gives the location of the protected premises.

If a single open fault or single ground fault impairs the circuit between the protected premises and the supervising station, then the signal produced by the turning of the code wheel transmits through earth ground. Operators must respond to a trouble signal generated by the fault. They then manually, or by a computer-based automation system, recondition the circuit to receive the signals through earth ground.

4-5.3.3.1 Transmitters.

4-5.3.3.1.1 A coded alarm signal from a transmitter shall consist of not less than three complete rounds of the number or code transmitted.

4-5.3.3.1.2* A coded fire alarm box shall produce not less than three signal impulses for each revolution of the coded signal wheel or equivalent device.

A-4-5.3.3.1.2 The following recommended coded signal designations for a building having four floors and basements are provided in Table A-4-5.3.3.1.2:

Table A-4-5.3.3.1.2 Recommended Coded Signal Designations

Location	Coded Signal
4th floor	2-4
3rd floor	2-3
2nd floor	2-2
1st floor	2-1
Basement	3-1
Subbasement	3-2

4-5.3.3.1.3 Circuit-adjusting means for emergency operating shall be permitted to either be automatic or be provided through manual operation upon receipt of a trouble signal.

Original McCulloh supervising stations required operator action to condition a circuit impaired by either an open fault or a ground fault to receive subsequent signals. As these stations have become equipped with computer-based automation systems, the interface equipment now often conditions the circuits automatically.

4-5.3.3.1.4 Equipment shall be provided at the supervising or subsidiary station on all circuits extending from the supervising or subsidiary station utilized for McCulloh systems for making the following tests:

(a) Current on each circuit under normal conditions;

(b) Current on each side of the circuit with the receiving equipment conditioned for an open circuit.

NOTE: The current readings in accordance with 4-5.3.3.1.4(a) should be compared with the normal readings to determine if a change in the circuit condition has occurred. A zero current reading in accordance with 4-5.3.3.1.4(b) indicates that the circuit is clear of a foreign ground.

By taking a current reading on each McCulloh circuit, operators may, sometimes, detect a "strap" across the circuit (short circuit). They have less success when the strap exists close to or at the protected premises. If sufficient resistance exists between the location of the strap and the protected premises, the loss of that resistance when someone applies the strap will result in an increase in current. The operator taking the reading should detect the increase in current.

In most cases a strap alone will not disable the transmission of signals. The McCulloh transmitter will also connect the circuit to ground with each pulse of the code wheel. However, the person placing the strap across the circuit may also disconnect both sides of the circuit beyond the strap. This action will prevent the McCulloh transmitters, located beyond the strap and the open circuit fault, from transmitting signals to the supervising station.

A foreign ground—an unintentional connection of the current to earth ground—can adversely affect the circuit's ability to transmit a signal. These current readings also help to detect the presence of foreign grounds. Once detected, a technician can locate and clear the foreign ground.

Typical foreign grounds include a tree branch rubbing against an aerial portion of the circuit, or water filling a conduit containing a portion of the circuit where the insulation has degraded.

4-5.3.3.2 **Transmission Channels.**

4-5.3.3.2.1 Circuits between the protected premises and the supervising or subsidiary station that are essential to the actuation or operation of devices initiating a signal indicative of fire shall be so arranged that the occurrence of a single break or single ground fault does not prevent transmission of an alarm.

Exception No. 1:   Circuits wholly within the supervising or subsidiary station.

Exception No. 2:   The carrier system portion of circuits.

A McCulloh system can continue to function even with a single open fault or a single ground fault on the circuit. Once the supervising station receives a trouble signal that indicates a single open fault or single ground fault, operators must respond to a trouble signal generated by the fault. They then manually, or by a computer-based automation system, recondition the circuit to receive the signals through earth ground.

Exceptions No. 1 and No. 2 exclude this requirement from circuits completely within the supervising station and from the portion of the circuit, described in subsection 4-5.3.3.2.6, that does not have continuous metallic continuity.

4-5.3.3.2.2 The occurrence of a single break or a single ground fault on any circuit shall not of itself cause a false signal that could be interpreted as an alarm of fire. Where such single fault prevents the normal functioning of any circuit, its occurrence shall be indicated automatically at the supervising station by a trouble signal compelling attention and readily distinguishable from signals other than those indicative of an abnormal condition of supervised parts of a fire suppression system(s).

This classic requirement dictates that the signals produced by a single open fault or a single ground fault on any circuit associated with the McCulloh system must not produce a false fire alarm signal. It also requires that such faults produce a trouble signal. While subsection 1-5.4.7(b) requires a distinct supervisory off-normal signal, subsection 4-5.3.3.2.2 permits a trouble signal to indicate both a fault and a fire extinguishing system supervisory off-normal condition. The code limits this combining of trouble and supervisory off-normal signals to McCulloh systems.

4-5.3.3.2.3 The circuits and devices shall be arranged to receive and record a signal readily identifiable as to location of origin, and provisions shall be made for equally identifiable transmission to the public fire service communications center.

4-5.3.3.2.4 Multipoint transmission channels between the protected premises and the supervising or subsidiary station and within the protected premises, consisting of one or more coded transmitters and an associated system unit(s), shall meet the requirements of either 4-5.3.3.2.5 or 4-5.3.3.2.6.

4-5.3.3.2.5 Where end-to-end metallic continuity is present, proper signals shall be received from other points under any one of the following transmission channel fault conditions at one point on the line:

(a) Open; or

(b) Ground; or

(c)* Wire-to-wire short; or

A-4-5.3.3.2.5(c) Though rare, it is understood that the occurrence of a wire-to-wire short on the primary trunk facility near the supervising station could disable the transmission system without immediate detection.

(d) Open and ground.

The traditional McCulloh system has end-to-end metallic continuity. Most often the subscriber leases the circuit between the protected premises and the supervising station from the public telephone utility. The telephone utility does not supply any power for the circuit. Rather, it simply provides a pair of wires. The telephone utility usually refers to such a circuit as a "PL circuit" (private line circuit).

The fact that the circuit must maintain end-to-end metallic continuity somewhat limits the electrical distance (resistance) between the protected premises and the supervising station. The resistance must not exceed the power to operate the system that the supervising station can supply.

4-5.3.3.2.6 Where end-to-end metallic continuity is not present, the nonmetallic portion of transmission channels shall meet all of the following requirements:

(a) Two nonmetallic channels or one channel plus a means for immediate transfer to a standby channel shall be provided for each transmission channel, with a maximum of eight transmission channels being associated with each standby channel, or shall be provided over one channel, provided that service is limited to one plant.

(b) The two nonmetallic channels (or one channel with standby arrangement) for each transmission channel shall be

provided by one of the following means, shown in descending order of preference:

1. Over separate facilities and separate routes; or

2. Over separate facilities in the same route; or

3. Over the same facilities in the same route.

(c) Failure of a nonmetallic channel or any portion thereof shall be indicated immediately and automatically in the supervising station.

(d) Proper signals shall be received from other points under any one of the following fault conditions at one point on the metallic portion of the transmission channel:

1. Open; or

2. Ground; or

3.* Wire-to-wire short.

As public telephone utilities have moved away from communications technology that uses end-to-end metallic continuity, the availability of PL circuits has significantly diminished. When the utility schedules the elimination of such circuits between certain telephone utility wire centers, the utility will sometimes provide an alternative. Such an alternate circuit must have the features described in this subsection. Alarm service providers sometimes also use this method to transport signals from remote areas where PL lines are not generally available to the supervising station.

A-4-5.3.3.2.6(d)3 Though rare, it is understood that the occurrence of a wire-to-wire short on the primary trunk facility near the supervising station could disable the transmission system without immediate detection.

4-5.3.3.3 Loading Capacity of McCulloh Circuits.

The loading capacities discussed in this subsection have been part of the NFPA signaling standards for more than 65 years. The capacities limit the number of signals lost under various adverse conditions. Such adverse conditions include open and ground faults and a clash of simultaneous signals coming from two or more protected premises on the same McCulloh circuit. Virtually every loading requirement contained in the *National Fire Alarm Code* has its root in these numbers for McCulloh systems.

4-5.3.3.3.1 The number of transmitters connected to any transmission channel shall be limited to avoid interference. The total number of code wheels or equivalent connected to a single transmission channel shall not exceed 250. Alarm signal transmission channels shall be reserved exclusively for fire alarm signal transmitting service.

Exception: As provided in 4-5.3.3.3.4.

4-5.3.3.3.2 The number of waterflow switches permitted to be connected to actuate a single transmitter shall not exceed five switches.

4-5.3.3.3.3 The number of supervisory switches permitted to be connected to actuate a single transmitter shall not exceed 20.

4-5.3.3.3.4 Combined alarm and supervisory transmission channels shall comply with the following:

(a) Where both sprinkler supervisory signals and fire or waterflow alarm signals are transmitted over the same transmission channel, provision shall be made to obtain either alarm signal precedence or sufficient repetition of the alarm signal to prevent the loss of any alarm signal.

(b) Other signal transmitters (e.g., burglar, industrial processes) on an alarm transmission channel shall not exceed five.

4-5.3.3.3.5* Where signals from manual fire alarm boxes and waterflow alarm transmitters within a building are transmitted over the same transmission channel and are operating at the same time, there shall be no interference with the fire box signals. Provision of the shunt noninterfering method of operation shall be permitted for this performance.

With a McCulloh coded-type manual fire alarm box, connected electrically first on the McCulloh circuit, operation of the box places a short circuit or shunt across the McCulloh circuit. It also disconnects the McCulloh transmitters connected electrically downstream from the manual fire alarm box. This effectively, if somewhat crudely, prevents another transmitter from interfering with the signal produced by the manual box.

A-4-5.3.3.3.5 At the time of system acceptance, verification should be made that manual fire alarm box signals are free of transmission channel interference.

4-5.3.3.3.6 One alarm transmission channel shall serve not more than 25 plants. A plant can consist of one or more buildings under the same ownership, and the circuit arrangement shall be such that an alarm signal cannot be received from more than one transmitter at a time within a plant. Where such noninterference is not provided, each building shall be considered a plant.

The routing of the McCulloh circuit throughout a large, multiple-building facility could have a significant effect on the loading of the circuit. A designer can extend the number of buildings served by a single McCulloh circuit by using a non-interfering shunt arrangement and very carefully routing the circuit throughout the facility. The circuit should begin at the most important building and extend to the least important building.

4-5.3.3.3.7 One sprinkler supervisory transmission channel circuit shall serve not more than 25 plants. A plant can consist of one or more buildings under the same ownership.

4-5.3.3.3.8 Connections to a guard supervisory transmission channel or to a combination manual fire alarm and guard transmission channel shall be limited so that not more than 60 scheduled guard report signals are transmitted in any 1-hour period. Patrol scheduling shall be such as to avoid interference between guard report signals.

This requirement presumes that an operator must manually record the guard supervisory signals.

4-5.3.4 Two-Way Radio Frequency (RF) Multiplex Systems.

A two-way radio frequency (RF) multiplex system consists of a traditional multiplex fire alarm system that uses a licensed two-way radio system to transmit signals from the protected premises to the supervising station. Essentially, the fire alarm system operates transparently over the radio portion of the system. The Code states requirements for two-way radio frequency (RF) multiplex systems identical to those for active multiplex systems.

4-5.3.4.1 The maximum end-to-end operating time parameters allowed for a two-way RF multiplex system are as follows:

(a) The maximum allowable time lapse from the initiation of a single fire alarm signal until it is recorded at the supervising station shall not exceed 90 seconds. When any number of subsequent fire alarm signals occur at any rate, they shall be recorded at a rate no slower than one every additional 10 seconds.

This subsection ensures that the system will complete an interrogation and response sequence at least every 90 seconds. As an alternative, the system may provide some other means to ensure alarm receipt within that time frame. For example, a designer could devise equipment that transmits alarm signals from the protected premises at a time other than during the normal interrogation and response sequence. However, most systems complete the interrogation and response sequence within 90-seconds.

(b) The maximum allowable time lapse from the occurrence of an adverse condition in any transmission channel until recording of the adverse condition is started shall not exceed 90 seconds for Type 4 and Type 5 systems. (*See 4-5.3.4.4.*)

As in subsection 4-5.3.4.1(a), subsection 4-5.3.4.1(b) also ensures that an interrogation and response sequence will be completed at least every 90 seconds.

(c) In addition to the maximum operating time allowed for fire alarm signals, the requirements of one of the following paragraphs shall be met:

1. A system unit having more than 500 initiating device circuits shall be able to record not less than 50 simultaneous status changes in 90 seconds.

2. A system unit having fewer than 500 initiating device circuits shall be able to record not less than 10 percent of the total number of simultaneous status changes within 90 seconds.

These requirements ensure that the portion of the multiplex system that processes and records status changes can do so with sufficient speed to handle a reasonable volume of signal traffic.

4-5.3.4.2 Facilities shall be provided at the supervising station for the following supervisory and control functions of the supervising or subsidiary station and the repeater station radio transmitting and receiving equipment. This shall be accomplished via a supervised circuit where the radio equipment is remotely located from the system unit.

(a) The following conditions shall be supervised at the supervising station:

1. RF transmitter in use (radiating);

2. Failure of ac power supplying the radio equipment;

3. RF receiver malfunction;

4. Indication of automatic switchover.

(b) Independent deactivation of either RF transmitter shall be controlled from the supervising station.

These supervisory functions help to ensure the continuity of signal transmission between the protected premises and the supervising station. In addition, the system also transmits an interrogation and response sequence between the protected premises and the supervising station every 90 seconds.

4-5.3.4.3 Transmission Channel.

4-5.3.4.3.1 The RF multiplex transmission channel shall terminate in a RF transmitter/receiver at the protected premises and in a system unit at the supervising or subsidiary station.

With this system, each protected premises will have its own radio frequency transmitter/receiver unit. The supervising station also has a radio frequency transmitter/receiver unit. The interrogation and response sequence takes place between these units. This system is similar to an active multiplex system where each protected premises has a transponder, and the supervising station has an active multiplex system control unit.

4-5.3.4.3.2 Operation of the transmission channel shall conform to the requirements of this code whether channels are private facilities, such as microwave, or leased facilities furnished by a communications utility company. Where private signal transmission facilities are utilized, the equipment necessary to transmit signals shall also comply with requirements for duplicate equipment or replacement of critical components, as described in 4-5.4.2.

This requirement ensures that the system will comply with the requirements of the Code, even if the facilities are leased from a communication utility company. It further ensures continuity of operations by requiring either redundant critical assemblies or replacement with on-premises spares. Either action must restore service within 30 minutes.

4-5.3.4.4* Two-way RF multiplex systems are divided into two categories based upon their ability to perform under adverse conditions. System classifications are of two types.

(a) A Type 4 system shall have two or more control sites configured as follows:

1. Each site shall have a RF receiver interconnected to the supervising or subsidiary station by a separate channel.

2. The RF transmitter/receiver located at the protected premises shall be within transmission range of at least two RF receiving sites.

3. The system shall contain two RF transmitters that are either:

a. Located at one site with the capability of interrogating all of the RF transmitters/receivers on the premises; or

b. Dispersed with all of the RF transmitters/receivers on the premises having the capability to be interrogated by two different RF transmitters.

4. Each RF transmitter shall maintain a status that allows immediate use at all times. Facilities shall be provided in the supervising or subsidiary station to operate any off-line RF transmitter at least once every 8 hours.

5. Any failure of one of the RF receivers shall in no way interfere with the operation of the system from the other RF receiver. Failure of any receiver shall be annunciated at the supervising station.

6. A physically separate channel shall be required between each RF transmitter or RF receiver site, or both, and the system unit.

These requirements essentially create a two-way radio frequency (RF) multiplex system that has redundancy of critical components. An authority having jurisdiction, or system designer, who expects a high volume of traffic or unusual transient radio frequency propagation problems would use such a system.

(b) A Type 5 system shall have a single control site configured as follows:

1. A minimum of one RF receiving site;

2. A minimum of one RF transmitting site.

NOTE: The sites above can be collocated.

A-4-5.3.4.4   The intent of the plurality of control sites is to safeguard against damage caused by lightning and to minimize the effect of interference on the receipt of signals.

A Type 4 two-way radio frequency (RF) multiplex system must have a plurality of control sites. Each site contains a transmitter/receiver unit.

4-5.3.4.5   Loading Capacities.

4-5.3.4.5.1   The loading capacities of two-way RF multiplex systems are based on the overall reliability of the signal receiving, processing, display, and recording equipment at the supervising or subsidiary station and the capability to transmit signals during adverse conditions of the transmission channels. Table 4-5.3.4.5.1 establishes the allowable loading capacities.

The loading of a two-way radio frequency (RF) multiplex system depends on the capability of the type of system. Since a Type 4 system has a redundant transmitter/receiver at different locations exerting control over the interrogation and response sequence between the protected premises and the supervising station, it has the greatest permitted system loading. A Type 5 system does not have dual transmitters/receivers in control of the system, so its trunk capacity is more limited than that of a Type 4 system.

4-5.3.4.5.2   **Exceptions to Loading Capacities Listed in Table 4-5.3.4.5.1.**   Where the signal receiving, processing, display, and recording equipment are duplicated at the

Table 4-5.3.4.5.1   Loading Capacities for Two-Way RF Multiplex Systems

	System Type	
	Type 4	Type 5
A. Trunks		
Maximum number of fire alarm service initiating device circuits per primary trunk facility	5120	1280
Maximum number of leg facilities for fire alarm service per primary trunk facility	512	128
Maximum number of leg facilities for all types of fire alarm service per secondary trunk facility[1]	128	128
Maximum number of all types of initiating device circuits per primary trunk facility in any combination	10,240	2560
Maximum number of leg facilities for types of fire alarm service per primary trunk facility in any combination[1]	1024	256
B. System Units at the Supervising Station		
Maximum number of all types of initiating device circuits per system unit[1]	10,240[2]	10,240[2]
Maximum number of fire protected buildings and premises per system unit	512[2]	512[2]
Maximum number of fire alarm service initiating device circuits per system	5120[2]	5120[2]
C. Systems Emitting from Subsidiary Station	Same as B	Same as B

[1]Includes every initiating device circuit (e.g., waterflow, fire alarm supervisory, guard, burglary, hold-up).
[2]Paragraph 4-5.3.4.5.2 applies.

supervising station and a switchover can be accomplished in not more than 30 seconds with no loss of signals during this period, the capacity of a system unit shall be unlimited.

This subsection modifies Part B of Table 4-5.3.4.5.1. However, to meet this requirement a two-way radio frequency (RF) multiplex system would have to employ complete redundancy of all critical components, and complete a switch-over in 30 seconds with no loss of signals. Systems meeting this requirement generally process all incoming signals in tandem. That is, the standby unit functions fully at all times and simply continues to function normally when the main unit fails. Operators would actually "change-over" only those incidental peripheral devices that had no required redundancy.

4-5.3.5 One-Way Private Radio Alarm Systems.

To create the requirements for a radio frequency transmission system that does not have an interrogation and response sequence to monitor the integrity of the transmission of signals between the protected premises and the supervising station, the Technical Committee borrowed heavily from the requirements for digital alarm communicator systems.

4-5.3.5.1 The requirements of 4-5.3.5 for a radio alarm repeater station receiver (RARSR) shall be satisfied where signals from each radio alarm transmitter (RAT) are received and supervised, in accordance with this chapter, by at least two independently powered, independently operating, and separately located RARSR.

The one-way radio alarm system consists of a radio frequency transmitter at the protected premises that connects to the protected premises control unit. This unit can transmit alarm, supervisory, and trouble signals to at least two receivers. The receivers relay the received signal to the supervising station by radio frequency or wired transmission means. This system allows the use of either a private system operated by a single alarm service provider, or a multi-user system operated by a one-way radio network provider. Most systems communicate through a multi-user network.

4-5.3.5.2* The end-to-end operating time parameters allowed for a one-way radio alarm system shall be as follows:

(a) There shall be a 90 percent probability that the time between the initiation of a single fire alarm signal until it is

recorded at the supervising station will not exceed 90 seconds.

(b) There shall be a 99 percent probability that the time between the initiation of a single fire alarm signal until it is recorded at the supervising station will not exceed 180 seconds.

(c) There shall be a 99.999 percent probability that the time between the initiation of a single fire alarm signal until it is recorded at the supervising station will not exceed 7.5 minutes (450 seconds), at which time the RAT shall cease transmitting.

When any number of subsequent fire alarm signals occur at any rate, they shall be recorded at an average rate no slower than one every additional 10 seconds.

(d) In addition to the maximum operating time allowed for fire signals, the system shall be able to record not less than 12 simultaneous status changes within 90 seconds at the supervising station.

A-4-5.3.5.2 It is intended that each RAT communicate with two or more independently located RARSRs. The location of such RARSRs should be such that they do not share common facilities.

NOTE: All probability calculations required for the purposes of Chapter 4 should be made in accordance with established communications procedures, should assume the maximum channel loading parameters specified, and should further assume that 25 RATs are actively in alarm and are being received by each RARSR.

Because this system does not have an interrogation and response sequence to verify the operating capability of the communication channel and all equipment associated with it, the system must rely on other means to achieve an acceptable level of operational integrity. The probabilities specified in subsection 4-5.3.5.2 help to ensure that level of integrity.

To achieve the probabilities, the system functions similarly to a digital alarm communicator transmitter. It makes a given number of attempts to reach one or both of the two receivers. If it does not succeed, it stops transmitting, so as to not tie up the receiver.

4-5.3.5.3 Supervision.

4-5.3.5.3.1 Equipment shall be provided at the supervising station for the supervisory and control functions of the supervising or subsidiary station and for the repeater station radio transmitting and receiving equipment. This shall be accomplished via a supervised circuit where the radio equipment is remotely located from the system unit. The following conditions shall be supervised at the supervising station:

(a) Failure of ac power supplying the radio equipment;

(b) RF receiver malfunction;

(c) Indication of automatic switchover (where applicable).

The specified supervisory functions help to ensure the continuity of signal transmission between the protected premises and the supervising station.

4-5.3.5.3.2 Protected Premises.

4-5.3.5.3.2.1 Interconnections between elements of transmitting equipment, including any antennas, shall be supervised either to cause an indication of failure at the protected premises or to transmit a trouble signal to the supervising station.

4-5.3.5.3.2.2 Where elements of transmitting equipment are physically separated, the wiring or cabling between them shall be protected by conduit.

Subsections 4-5.3.5.3.2.2.1 and 4-5.3.5.3.2.2 address two serious points of potential failure. Either the loss of the antenna, or the loss of connection between the transmitter and the antenna, would impair transmission.

In some systems, the transmitter connects directly to the antenna. In others, the installer locates the antenna at a point in the building more advantageous for successful transmission of a signal. These requirements ensure that a trouble signal resulting from the loss of the antenna or its connection will at least annunciate locally. They further require mechanical protection in conduit for the conductors to a remote antenna.

4-5.3.5.4 Transmission Channels.

4-5.3.5.4.1 The one-way RF transmission channel shall originate with a one-way RF transmitting device at the protected premises and shall terminate at the RF receiving system of an RARSR capable of receiving transmissions from such transmitting devices.

4-5.3.5.4.2 A receiving network transmission channel shall terminate at an RARSR at one end and with either another RARSR or a radio alarm supervising station receiver (RASSR) at the other end.

This subsection permits the architecture necessary to develop a network to handle a large number of radio alarm transmitters. The network interconnections can use multiple radio alarm repeater station receivers (RARSRs) that in turn repeat the received signals to other RARSRs until the signals ultimately reach a radio alarm supervising station receiver (RASSR). Along each segment of the transmission path, at least two RARSR's must always receive the signal.

4-5.3.5.4.3 Operation of receiving network transmission channels shall conform to the requirements of this code whether channels are private facilities, such as microwave, or leased facilities furnished by a communications utility company. Where private signal transmission facilities are utilized, the equipment necessary to transmit signals shall also comply with requirements for duplicate equipment or replacement of critical components as described in 4-5.4.2.

4-5.3.5.4.4 The system shall provide information indicating the quality of the received signal for each RARSR supervising each RAT in accordance with 4-5.3.5 and shall provide information at the supervising station when such signal quality falls below the minimum signal quality levels set forth in 4-5.3.5.

4-5.3.5.4.5 Each RAT shall be installed in such a manner so as to provide a signal quality over at least two independent one-way RF transmission channels, of the minimum quality level specified, that satisfies the performance requirements in 4-5.2.2 and 4-5.4.

This requirement ensures that the system will comply with the Code, even if the installer leases facilities from a communications utility company or other one-way radio network service provider. It further ensures continuity of operations by requiring either redundant critical assemblies, or that technicians can replace critical assemblies with on-premises spares and restore service within 30 minutes.

The system must also monitor the quality of the transmitted signal, including the various operating time parameters specified in subsection 4-5.3.5.2. One design provides each RAT with a clock. Each transmitted signal includes the time of first transmission and the current time, along with the alarm, supervisory, or trouble data.

4-5.3.5.5 Nonpublic one-way radio alarm systems shall be divided into two categories based upon the following number of RASSRs present in the system:

(a) A Type 6 system shall have one RASSR and at least two RARSRs.

(b) A Type 7 system shall have more than one RASSR and at least two RARSRs.

In a Type 7 system, when more than one RARSR is out of service and, as a result, any RATs are no longer being supervised, the affected supervising station shall be notified.

In a Type 6 system, when any RARSR is out of service, a trouble signal shall be annunciated at the supervising station.

The Type 6 system serves a single supervising station. The Type 7 system serves more than one supervising station. A multi-user network most closely fits the Type 7 system description.

4-5.3.5.6 The loading capacities of one-way radio alarm systems are based on the overall reliability of the signal receiving, processing, display, and recording equipment at the supervising or subsidiary station and the capability to transmit signals during adverse conditions of the transmission channels. Table 4-5.3.5.6 establishes the allowable loading capacities.

4-5.3.5.7 Exceptions to Loading Capacities Listed in Table 4-5.3.5.6. Where the signal receiving, processing,

Table 4-5.3.5.6 Loading Capacities of One-Way Radio Alarm Systems

	System Type	
	Type 6	Type 7
A. Radio Alarm Repeater Station Receiver (RARSR)		
Maximum number of fire alarm service initiating device circuits per RARSR	5120	5120
Maximum number of RATs for fire	512	512
Maximum number of all types of initiating device circuits per RARSR in any combination[1]	10,240	10,240
Maximum number of RATs for all types of fire alarm service per RARSR in any combination[1]†	1024	1024
B. System Units at the Supervising Station		
Maximum number of all types of initiating device circuits per system unit[1]	10,240[2]	10,240[2]
Maximum number of fire protected buildings and premises per system unit	512[2]	512[2]
Maximum number of fire alarm service initiating device circuits per system unit	5120[2]	5120[2]

[1]Includes every initiating device circuit (e.g., waterflow, fire alarm, supervisory, guard, burglary, hold-up).
[2]Paragraph 4-5.3.5.7 applies.
†Each supervised BA (open/close) or each suppressed guard tour transmitter shall reduce the allowable RATs by five.
Each guard tour transmitter shall reduce the allowable RATs by 15.
Each two-way protected premises radio transmitter shall reduce the allowable RATs by two.

display, and recording equipment are duplicated at the supervising station and a switchover can be accomplished in not more than 30 seconds with no loss of signals during this period, the capacity of a system unit is unlimited.

This subsection modifies Part B of Table 4-5.3.5.6. However, to meet this requirement a one-way radio alarm system would have to employ complete redundancy of all critical components, and complete a switchover in 30 seconds with no loss of signals. Systems meeting this requirement generally process all incoming signals in tandem. That is, the standby unit functions fully at all times and simply continues to function normally when the main unit fails. Operators would actually "change-over" only those incidental peripheral devices that had no required redundancy.

4-5.3.6 Directly-Connected Noncoded Systems.

4-5.3.6.1 Circuits for transmission of alarm signals between the fire alarm control unit or the transmitter in the protected premises and the supervising station shall be arranged so as to comply with either of the following provisions:

(a) These circuits shall be arranged so that the occurrence of a single break or single ground fault does not prevent the transmission of an alarm signal. Circuits complying with this paragraph shall be automatically self-adjusting in the event of either a single break or a single ground fault and shall be automatically self-restoring in the event that the break or fault is corrected.

Only one manufacturer offers a system that meets the requirements of subsection 4-5.3.6.1(a). This system powers the circuit by means of a float-charged set of batteries with a center tap connection to earth ground. Two sets of alarm relays are connected across the circuit at the supervising station and reference to earth ground. Loss of one side of the circuit would not prevent the transmission of the signal over the remaining side of the circuit and earth ground.

(b) These circuits shall be arranged so that they are normally isolated from ground (except for reference ground detection) and so that a single ground fault does not prevent the transmission of an alarm signal. Circuits complying with this paragraph shall be provided with a ground reference circuit so as to detect and indicate automatically the existence of a single ground fault, unless a multiple ground-fault condition that would prevent alarm operation is to be indicated by an alarm or by a trouble signal.

The vast majority of the circuits employed for transmitting signals from a protected premises to a supervising station over directly connected noncoded systems meet the requirements of subsection 4-5.3.6.1(b). In most cases, the protected premises or alarm service provider leases the circuits for directly-connected noncoded systems from the public telephone utility. The telephone utility does not supply any power for the circuit. Rather, it simply provides a pair of wires. The telephone utility usually refers to such a circuit as a "PL circuit" (private line circuit).

The fact that the circuit must maintain end-to-end metallic continuity somewhat limits the electrical distance (resistance) between the protected premises and the supervising station. The resistance must not exceed that for which either the protected premises or the supervising station can supply power to operate the system.

4-5.3.6.2 Circuits for transmission of supervisory signals shall be separate from alarm circuits. These circuits within the protected premises and between the protected premises and the supervising station shall be arranged as described in 4-5.3.6.1(a) or (b).

Exception: Where the reception of alarm signals and supervisory signals at the same supervising station is permitted by the authority having jurisdiction, the supervisory signals do not interfere with the alarm signals, and alarm signals have priority, the same circuit between the protected premises and the supervising station shall be permitted to be used for alarm and supervisory signals.

4-5.3.6.3 The occurrence of a single break or a single ground fault on any circuit shall not of itself cause a false signal that could be interpreted as an alarm of fire.

4-5.3.6.4 The requirements of 4-5.3.6.1 and 4-5.3.6.2 shall not apply to the following circuits:

(a) Circuits wholly within the supervising station;

(b) Circuits wholly within the protected premises extending from one or more automatic fire detectors or other noncoded initiating devices other than waterflow devices to a transmitter or control unit; or

(c) Power supply leads wholly within the building or buildings protected.

These requirements clarify that the named circuits need not have the operational capability of the circuits extending between the protected premises and the supervising station.

4-5.3.6.5 **Loading Capacity of Circuits.**

4-5.3.6.5.1 The number of initiating devices connected to any signaling circuit and the number of plants that shall be permitted to be served by a signal circuit shall be determined by the authority having jurisdiction and shall not exceed the limitations specified in 4-5.3.6.5.

NOTE: A plant can consist of one or more buildings under the same ownership.

4-5.3.6.5.2 A single circuit shall not serve more than one plant.

NOTE: Where a single plant involves more than one gate entrance or involves a number of buildings, separate circuits might be required so that the alarm to the supervising station indicates the area to which the fire department is to be dispatched.

Unique among transmission technologies, a directly-connected noncoded system may serve only one plant.

4-5.3.7 **Private Microwave Radio Systems.**

The Technical Committee originally consulted with AT&T in developing the requirements in subsection 4-5.3.7. Based on standardized microwave relay link requirements, these requirements help to ensure normal network reliability.

In most cases, a private microwave radio system would transport a rather high volume of signal traffic from a subsidiary station to a supervising station.

4-5.3.7.1* Where a private microwave radio is used as the transmission channel and communications channel, appropriate supervised transmitting and receiving equipment shall be provided at supervising, subsidiary, and repeater stations.

A-4-5.3.7.1 A private microwave radio can be used either as a transmission channel, to connect a transmitter to a supervising station or subsidiary station, or as a communications channel to connect a subsidiary station(s) to a supervising station(s). This can be done independently or in conjunction with wireline facilities.

4-5.3.7.2 Where more than five protected buildings or premises or 50 initiating devices or initiating device circuits are being serviced by a private radio carrier, the supervising, subsidiary, and repeater station radio facilities shall meet all of the following:

(a) Dual supervised transmitters, arranged for automatic switching from one to the other in case of trouble, shall be installed. Where the transmitters are located where someone is always on duty, switchboard facilities shall be permitted to be manually operated, provided the switching can be carried out within 30 seconds. Where the transmitters are located where no one is normally on duty, the circuit extending between the supervising station and the transmitters shall be a supervised circuit.

(b)* Transmitters shall be operated on a time ratio of 2:1 within each 24 hours.

A-4-5.3.7.2(b) Transmitters should be operated alternately, 16 hours on and 16 hours off.

(c) Dual receivers shall be installed with a means for selecting a usable output from one of the two receivers. The failure of one shall in no way interfere with the operation of the other. Failure of either receiver shall be annunciated.

4-5.3.7.3 Means shall be provided at the supervising station for the supervision and control of supervising, subsidiary, and repeater station radio transmitting and receiving equipment. This shall be accomplished via a supervised circuit where the radio equipment is remote from the supervising station.

(a) The following conditions shall be supervised at the supervising station:

1. Transmitter in use (radiating);
2. Failure of ac power supplying the radio equipment;
3. Receiver malfunction;
4. Indication of automatic switchover.

(b) It shall be possible to independently deactivate either transmitter from the supervising station.

4-5.4 Display and Recording.

These requirements specify the content and nature of the display and recording of signals received at a supervising station. The requirements take into account a reasonable quantity of signal traffic. They also consider certain ergonomic necessities for interfacing electronically reproduced signals with one or more human operators.

4-5.4.1* Any status changes (including the initiation or restoration to normal of a trouble condition) that occur in an initiating device or in any interconnecting circuits or equipment, including the local protected premises controls from the location of the initiating device(s) to the supervising station, shall be presented in a form to expedite prompt operator interpretation. Status change signals shall provide the following information:

(a) *Type of Signal.* Identification of the type of signal to show whether it is an alarm, supervisory, delinquency, or trouble signal;

(b) *Condition.* Identification of the signal to differentiate between an initiation of an alarm, supervisory, delinquency, or trouble signal and a restoration to normal from one or more of these conditions;

(c) *Location.* Identification of the point of origin of each status change signal.

A-4-5.4.1 The signal information may be permitted to be provided in coded form. Records may be permitted to be used to interpret these codes.

4-5.4.2* Where duplicate equipment for signal receiving, processing, display, and recording is not provided, the installed equipment shall be so designed that any critical assembly can be replaced from on-premises spares and the system can be restored to service within 30 minutes. A critical assembly is an assembly in which a malfunction prevents the receipt and interpretation of signals by the supervising station operator.

Exception: Proprietary and remote station systems.

A-4-5.4.2 In order to expedite repairs, it is recommended that spare modules, such as printed circuit boards, CRT displays, or printers, be stocked at the supervising station.

This requirement ensures that a technician will properly repair any malfunction in a critical assembly. For a description of critical assembly, see subsection 4-5.4.2. The technician may repair the defective assembly but, most often, the technician will replace the defective assembly with an on-premises spare. Any assembly too complex for a technician to readily repair would require a duplicate.

4-5.4.3* Any method of recording and display or indication of change of status signals shall be permitted, provided all of the following conditions are met:

(a) Each change of status signal requiring action to be taken by the operator shall result in an audible signal and not less than two independent methods of identifying the type, condition, and location of the status change.

(b) Each change of status signal shall be automatically recorded. The record shall provide the type of signal, condition, and location as required by 4-5.4.1 in addition to the time and date the signal was received.

(c) Failure of an operator to acknowledge or act upon a change of status signal shall not prevent subsequent alarm signals from being received, indicated or displayed, and recorded.

(d) Change of status signals requiring action to be taken by the operator shall be displayed or indicated in a manner that clearly differentiates them from those that have been acted upon and acknowledged.

(e) Each incoming signal to a DACR or DARR shall cause an audible signal that persists until manually acknowledged.

Exception: Test signals (see 4-5.3.2.1.10) received at a DACR or a DARR.

A-4-5.4.3 For all forms of transmission, the maximum time to process an alarm signal should be 90 seconds. The maximum time to process a supervisory signal should be 4 minutes. The time to process an alarm or supervisory signal is defined as that time measured from receipt of a signal until retransmission or subscriber contact is initiated.

When the level of traffic in a supervising station system reaches a magnitude such that delayed response is possible, even where the loading tables or loading formulas of this code are not exceeded, it is envisioned that it will be necessary to employ an enhanced method of processing.

For example, in a system where a single DACR instrument provided with fire and burglar alarm service is connected to multiple telephone lines, it is conceivable that, during certain periods of the day, fire alarm signals could be delayed by the security signaling traffic, such as opening and closing signals. Such an enhanced system would, upon receipt of a signal:

(a) Automatically process the signals, differentiating between those that require immediate response by supervising station personnel and those that need only be logged.

(b) Automatically provide relevant subscriber information to assist supervising station personnel in their response.

(c) Maintain a timed, unalterable log of the signals received and the response of supervising station personnel to such signals.

4-5.5 **Testing and Maintenance.** Testing and maintenance of communications methods shall be in accordance with the requirements of Chapter 7.

4-6   Public Fire Alarm Reporting Systems.

NOTE: The requirements of Chapters 1, 5, and 7 apply to public fire alarm reporting systems, unless they conflict with the requirements of this section.

4-6.1 **Scope.** This section covers the general requirements and use of public fire alarm reporting systems. These systems include the equipment necessary to effect the trans-

mission and reception of fire alarms or other emergency calls from the public.

Section 4-6 deals with the means provided by a municipality or other governmental entity to allow the public to initiate a fire alarm signal. The public fire reporting system would transmit this signal to a public fire service communication center.

This system transmits only fire alarm signals or trouble signals relating to the system itself. The requirements of this section do not contemplate the transmission of supervisory signals.

A municipality or other governmental entity may exercise its right as an authority having jurisdiction and permit the transmission of supervisory signals from buildings the entity owns. However, this section of the Code does not contain appropriate requirements for transmission of supervisory signals.

The public initiates a fire alarm signal at a publicly accessible manual fire alarm box. Section 4-7 provides requirements for interfacing a protected premises fire alarm system with a public fire reporting system. The protected premises fire alarm system would meet the requirements of Chapters 1, 3, 5,. 6 and 7. The Code calls such an interface an Auxiliary fire alarm system. See Figures 4.8 and 4.9.

4-6.2 **General Fundamentals.**

4-6.2.1* Where implemented at the option of the authority having jurisdiction, a public fire alarm reporting system shall be designed, installed, operated, and maintained to provide the maximum practicable reliability for transmission and receipt of fire alarms.

A-4-6.2.1 Where choosing from available options to implement a public fire alarm reporting system, the operating agency should consider which of the choices would facilitate the maximum reliability of the system where such a choice is not cost prohibitive.

4-6.2.2 It shall be permitted for a public fire alarm reporting system, as described herein, to be used for the transmission of other signals or calls of a public emergency nature, provided such transmission does not interfere with the transmission and receipt of fire alarms.

In addition to transmitting a fire alarm signal to the public fire service communication center, this subsection permits a system to transmit a voice or electronic request for emergency medical or police response. The Code

Figure 4.8 *Typical public fire service communications center. Photograph courtesy of Manchester Fire Department, Manchester, NH.*

Figure 4.9 *Typical manual fire alarm box used in a Public Fire Reporting System.*

permits these non-fire transmissions as long as they do not interfere with the transmission of fire alarm signals.

4-6.2.3 Alarm systems shall be Type A or Type B. A Type A system shall be provided where the number of all alarms required to be transmitted over the dispatch circuits exceeds 2500 per year.

NOTE: Where a Type A system is required, automatic transmission of alarms from boxes by use of electronic equipment is permitted, provided the following conditions are satisfied:

(a) Reliable facilities are provided for the automatic receipt, storage, retrieval, and transmission of alarms in the order received; and

(b) Override capability is provided to the operator(s) so that manual transmission and dispatch facilities are instantly available.

In a Type A system, operators at the public fire service communication center receive signals from the public fire reporting system. They then retransmit these signals to the fire stations designated to respond. Subsection

4-6.2.3 requires systems transmitting more than 2500 alarms on the dispatch circuits to be Type A systems.

In a Type B system, equipment at the public fire service communication center automatically retransmits the signals to all fire stations and other locations connected to the system.

4-6.3   Management and Maintenance.   See Chapter 7.

4-6.4   Equipment and Installation.

4-6.4.1   Means for actuation of alarms by the public shall be conspicuous and readily accessible for easy operation.

4-6.4.2   Public fire alarm reporting systems as defined in this chapter, shall, in their entirety, be subject to a complete operational acceptance test upon completion of system installation. This test(s) shall be made in accordance with the requirements of the authority having jurisdiction. However, in no case shall the operational functions tested be less than those stipulated in Chapter 7. Similar tests shall be performed on any alarm reporting devices as identified in this chapter that are added subsequent to the installation of the initial system.

4-6.4.3   Publicly accessible boxes shall be recognizable as such. Boxes shall have operating instructions plainly marked on the exterior surface.

4-6.4.4   The actuating device shall be readily available and of such design and so located as to make the method of its use apparent.

4-6.4.5   Publicly accessible boxes shall be as conspicuous as possible. Their color shall be distinctive.

4-6.4.6   All publicly accessible boxes mounted on support poles shall be identified by a wide band of distinctive colors or adequate signs placed 8 ft (2.44 m) above the ground and visible from all directions wherever possible.

4-6.4.7*   Indicating lights of a distinctive color, visible for at least 1500 ft (460 m), shall be installed over publicly accessible boxes in mercantile and manufacturing areas. Equipping the street light nearest the box with a distinctively colored light shall be considered as meeting this requirement.

A-4-6.4.7   Indicating Lights.

(a) The current supply for designating lamps at street boxes should be secured at lamp locations from the local electric utility company.

(b) Alternating current power may be permitted to be superimposed on metallic fire alarm circuits for supplying designating lamps or for control or actuation of equipment devices for fire alarm or other emergency signals, provided the following conditions exist:

1. Voltages between any wire and ground or between one wire and any other wire of the system should not exceed 150 volts; the total resultant current in any line circuit should not exceed $1/4$ ampere.

2. Components such as coupling capacitors, transformers, chokes, or coils should be rated for 600-volt working voltage and have a breakdown voltage of at least twice the working voltage plus 1000 volts.

3. There is no interference with fire alarm service under any conditions.

Superimposing box light power on the fire alarm circuit was popular in the 1930s and 1940s, but system operators seldom use this method today. Another somewhat popular building fire alarm scheme of yesteryear imposed alternating current for audible fire alarm notification appliances on the direct current manual fire alarm initiating device circuit of a shunt-type master fire alarm box. The Gamewell Company marketed this system under the trade name of "dual alarm." Such a system would no longer meet the requirements of Chapters 1, 3, 5, 6 and 7.

4-6.4.8   Boxes shall be securely mounted on poles, pedestals, or structural surfaces as directed by the authority having jurisdiction.

4-6.4.9   Concurrent operation of at least four boxes shall not result in the loss of an alarm.

To meet this requirement, the box must sense that a second box has begun to transmit over the common box circuit. The first box retains its signal until it senses a clear circuit. Then it transmits its signal. In meeting this requirement, manufacturers must create non-interfering and successive boxes, or the equivalent.

4-6.5   Design of Boxes.   See Chapter 5.

For box design information, refer to subsection 5-8.2.

4-6.6*   **Location of Boxes.**   The location of publicly accessible boxes shall be designated by the authority having jurisdiction. Schools, hospitals, nursing homes, and places of public assembly shall have a box located at or near the main entrance, as directed by the authority having jurisdiction.

In most cases, the municipal fire officials will serve as the authority having jurisdiction. The Municipal Grading Schedule of the Insurance Services Office may also influence their decisions.

A-4-6.6 Where the intent is for complete coverage, it should not be necessary to travel in excess of one block or 500 ft (150 m) to reach a box. In residential areas, it should not be necessary to travel in excess of two blocks or 800 ft (240 m) to reach a box.

4-6.7 Power Supply.

4-6.7.1 General.

4-6.7.1.1 Batteries, motor-generators, or rectifiers shall be sufficient to supply all connected circuits without exceeding the capacity of any battery or overloading any generator or rectifier, so that circuits developing grounds or crosses with other circuits each can be supplied by an independent source to the extent required by 4-6.7.1.8(b).

4-6.7.1.2 Provision shall be made in the operating room for supplying any circuit from any battery, generator, or rectifier. Enclosed fuses shall be provided at points where supplies for individual circuits are taken from common leads. Necessary switches, testing, and signal-transmitting and receiving devices shall be provided to allow the isolation, control, and test of each circuit up to at least 10 percent of the total number of box and dispatch circuits, but never less than two circuits.

These requirements help to ensure maximum reliability for the public fire alarm reporting system. The last phrase intends that the system will have enough equipment, such as transmitting and receiving equipment, so that no one device or appliance serves more than 10 percent of the circuits. The stock of spare parts will always have at least two of every device or appliance on hand.

4-6.7.1.3 Where common-current source systems are grounded, the ground shall not exceed 10 percent of resistance of any connected circuit and shall be located at one side of the battery. Visual and audible indicating devices shall be provided for each box and dispatch circuit to give immediate warning of ground leakage endangering operability.

The system installers have set up a large percentage of the cable plant for the public fire reporting system as either aerial or underground construction. Foreign grounds on a circuit will render a portion inoperable. For this reason, operators must give much attention to the prompt discovery of excess current leakage to ground.

4-6.7.1.4 Local circuits at communications centers shall be supplied either in common with box circuits or coded radio-receiving system circuits or by a separate power source. The source of power for local circuits required to operate the essential features of the system shall be supervised.

The system must monitor the integrity of the power for circuits and equipment within the public fire service communication center. The loss of this power must cause a trouble signal.

4-6.7.1.5 Visual and audible means to indicate a 15 percent or greater reduction of normal power supply (rated voltage) shall be provided.

When power for the public fire reporting system or for local circuits drops 15 percent or more below the normal rated voltage, such reduction must initiate a trouble signal.

4-6.7.1.6 The forms and arrangements of power supply shall be classified as described in 4-6.7.1.7 through 4-6.7.1.9.

NOTE: Where the electrical service/capacity of the equipment required under NFPA 1221, *Standard for the Installation, Maintenance, and Use of Public Fire Service Communication Systems*, 2-1.6, is adequate to satisfy the needs of equipment in Section 4-6, such equipment is not required to be duplicated.

The wide range of alternatives for power supplies offered by the forms described in subsections 4-6.7.1.7, 4-6.7.1.8, and 4-6.7.1.9 gives the designer of a public fire alarm reporting system a great deal of flexibility in design execution.

4-6.7.1.7 Form 2. Form 2 shall be permitted for Type A systems only. Box circuits shall be served in multiple by:

(a)* *Form 2A.* A rectifier or motor-generator powered from a single source of alternating current, with a floating storage battery having a 24-hour standby capacity.

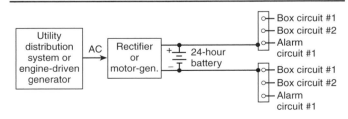

Figure A-4-6.7.1.7(a) *Form 2A.*

(b)* *Form 2B.* A rectifier or motor-generator powered from two sources of alternating current, with a floating storage battery having a 4-hour standby capacity.

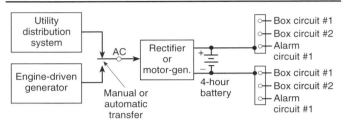

Figure A-4-6.7.1.7(b)(1) *Form 2B-1.*

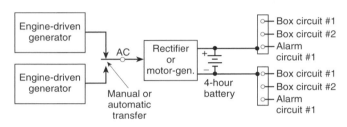

Figure A-4-6.7.1.7(b)(2) *Form 2B-2.*

(c)* *Form 2C.* A duplicate rectifier or motor-generator powered from two sources of alternating current with transfer facilities to apply power from the secondary source to the system within 30 seconds (*see* NFPA 1221, *Standard for the Installation, Maintenance, and Use of Public Fire Service Communication Systems*). Each rectifier or motor-generator shall be capable of powering the entire system.

NOTE: For Forms 2A, 2B, and 2C, these arrangements are permitted but are not recommended where circuits are wholly or partly open-wire because of the possibility of trouble from multiple grounds.

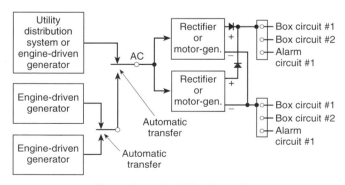

Figure A-4-6.7.1.7(c) *Form 2C.*

4-6.7.1.8   Form 3. Each box circuit or coded radio receiving system shall be served by:

(a)* *Form 3A.* A rectifier or motor-generator powered from a single source of alternating current with a floating storage battery having a 60-hour standby capacity.

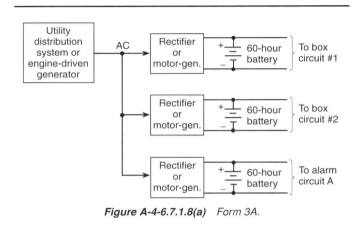

Figure A-4-6.7.1.8(a) *Form 3A.*

(b)* *Form 3B.* A rectifier or motor-generator powered from two sources of alternating current with a floating storage battery having a 24-hour standby capacity.

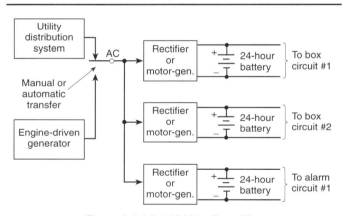

Figure A-4-6.7.1.8(b)(1) *Form 3B-1.*

4-6.7.1.9   Form 4. Each box circuit or coded radio receiving system shall be served by:

(a)* *Form 4A.* An inverter powered from a common rectifier receiving power by a single source of alternating current with a floating storage battery having a 24-hour standby capacity.

(b)* *Form 4B.* An inverter powered from a common rectifier receiving power from two sources of alternating current with a floating storage battery having a 4-hour standby capacity.

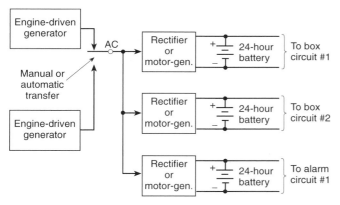

Figure A-4-6.7.1.8(b)(2)   *Form 3B-2.*

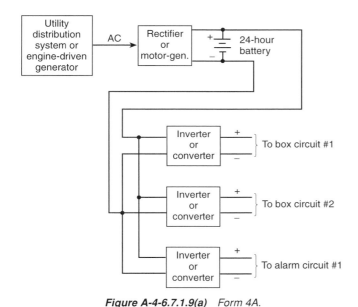

Figure A-4-6.7.1.9(a)   *Form 4A.*

NOTE: For Form 4A and Form 4B, it is permitted to distribute the system load between two or more common rectifiers and batteries.

(c)* *Form 4C.* A rectifier, converter, or motor-generator receiving power from two sources of alternating current with transfer facilities to apply power from the secondary source to the system within 30 seconds. *(See* NFPA 1221, *Standard for the Installation, Maintenance, and Use of Public Fire Service Communication Systems.) (See figure on next page.)*

4-6.7.2 Rectifiers, Converters, Inverters, and Motor-Generators.

4-6.7.2.1 Rectifiers shall be supplied through an isolating transformer taking energy from a circuit not to exceed 250 volts.

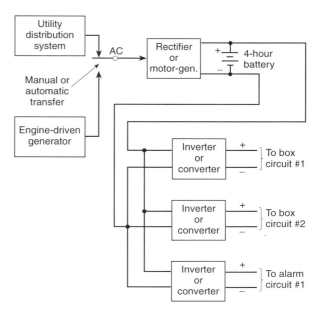

Figure A-4-6.7.1.9(b)(1)   *Form 4B-1.*

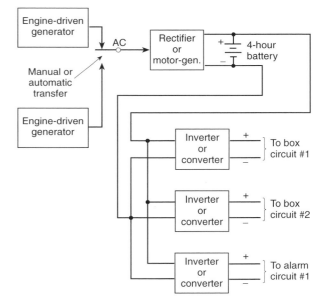

Figure A-4-6.7.1.9(b)(2)   *Form 4B-2.*

4-6.7.2.2 Complete, ready-to-use spare units or spare parts shall be available in reserve.

4-6.7.2.3 One spare rectifier shall be provided for each 10 required for operation, but in no case shall less than one be available.

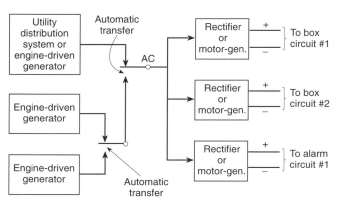

Figure A-4-6.7.1.9(c)   Form 4C.

4-6.7.2.4 Leads from rectifiers or motor-generators, with storage battery floating, shall have fuses rated at not less than 1 ampere and not more than 200 percent of maximum connected load. Where not provided with battery floating, the fuse shall be not less than 3 amperes.

These fuse sizing requirements provide circuits with sufficient protection without being overly sensitive. Too frequent operation of the protection would reduce the overall reliability of the public fire alarm reporting system.

4-6.7.3 Engine-Driven Generator Sets.

The requirements in this subsection help to ensure the continuity of supplied power. This helps to assure the reliability of the public fire alarm reporting system.

4-6.7.3.1 The provisions of 4-6.7.3 shall apply to generators driven by internal combustion engines.

4-6.7.3.2 The installation of such units shall conform to the provisions of NFPA 37, *Standard for the Installation and Use of Stationary Combustion Engines and Gas Turbines*, and NFPA 110, *Standard for Emergency and Standby Power Systems*.

Exception: Where restricted by the provisions of 4-6.7.3.

4-6.7.3.3 The engine-driven generator shall be located in an adequately ventilated cutoff area of the building housing the communications center equipment. The area housing the unit shall be used for no other purpose other than for storage of spare parts or equipment. Exhaust fumes shall be discharged directly outside the building.

4-6.7.3.4 Liquid fuel shall be stored in outside underground tanks and gravity feed shall not be used. Sufficient fuel shall be available for 24 hours of operation at full load where a reliable source of fuel supply is available, at any time, on 2-hours' notice. Where a source of supply is not reliable or readily available, or where special arrangements need to be made for refueling as necessary, a supply sufficient for 48 hours of operation at full load shall be maintained.

4-6.7.3.5 Liquefied petroleum gas and natural gas installations shall meet the requirements of NFPA 54, *National Fuel Gas Code*, and NFPA 58, *Standard for the Storage and Handling of Liquefied Petroleum Gases*.

4-6.7.3.6 The unit, as a minimum, shall be of sufficient capacity to supply power to operate all fire alarm facilities and emergency lighting of the operating rooms or communications building.

4-6.7.3.7 A separate storage battery on automatic float charger shall be provided for starting the engine-driven generator.

4-6.7.3.8 Where more than one engine-driven generator is provided, each shall be provided with a separate fuel line and transfer pump.

4-6.7.4 Float-Charged Batteries.

The requirements in subsection 4-6.7.4 help to ensure the continuity of the supplied power. This helps to assure the reliability of the public fire alarm reporting system.

4-6.7.4.1 Batteries shall be of the storage type; primary batteries (dry cells) shall not be used. All cells shall be of the sealed type. Lead-acid batteries shall be in jars of glass or other suitable transparent materials; other types of batteries shall be in containers suitable for the purpose.

4-6.7.4.2 Batteries shall be located in the same building as the operating equipment, preferably on the same floor, and readily accessible for maintenance and inspection. The battery room shall be aboveground and shall be ventilated to prevent accumulation of explosive gas mixtures; special ventilation shall be required only for unsealed cells.

The installer must select an above ground location, unless he or she constructs a location specifically for use below grade.

Exception: Where permitted by NFPA 1221, Standard for the Installation, Maintenance, and Use of Public Fire Service

Communication Systems, 2-11.2, the battery room shall not be required to be aboveground.

4-6.7.4.3 Batteries shall be mounted to provide effective insulation from the ground and from other batteries. The mounting shall be suitably protected against deterioration, and consideration shall be given to stability, especially in geographic areas subject to seismic disturbance.

As stated previously, the requirements for power supplies in 4-6.7.1.6 through 4-6.7.4.3 all help to ensure the overall reliability of the power supply portion of the public fire alarm reporting system.

4-6.8 **Requirements for Metallic Systems and Metallic Interconnections.**

4-6.8.1 **Circuit Conductors.**

4-6.8.1.1 Wires shall be terminated so as to provide good electrical conductivity and to prevent breaking from vibration or stress.

4-6.8.1.2 Circuit conductors on terminal racks shall be identified and isolated from conductors of other systems wherever possible and shall be suitably protected from mechanical injury.

4-6.8.1.3 Exterior cable and wire shall conform to International Municipal Signal Association specifications or their equivalent.

Exception: Where otherwise provided herein.

The International Municipal Signal Association (IMSA) publishes wire and cable specifications for use in the installation of public fire reporting systems. IMSA can be reached at P.O. Box 539, Newark, NY 14513, (315) 331-2182.

4-6.8.1.4 Where a public box is installed inside a building, it shall be placed as close as practical to the point of entrance of the circuit. The circuit from the point of entrance to the public box shall be installed in conduit or electrical metallic tubing in accordance with Chapter 3 of NFPA 70, *National Electrical Code.*

Exception: This requirement shall not apply to coded radio box systems.

The intent of subsection 4-6.8.1.4 and of subsection 4-6.9.1.1.2 is to limit the exposure of the public fire

alarm reporting system circuit to a hostile fire within a building. If an installer runs that circuit extensively throughout the building, a fire could more likely burn through a portion of the circuit before the system transmits and alarm signal. This would render the protection system useless.

4-6.8.2 **Cables.**

4-6.8.2.1 **General.**

4-6.8.2.1.1 Cables that meet the requirements of NFPA 70, *National Electrical Code*, Article 310, for installation in wet locations shall be satisfactory for overhead or underground installation.

Exception: Direct-burial cable shall be specifically approved for this purpose.

4-6.8.2.1.2 Paper or pressed pulp insulation shall not be considered satisfactory for an emergency service such as a fire alarm system, except that cables containing conductors with such insulation shall be permitted where pressurized with dry air or nitrogen. Loss of pressure in cables shall be indicated by a visual or audible warning system located where an individual who can interpret the pressure readings and who has authority to have the indicated abnormal condition corrected is in constant attendance.

Certain cable constructions installed in the 1940s, particularly for telephone and other communications purposes, used copper wire insulated with paper or pressed pulp materials. These cables must remain dry or the insulation materials will degrade. To ensure dryness, installers commonly pressurize the cable with dry nitrogen and monitor the cable for leakage. Installers would not normally use such cable in new installations.

4-6.8.2.1.3 Natural rubber-sheathed cable shall not be used where it could be exposed to oil, grease, or other substances or conditions that tend to deteriorate the cable sheath. Braided-sheathed cable shall be used only inside of buildings where run in conduit or metal raceways.

4-6.8.2.1.4 Other municipally controlled signal wires shall be permitted to be installed in the same cable with fire alarm wires. Cables controlled by or containing wires of private signaling organizations shall be permitted to be used for fire alarm purposes only by permission of the authority having jurisdiction.

Occasionally, municipalities that maintain their own governmental service telephone system will run the wiring for that service in the same cable as the public fire alarm reporting system. Conversely, parallel telephone-type systems and, on rare occasions, other types of public fire reporting systems may consist of interconnecting wiring leased from the public telephone utility, or even from a private agency, such as Western Union. The requirements of this subsection apply to such cases.

4-6.8.2.1.5 Signaling wires that, because of the source of current supply, might introduce a hazard shall be protected and supplied as required for lighting circuits.

See NFPA 70, *National Electrical Code®*, Articles 760 and 800 for protection requirements.

4-6.8.2.1.6 All cables with all taps and splices made shall be tested for insulation resistance when installed, but before connection to terminals. Such tests shall indicate an insulation resistance of at least 200 megohms per mile between any one conductor and all other conductors, the sheath, and the ground.

This subsection requires installers to test cables and splices with a megohm meter (megger) to ensure the dielectric strength of the insulation. Installers must conduct this test before they connect any devices or appliances to the cable plant.

4-6.8.2.2 Underground Cables.

4-6.8.2.2.1 Underground cables in duct or direct burial shall be brought aboveground only at points where liability of mechanical injury or of disablement from heat incident to fires in adjacent buildings is minimized.

4-6.8.2.2.2 Cables shall be permitted in duct systems and manholes containing low-tension fire alarm system conductors only, except low-tension secondary power cables shall be permitted. Where in duct systems or manholes containing power circuit conductors in excess of 250 volts to ground, fire alarm cables shall be located as far as possible from such power cables and shall be separated from them by a noncombustible barrier or by such other means as is practicable to protect the fire alarm cables from injury.

4-6.8.2.2.3 All cables installed in manholes shall be properly racked and marked for identification.

4-6.8.2.2.4 All conduits or ducts entering buildings from underground duct systems shall be effectively sealed against moisture or gases entering the building.

4-6.8.2.2.5 Cable joints shall be located only in manholes, fire stations, and other locations where proper accessibility is provided and where there is little liability of injury to the cable due to either falling walls or operations in the buildings. Cable joints shall be made to provide and maintain conductivity, insulation, and protection at least equal to that afforded by the cables that are joined. Cable ends shall be sealed against moisture.

4-6.8.2.2.6 Direct-burial cable, without enclosure in ducts, shall be laid in grass plots, under sidewalks, or in other places where the ground is not likely to be opened for other underground construction. If splices are made, such splices shall, where practicable, be accessible for inspection and tests. Such cables shall be buried at least 18 in. (0.5 m) deep and, where crossing streets or other areas likely to be opened for other underground construction, shall be in duct or conduit or be covered by creosoted planking of at least 2 in. × 4 in. (50 mm × 100 mm) with half-round grooves, spiked or banded together after the cable is installed.

To maintain the overall operational integrity of the public fire alarm reporting system, the requirements of subsections 4-6.8.2.2.1 through 4-6.8.2.2.6 help to ensure that underground cables will not be exposed to undue potential for mechanical injury.

4-6.8.2.3 Aerial Construction.

4-6.8.2.3.1 Fire alarm wires shall be run under all other wires except communications wires. Suitable precautions shall be provided where passing through trees, under bridges, over railroads, and at other places where injury or deterioration is possible. Wires and cables shall not be attached to a crossarm carrying electric light and power wires, except circuits carrying up to 220 volts for municipal communications use. Such 220-volt circuits shall be tagged or otherwise identified.

4-6.8.2.3.2 Aerial cable shall be supported by messenger wire of adequate tensile strength.

Exception: Where permitted in 4-6.8.2.3.3.

4-6.8.2.3.3 Two-conductor cable shall be messenger-supported.

Exception: Where two-conductor cable has conductors of No. 20 AWG or larger size and has mechanical strength equivalent to No. 10 AWG hard-drawn copper.

4-6.8.2.3.4 Single wire shall meet International Municipal Signal Association specifications and shall not be smaller than No. 10 Roebling gauge where of galvanized iron or steel, No. 10 AWG where of hard-drawn copper, No. 12 AWG where of approved copper-covered steel, or No. 6 AWG where of aluminum. Span lengths shall not exceed manufacturers' recommendations.

4-6.8.2.3.5 Wires to buildings shall contact only intended supports and shall enter through an approved weatherhead or suitable sleeves slanting upward and inward. Drip loops shall be formed on wires outside of buildings.

4-6.8.2.4 Leads Down Poles.

4-6.8.2.4.1 Leads down poles shall be protected against mechanical injury. Any metallic covering shall form a continuous conducting path to ground. Installation, in all cases, shall prevent water from entering the conduit or box.

4-6.8.2.4.2 Leads to boxes shall have 600-volt insulation approved for wet locations, as defined in NFPA 70, *National Electrical Code*.

To maintain the overall operational integrity of the public fire alarm reporting system, the requirements of subsections 4-6.8.2.3, 4-6.8.2.4, and their related subsections help to ensure that nothing exposes aerial cables and leads down poles, reducing the risk of both mechanical injury and electrical failure.

4-6.8.2.5 Wiring Inside Buildings.

4-6.8.2.5.1 At the communications center, conductors shall extend as directly as possible to the operating room in conduits, ducts, shafts, raceways, or overhead racks and troughs of a type of construction affording protection against fire and mechanical injury.

4-6.8.2.5.2 All conductors inside buildings shall be in conduit, electrical tubing, metal molding, or raceways. Installation shall be in accordance with NFPA 70, *National Electrical Code*.

4-6.8.2.5.3 Conductors shall have an approved insulation; the insulation or other outer covering shall be flame-retardant and moisture-resistant.

4-6.8.2.5.4 Conductors shall be installed as far as possible without joints. Splices shall be permitted only in junction or terminal boxes. Wire terminals, splices, and joints shall conform to NFPA 70, *National Electrical Code*.

4-6.8.2.5.5 Conductors bunched together in a vertical run connecting two or more floors shall have a flame-retardant covering sufficient to prevent the carrying of fire from floor to floor.

Exception: This requirement shall not apply where the conductors are encased in a metallic conduit or located in a fire-resistive shaft having fire stops at each floor.

4-6.8.2.5.6 Where cable or wiring is exposed to unusual fire hazards, it shall be properly protected.

4-6.8.2.5.7 Cable terminals and cross-connecting facilities shall be located in or adjacent to the operations room.

4-6.8.2.5.8 Where signal conductors and electric light and power wires are run in the same shaft, they shall be separated by at least 2.0 in. (51 mm), or either system shall be encased in a noncombustible enclosure.

To maintain the overall operational integrity of the public fire alarm reporting system, the requirements of subsection 4-6.8.2.5 and its related subsections help to ensure that nothing exposes wiring inside a building, preventing both the risk of mechanical injury or electrical failure. The requirements also make certain that the wiring will not contribute to the spread of a fire in a building.

4-6.9 Facilities for Signal Transmission.

4-6.9.1 Circuits.

4-6.9.1.1 General.

4-6.9.1.1.1 ANSI/IEEE C2, *National Electrical Safety Code*, shall be used as a guide for the installation of outdoor circuitry.

Public and private electric company utilities, public and private telephone utilities, and public and private community antenna television company utilities use this document. A committee from the Institute of Electrical and Electronic Engineers, developed the document. ANSI/IEEE C2 describes the placement and spacing of outdoor aerial cable installations to ensure safe operation of the associated systems.

4-6.9.1.1.2   In all installations, first consideration shall be given to continuity of service. Particular attention shall be given to liability of mechanical injury; disablement from heat that is incident to a fire; injury by falling walls; and damage by floods, corrosive vapors, or other causes.

4-6.9.1.1.3   Open local circuits within single buildings shall be permitted in accordance with Chapter 3.

Fire alarm system circuits that are not a part of the public fire alarm reporting system must be installed in accordance with Chapter 3 of the Code.

4-6.9.1.1.4   All circuits shall be so routed as to allow ready tracing of circuits for trouble.

4-6.9.1.1.5   Circuits shall not pass over, under, through, or be attached to buildings or property not owned by or under the control of the authority having jurisdiction or the agency responsible for maintaining the system.

Exception: Where the circuit is terminated in a box on the premises.

Installers strung the circuits for many of the original public fire alarm reporting systems in various East Coast municipalities throughout a city from building to building. Engineers soon learned that fires in those buildings would damage the circuits, thereby placing the operational integrity of the public fire alarm reporting system in jeopardy.

4-6.9.1.2   Box Circuits.   Accessible and reliable means, available only to the authority having jurisdiction or the agency responsible for maintaining the public fire alarm reporting system, shall be provided for disconnecting the auxiliary loop to the box inside the building, and definite notification shall be given to occupants of the building when the interior box is not in service.

If an installer makes a connection to the public fire alarm reporting system in accordance with the requirements of Section 4-7, he or she must provide a means to disconnect the protected premises connection. Only the authority having jurisdiction over the public fire alarm reporting system may have access to this disconnecting means.

4-6.9.1.3   Tie Circuits.

Tie circuits connect the public fire service communication center with a subsidiary communication center. For example, in a large municipality, a subsidiary communication center concentrates signals from a particular neighborhood before transmitting them to the public fire service communication center. Also, in some cities where several boroughs have their own public fire service communication center (New York City, for example), the system may use tie circuits to interconnect the centers. This allows the centers to handle signals even with one of the centers impaired.

4-6.9.1.3.1   A separate tie circuit shall be provided from the communications center to each subsidiary communications center.

4-6.9.1.3.2   The tie circuit between the communications center and the subsidiary communications center shall not be used for any other purpose.

4-6.9.1.3.3   In a Type B wire system, where all boxes in the system are of the succession type, it shall be permitted to use the tie circuit as a dispatch circuit to the extent permitted by NFPA 1221, *Standard for the Installation, Maintenance, and Use of Public Fire Service Communication Systems.*

4-6.9.1.4*   **Circuit Protection.**

A-4-6.9.1.4   All requirements for circuit protection do not apply to coded radio reporting systems. These systems do not use metallic circuits.

Circuit protection limits equipment damage caused when transient currents are applied to the circuits of the public fire alarm reporting system. Lightning is as one source of such transients.

4-6.9.1.4.1   General.

4-6.9.1.4.1.1   The protective devices shall be located close to or be combined with the cable terminals.

4-6.9.1.4.1.2   Lightning arresters suitable for the purpose shall be provided. Lightning arresters shall be marked with the name of the manufacturer and model designation.

4-6.9.1.4.1.3   All lightning arresters shall be connected to a suitable ground in accordance with NFPA 70, *National Electrical Code.*

4-6.9.1.4.1.4   All fuses shall be plainly marked with their rated ampere capacity. All fuses rated over 2 amperes shall be of the enclosed type.

4-6.9.1.4.1.5 Circuit protection required at the communications center shall be provided in every building housing communications center equipment.

4-6.9.1.4.1.6 Each conductor entering a fire station from partially or entirely aerial lines shall be protected by a lightning arrester.

4-6.9.1.4.2 Communications Center.

4-6.9.1.4.2.1 All conductors entering the communications center shall be protected by the following devices, in the order named, starting from the exterior circuit:

(a) A fuse rated at 3 amperes minimum to 7 amperes maximum, and not less than 2000 volts;

(b) A lightning arrester;

(c) A fuse or circuit breaker, rated at $1/2$ ampere.

4-6.9.1.4.2.2 The $1/2$-ampere protection on the tie circuits shall be omitted at subsidiary communications centers.

4-6.9.1.4.3 Protection on Aerial Construction.

4-6.9.1.4.3.1 At junction points of open aerial conductors and cable, each conductor shall be protected by a lightning arrester of the weatherproof type. There also shall be a connection between the lightning arrester ground, any metallic sheath, and messenger wire.

4-6.9.1.4.3.2 Aerial open-wire and non-messenger-supported two-conductor cable circuits shall be protected by a lightning arrester at intervals of approximately 2000 ft (610 m).

4-6.9.1.4.3.3 Lightning arresters, other than of the air-gap or self-restoring type, shall not be installed in fire alarm circuits.

4-6.9.1.4.3.4 All protective devices shall be accessible for maintenance and inspection.

4-6.10 Power.

4-6.10.1 Requirements for Constant-Current Systems.

The coded wired public fire alarm reporting system operates at a constant current of nominally 100 milliamperes. The requirements of subsections 4-6.10.1.1 through 4-6.10.1.4 help to ensure that such a system will maintain a high level of operational integrity.

4-6.10.1.1 Means shall be provided for manually regulating the current in box circuits so that the operating current is maintained within 10 percent of normal throughout changes in external circuit resistance from 20 percent above to 50 percent below normal.

4-6.10.1.2 The voltage supplied to maintain normal line current on box circuits shall not exceed 150 volts, measured under no-load conditions, and shall be such that the line current cannot be reduced below safe operating value by the simultaneous operation of four boxes.

4-6.10.1.3 Visual and audible means to indicate a 20 percent or greater reduction in the normal current in any alarm circuit shall be provided. All devices connected in series with any alarm circuit shall function properly when the alarm circuit current is reduced to 70 percent of normal.

4-6.10.1.4 Sufficient meters shall be provided to indicate the current in any box circuit and the voltage of any power source. Meters used in common for several circuits shall be provided with cut-in devices designed to reduce the probability of cross-connecting circuits.

4-6.11 Receiving Equipment—Facilities for Receipt of Box Alarms.

The Technical Committee based the requirements of the following subsections on the premise that the public fire service communication center will receive the signals transmitted over the public fire alarm reporting system and will automatically record them in a manner that provides a permanent visual record of the signals. At the same time, an audible notification appliance will alert the operators to incoming signals.

The signal will indicate the exact location of its origin. This indication comes from a unique number assigned to each public fire alarm box. An operator reading a manual chart, or an interface to a computer-aided dispatching system—NFPA 1221 covers these requirements—will translate the box number to an exact location.

4-6.11.1 General.

4-6.11.1.1 Alarms from boxes shall be automatically received and recorded at the communications center.

4-6.11.1.2 A permanent visual record and an audible signal shall be required to indicate the receipt of an alarm. The permanent record shall indicate the exact location from which the alarm is being transmitted.

NOTE: The audible signal device can be common to several box circuits and arranged so that the fire alarm operator can manually silence the signal temporarily by a self-restoring switch.

4-6.11.1.3 Facilities shall be provided that automatically record the date and time of receipt of each alarm.

Exception: Only the time shall be required to be automatically recorded for voice recordings.

4-6.11.2 Visual Recording Devices.

4-6.11.2.1 A device for producing a permanent graphic recording of all alarm, supervisory, trouble, and test signals received or retransmitted, or both, shall be provided at each communications center for each alarm circuit and tie circuit. Where each circuit is served by a dedicated recording device, the number of reserve recording devices required on site shall be equal to at least 5 percent of the circuits in service and in no case less than 1 percent. Where two or more circuits are served by a common recording device, a reserve recording device shall be available on site for each circuit connected to a common recorder.

4-6.11.2.2 In a Type B wire system, one such recording device shall be installed in each fire station and at least one shall be installed in the communications center.

4-6.12 Supervision.

The requirements of subsections 4-6.12.1 through 4-6.12.5 help to ensure that any impairment to the circuits or power supplies of the public fire alarm reporting system will result in a trouble signal at the public fire service communication center.

4-6.12.1 To ensure reliability, wired circuits upon which transmission and receipt of alarms depend shall be under constant electrical supervision to give prompt warning of conditions adversely affecting reliability.

4-6.12.2 The power supplied to all required circuits and devices of the system shall be supervised.

4-6.12.3 Trouble signals shall actuate a sounding device located where there is a responsible person on duty at all times.

4-6.12.4 Trouble signals shall be distinct from alarm signals and shall be indicated by both a visual light and an audible signal.

NOTE 1: The audible signal can be common to several supervised circuits.

NOTE 2: A switch for silencing the audible trouble signal is permitted, provided the visual signal remains operating until the silencing switch is restored to its normal position.

4-6.12.5 The audible signal shall be responsive to faults on any other circuits that occur prior to restoration of the silencing switch to normal.

4-6.13 Coded Wired Reporting Systems.

The following subsections (4-6.13, 4-6.14, 4-6.15, and 4-6.16) give the requirements for each of the four types of public fire alarm reporting systems: coded wired, coded radio, telephone (series), and telephone (parallel).

4-6.13.1 For a Type B system, the effectiveness of noninterference and succession functions between box circuits shall be no less than between boxes in any one circuit. The disablement of any metallic box circuit shall cause a warning signal in all other circuits, and, thereafter, the circuit or circuits not otherwise broken shall be automatically restored to operative condition.

In a Type B coded wired system, the system "repeats" signals on one box circuit or alarm circuit on to the other box and alarm circuits. By repeating these signals, other boxes will sense the busy circuit and wait for a clear circuit before transmitting. This ensures the proper functioning of the non-interfering and successive features.

4-6.13.2 Box circuits shall be sufficient in number and so laid out that the areas that would be left without box protection in case of disruption of a circuit do not exceed those covered by 20 properly spaced boxes where all or any part of the circuit is of aerial open-wire, or 30 properly spaced boxes where the circuit is entirely in underground or messenger-supported cable.

4-6.13.3 Where all boxes on any individual circuit and associated equipment are designed and installed to provide for receipt of alarms through the ground in the event of a break in the circuit, the circuit shall be permitted to serve twice the number of aerial open-wire and cable circuits, respectively, as are specified in 4-6.13.2.

In most coded wired systems, when a fire alarm box senses that the circuit has an open fault, it will idle for one round and then connect the box to earth ground. Sensing an open circuit, the receiving equipment at the public fire service communication center will also connect itself to earth ground. This "conditioning" of the circuit will allow the box to transmit its signal through earth ground. Of course, if two open faults occur on the circuit, the boxes isolated between the faults cannot transmit a signal.

4-6.13.4   The installation of additional boxes in an area served by the number of properly spaced boxes indicated above shall not constitute geographical overloading of a circuit.

Once an installer has properly spaced boxes throughout an area, adding more boxes will not overload the circuit. An installer might add more boxes when he or she installs master fire alarm boxes to allow the connection of protected premises fire alarm systems to the public fire alarm reporting system.

4-6.13.5   Sounding devices for signals shall be provided for box circuits.

NOTE 1:   In a Type A system, it is satisfactory to use a common sounding device for more than one circuit, and it should be installed at the communications center.

NOTE 2:   In a Type B system, a sounding device is to be installed in each fire station at the same location as the recording device for that circuit, unless installed at the communications center, where a common sounding device is permitted.

4-6.14   **Coded Radio Reporting Systems.**

4-6.14.1   **Radio Box Channel (Frequency).**

4-6.14.1.1   The number of boxes permitted on a single frequency shall be governed by the following:

(a)   For systems utilizing one-way transmission in which the individual box automatically initiates the required message (*see* 4-6.14.6.3) using circuitry integral to the boxes, not more than 500 boxes shall be permitted on a single frequency.

(b)   For systems utilizing a two-way concept in which interrogation signals (*see 4-6.14.6.3*) are transmitted to the individual boxes from the communications center on the same frequency used for receipt of alarms, not more than 250 boxes shall be permitted on a single frequency. Where interrogation signals are transmitted on a frequency that differs from that used for receipt of alarms, not more than 500 boxes shall be permitted on a single frequency.

(c)   A specific frequency shall be designated for both fire and other fire-related or public safety alarm signals, and supervisory signals (test and tamper). All acknowledgment and other signals shall utilize a separate frequency.

This requirement prevents the public fire service communication center from using the radio frequency assigned to the boxes for normal two-way or one-way radio communication. Such use might inadvertently interfere with receipt of signals from the boxes.

4-6.14.1.2   Where box message signals to the communications center or acknowledgment of message receipt signals from the communications center to the box are repeated, associated repeating facilities shall conform to the requirements indicated in NFPA 1221, *Standard for the Installation, Maintenance, and Use of Public Fire Service Communication Systems, 3-4.1.2.*

4-6.14.2   **Metallic Interconnections.** Accessible and reliable means, available only to the agency responsible for maintaining the public fire alarm reporting system, shall be provided for disconnecting the auxiliary loop to the box inside the building, and definite notification shall be given to occupants of the building when the interior box is not in service.

If an installer makes a connection to the public fire alarm reporting system in accordance with the requirements of section 4-7, he or she must provide a means to disconnect the protected premises connection. Only the authority having jurisdiction over the public fire alarm reporting system may have access to this disconnecting means.

4-6.14.3   **Receiving Equipment—Facilities for Receipt of Box Alarms.**

4-6.14.3.1   **Type A System.**

4-6.14.3.1.1*   For each frequency used, two separate receiving networks, each including an antenna, an audible alerting device, a receiver, a power supply, signal processing equipment, a means of providing a permanent graphic recording of the incoming message that is both timed and dated, and other associated equipment shall be provided and shall be installed at the communications center. Facilities shall be so arranged that a failure of either receiving network cannot affect the receipt of messages from boxes. *(See figure on next page.)*

These requirements for redundant receiving equipment help to ensure the overall reliability of the coded radio public fire alarm reporting system.

Polling required for transpondence-(two-way) type systems only.

Figure A-4-6.14.3.1.1   *Type A system receiving networks.*

4-6.14.3.1.2 Where the system configuration is such that a polling device is incorporated into the receiving network to allow remote/selective initiation of box tests (*see Chapter 7*), a separate such device shall be included in each of the two required receiving networks. Furthermore, the polling devices shall be configured for automatic cycle initiation in their primary operating mode, capable of continuous self-monitoring, and integrated into the network(s) to provide automatic switchover and operational continuity in the event of failure of either device.

Some coded radio systems provide for an interrogation and response sequence initiated from the public fire service communication center to ensure the operational integrity of the system. When the system provides such an arrangement, the requirements of subsection 4-6.14.3.1.2 indicate the need for redundancy to further ensure the integrity of the system.

4-6.14.3.1.3 Test signals from boxes shall not be required to include the date as part of their permanent recording, provided that the date is automatically printed on the recording tape at the beginning of each calendar day.

4-6.14.3.2 **Type B System.**

4-6.14.3.2.1 For each frequency used, a single, complete receiving network shall be permitted in each fire station, provided the communications center conforms to 4-6.14.3.1.1. Where the jurisdiction maintains two or more alarm reception points in operation, one receiving network shall be permitted to be at each alarm reception point.

4-6.14.3.2.2 Where alarm signals are transmitted to a fire station from the communications center using the coded radio-type receiving equipment in the fire station to receive and record the alarm message, a second receiving network conforming to 4-6.14.3.2.1 shall be provided at each fire station, and that receiving network shall employ a frequency other than that used for the receipt of box messages.

4-6.14.4 **Power.** Power shall be provided in accordance with 4-6.7.

4-6.14.5 **Testing.** See Chapter 7.

4-6.14.6 **Supervision.**

4-6.14.6.1 All coded radio box systems shall provide constant monitoring of the frequency in use. Both an audible and a visual indication of any sustained carrier signal, where in excess of 15 seconds' duration, shall be provided for each receiving system at the communications center.

An open fault or ground fault on a coded wired public fire alarm reporting system would interfere with the transmission of signals. Similarly, a sustained carrier would interfere with the transmission of signals from the boxes on a coded radio public fire alarm reporting system.

4-6.14.6.2 The power supplied to all required circuits and devices of the system shall be supervised.

4-6.14.6.3 Each coded radio box shall automatically transmit a test message at least once in each 24-hour period.

The 24-hour test signal safeguards against the catastrophic failure of a single fire alarm box.

4-6.14.6.4 Receiving equipment associated with coded radio-type systems, including any related repeater(s), shall be tested at least hourly. The receipt of test messages shall be considered sufficient to comply with this requirement, provided at least one such message is received each hour.

4-6.14.6.5 Radio repeaters upon which receipt of alarms depend shall be provided with dual receivers and transmitters. Failure of the primary transmitter or receiver shall cause an automatic switchover to the secondary receiver and transmitter.

Exception: Where the repeater controls are located where an individual is always on duty, manual switchover shall be permitted, provided it can be completed within 30 seconds.

4-6.14.6.6 Trouble signals shall actuate a sounding device located where there is always a responsible person on duty.

4-6.14.6.7 Trouble signals shall be distinct from alarm signals and shall be indicated by both a visual light and an audible signal.

NOTE 1:   The audible signal may be permitted to be common to several supervised circuits.

NOTE 2:   A switch for silencing the audible trouble signal may be permitted where the visual signal remains operating until the silencing switch is restored to its normal position.

4-6.14.6.8 The audible signal shall be responsive to faults on any other circuits that might occur prior to restoration of the silencing switch to normal.

4-6.15 Telephone (Series) Reporting Systems.

Sometimes installers add components to all or a portion of an existing coded wired reporting system to give that system the capability of transmitting and receiving voice alarms. In these cases, the telephone (series) system uses the same cable plant as the coded wired system.

4-6.15.1 A permanent visual recording device installed in the communications center shall be provided to record all incoming box signals. A spare recording device shall be provided for five or more box circuits.

This permanent visual recording device records the date, time, and box number, but not the content of the voice message. See subsection 4-6.15.4.

4-6.15.2 A second visual means of identifying the calling box shall be provided.

4-6.15.3 Audible signals shall indicate all incoming calls from box circuits.

4-6.15.4 All voice transmissions from boxes for emergencies shall be recorded with the capability of instant playback.

Special audio recording equipment has been designed that not only provides an audio log of signal content from the boxes, but also allows an operator to instantly recycle to the beginning of each message. This allows operators at the public fire service communication center to review messages with unclear content.

4-6.15.5 A voice recording facility shall be provided for each operator handling incoming alarms to eliminate the possibility of interference.

4-6.15.6 Box circuits shall be sufficient in number and so laid out that the areas that would be left without box protection in case of disruption of a circuit do not exceed those covered by 20 properly spaced boxes where all or any part of the circuit is of aerial open-wire, or 30 properly spaced boxes where the circuit is entirely in underground or messenger-supported cable.

4-6.15.7 Where all boxes on any individual circuit and associated equipment are designed and installed to provide for receipt of alarms through the ground in the event of a break in the circuit, the circuit shall be permitted to serve twice the number of aerial open-wire and cable circuits, respectively, as are specified in 4-6.15.6.

In some telephone (series) systems, when a fire alarm box senses that the circuit has an open fault, it will idle for one round and then connect the box to earth ground. Sensing an open circuit, the receiving equipment at the public fire service communication center will also connect itself to earth ground. This "conditioning" of the circuit will allow the box to transmit its signal through earth ground. Of course, if two open faults occur on the circuit, the boxes isolated between the faults cannot transmit a signal.

4-6.15.8 The installation of additional boxes in an area served by the number of properly spaced boxes indicated above shall not constitute geographical overloading of a circuit.

Once an installer has properly spaced boxes throughout an area, adding more boxes will not overload the circuit. An installer might add more boxes when he or she installs master fire alarm boxes to allow the connection of protected premises fire alarm systems to the public fire alarm reporting system.

4-6.16 Telephone (Parallel) Reporting Systems.

Owners of systems typically lease telephone (parallel) public fire alarm reporting systems from the public telephone utility. In most systems, each box has its own circuit that leads back through the public telephone utility wire centers to the public fire service communication center. However, in some cases, each box connects to a concentrator-identifier. This concentrator-identifier, in turn, connects to the public fire service communication center.

4-6.16.1 Box Circuits.

4-6.16.1.1 Where a municipal box is installed inside a building, it shall be placed as close as practical to the point of entrance of the circuit, and the exterior wire shall be installed in conduit or electrical metallic tubing, in accordance with Chapter 3 of NFPA 70, *National Electrical Code.*

4-6.16.1.2 Accessible and reliable means, available only to the authority having jurisdiction or the agency responsible for maintaining the public fire alarm reporting system, shall be provided for disconnecting the box inside the building, and definite notification shall be given to occupants of the building when the interior box is not in service.

If an installer makes a connection to the public fire alarm reporting system in accordance with the requirements of section 4-7, he or she must provide a means to disconnect the protected premises connection. Only the authority having jurisdiction over the public fire alarm reporting system may have access to this disconnecting means.

4-6.16.1.3 A separate circuit shall be provided for each box.

4-6.16.1.4 Where a concentrator-identifier or similar device is employed, at least two tie circuits for the first 40 boxes connected shall be provided to the communications center. A tie circuit shall be provided for each 40 additional boxes, or fraction thereof, connected to the concentrator-identifier.

NOTE: These tie circuits are not to be used for any other purpose or function.

A concentrator-identifier connects a group of boxes in a particular geographic area to the public fire service communication center. Systems use this device where an excessively long circuit would extend from each box to the communication center. The concentrator-identifier connects to the communication center by means of at least two tie circuits. The concentrator-identifier precisely indicates at the communication center exactly which box has been operated.

4-6.16.1.5 Power shall be provided in accordance with Section 4-6.7.

4-6.16.2 Receiving Equipment—Facilities for Receipt of Box Alarms.

4-6.16.2.1 The box circuits shall be terminated:

(a) Directly on a console or switchboard located in the communications center; or

(b) In concentrator-identifier equipment located in a subsidiary communications center.

NOTE: The audible signal device can be common to several box circuits and arranged so that the operator can manually silence the signal temporarily with a self-restoring switch.

4-6.16.2.2 All voice transmissions from boxes for emergencies shall be recorded with the capability of instant playback.

Special audio recording equipment has been designed that not only provides an audio log of signal content from the boxes, but also allows an operator to instantly recycle to the beginning of each message. This allows operators at the public fire service communication center to review messages with unclear content.

4-6.16.2.3 A means of voice recording shall be provided for each operator handling incoming alarms to eliminate the possibility of interference.

4-6.16.2.4 Either a continuous line test or periodic (up to 6 minutes) automatic line tests shall detect an open, short, ground, or leakage condition. When one of these conditions occurs, a visual and audible trouble signal shall be actuated where there is an operator on duty.

The equipment supplied by the public telephone utility either continuously monitors the integrity of each circuit for the faults stated in 4-6.16.2.4 or it provides a test for these faults no less frequently than every 6 minutes.

4-7 Auxiliary Fire Alarm Systems.

NOTE: The requirements of Chapters 1 and 7 apply to auxiliary fire alarm systems, unless they conflict with the requirements of this section.

4-7.1   Scope. This section describes the equipment and circuits necessary to connect a protected premises (*see* Chapter 3) to a public fire alarm reporting system (*see Section* 4-6).

4-7.2   General.

4-7.2.1 An auxiliary fire alarm system shall be used only in connection with a public fire alarm reporting system that is suitable for the service. A system satisfactory to the authority having jurisdiction shall be considered as meeting this requirement.

If a community has not provided a public fire alarm reporting system, no auxiliary fire alarm system can exist. An auxiliary system depends on the public fire alarm reporting system to transmit signals from the protected premises to the public fire service communication center.

4-7.2.2 Permission for the connection of an auxiliary fire alarm system to a public fire alarm reporting system, and acceptance of the type of auxiliary transmitter and its actuating mechanism, circuits, and components connected thereto, shall be obtained from the authority having jurisdiction.

4-7.2.3 An auxiliary fire alarm system shall be maintained and supervised by a responsible person or corporation.

Proper maintenance of an auxiliary system requires careful coordination with those who operate and maintain the public fire alarm reporting system.

4-7.2.4 Section 4-7 does not require the use of audible alarm signals other than those necessary to operate the auxiliary fire alarm system. Where it is desired to provide fire alarm evacuation signals in the protected property, the alarms, circuits, and controls shall comply with the provisions of Chapter 3, in addition to the provisions of Section 4-7.

An auxiliary system will not, itself, notify occupants of a fire alarm. If the authority having jurisdiction requires such notification, then a protected premises fire alarm system with notification appliances must be installed in accordance with Chapters 1, 3, 5, 6 and 7.

4-7.3   Communications Center Facilities. The communications center facilities shall be in accordance with the requirements of Section 4-6.

4-7.4   Equipment.

4-7.4.1   Types of Systems. There are three types of auxiliary fire alarm systems as follows:

(a)* *Local Energy Type.*

A-4-7.4.1(a) The local energy-type system [*see* Figures *A-4-7.4.1(a)(1) and A-4-7.4.1(a)(2)*] is electrically isolated from the public fire alarm reporting system and has its own power supply. The tripping of the transmitting device does not depend on the current in the system. In a wired circuit, receipt of the alarm by the communications center when the circuit is accidentally opened depends on the design of the transmitting device and the associated communications center equipment (i.e., whether or not the system is designed to receive alarms through manual or automatic ground operational facilities). In a radio box-type system, receipt of the alarm by the communications center depends on the proper operation of the radio transmitting and receiving equipment.

1. Local energy systems shall be permitted to be of the coded or noncoded type.

2. Power supply sources for local energy systems shall conform to Chapter 1.

(b)* *Shunt Type.*

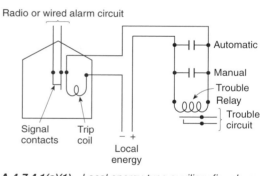

Figure A-4-7.4.1(a)(1)   *Local energy-type auxiliary fire alarm system.*

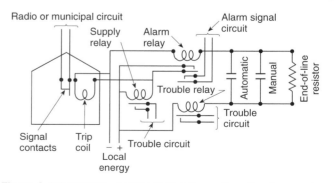

Figure A-4-7.4.1(a)(2)   *Local energy-type auxiliary fire alarm system.*

A-4-7.4.1(b) The shunt-type [*see Figures A-4-7.4.1(b) and A-4-7.4.1(b)8*] is electrically connected to, and is an integral part of, the public fire alarm reporting system. A ground fault on the auxiliary circuit is a fault on the public fire alarm reporting system circuit, and an accidental opening of the auxiliary circuit sends a needless (or false) alarm to the communications center. An open circuit in the transmitting device trip coil is not indicated either at the protected property or at the communications center; also, if an initiating device is operated, an alarm is not transmitted, but an open circuit indication is given at the communications center. If a public fire alarm reporting system circuit is open when a connected shunt-type system is operated, the transmitting device does not trip until the public fire alarm reporting system circuit returns to normal, at which time the alarm is transmitted, unless the auxiliary circuit is first returned to a normal condition.

Additional design restrictions for shunt systems are found in laws or ordinances.

1. Shunt systems shall be noncoded with respect to any remote electrical tripping or actuating devices.

2. All conductors of the shunt circuit shall be installed in accordance with NFPA 70, *National Electrical Code*, Article 346, for rigid conduit, or Article 348, for electrical metallic tubing.

3. Both sides of the shunt circuit shall be in the same conduit.

4. Where an auxiliary transmitter is located within a private premises, it shall be installed in accordance with 4-6.9.1.

5. Where a shunt loop is used, it shall not exceed a length of 750 ft (230 m) and shall be in conduit.

6. Conductors of the shunt circuits shall not be smaller than No. 14 AWG and shall be insulated as prescribed in NFPA 70, *National Electrical Code*, Article 310.

7. The power for shunt-type systems shall be provided by the public fire alarm reporting system.

8. A local system made to an auxiliary system by the addition of a relay whose coil is energized by a local power supply and whose normally closed contacts trip a shunt-type master box shall not be permitted. [*See Figure A-4-7.4.1(b)8.*]

(c)* *Parallel Telephone Type.*

A-4-7.4.1(c) A parallel telephone-type system [*see Figure A-4-7.4.1(c)*] is a system in which alarms are transmitted over a circuit directly connected to the annunciating switchboard at the public fire service communications center and terminated at the protected property by an end-of-line device.

Such auxiliary systems are for connection to public fire alarm reporting systems of the type in which each alarm box annunciates at the communications center by individual circuit.

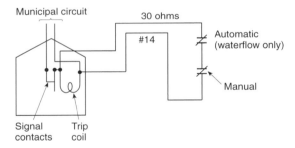

Figure A-4-7.4.1(b) *Shunt-type auxiliary fire alarm system.*

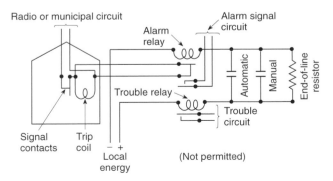

Figure A-4-7.4.1(b)8 *Shunt-type auxiliary fire alarm system.*

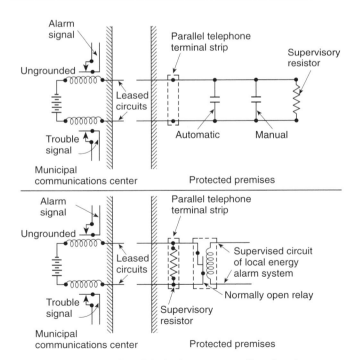

Figure A-4-7.4.1(c) *Parallel telephone-type auxiliary fire alarm system.*

NOTE: The essential difference between the local energy or parallel telephone types and the shunt-type system is that accidental opening of the alarm-initiating circuits causes an alarm on the shunt-type system only.

1. Parallel telephone systems shall be noncoded with respect to any remote electrical tripping or actuating devices.

2. Two methods of parallel telephone systems shall be permitted to be used as follows:

a.The circuits are extended beyond the entrance termination point to actuating devices, with the supervisory device beyond the last actuating device in the circuit; or

b.The supervisory device for the circuit is located at the entrance termination point. The tripping relay shall be located immediately adjacent to the supervisory device and shall be connected thereto with conductors not smaller than No. 14 AWG in conduit.

3. Nonvoice circuits connected to a parallel telephone system shall be indicated with a color that is both distinctive and different from that of voice circuits and shall be grouped in a reserved separate section of the receiving equipment with adequate written warning that no voice is to be expected on these alarms and that the fire department is to be dispatched on alarm light indications.

The detailed requirements of subsections 4-7.4.1(a) through (c) describe the three types of auxiliary systems: local energy, shunt, and parallel telephone. In each case, the requirements help ensure the operational integrity of the particular type of system. Most authorities having jurisdiction consider the shunt system the least desirable. Also, see definitions and commentary for each type of system in Section 1-4.

4-7.4.2   The interface of the three types of auxiliary fire alarm systems with the four types of public fire alarm reporting systems shall be in accordance with Table 4-7.4.2.

4-7.4.3   The application of the three types of auxiliary fire alarm systems shall be limited to the initiating devices specified in Table 4-7.4.3.

Table 4-7.4.2   Application of Public Fire Alarm Reporting Systems with Auxiliary Fire Alarm Systems

Reporting Systems	Local Energy Type	Shunt Type	Parallel Type
Coded wired	Yes	Yes	No
Coded radio	Yes	No	No
Telephone series	Yes	No	No
Telephone parallel	No	No	Yes

Table 4-7.4.3   Application of Initiating Device with Auxiliary Fire Alarm Systems

Initiating Devices	Local Energy Type	Shunt Type	Parallel Type
Manual fire alarm	Yes	Yes	Yes
Waterflow or actuation of the fire extinguishing system(s) or suppression system(s)	Yes	Yes	Yes
Automatic detection devices	Yes	No	Yes

4-7.4.4   Location of Transmitting Devices.

4-7.4.4.1   Shunt-type auxiliary systems shall be arranged so that one auxiliary transmitter does not serve more than 100,000 ft^2 (9290 m^2) total area.

Exception: Where otherwise permitted by the authority having jurisdiction.

4-7.4.4.2   A separate auxiliary transmitter shall be provided for each building or where permitted by the authority having jurisdiction for each group of buildings of single ownership or occupancy.

4-7.4.4.3   The same box shall be permitted to be used as a public fire alarm reporting system box and as a transmitting device for an auxiliary system where permitted by the authority having jurisdiction, provided that the box is located at the outside of the entrance to the protected property.

NOTE: The fire department might require the box to be equipped with a signal light to differentiate between automatic and manual operation, unless local outside alarms at the protected property serve the same purpose.

4-7.4.4.4   The transmitting device shall be located as required by the authority having jurisdiction.

4-7.4.4.5   The system shall be so designed and arranged that a single fault on the auxiliary system shall not jeopardize operation of the public fire alarm reporting system and shall not, in case of a single fault on either the auxiliary or public fire alarm reporting system, transmit a false alarm on either system.

Exception: Shunt systems. [See 4-7.4.1(b).]

4-7.5   **Personnel.**   Personnel necessary to receive and act on signals from auxiliary fire alarm systems shall be in accordance with the requirements of Section 4-6 and NFPA 1221,

Standard for the Installation, Maintenance, and Use of Public Fire Service Communication Systems.

4-7.6 **Operations.** Operations for auxiliary fire alarm systems shall be in accordance with the requirements of Section 4-6 and NFPA 1221, *Standard for the Installation, Maintenance, and Use of Public Fire Service Communication Systems.*

4-7.7 **Testing and Maintenance.** Testing and maintenance of auxiliary fire alarm systems shall be in accordance with the requirements of Chapter 7.

5

Initiating Devices

Contents, Chapter 5

This chapter deals with the sensors that provide input to the fire alarm system control unit. Fire detectors initiate the fire alarm system response to the fire. The term initiating device applies to all types of sensors, from manually operated fire alarm boxes to fire suppression system actuation monitoring switches, that provide an in-coming signal into the fire alarm system control unit.

Initiating devices are ranked by speed of response. Because small fires are easier to extinguish and produce less damage than large fires, the sooner the fire is detected, the better. However, with increased sensitivity and speed of response comes reduced stability. The criteria established here for each general type of detector reflect an effort to balance the speed and stability trade-off. Every requirement in Chapter 5 of the National Fire Alarm Code stems from the need for speed and surety of response to a fire with minimal probability that an alarm signal will result from some non-fire circumstance.

Fire detection devices do not actually respond to the fire, itself, but to some change in the ambient conditions created by the fire in the immediate vicinity of the detector. A heat detector responds to an increase in the ambient temperature in its immediate vicinity. A smoke detector responds to the presence of smoke in the air in its immediate vicinity. A flame detector responds to the influx of radiant energy that has traveled from the fire to the detector. In each case, either heat, smoke, or light must travel from the fire to the detector.

The transmission of radiant energy is relatively straight forward. However, the placement and spacing of both smoke and heat detectors is dependent upon the transfer of combustion products (heat, smoke, etc.) from the location of the fire to the detector. This transfer of smoke aerosol or heated combustion product gases and air, is described in a set of physical principals generally called "fire plume dynamics." The combustion reactions of the fire heat the air immediately above it as hot combustion product gases and radiant energy are released. The hot air and combustion product gas mixture expands and rises in an expanding column from the fire itself until it impacts the ceiling. As more hot air and combustion product gases continue to be produced by the fire below and flow upward, the plume turns and forms a "ceiling jet". This jet is a layer of hot air and combustion gases that expand radially away from the fire plume center-line as shown in Figure 5-1. This ceiling jet carries the heat and combustion product gases (smoke) to the heat detector or smoke detector mounted on the ceiling.

Most of the placement and spacing criteria established by this Code for automatic fire detectors are intended to place a sufficient number of detectors in the proper location so that there is a detector in the path of heat, smoke, or radiant energy regardless of where the fire is located within the protected area. The criteria established for the installation of each type of detection device are based upon years of experience and the best research data currently available.

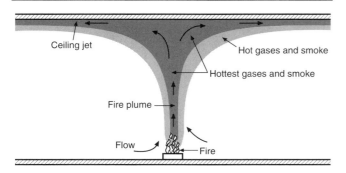

Figure 5.1 *The location and placement of heat and smoke detectors is determined by fire plume dynamics. It is the fire plume that carries the heat and smoke to the detector. Drawing courtesy of J.M.Cholin Consultants, Inc., Oakland, NJ.*

5-1 Introduction.

5-1.1 Scope. This chapter covers minimum requirements for performance, selection, use, and location of automatic fire detection devices, sprinkler waterflow detectors, manually activated fire alarm stations, and supervisory signal initiating devices, including guard tour reporting used to ensure timely warning for the purposes of life safety and the protection of a building, space, structure, area, or object.

NOTE: For detector requirements in a household system, refer to Chapter 2.

It is crucial to understand the scope limitations of Chapter 5. When this Code is referenced in laws or ordinances, the requirements of the Code assume the "effect of law." The scope statement determines whether the provisions of this chapter are applicable to a device in question.

As used here, the term "initiating device" has been broadened to cover not only traditional fire detection devices, but also other devices that monitor some condition related to fire safety. This includes sprinkler system flow switches, pressure switches, valve tamper switches, manual alarm stations, municipal fire alarm boxes, and any signaling switches used to monitor special extinguishing systems. The requirements in Section 5-1 are applied to all monitoring devices covered in 5-1.1 that provide information, either in the form of a switch transition or electronic code, to a fire alarm control unit.

5-1.2 Purpose.

5-1.2.1 The material in this chapter is intended for use by persons knowledgeable in the application of fire detection and fire alarm systems/services.

5-1.2.2 Automatic and manual initiating devices contribute to life safety, fire protection, and property conservation only where used in conjunction with other equipment. The interconnection of these devices with control equipment configurations and power supplies, or with output systems responding to external actuation, is detailed elsewhere in this code or other appropriate NFPA codes and standards.

Chapter 5 of the *National Fire Alarm Code* covers the requirements relevant to the installation of fire alarm and supervisory initiating devices when they are required by some other code or standard, such as NFPA 101, *Life Safety Code.*® The requirements also apply to any detection devices installed as part of a new or existing "non-required" fire alarm system. Chapter 5 establishes the selection and placement criteria that determine the necessary number and type of detectors, but does not address which types of facilities need initiating devices. The requirement for detection or some form of initiating device is established in the codes and standards that cover a specific class of occupancy or, in some cases, a class of fire protection system. Detection requirements may also be initiated by the property owner, property insurance carrier or other authorities having jurisdiction. Once this requirement has been established for detection devices to be installed in a property, the reader could refer to this Code for the specifics of selection and placement.

For example, NFPA 664, *Standard for the Prevention of Fires and Explosions in Wood Processing and Woodworking Facilities*, requires the use of spark/ember detectors in certain specific instances. The reader of NFPA 664 must then refer to this chapter of the *National Fire Alarm Code* for the relevant installation requirements for spark/ember detectors. In a second example, Section 28 of NFPA 101, which covers industrial facilities, requires a fire alarm system when there are more than 100 people on site or more than 25 people on a floor other than the ground floor. The fire protection designer of such an industrial site must then refer to NFPA 72 for the relevant requirements for that fire alarm system. The designer must refer to this chapter for the determination of the type, quantity, and placement of the fire detection devices. The designer must also refer to this chapter for the installation requirements for flow switches, pressure switches, and other initiating devices that may be required by NFPA 13, *Standard for the Installation of Sprinkler Systems*, or NFPA 12, *Standard on Carbon Dioxide Extinguishing Systems*, or other such standards. When a designer is placing detection in a specific area or to protect from a specific hazard, the detection devices to be installed must follow the requirements outlined in this Code.

5-1.3 Installation and Required Location of Initiating Devices.

This portion of Chapter 5 begins with some basic requirements that apply to all initiating devices, regardless of type. These requirements have come from years of experience relating to the installation of heat, smoke, or radiant energy-sensing detectors. However, with the broadening of the scope of this chapter to cover supervisory switches, manual fire alarm boxes, and other types of initiating devices, these general requirements are applicable to those types of initiating devices also.

5-1.3.1   Where subject to mechanical damage, an initiating device shall be protected. A mechanical guard used to protect a smoke or heat detector shall be listed for use with the detector being used.

A prudent designer and installer would apply this requirement to every component of the fire alarm system. The cause of many unwanted alarms as well as system failures has been found to be the result of damage to a detector or other initiating device. See Figures 5.2 and 5.3 for examples of protected detectors.

Mechanical damage is not necessarily limited to catastrophic destruction. Mechanical damage can occur over an extended period of time from vibration, extremes in temperature, corrosive atmospheres, other chemical reactions or excessive humidity. The designer and installer must be sure that the initiating device will be appropriate for the environment in which it is to be installed.

5-1.3.2   In all cases, initiating devices shall be supported independently of their attachment to the circuit conductors.

This requirement is to be applied to all types of initiating devices. The copper used in the wiring conductors is not formulated to serve as a mechanical support. Copper, by its very nature, fatigues over time if placed under a mechanical stress. This fatigue results in increasing brittleness and increasing electrical resistance over time. Ultimately, the fatigued conductor either breaks or its

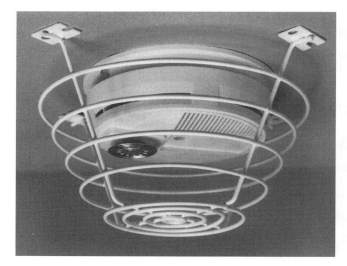

Figure 5.2   *Combination smoke and heat detector with protective cover. Photograph courtesy of Mammoth Fire Alarms (ESL Sentrol, Inc.), Lowell, MA.*

Figure 5.3   Rate-of-rise heat detector with protective cover. Photograph courtesy of Mammoth Fire Alarms (Chemetron), Lowell, MA.

resistance becomes too high to allow the initiating device circuit to function properly. In either case, the operation of the circuit is impaired, and a tragic loss of life could conceivably result because of fire alarm system failure.

Initiating devices should always be mounted in the manner shown in the manufacturer's instructions. The requirements for listing include a method for mounting that adequately supports the initiating device so that no mechanical stresses are applied to the circuit conductors. Furthermore, listing also requires that no electrical shock hazard exists when mounted pursuant to the instructions. When the instructions show use of a backbox, then it is a requirement of the listing, and the specific type of backbox shown must be used. If not shown, the use of a backbox is determined by field conditions and the requirements of NFPA 70, *National Electrical Code.*®

5-1.3.3 Initiating devices shall be installed in all areas where required by the appropriate NFPA standard or the authority having jurisdiction. Each installed initiating device shall be accessible for periodic maintenance and testing.

The first requirement of 5-1.3.3 provides correlation between the *National Fire Alarm Code* and other codes and standards. Initiating devices must be used wherever they are required by another code or standard. The question of how many and how they should be installed is established by this chapter. Note that the authority having jurisdiction may require initiating devices in areas where they are not necessarily required by other codes or standards. If the authority having jurisdiction makes

such a requirement, those initiating devices must also be installed in a manner consistent with this code.

The second requirement addresses the issue of accessibility of initiating devices. The prudent designer or installer should apply this to all system components. The term "accessible" is often subject to debate. For example, if initiating devices such as smoke detectors are mounted on the ceiling of an auditorium, one individual can assert that they are accessible with a scaffold while another can disagree, asserting that erecting a scaffold precludes the use of the facility for its intended purpose and is, therefore, not a viable alternative. The accessibility of a detector or other initiating device will ultimately be reflected in the ability of service personnel to perform maintenance at the required frequency, as outlined in Chapter 7.

5-1.3.4*   **Connection to the Fire Alarm System.**   Duplicate terminals or leads, or their equivalent, shall be provided on each initiating device for the express purpose of connecting into the fire alarm system to provide supervision of the connections. Such terminals or leads are necessary to ensure that the wire run is broken and that the individual connections are made to the incoming and outgoing leads or other terminals for signaling and power.

Exception: Initiating devices that provide equivalent supervision.

Traditionally, fire alarm system control units have used a small supervisory current to recognize a break in a conductor or the removal of a detector from the circuit. Under normal conditions, the supervisory current flows through the circuit. When a detector is removed or a conductor is broken, the current path is interrupted and the flow of current stops. The control unit translates this into a trouble signal.

It has been common practice in the electrical trade when installing initiating devices to remove a short section of insulation from the conductor and to loop the wire beneath the screw terminal without ever cutting the conductor. This is an unacceptable method of installation. If this method is employed, the connection to the initiating device (detector) could loosen over time, and the control unit would not be able to recognize this as a break in the circuit. Subsection 5-1.3.4 was incorporated into the code to preclude this practice.

Recently, systems using "smart" detectors and even "smarter" control units have been introduced. A microcomputer in the control unit maintains a list of the "names" of all of the initiating devices in the system. It sequentially addresses each device by name and verifies the response from that device. In this way, the control unit recognizes when an initiating device fails to respond, indicating either a device failure or a break in the wiring. This method does not depend upon the continuous flow of current. Consequently, these systems are exempt from the duplicate terminal requirement. See Figures 5.4 and 5.5.

A-5-1.3.4   Refer to Figures A-5-1.3.4(a) and (b) on pages 119–120 for proper connections of automatic fire detectors to fire alarm system initiating device circuits and power supply circuits.

It is helpful to review the equivalent circuit inside the detector when considering Figures A-5-1.3.4(a) and (b). The "generic" 4-wire detector has power supply terminals, terminals for a normally closed trouble contact, and terminals for a normally open alarm contact. Figure 5.6 shows the equivalent schematic of the detector (initiating device). The numbering of the terminals shown here is strictly illustrative and will probably not be consistent with the numbering of commercially available detectors.

Using the designations in Figure 5.6, the operating potential (voltage) for the detector is supplied to terminals 7 and 8. Within the detector there is a connection from terminal 7 to terminal 4, and from terminal 8 to terminal 3. Terminals 4 and 3 are wired to terminals 7 and 8, respectively, of the next detector on the circuit, providing operating potential (voltage) to the subsequent detectors. The application of operating potential (voltage) in the proper polarity closes a normally closed trouble contact (n.c.) between terminals 5 and 6. Within the detector there is a jumper between terminals 1 and 2. Thus, under normal operational conditions, terminals 1,

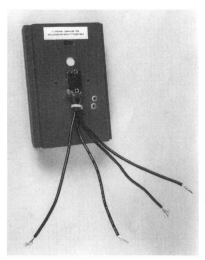

Figure 5.5   Manual fire alarm box showing incoming and outgoing leads. Photograph courtesy of Mammoth Fire Alarms (Fire Control Instruments), Lowell, MA.

2, 5, and 6 provide a circuit path for the supervisory current. Between terminals 1 and 6 there is a normally open alarm contact (n.o.) that closes when the detector senses the by-products of fire.

The normally closed contacts allow a supervisory current to flow from the control unit into terminal 6, out terminal 5, on through each detector, through the end-of-line device, and back through terminals 1 and 2 of each detector to the control unit. If an initiating device (detector) loses its source of operating potential (voltage), its trouble

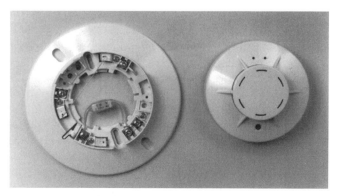

Figure 5.4   Smoke detector with base showing incoming and outgoing terminals. Photograph courtesy of Mammoth Fire Alarms (Kidde-Fenwal), Lowell, MA.

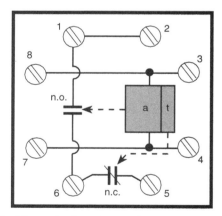

Figure 5.6   The equivalent circuit of a generic 4-wire detector. The circuitry (shaded area) is divided into two parts, a sensing part, a, which operates the alarm operated contact and a trouble indicating part, t, which operates the trouble operated contact. See commentary following A-5-1.3.4 for explanation. Drawing courtesy of J. M. Cholin Consultants, Inc. Oakland, NJ.

contact between terminals 5 and 6 opens, interrupting the current flow. If an initiating device senses a fire, the alarm contact between terminals 1 and 6 closes, bypassing the end-of-line device, which increases the current flowing through the initiating device circuit. The control unit interprets the larger flow of current as a fire alarm.

The ability to use the three-wire format or the four-wire format is determined by the initiating device input circuit of the fire alarm control unit, not the detector. Some control units use one side of the power supply as part of the initiating device circuit, others do not. The system must be wired according to the instructions provided by the manufacturer of the fire alarm system control unit. In addition, the only detectors that may be connected to a fire alarm control unit are those that have been listed as being compatible with the specific make and model control unit.

5-1.4 Requirements for Smoke and Heat Detectors.

5-1.4.1 Detectors shall not be recessed into the mounting surface in any manner.

Exception: Where tested and listed for such recessed mounting.

Recessing fire detectors has an adverse effect on the ability of the detector to perform as intended. Consider first the case of a heat detector that has been recessed, contrary to this Code. A heat detector must absorb heat from the hot gases of the ceiling jet shown in Figure 5-1 before it can respond. Approximately 95-98 percent of the heat a detector receives is carried to the detector in the hot air and combustion product gases of the ceiling jet, created by the fire. This is called convection or

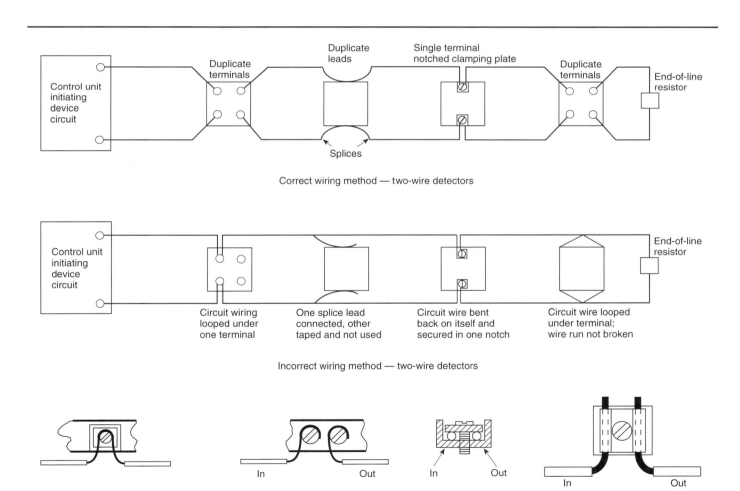

Figure A-5-1.3.4(a)   Correct wiring methods — four-wire detectors with separate power supply.

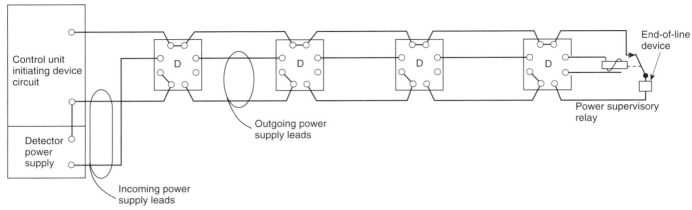

D = Detector

Illustrates four-wire smoke detector employing a three-wire connecting arrangement. One side of power supply is connected to one side of initiating device circuit. Wire run broken at each connection to smoke detector to provide supervision.

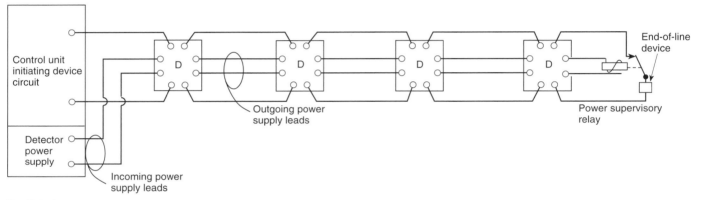

D = Detector

Illustrates four-wire smoke detector employing a four-wire connecting arrangement. Incoming and outgoing leads or terminals for both initiating device and power supply connections. Wire run broken at each connection to provide supervision.

Figure A-5-1.3.4(b)   *Wiring arrangements for three- and four-wire detectors.*

convective heat transfer. When a detector is recessed, it is removed from the flow of air, and, consequently, the quantity of heat it receives per unit of time is reduced. This slows down response, allowing the fire to grow larger before it is detected. A heat detector also receives a small percentage of radiated heat. If the detector is recessed, this radiated heat energy cannot strike the detector. The result is very slow response and a fire that grows very large before it is detected.

Other modes of fire detection are also less sensitive if they are recessed. Smoke detectors depend upon air movement to convey smoke from the fire to the detector. Usually this air movement is the ceiling jet from the fire. Smoke detectors are typically mounted on the ceiling to take advantage of the fire plume dynamics and

the ceiling jet. However, because there is frictional energy loss between the ceiling jet and the ceiling surface there is a very thin layer of air immediately below the ceiling surface that is not as involved in the ceiling jet flow created by the fire plume and, consequently, contains less smoke. If a smoke detector is recessed, this more slowly moving, relatively clean air immediately below the ceiling surface could impede the flow of smoke into the detector sensing chamber, again retarding response to the fire.

5-1.4.2 Where required, total coverage shall include all rooms, halls, storage areas, basements, attics, lofts, spaces above suspended ceilings, and other subdivisions and accessible spaces; and the inside of all closets, elevator shafts, enclosed stairways, dumbwaiter shafts, and chutes. Inaccessible areas shall not be required to be protected by detectors.

Exception No. 1: Where inaccessible areas contain combustible material, they shall be made accessible and shall be protected by a detector(s).

Exception No. 2: Detectors shall not be required in combustible blind spaces where any of the following conditions exist:

(a) Where the ceiling is attached directly to the underside of the supporting beams of a combustible roof or floor deck;

(b) Where the concealed space is entirely filled with a noncombustible insulation. In solid joist construction, the insulation shall be required to fill only the space from the ceiling to the bottom edge of the joist of the roof or floor deck;

(c) Where there are small, concealed spaces over rooms, provided any space in question does not exceed 50 ft² (4.6 m²) in area ;

(d) In spaces formed by sets of facing studs or solid joists in walls, floors, or ceilings where the distance between the facing studs or solid joists is less than 6 in. (150 mm).

Exception No. 3: Detectors shall not be required below open grid ceilings where all of the following conditions exist:

(a) The openings of the grid are ¹/₄ in. (6.4 mm) or larger in the least dimension.

(b) The thickness of the material does not exceed the least dimension.

(c) The openings constitute at least 70 percent of the area of the ceiling material.

Exception No. 4: Concealed, accessible spaces above suspended ceilings, used as a return air plenum meeting the requirements of NFPA 90A, Standard for the Installation of Air Conditioning and Ventilating Systems, where equipped with smoke detection at each connection from the plenum to the central air-handling system.

This paragraph deals with the concept of total coverage. The terms "total coverage" and "partial coverage" were introduced by NFPA 101, and by other codes and standards with the phrases "complete smoke detection system" and "partial smoke detection system." Total coverage, a complete smoke detection system, is required by the *Life Safety Code®* for a number of types of occupancies. However, it is this subsection of the *National Fire Alarm Code* that defines exactly what total coverage entails.

This section requires that total coverage entails placing detectors in all accessible spaces. The underlying premise is that if an enclosed space is accessible, it may, at some time, be used to store combustible materials. Notice that there is a parallel between the language of this paragraph and an that of paragraph 4-5.1 of NFPA 13, *Standard on the Installation of Sprinkler systems.* The concepts are analogous.

Inaccessible, noncombustible spaces do not need detectors. This includes a number of blind, boxed-in spaces that are common in stud-wall, curtain-wall, and frame construction, which, if not excepted from this requirement, would result in detectors being placed within walls, etc. The basis for this exception is that these spaces contain limited combustibles, and the probability of an ignition originating in these spaces is remote.

Exception 1 requires that inaccessible, combustible spaces must be made accessible and be provided with detection. Applying the criteria in Exception 1 requires engineering judgment.

Exception 2 covers boxed-in spaces that often occur in modern construction and renovation where a void space results. Experience has shown that such spaces are not significant as a fire ignition location and hence do not warrant detectors.

The third exception pertains to open-grid ceilings where all of the listed criteria are met. Where true open grid ceilings exist, the ceiling does not represent a significant barrier to the movement of smoke and fire gases. In most facilities where there are suspended ceilings, the above-ceiling space contains combustibles and does not comply with Exception No. 3. These spaces must, therefore, be equipped with detection when total coverage is required. If the open-grid ceiling does not meet all of the criteria in Exception No. 3 in a site where complete coverage is required, detectors must be placed in the above-ceiling space.

The fourth exception addresses above-ceiling spaces that are used as a return air plenum. When above-ceiling air plenums are in compliance with the requirements of NFPA 90A and are equipped with smoke detection at each of the exhaust duct connections, again per NFPA 90A, then the requirement for detectors throughout the above-ceiling space is waived. This exception does not apply only when detectors have been installed to comply with NFPA 90A, but is an alternative to regularly spaced area detectors in the above-ceiling space. The detectors in the exhaust duct will not be effective under "no airflow" conditions, and area detection at the ceiling plane is still required. The relevant sections of NFPA 90A limit the types and quantities of combustible materials that may be included within the return air plenum.

Formal Interpretation 78-2
Reference: 5-1.3.4

In 5-1.3.4 it states "Where total coverage is required, this shall include all rooms, . . . spaces above suspended ceilings, . . . and chutes." I am interested in the "spaces above suspended ceilings."

There are many buildings which have suspended ceilings—acoustic tile exposed T-bar ceilings.

Question 1: Does the above mean that detectors are required on the underside of the suspended ceiling for area protection of the room and also detectors required above the ceiling to protect space between ceiling and roof or space between ceiling and floor above?
Answer: Yes.

Question 2: Is there any criteria to consider that would not require detectors in the spaces above suspended ceilings?
Answer: Yes. If the space contained no combustible material as defined by NFPA 220 and the ceiling tiles were secured to their T-bar by clips or other methods of fixing such as in an approved fire resistant ceiling-roof assembly or, if the authority having jurisdiction does not require total coverage.

Issue Edition: NFPA 72E-1978
Reference: 2-6.5
Issue Date: September 1980 ∎

5-1.4.3* Detectors shall be required underneath open loading docks or platforms and their covers and for accessible underfloor spaces.

Exception: Where permitted by the authority having jurisdiction, detectors shall not be required where all of the following conditions exist:

(a) The space is not accessible for storage purposes or entrance of unauthorized persons and is protected against the accumulation of windborne debris.

(b) The space contains no equipment such as steam pipes, electric wiring, shafting, or conveyors.

(c) The floor over the space is tight.

(d) No flammable liquids are processed, handled, or stored on the floor above.

There have been fire losses that were far worse than they should have been simply because detectors were not

Formal Interpretation 78-3
Reference: 5-1.3.4

Question 1: Does Exception No. 1 apply only to the last sentence of 5-1.3.4, referring to "inaccessible areas which contain combustible material"?
Answer: Yes.

Question 2: If the answer to Question 1 is "yes," does this indicate that inaccessible areas which do not contain combustible material need not have detectors?
Answer: Yes.

Question 3: If the answer to Question 2 is "yes," does the installation of small access doors for servicing of smoke or fire dampers make the space above an otherwise non-accessible ceiling accessible within the intent of the code?
Answer: Yes.

Issue Edition: NFPA 72E-1978
Reference: 2-6.5
Issue Date: September 1980 ∎

placed in the types of spaces referenced here. The exception allows the authority having jurisdiction to waive this requirement only when all of the referenced conditions in (a) through (d) exist.

A-5-1.4.3 Detectors might be required under large benches, shelves, or table, and inside cupboards or other enclosures.

The importance of the requirement in 5-1.4.3 and this appendix material becomes obvious when one considers the role of the fire plume in transporting the heat and combustion product gases from the fire to the detectors. A floor deck or other large horizontal surface between the fire and the detectors can delay response by interfering with the flow produced by the fire plume.

5-1.4.4* Where codes, standards, laws, or authorities having jurisdiction require the protection of selected areas only, the specified areas shall be protected in accordance with this code.

This paragraph establishes that whenever and wherever an initiating device is installed, regardless of its ultimate purpose or function, it must be installed in accordance with this Code.

A-5-1.4.4 If there are no detectors in the room or area of origin, the fire could be too large to control where detected by remotely located detectors.

This appendix item was added to the Code to advise the user that partial detection provides less warning of impending fire than complete or total detection coverage. If early warning is the designer's goal, then the detector must be in the area of the fire to meet that goal.

5-1.4.5* **Stratification.** The possible effect of stratification below the ceiling shall be considered. (*Also see Appendix B for additional guidance.*)

A-5-1.4.5 **Stratification.** Stratification of air in a room can hinder air containing smoke particles or gaseous combustion products from reaching ceiling-mounted smoke or fire-gas detectors.

Stratification occurs when air containing smoke particles or gaseous combustion products is heated by smoldering or burning material and, becoming less dense than the surrounding cooler air, rises until it reaches a level at which there is no longer a difference in temperature between it and the surrounding air.

Stratification also can occur when evaporative coolers are used, because moisture introduced by these devices can condense on smoke, causing it to fall toward the floor. Therefore, to ensure rapid response, it might be necessary to install smoke detectors on sidewalls or at locations below the ceiling.

In installations where detection of smoldering or small fires is desired and where the possibility of stratification exists, consideration should be given to mounting a portion of the detectors below the ceiling. In high ceiling areas, projected beam-type or air sampling-type detectors at different levels also should be considered. *See Figure A-5-1.4.5.*

When the combustion product gases (smoke) are formed in a fire they are hot and, consequently, begin expanding. These expanded gases are less dense than the surrounding air and are buoyed upward. Because the gases are still hot they continue to expand and a V-shaped fire plume results, small at the bottom and ever larger the higher it goes. The Ideal Gas Law requires that as a gas expands it loses heat, becoming cooler. In addition, gas mixes with the surrounding air as it rises. Consequently, the combustion product gases in the fire plume eventually decrease in temperature due to the volumetric expansion and entrainment of ambient air until they are no longer hotter than the surrounding air. At this height there is no longer an upward push on the plume and it spreads out in a layer. If this height is reached before the fire plume impinges upon the ceiling there is no force causing the fire plume to turn and form a ceiling jet. All of the spacing criteria for smoke and heat detectors is based upon the existence of a ceiling jet

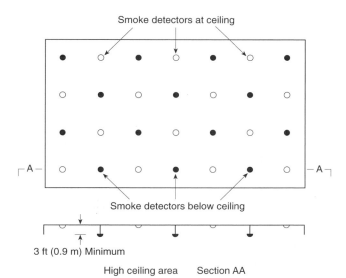

Figure A-5-1.4.5 *Smoke detector layout accounting for stratification.*

that moves the smoke and heat horizontally across the ceiling. Stratification impacts the performance of the detection system and is most likely to occur when the fires are small and the ceilings are high. However, HVAC systems designed to form a layer of cool air at some given distance above the floor can create exactly the same conditions as naturally occurring stratification with the same profound effects on the performance of a detection system.

5-2 Heat-Sensing Fire Detectors.

Heat detectors shall be installed in all areas where required either by the appropriate NFPA standard or the authority having jurisdiction.

We must understand the relationship between heat and temperature if we are to apply heat-sensing detectors properly. Heat is energy and can be quantified in terms of an amount and temperature is a measure of intensity and is quantified in terms of extent. Heat detectors are devices that change in some way when the temperature at the detector achieves a particular level, such as a rate-of-rise type, or a set-point, such as a fixed temperature type. Increase in temperature is due to the absorption of heat from a fire, primarily by convective heat transfer, and, to a much lesser degree, radiation.

Heat detectors are available in two general types: spot-type, which are devices that occupy a specific spot or point, and line-type, which are linear devices that extend over a distance, sensing temperature along their entire length.

Heat detectors operate on one or more of three different principles. These operating principles are categorized as fixed temperature, rate compensation, and rate-of-rise. Each principle has its performance advantages and can be employed in either a spot-type device or a line-type device.

See Figures 5.7 through 5.14.

There are a number of different technologies that can be used to detect the heat from a fire. These technologies include:

1. Expanding bimetallic components
2. Eutectic solders
3. Eutectic salts
4. Melting insulators
5. Thermistors
6. Temperature-sensitive semiconductors
7. Expanding air volume
8. Expanding liquid volume
9. Temperature-sensitive resistors
10. Thermopiles

The code has been written to allow the development and use of new technologies. The reader must make special note not to confuse the terms "type" and "principle" with "technology" —the method employed to achieve heat detection.

A precise definition and explanation of the mode of operation for each type of heat detector can be found in Section 1-4.

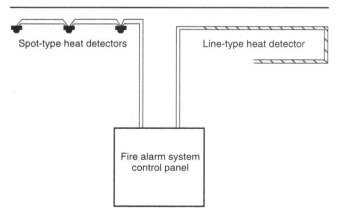

Figure 5.7   *The two types of heat detector, spot-type and line-type. Drawing courtesy of J. M. Cholin Consultants, Inc., Oakland, NJ.*

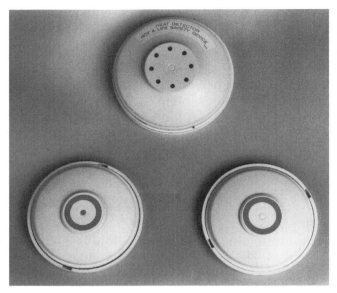

Figure 5.8   *Typical "low-profile" rate-of-rise and fixed-temperature heat detectors. Photograph courtesy of Mammoth Fire Alarms (Edwards System Technology, Chemetron), Lowell, MA.*

5-2.1 Temperature Classification.

The performance of a heat detector is dependent upon two parameters: its temperature classification and its time dependent thermal response characteristics. Traditionally, the temperature classification has been the principal parameter used in selecting the proper detector for a given site.

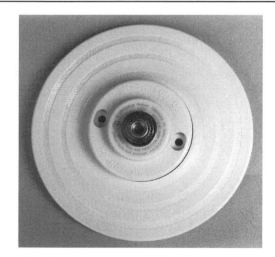

Figure 5.9   *Typical spot-type fixed-temperature heat detector. Photograph courtesy of Mammoth Fire Alarms (Chemetron), Lowell, MA.*

Figure 5.10 *Line-type heat detectors. Photograph courtesy of Protectowire Company, Hanover, MA.*

Figure 5.12 *Rate compensation heat detector - horizontal mounting. Photograph courtesy of Kidde-Fenwal Protection Systems, Ashland, MA.*

Figure 5.11 *Typical line-type heat detector installed in cable tray applications. Photograph courtesy of Protectowire Company, Hanover, MA.*

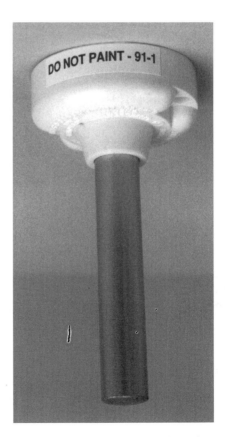

Figure 5.13 *Typical rate compensation heat detector - vertical mounting. Photograph courtesy of Mammoth Fire Alarms (Thermotech), Lowell, MA.*

Figure 5.14 *Typical spot-type combination rate-of-rise fixed-temperature heat detector. Photograph courtesy of Mammoth Fire Alarms (Chemetron), Lowell, MA.*

5-2.1.1   Color Coding.

5-2.1.1.1   Heat detectors of the fixed-temperature or rate-compensated, spot-pattern type shall be classified as to the temperature of operation and marked with the appropriate color code. *See Table 5-2.1.1.1.*

Spot-type heat detectors are the most popular type of heat detector in general use. In Table 5-2.1.1.1 there are specific criteria for the nominal temperature classification versus the maximum expected normal temperature for the location of the detector. Note that it is necessary to provide for at least a 20°F difference between the temperature classification of the detector and the maximum expected normal temperature. This requirement technically applies only to low temperature detectors, not to detectors with other temperature classifications. Nevertheless, it is still good practice to provide for at least a 20°F difference between maximum ambient and detector temperature classification when using heat detectors with the higher temperature classifications. Also note that it is crucial not to select a detector temperature classification any higher than necessary. The higher the temperature classification, the longer it will take to achieve an alarm because a larger fire is needed to produce the higher temperatures at the detector location.

The unified color coding of heat detectors facilitates inspections, making it possible to identify the temperature rating of a ceiling-mounted heat detector while standing on the floor. The color code for heat detectors is very similar to that used for sprinkler heads, as described in NFPA 13-1994 paragraph 2-2.3.1. The manufacturer also provides this information in their data sheets for each type of heat detector.

Exception: Heat detectors where the alarm threshold is field adjustable and that are marked with the temperature range.

5-2.1.1.2   Where the overall color of a detector is the same as the color code marking required for that detector, one of the following arrangements, applied in a contrasting color and visible after installation, shall be employed:

(a) A ring on the surface of the detector; or

(b) The temperature rating in numerals at least 3/8 in. (9.5 mm) high.

See Figures 5.15 and 5.16.

Table 5-2.1.1.1   Temperature Classification for Heat Detectors

Temperature Classification	Temp. Rating Range (°F)	Temp. Rating Range (°C)	Max. Ceiling Temp. (°F)	Max. Ceiling Temp. (°C)	Color Code
Low[1]	100 to 134	39 to 57	20 below[2]	11 below[2]	Uncolored
Ordinary	135 to 174	58 to 79	100	38	Uncolored
Intermediate	175 to 249	80 to 121	150	66	White
High	250 to 324	122 to 162	225	107	Blue
Extra high	325 to 399	163 to 204	300	149	Red
Very extra high	400 to 499	205 to 259	375	191	Green
Ultra high	500 to 575	260 to 302	475	246	Orange

[1]Intended only for installation in controlled ambient areas. Units shall be marked to indicate maximum ambient installation temperature.
[2]Maximum ceiling temperature has to be 20°F (11°C) or more below detector rated temperature.
NOTE: The difference between the rated temperature and the maximum ambient should be as small as possible to minimize the response time.

Figure 5.15 *Heat detectors with color coded rings. Photograph courtesy of Mammoth Fire Alarms (Chemetron), Lowell, MA.*

Figure 5.16 *Heat detector with temperature marked with numerals. Photograph courtesy of Mammoth Fire Alarms (Edwards Systems Technology/EST), Lowell, MA.*

5-2.1.2* A heat detector integrally mounted on a smoke detector shall be listed or approved for not less than 50-ft (15-m) spacing.

A-5-2.1.2 The linear space rating is the maximum allowable distance between heat detectors. The linear space rating is also a measure of the heat detector response time to a standard test fire where tested at the same distance. The higher the rating, the faster the response time. This code recognizes only those heat detectors with ratings of 50 ft (15 m) or more.

This paragraph originated in NFPA 74, *Standard for the Installation, Maintenance, and Use of Household Fire Warning Equipment.* There are many common smoke detectors designed for household applications that are equipped with an integral heat sensor. In order

for the heat sensor portion of the detector to comply with the Code, it must have a 50-foot spacing factor. See Figure 5.17.

5-2.2 Location.

This section of the Code prescribes the proper location of heat detectors for general purpose, open area protection. The location stipulated is intended to take maximum benefit of the ceiling jet produced by a fire. Since the occurrence of a ceiling jet causes the hot combustion product gases to flow radially away from the fire plume center-line, a ceiling location provides for the maximum flow across the detector and hence the maximum speed of response to a growing fire.

5-2.2.1* Spot-type heat detectors shall be located on the ceiling not less than 4 in. (100 mm) from the sidewall or on the sidewalls between 4 in. and 12 in. (100 mm and 300 mm) from the ceiling. *See Figure A-5-2.2.1.*

Exception No. 1:   In the case of solid open joist construction, detectors shall be mounted at the bottom of the joists.

Exception No. 2:   In the case of beam construction where beams are less than 12 in. (300 mm) in depth and less than 8 ft (2.4 m) on center, detectors shall be permitted to be installed on the bottom of beams.

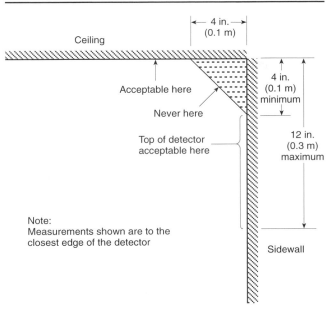

Figure A-5-2.2.1 *Example of proper mounting for detectors.*

Figure 5.17   Smoke detector with 50-foot listed heat detection. Photograph courtesy of Mammoth Fire Alarms (ESL/Sentrol, Inc.), Lowell, MA.

This paragraph is applicable only to spot-type heat detectors. The terms ceiling, joist and beam are defined in Section 1-4 for use in this context.

Section 1-4 defines a ceiling as the upper surface of a space, regardless of height. Subsection 5-1.4.2 identifies those spaces where detectors must be placed when total coverage is required. Spot-type heat detectors must be located on the ceiling of those spaces, at a distance 4 inches or more from a vertical side wall, or on the side walls within 12 inches of the ceiling. These locations derive maximum benefit from the upward flow of the fire plume and the flow of the ceiling jet beneath the ceiling plane. The best currently available research data support the existence of a dead air space where the walls meet the ceiling in a typical room. Figure A-5-2.2.1 shows this dead air space extending 4 inches in from the wall and 4 inches down from the ceiling. Consequently, the Code excludes detectors from that area.

The applicability of the exceptions are determined by the type of ceiling configuration in the protected space. But these definitions of the terms "joist" and "beam" must be inferred from the definition of "Solid Joist Construction" and "Beam Construction" in Section 1-4. Joists are solid projections extending downward from the ceiling, whether structural or not, which are more than 4.0 inches in depth and are spaced on 3.0 foot centers or less. The commonly encountered 2"x10" installed on 16 inch centers supporting a roof deck is

typical of solid joist construction. The structural component commonly called a "bar-joist" is actually an open web beam. If the upper web member of an open web beam is less than 4 inches deep, the beam is ignored; if it is more than 4 inches deep, it is deemed either a joist or a beam depending upon the center to center spacing.

The narrow spacing between joists (usually 16 inches) creates air pockets between them. These air pockets have two effects on the flow of the ceiling jet. First they tend to slow down the ceiling jet. Secondly, they force the ceiling jet to flow across the bottoms of the joists. See Figure 5.18. Exception 1 requires that heat detectors be placed on the bottoms of joists rather than up in the pockets between them. This puts the detectors in the region of maximum ceiling jet flow.

Beams are defined as being solid projections extending downward from the ceiling, whether structural or not, which are more than 4.0 inches in depth and are spaced on centers of more than 3.0 feet. The principal distinction between a joist and a beam in the context of the Code is the center to center spacing. Exception 2 allows the detectors to be placed on the beam bottoms only when they are less than 12 inches deep, and only when they are on centers of less than 8 feet. If they are more than 12 inches deep or if they are spaced more than 8 feet apart, the detectors must be placed on the ceiling surface between the beams.

Finally, notice that the only permitted location for spot-type heat detectors is at the ceiling plane, consistent with the stipulations in this paragraph. There is no research to provide guidance for detector placement in areas without ceilings. By inference, if there is no ceiling, there cannot be heat detectors installed in compliance with this Code.

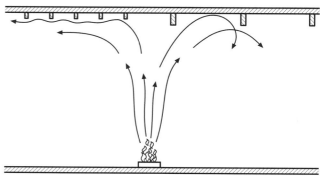

Joists: Less than 3 ft on center, more than 4 in. (100 mm) deep
Beams: More than 3 ft on center, more than 4 in. (100 mm) deep

Figure 5.18   The effect of joists and beams on the ceiling jet. Note: Joists—less than 3 ft. on center, more than 4 in. deep; Beams—more than 3 ft. on center, more than 4 in. deep. Drawing courtesy of J.M.Cholin Consultants, Inc., Oakland, NJ.

5-2.2.2 Line-type heat detectors shall be located on the ceiling or on the sidewalls not more than 20 in. (500 mm) from the ceiling.

Exception No. 1: In the case of solid open joist construction, detectors shall be mounted at the bottom of the joists.

Exception No. 2: In the case of beam construction where beams are less than 12 in. (300 mm) in depth and less than 8 ft (2.4 m) on center, detectors shall be permitted to be installed on the bottom of beams.

Exception No. 3: Where a line-type detector is used in an application other than open area protection, the manufacturer's installation instructions shall be followed.

This paragraph only applies to linear heat detectors. It does not prohibit the installation of a line-type detector in the portion of the ceiling within 4 inches of a vertical wall or on the vertical wall within 4 inches of the ceiling; (the area often referred to as the dead air space). Although installation in this area is not prohibited, the prudent designer will avoid the ceiling/wall corner, even when using a line-type heat detector.

Also note that Exception 3 recognizes the installation of linear heat detection for uses other than open area protection. Linear heat detection can be used for special application purposes installation is consistent with the manufacturer's instructions. When linear heat detection is used in this way, keep in mind that there is no ceiling jet moving the hot combustion product gases horizontally to the detector location. Consequently, the spacing criteria provided later in the Code will not be applicable.

5-2.3* **Temperature.** Detectors having fixed-temperature or rate-compensated elements shall be selected in accordance with Table 5-2.1.1.1 for the maximum ceiling temperature that can be expected.

A-5-2.3 A heat detector with a temperature rating somewhat in excess of the highest normally expected ambient temperature is specified in order to avoid the possibility of premature operation of the heat detector to nonfire conditions.

5-2.4* **Spacing.**

A-5-2.4 In addition to the special requirements for heat detectors installed on ceilings with exposed joists, reduced spacing also might be required due to other structural characteristics of the protected area, possible drafts, or other conditions that could affect detector operation.

The spacing criteria established by 5-2.4.1 determine how many detectors of a given type are necessary to properly protect a given area. To compensate for the impact that variations in the specific hazard area can have on the temperature and velocity of the ceiling jet, the spacing of detectors of known performance (listed spacing) is adjusted. These spacing adjustments are intended to compensate for environmental impacts on the detector performance and provide response roughly equivalent to that attainable from the same detectors when installed on smooth level ceilings.

The number of detectors required is a function of the spacing factor, "S", of the detector to be used. The spacing is established through a series of fire tests and is indicative of the relative sensitivity of the detector. The spacing derived from the fire tests relate heat detectors to the response of a special test sprinkler head. The test fire is situated at the center of an array of special test sprinkler heads, installed on 10 foot by 10 foot centers. This places the test fire 7.07 feet from the test sprinklers. Detector performance is defined relative to the distance at which it could detect the same fire that fused the test sprinkler head in 2 minutes ± 10 seconds. For example, a heat detector receives a 50 foot listed spacing if it responds, when installed on a 50 foot by 50 foot array, to the test fire just before the special test sprinkler head operates.

Alternatively, the designer may use the analytical method described in Appendix B of this Code. Because Appendix B is optional, it is recommended that the designer obtain the approval or acceptance of the authority having jurisdiction prior to using Appendix B for the design.

The number of detectors necessary for a given application also depends on the ceiling height, the type of ceiling (whether it has exposed joists or beams), and other features that may affect the flow of air or the accumulation of heat from a fire. All of these factors enter into the spacing design rules that follow.

5-2.4.1* **Smooth Ceiling Spacing.** One of the following requirements shall apply:

A-5-2.4.1 Maximum linear spacings on smooth ceilings for spot-type heat detectors are determined by full-scale fire tests. These tests assume that the detectors are to be installed in a pattern of one or more squares, each side of which equals the maximum spacing as determined in the test. This is illustrated in Figure A-5-2.4.1(a). The detector to be tested is placed at a corner of the square so that it is positioned at the farthest possible distance from the fire while remaining within the square. Thus, the distance from the detector ("D") to the fire ("F") is always the test spacing multiplied by 0.7 and can be set up as shown in Table A-5-2.4.1.

Table A-5-2.4.1   Test Spacing for Spot-Type Heat Detectors

Test Spacing		Maximum Test Distance from Fire to Detector (0.7 × D)	
(ft)	(m)	(ft)	(m)
50 × 50	15.24 × 15.24	35	10.67
40 × 40	12.19 × 12.19	28	8.53
30 × 30	9.1 × 9.1	21	6.40
25 × 25	7.62 × 7.62	17.5	5.33
20 × 20	6.10 × 6.10	14	4.27
15 × 15	4.57 × 4.57	10.5	3.20

Once the correct maximum test distance has been determined, it is valid to interchange the positions of the fire ("F") and the detector ("D"). The detector is now in the middle of the square, and the listing specifies that the detector is adequate to detect a fire that occurs anywhere within that square—even out to the farthest corner.

In laying out detector installations, designers work in terms of rectangles, as building areas are generally rectangular in shape. The pattern of heat spread from a fire source, however, is not rectangular in shape. On a smooth ceiling, heat spreads out in all directions in an ever-expanding circle. Thus, the coverage of a detector is not, in fact, a square, but rather a circle whose radius is the linear spacing multiplied by 0.7

This is graphically illustrated in Figure A-5-2.4.1(b). With the detector as the center, by rotating the square, an infinite number of squares can be laid out, the corners of which create the plot of a circle whose radius is 0.7 times the listed spacing. The detector will cover any of these squares and, consequently, any point within the confines of the circle.

So far this explanation has considered squares and circles. In practical applications, very few areas turn out to be exactly square, and circular areas are extremely rare. Designers deal generally with rectangles of odd dimensions and corners of rooms or areas formed by wall intercepts, where spacing to one wall is less than ½ the listed spacing. To simplify the rest of this explanation, the use of a detector with a listed spacing of 30 ft × 30 ft (9.1 m × 9.1 m) should be considered. The principles derived are equally applicable to other types.

Figure A-5-2.4.1(c) illustrates the derivation of this concept. A detector is placed in the center of a circle with a radius of 21 ft (0.7 × 30 ft) [6.4 m (0.7 × 9.1 m)]. A series of rectangles with one dimension less than the permitted maximum of 30 ft (9.1 m) is constructed within the circle. The following conclusions can be drawn:

(a) As the smaller dimension decreases, the longer dimension can be increased beyond the linear maximum spacing of the detector with no loss in detection efficiency.

(b) A single detector covers any area that fits within the circle. For a rectangle, a single, properly located detector may

be permitted, provided the diagonal of the rectangle does not exceed the diameter of the circle.

(c) Relative detector efficiency actually is increased, because the area coverage in square feet is always less than the 900 ft^2 (83.6 m^2) permitted if the full 30 ft × 30 ft (9.1 m × 9.1 m) square were to be utilized. The principle illustrated here allows equal linear spacing between the detector and the fire, with no recognition for the effect of reflection from walls or partitions, which in narrow rooms or corridors is of additional benefit. For detectors that are not centered, the longer dimension should always be used in laying out the radius of coverage.

Areas so large that they exceed the rectangular dimensions given in Figure A-5-2.4.1(c) require additional detectors. Often proper placement of detectors can be facilitated by breaking down the area into multiple rectangles of the dimensions that fit most appropriately [see Figure A-5-2.4.1(d)]. For example, see Figure A-5-2.4.1(c). A corridor 10 ft (3 m) wide and up to 82 ft (25 m) long can be covered with two 30-ft (9.1-m) detectors. An area 40 ft (12.2 m) wide and up to 74 ft (22.6 m) long can be covered with four detectors. Irregular areas need more careful planning to make certain that no spot on the ceiling is more than 21 ft (6.4 m) away from a detector. These points can be determined by striking arcs from the remote corner. Where any part of the area lies beyond the circle with a radius of 0.7 times the listed spacings, additional detectors are required.

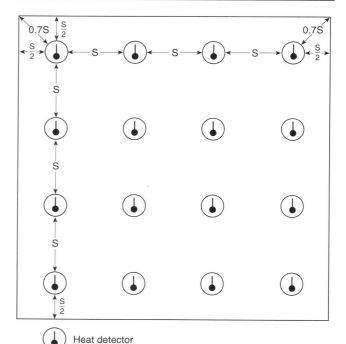

⊙ Heat detector

S   Spacing between detectors

Figure A-5-2.4.1(a)   *Spot-type heat detectors.*

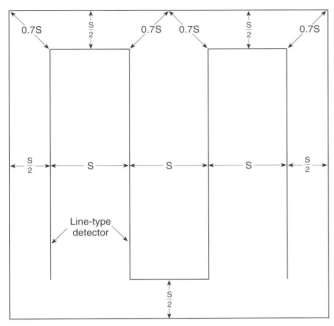

Figure A-5-2.4.1(b)   *Line-type detectors — spacing layouts, smooth ceiling.*

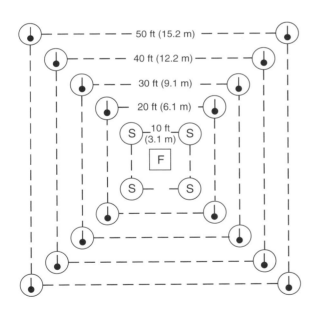

Legend

F   –Test fire, denatured alcohol, 190 proof. Pan located approximately 3 ft (0.9 m) above floor.

(S)   –Indicates normal sprinkler spacings on 10-ft (3.1-m) schedules.

●   –Indicates normal heat detector spacing on various spacing schedules.

Figure A-5-2.4.1(c)   *Fire test layout.*

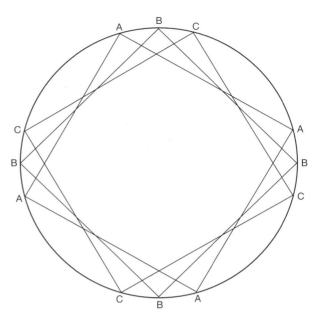

Figure A-5-2.4.1(d)   *A detector will cover any square laid out in the confines of a circle whose radius is 0.7 times the listed spacing.*

(a)   The distance between detectors shall not exceed their listed spacing, and there shall be detectors within a distance of ¹/₂ the listed spacing, measured at a right angle, from all walls or partitions extending to within 18 in. (460 mm) of the ceiling; or

(b)   All points on the ceiling shall have a detector within a distance equal to 0.7 times the listed spacing (0.7S). This is useful in calculating locations in corridors or irregular areas.

5-2.4.1.1*   Irregular Areas.   For irregularly shaped areas, the spacing between detectors shall be permitted to be greater than the listed spacing, provided the maximum spacing from a detector to the farthest point of a sidewall or corner within its zone of protection is not greater than 0.7 times the listed spacing. *See Figure A-5-2.4.1.1.*

5-2.4.1.2*   High Ceilings.   On ceilings 10 ft to 30 ft (3 m to 9.1 m) high, heat detector linear spacing shall be reduced in accordance with Table 5-2.4.1.2.

Exception: Table 5-2.4.1.2 shall not apply to the following detectors, which rely on the integration effect:

(a) Line-type electrical conductivity detectors [see A-1-4, "Fixed Temperature Detector," (b), "Electrical Conductivity"];

(b) Pneumatic rate-of-rise tubing [see A-1-4, "Rate-of-Rise Detector," (a), "Pneumatic Rate-of-Rise Tubing"];

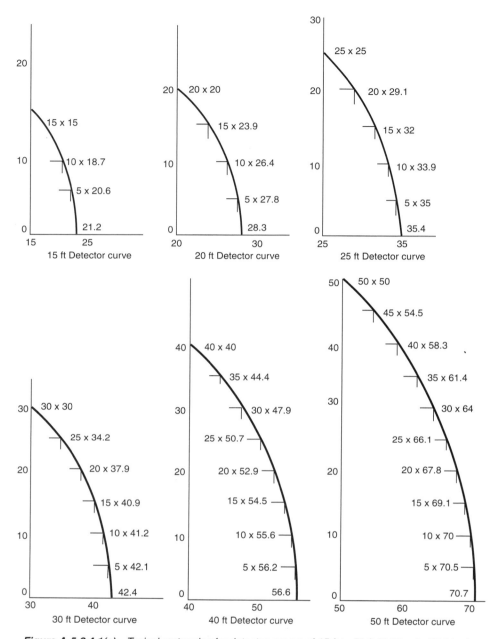

Figure A-5-2.4.1(e)   *Typical rectangles for detector curves of 15 ft to 50 ft (4.57 m to 15.24 m).*

(c) Series connected thermoelectric effect detectors [see A-1-4, "Rate-of-Rise Detector," (c), "Thermoelectric Effect Detector"].

In these cases, the manufacturer's recommendations shall be followed for appropriate alarm point and spacing.

NOTE: Table 5-2.4.1.2 provides for spacing modifications to take into account different ceiling heights for generalized fire conditions. An alternative design method that allows a designer to take into account ceiling height, fire size, and ambient temperature is provided in Appendix B.

A-5-2.4.1.2   Both 5-2.4.1.2 and Table 5-2.4.1.2 are constructed to provide detector performance on higher ceilings [to 30 ft (9.1 m) high)] that is essentially equivalent to that which would exist with detectors on a 10-ft (3-m) ceiling.

The Fire Detection Institute Fire Test Report (*see references in Appendix C*), used as a basis for Table 5-2.4.1.2, does not include data on integration-type detectors.

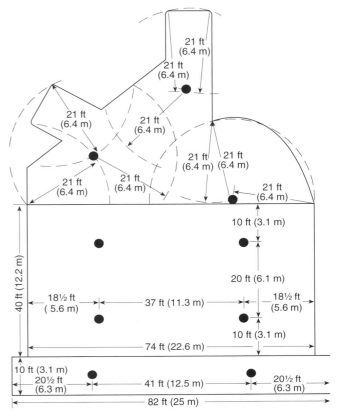

S – Detector spacing
● – Smoke detector or heat detector

Figure A-5-2.4.1.1 *Smoke or heat detector spacing layout, irregular areas.*

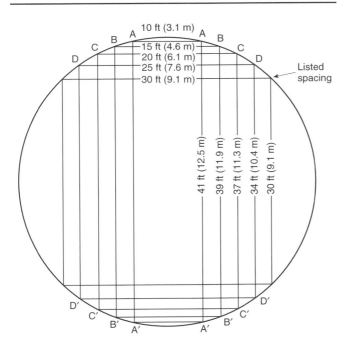

Rectangle A = 10 ft x 41 ft = 410 ft² (3.1 m x 12.5 m = 38.1 m²)
          B = 15 ft x 39 ft = 585 ft² (4.6 m x 11.9 m = 54.3 m²)
          C = 20 ft x 37 ft = 740 ft² (6.1 m x 11.3 m = 68.8 m²)
          D = 25 ft x 34 ft = 850 ft² (7.6 m x 10.4 m = 78.9 m²)
Listed spacing = 30 ft x 30 ft = 900 ft² (9.1 m x 9.1 m = 83.6 m²)

Figure A-5-2.4.1(f) *Detector spacing, rectangular areas.*

Pending development of such data, the manufacturer's recommendations provide guidance.

The spacing factor for a given detector is a rough measure of how far the ceiling jet can travel from the test fire (used in the listing evaluation) before it has cooled and slowed down too much to provide reliable detection in the required time period. As hot combustion product gases in the fire plume rise from the fire, the gases expand, giving off energy and actually cooling. This leaves less energy available to accelerate the ceiling jet once the fire plume impinges upon the ceiling plane. Increasing the ceiling height has a very significant effect on the ceiling jet temperature and velocity. The reduction of spacing with increased ceiling height places detectors closer to the fire plume center-line, thus allowing the hot combustion product gas, air and radiated heat to travel a shorter distance before encountering a detector.

Table 5-2.4.1.2   Heat Detector Spacing Reduction Based on Ceiling Height

Ceiling Height Above		Up to		Percent of Listed Spacing
(ft)	(m)	(ft)	(m)	
0	0	10	3.05	100
10	3.05	12	3.66	91
12	3.66	14	4.27	84
14	4.27	16	4.88	77
16	4.88	18	5.49	71
18	5.49	20	6.10	64
20	6.10	22	6.71	58
22	6.71	24	7.32	52
24	7.32	26	7.93	46
26	7.93	28	8.54	40
28	8.54	30	9.14	34

The inverse square law predicts that when the distance between the fire and the detector is doubled, the amount of radiated heat reaching the detector will be reduced by a factor of 4. This also contributes to the need to reduce the spacing as the ceiling height is increased.

It is important to note that Table 5-2.4.1.2 covers ceiling heights up to 30 feet. This is the highest ceiling for which the Technical Committee had test data. (See references in Appendix B.) Where ceilings higher than 30 feet are encountered, the designer must act with the knowledge that those conditions are beyond the limits of the testing that provided the basis for the requirements of the Code. There is the temptation to extrapolate for higher ceiling heights. However, a theoretical basis for doing so has yet to be reviewed by the Technical Committee. This is currently an area of considerable research that may yield important new insights in the near future.

The Code neither prohibits nor permits the use of heat detectors on ceilings higher than 30 feet. Computer models such as FPETool and Hazard 1 have been used to predict performance at higher ceiling heights and there have been some studies that have confirmed the predictions derived from these models. However, it should be understood that in the context of an exponentially growing fire, a much larger fire will be necessary to activate the detectors on higher ceilings, and the fire allowed by the delayed detection may be considerably larger than that normally assumed. In some cases, this will mean that the detection system will not meet the protection goals of the owner or the intent of the Code. The final decision as to whether or not a design is acceptable rests with the authority having jurisdiction.

5-2.4.2*   Solid Joist Construction.   The spacing of heat detectors, where measured at right angles to the solid joists, shall not exceed 50 percent of the smooth ceiling spacing permitted under 5-2.4.1 and 5-2.4.1.1. *See Figure A-5-2.4.2.*

Subsection 5-2.2 establishes the requirement to locate heat detectors on the bottom of joists. In subsection 5-2.4.2, the effect of joists on detector spacing is established. The hot combustion product gases and smoke from a fire rise vertically in a plume until it impinges upon the ceiling. There, the hot combustion product gases and entrained air of the fire plume change direction and move horizontally across the ceiling, becoming a ceiling jet. When the joists are running parallel to the direction of travel of the ceiling jet, they have little effect on the speed with which the hot gases of the ceiling jet move across the ceiling. However, when the joists are perpendicular to the direction of gas travel from the fire to the detector, they produce turbulence and thus reduce the ceiling jet velocity. This necessitates a closer spacing in the direction perpendicular to the joists if uniform performance is to be attained.

Remember that joists are solid members extending more than 4 inches down from the ceiling and are installed on centers of less than 3 feet (1 m). If the solid members extending down from the ceiling are on 3-foot centers or larger, they are beams. Also remember that bar-joists have no effect on spacing unless the top cord is greater than 4 inches (0.1 m or 100 cm).

5-2.4.3*   Beam Construction.   A ceiling shall be treated as a smooth ceiling where the beams project no more than 4 in. (100 mm) below the ceiling. Where the beams project

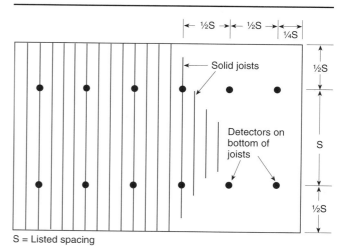

Figure A-5-2.4.2   *Detector spacing layout, solid joist construction.*

more than 4 in. (100 mm) below the ceiling, the spacing of spot-type heat detectors at right angles to the direction of beam travel shall be not more than 2/3 of the smooth ceiling spacing permitted under 5-2.4.1 and 5-2.4.1.1. Where the beams project more than 18 in. (460 mm) below the ceiling and are more than 8 ft (2.4 m) on center, each bay formed by the beams shall be treated as a separate area.

Keeping mind that beams project more than 4 inches from the ceiling and are on center to center spacing greater than 3 feet, beams create barriers to the horizontal flow of the ceiling jet. In essence, the bay created by the beams and the walls at either end, or the cross beams extending from beam to beam, must fill up with smoke and hot combustion product gases before spilling into the next bay. This fill and spill progression of the ceiling jet is slower than the velocity attained on a smooth flat ceiling. Consequently, the detector spacing in the direction perpendicular to the beams must be reduced to compensate for this reduced ceiling jet velocity if consistent performance is to be attained.

Open-web beams and trusses have little effect on the passage of air currents caused by fire. Generally, they are not considered in determining the proper spacing of detectors unless the solid part of the top cord extends more than 4 inches down from the ceiling. See Figure 5.19.

A-5-2.4.3 Location and spacing of heat detectors should consider beam depth, ceiling height, beam spacing, and fire size.

(a) If the ratio of beam depth (D) to ceiling height (H) (D/H) is greater than 0.10 and the ratio of beam spacing (W) to ceiling height (H) (W/H) is greater than 0.40, heat detectors should be located in each beam pocket.

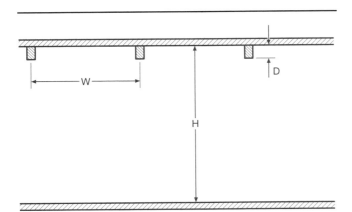

Figure 5.19 *The depth and center to center spacing of beams relative to the ceiling height affects the preferred location for heat detectors per A-5-2.4.3. Drawing courtesy of J.M.Cholin Consultants, Inc., Oakland, NJ.*

(b) If either the ratio of beam depth to ceiling height (D/H) is less than 0.10 or the ratio of beam spacing to ceiling height (W/H) is less than 0.40, heat detectors should be installed on the bottom of the beams.

The criteria included in this Appendix material consider the thickness of the ceiling jet under varied conditions. In general, research has shown that the ceiling jet occupies the upper 10% of the compartment volume. If the downward extension of the beams is less than 10% of the ceiling height, then the impact of the beams on the flow of the hot combustion product gases in the ceiling jet will be lessened as a significant portion of the ceiling jet will pass beneath the beams. Likewise, with the relatively narrow center-to-center spacing resulting in a width to height ratio of less than 0.4, the beam bay fills very rapidly, making the fill part of the fill and spill propagation of the ceiling jet a relatively short delay in time. Where the beam depths are relatively large, or the bay volumes, proportional to beam center-to-center spacing are large, the fill delay is significant and detectors must be located in each bay. See Figure 5.19.

5-2.4.4 Sloped Ceilings.

When the fire plume impinges upon a sloped ceiling, the development of the ceiling jet is affected by the slope of the ceiling. Because it takes less energy to turn the flow of combustion product gases and entrained air less than 90 degrees, the ceiling jet moves more rapidly up a sloped ceiling and more slowly across the slope than it would across a level ceiling. However, when the ceiling jet reaches the peak of the roof, its flow stops. This effects the placement and spacing of heat detectors. At the peak, the ceiling jet collides with a mass of the hottest air that normally exists beneath the ceiling. Furthermore, because the jet cannot continue flowing without moving down the opposite side of the roof, the ceiling jet stops. This creates a thermal lag of heat detection (controlled by the velocity of the ceiling jet flow across the detector) when the ceiling jet stops the thermal lag increases significantly. The spacing adjustment rules provided here are derived from the experience of the past and the study of ceiling jet flows under these ceilings.

5-2.4.4.1* **Peaked.** A row of detectors shall first be spaced and located at or within 3 ft (0.9 m) of the peak of the ceiling, measured horizontally. The number and spacing of additional detectors, if any, shall be based on the horizontal projection of the ceiling in accordance with the type of ceiling construction. *See Figure A-5-2.4.4.1.*

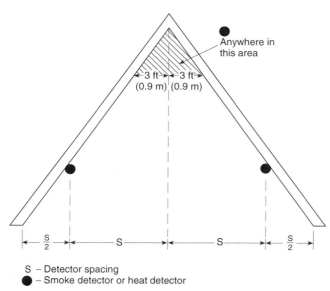

Figure A-5-2.4.4.1   *Smoke or heat detector spacing layout, sloped ceilings (peaked type).*

5-2.4.4.2*   **Shed.**   Sloped ceilings having a rise greater than 1 ft in 8 ft (1 m in 8 m) horizontally shall have a row of detectors located on the ceiling within 3 ft (0.9 m) of the high side of the ceiling measured horizontally, spaced in accordance with the type of construction. The remaining detectors, if any, shall be located in the remaining area on the basis of the horizontal projection of the ceiling. *See Figure A-5-2.4.4.2.*

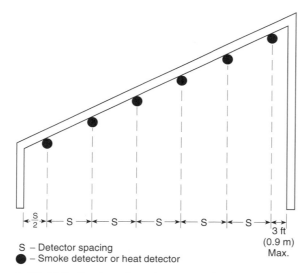

Figure A-5-2.4.4.2   *Smoke or heat detector spacing layout, sloped ceilings (shed type).*

5-2.4.4.3   For a roof slope of less than 30 degrees, all detectors shall be spaced utilizing the height at the peak. For a roof slope of greater than 30 degrees, the average slope height shall be used for all detectors other than those located in the peak.

5-3   Smoke-Sensing Fire Detectors.

5-3.1   General.

Precise definitions and descriptions of the mode of operation of each type of smoke detector can be found in Section 1-4, Definitions.

5-3.1.1*   The purpose of Section 5-3 is to provide information to assist in design and installation of reliable early warning smoke detection systems for protection of life and property.

A-5-3.1.1   The addition of a heat detector to a smoke detector does not enhance its performance as an early warning device.

The location and spacing criteria in Section 5-3 are based on smoke detectors that have been listed by a nationally recognized testing laboratory as having passed specific performance tests. To pass the tests, the detectors must respond to nominal smoke obscuration of 1 to 4 percent, depending on the type of smoke and type of detector. In most fires, these devices respond much sooner than either sprinkler systems or heat detectors. In flaming fire tests, smoke detectors activate long before typical heat detectors. The difference in the speed of response is even more dramatic with low-energy smoldering fires. This difference in the speed of response is the basis for concluding that the addition of a heat detector to a smoke detector adds little to overall fire detection performance when early warning is the design criteria.

5-3.1.2   Section 5-3 covers general area application of smoke detectors in ordinary indoor locations.

This paragraph limits the applicability of the requirements and recommendations of Section 5-3 to "general area application . . . in ordinary indoor locations." Whether or not a hazard area falls into this category is up to the authority having jurisdiction.

There are some authorities having jurisdiction that will establish additional requirements for specific types of occupancies above and beyond the requirements of

Section 5-3. The designer may also choose closer spacing in areas where there are extremely valuable assets, such as in a data center.

Finally, the common interpretation of 5-3.1.2 usually does not include special compartments, such as switch gear enclosures or aircraft lavatories. While detectors may be used in these and similar locations, the designer should use engineering judgment.

5-3.1.3 For information on use of smoke detectors for control of smoke spread, refer to Section 5-10.

Early in the second half of this century, there were several fires in high-rise buildings that demonstrated the futility of trying to evacuate all of the occupants. A new strategy of "protecting in place" or using "refuge zones" was developed. Concurrently, it became well-known that smoke inhalation was the principal cause of death associated with fires. If occupants were to be protected in place, it was critical to be able to automatically control the flow of smoke with the heating, ventilating, and air conditioning (HVAC) system. Smoke detectors are employed for that purpose, and their use for such applications is covered in Section 5-10.

5-3.1.4 For additional guidance in the application of smoke detectors for flaming fires of various sizes and growth rates in areas of various ceiling heights, refer to Appendix B.

Traditionally, the application of smoke detectors has been for early warning. Appendix B allows the use of smoke detectors for flaming fire detection. This limitation on the applicability of Appendix B for smoke detector design comes from the fact that the design curves are predicated upon smoke detector activation when a temperature rise of 13 degrees C (20 degrees F) occurs at the detector. This temperature to smoke density correlation is a simplifying assumption that was introduced in the early research and has persisted since the 1980s. It is considered to be an extremely conservative estimate. In some applications, the extension of the normal 30-foot spacing for smoke detectors for flaming fire detection allowed by Appendix B is appropriate. However, this spacing allowance should not be confused with the 30-foot spacing normally used in life safety or early warning applications.

Testing performed under the auspices of the Fire Detection Institute was the basis for the new method for predicting detector response. This testing gave rise to the computational procedure outlined in Appendix B for smoke detector applications in flaming fire detection scenarios. It provides a more analytical and precise method of determining detector spacing.

5-3.2* Smoke detectors shall be installed in all areas where required either by the appropriate NFPA standard or by the authority having jurisdiction.

A-5-3.2 The person designing an installation should keep in mind that, in order for a smoke detector to respond, the smoke has to travel from the point of origin to the detector. In evaluating any particular building or location, likely fire locations should be determined first. From each of these points of origin, paths of smoke travel should be determined. Wherever practical, actual field tests should be conducted. The most desired locations for smoke detectors are the common points of intersection of smoke travel from fire locations throughout the building.

NOTE: This is one of the reasons that specific spacing is not assigned to smoke detectors by the testing laboratories.

5-3.3 Sensitivity.

5-3.3.1 Smoke detectors shall be marked with their nominal production sensitivity (percent per foot obscuration), as required by the listing. The production tolerance around the nominal sensitivity also shall be indicated.

Because the mission of most smoke detection systems is the protection of human life, the response of a smoke detector is usually defined in human terms. The percent per foot obscuration method of measuring sensitivity relates to a person's ability to see well enough to escape from a fire. Smoke is composed of both visible and invisible particulate matter. However, the portion of the smoke that is invisible has little immediate impact on an individual's ability to escape.

It is not currently accepted practice to adjust detector spacing based upon the detector sensitivity. Often the selection of detector for a given purpose will include an analysis of the performance that can be expected from a detector of given sensitivity.

5-3.3.2 Smoke detectors that have provision for field adjustment of sensitivity shall have an adjustment range of not less than 0.6 percent per foot obscuration. Where the means of adjustment is on the detector, a method shall be available to restore the detector to its factory calibration. Detectors that have provision for program controlled adjustment of sensitivity shall be permitted to be marked with their programmable sensitivity range only.

The adjustment of detector sensitivity over a range of less than 0.6 percent per foot has little, if any, practical benefit. Even when smoke detectors are used for property conservation (as in data centers), the difference in response represented by an adjustment range of less than 0.6 percent per foot is minor.

There are some smoke detectors that have a feature allowing the adjustment of detector sensitivity to accommodate the immediate ambient conditions in the area of the detector. Other smoke detectors send a voltage or current back to the control unit that is proportional to the smoke-sensing signal in the detector. The trip point of the detector is a voltage or current level stored in the control unit memory. In either case, there may be occasion to adjust the detector sensitivity, either at the detector or control unit. The means to restore the detector to its factory sensitivity must be provided, and the detector must be labeled, showing the sensitivity range. In some cases, the adjustment feature may be used between cleaning intervals to maintain stability. Naturally, after the unit has been cleaned, it is desirable to restore it to its original design sensitivity. The maintenance of smoke detectors is covered in Chapter 7.

5-3.4   Location and Spacing.

5-3.4.1*   General.

A-5-3.4.1   For operation, all types of smoke detectors depend on smoke entering the sensing chamber or light beam. Where sufficient concentration is present, operation is obtained. Since the detectors are usually mounted on the ceiling, response time depends on the nature of the fire. A hot fire rapidly drives the smoke up to the ceiling. A smoldering fire, such as in a sofa, produces little heat; therefore, the time for smoke to reach the detector is increased.

5-3.4.1.1   The location and spacing of smoke detectors shall result from an evaluation based on the guidelines detailed in this code and on engineering judgment. Some of the conditions that shall be considered include:

(a) Ceiling shape and surface.

(b) Ceiling height.

(c) Configuration of contents in the area to be protected.

(d) Burning characteristics of the combustible materials present.

(e) Ventilation.

(f) Ambient environment.

These general criteria are far less specific than those established for heat detectors. The reason for this can be understood by reviewing the importance fire plume dynamics plays in the location and spacing of heat detectors versus smoke detectors.

Heat detectors depend upon the fire plume and ceiling jet to carry hot combustion product gases and entrained air to the detector where heat can flow from the ceiling jet into the detector, resulting in an alarm. The fire liberates significant quantities of energy and is the "engine" that creates its own air currents. The energy from the fire propels the hot air/smoke mixture across the ceiling. Under these circumstances, it is possible to model the flow of the fire plume and ceiling jet with computer programs that apply the rules of fluid flow physics and thermodynamics. The Fire Detection Institute has been an important guiding force in the development of these models. These models predict that smoke detectors provide response well before heat detectors. This prediction has been verified experimentally.

However, under smoldering, low-energy-output fire conditions the fire has not yet achieved an energy output (heat release rate) sufficient to serve as the primary source of propulsion for the smoke. The flow of smoke through the hazard area is dominated by existing air currents, with little, if any, contribution from the fire. This makes the prediction of flow far more dependent upon site-specific air flow variables, and detection is much more difficult. Consequently, the location and spacing of smoke detectors must be determined subject to the judgment of the designer on how the site-specific environment will effect the flow of smoke from these early stage, low-energy-output fires.

5-3.4.1.2   Where the intent is to protect against a specific hazard, the detector(s) shall be permitted to be installed closer to the hazard in a position where the detector can readily intercept the smoke.

The Code specifically allows the designer to add detectors where he/she expects the pre-existing, normal air currents to convey the smoke from an early-stage fire. Usually, the design process begins by locating detectors so that they will provide general area protection. Then, additional detectors are added, or positions adjusted, to take into account known or anticipated ignition sources and known air currents. This paragraph of the Code is generally cited when additional detectors are placed above switch-gear enclosures, power supplies and similar assets with known histories of ignition as well as high dollar value.

5-3.4.2   Air Sampling-Type Smoke Detector.   Each sampling port of an air sampling-type smoke detector shall be treated as a spot-type detector for the purpose of location and spacing. Maximum air sample transport time from the farthest sampling point shall not exceed 120 seconds.

An air sampling-type smoke detector is defined in Section 1-4. These detectors use a sampling tube and draw a sample of air from the hazard area to the detector where the presence of visible smoke or invisible combustion products is determined. The air transport time criterion places an effective limit on the design of the fan and the maximum distance from the detector to the farthest sampling port, as well as the size and layout of the sampling tubes. The manufacturer's listing and instructions will provide the detail on how the particular product must be used in order to comply with this limitation. Some air sampling-type smoke detectors have a means to detect changes in airflow, providing some measure of monitoring the integrity of the tubing or piping network. See Figure 5.20.

5-3.4.3*   Spot-Type Smoke Detectors.

A-5-3.4.3 In high ceiling areas, such as atriums, where spot-type smoke detectors are not accessible for periodic maintenance and testing, projected beam-type or air sampling-type detectors should be considered where access can be provided.

The issue of accessibility and the maintenance of a smoke detection system cannot be over-emphasized. The designer must exercise her/his judgment and discretion in order to provide a system that can be maintained pursuant to the criteria established in Chapter 7. Paragraph A-5-3.4.3 clarifies the 5-3.3.3.1 requirement to

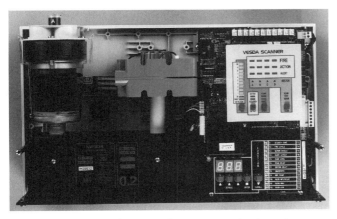

Figure 5.20   Typical air sampling-type smoke detection apparatus. Photograph courtesy of Vision Systems, Inc., Hingham, MA.

locate detectors on the ceiling or within 12 inches of it, and 5-1.4.2, which requires that all initiating devices, including smoke detectors, be installed in such a manner that they can be maintained. Atriums and other areas with exceptionally high ceilings (auditoriums, gymnasiums, exhibit halls, storage facilities, and some manufacturing facilities) represent very difficult situations for the use of spot-type smoke detection because of the problems that arise due to stratification, maintenance concerns, and smoke dissipation. This paragraph advises the designer to consider either air-sampling or projected beam-type photoelectric light obscuration smoke detection as alternatives.

5-3.4.3.1 Spot-type smoke detectors shall be located on the ceiling not less than 4 in. (100 mm) from a sidewall to the near edge or, where on a sidewall, between 4 in. and 12 in. (100 mm and 300 mm) down from the ceiling to the top of the detector. *See Figure A-5-2.2.1.*

Exception No. 1:   See 5-1.4.5.

Exception No. 2:   See 5-3.4.6.

The same fluid dynamics that predict that there will be dead-air space where the wall joins the ceiling in the context of heat detectors also predicts that this space will generally not participate in the normal ambient circulation of air within the space and hence will not provide a location where a smoke detector can be expected to perform. This location requirement is valid for both the low-energy incipient fire as well as a high-energy-output fire—one that is immediately life threatening. Either the normally existing air movements or the fire plume from the larger fire convey smoke to ceiling-mounted detectors. In order for the detectors to be able to respond they must be in the working air volume of the compartment. This logic is identical to that used in locating heat detectors. See 5-2.2.1.

Paragraph 5-3.4.3.1 instructs us to locate smoke detectors on the ceiling, not less than 4 inches from the side wall (or a beam greater than 8 inches in depth), or on the side walls within 12 inches of the ceiling but not less than 4 inches from the ceiling. These locations derive maximum benefit from the normal ambient air currents and the upward flow of smoke from the fire. The best currently available research data supports the theory that there is dead air space where the walls meet the ceiling in a typical room.

Figure A-5-2.5.1 shows this dead air space extending 4 inches in from the wall and 4 inches down from the ceiling. The code excludes detectors from that area.

While it is not explicitly stated in the Code it is clear that this criterion is intended to assure that the detector is in the "working air volume" of the room. Placing a detector in any location that is not in the working air volume of the room is not consistent with the general intent of the Code. For example, locating a smoke detector on the ceiling in the corner of the room, 4 inches from each wall is not specifically prohibited but is inconsistent with the general intent of this paragraph. The same would hold true for placing a detector on a side wall in the room corner.

Exception 1 provides for the precedence of paragraph 5-1.4.5 over this paragraph in the event of a conflict. Where stratification can be expected the location and spacing of smoke detectors must be adjusted. The design of a smoke detection system must address both the high-energy-output fire and the low-energy-output fire. In areas of high ceilings, this often necessitates layers of detectors or combining detectors to address all of the possible fire scenarios.

The objective of detecting the fire before it has achieved a high-energy output requires additional insight into the placement of detectors. The high-energy-output flaming fire produces a fire plume that propels smoke and hot air upward. The larger the fire, the higher the plume will extend and the greater the air velocity within the plume. In the low-energy-output smoldering fire [the type of ignition often encountered in residential (homes, hotels, apartments], institutional (hospitals, nursing homes, schools), and commercial (offices, stores) occupancies], significant quantities of smoke may be produced before the development of an energetic fire plume. This smoke may lack the energy to rise up to ceiling-mounted smoke detectors when the ceilings are higher than normally encountered. This situation must be addressed in any fire alarm system designed for residential, institutional, or commercial occupancies. The addition of smoke detectors at some distance below the ceiling does not eliminate the requirement for ceiling-mounted detectors.

Exception 2 provides for the precedence of paragraph 5-3.4.6 over the location criteria in this paragraph. Paragraph 5-3.4.6 addresses the impact of beams and joists on the performance of smoke detectors

5-3.4.3.2* To minimize dust contamination of smoke detectors where installed under raised room floors and similar spaces, they shall be mounted only in an orientation for which they have been listed. *See Figure A-5-3.4.3.2.*

A-5-3.4.3.2   See Figure A-5-3.4.3.2.

The fast-moving air in a data center underfloor space has sufficient energy to suspend dust. As that air enters the detector, it slows down and the suspended dust settles in the detector. The accumulation of dust within a smoke detector has a similar effect to that of smoke. In an ionization detector, the dust impedes the flow of current within the chamber. In a photoelectric detector, the dust increases the reflectance within the chamber. Thus, dust causes each type of detector to become more sensitive, increasing the likelihood of false alarms. The permitted orientations shown in Figure A-5-3.4.2.1 minimize the possibility of dust falling into the detector from the floor and also minimize the effect of air-conveyed dust on the detector.

There are other concerns that reinforce the benefits of positioning detectors as shown in Figure A-5-3.4.2.1. They place the detector in the upper half of the sub-floor volume. Since the sub-floor space is there to allow for the routing of cables between machines the floor is usually covered with cable. This cable has the same effect on the flow of air in the underfloor volume that joists have on air flow in a room. The cables create turbulence and force the flow to be concentrated in the upper half of the underfloor volume. Placing the detector in the upper half of the underfloor improves the system's ability to respond to an early-stage fire.

Another reason for positioning detectors as shown in Figure A-5-3.5.2.1 is that detectors mounted in the upper half of the underfloor volume are far less likely to be damaged as new cables are installed or old cables are rerouted through the underfloor space. If water-cooled computers are in use, the detectors are less likely to become wet if the computer cooling system leaks. Also, when there is no airflow, the detectors will be in the best orientation for detection. Finally, Figure A-5-3.4.2.1 shows the detectors in the orientation for which they have been tested and listed.

5-3.4.4 Projected Beam-Type Smoke Detectors. Projected beam-type smoke detectors (*see A-1-4, "Photoelectric Light Obscuration Smoke Detection"*) normally shall be located with their projected beams parallel to the ceiling and in accordance with the manufacturer's documented instructions.

Exception No. 1: See 5-1.4.5.

Exception No. 2: Beams shall be permitted to be installed vertically or at any angle needed to afford protection of the hazard involved (e.g., vertical beams through the open shaft area of a stairwell where there is a clear vertical space inside the handrails).

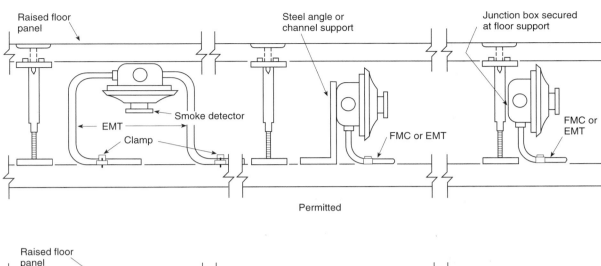

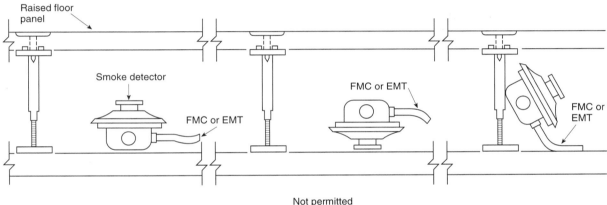

Figure A-5-3.4.3.2   *Mounting installations, permitted (top) and not permitted (bottom).*

5-3.4.4.1   The beam length shall not exceed the maximum permitted by the equipment listing.

Projected-beam smoke detectors have limitations on both the minimum and maximum beam length over which they will operate properly. The minimum beam length limitation is established by the lowest smoke concentration it can detect at that minimum beam length. The maximum beam length is determined by the maximum distance at which it can maintain its design stability even when some normal light obscuration is present. The projected-beam smoke detector must be able to identify a low concentration of smoke distributed along a substantial portion of the beam and a high concentration of smoke localized in a short segment of the beam. Each manufacturer obtains a listing from a nationally recognized testing laboratory that sets the upper and lower limits on the beam length. Failure to observe these limits could result in an unstable detector or the failure to detect a fire.

5-3.4.4.2   Where mirrors are used with projected beams, they shall be installed in accordance with the manufacturer's documented instructions.

5-3.4.5   Smooth Ceiling Spacing.

5-3.4.5.1   Spot-Type Detectors.

5-3.4.5.1.1   On smooth ceilings, spacing of 30 ft (9.1 m) shall be permitted to be used as a guide. In all cases, the manufacturer's documented instructions shall be followed. Other spacing shall be permitted to be used depending on ceiling height, different conditions, or response requirements. (*See Appendix B for detection of flaming fires.*)

Because of the difficulty in modeling the flow of smoke in low energy output fire scenarios, the Code cannot provide definitive spacing for smoke detectors at this time. This spacing guideline is based upon the fire tests conducted by the nationally recognized testing laboratories. Additional research is being funded and managed under the auspices of the National Fire Protection Research Foundation and the Fire Detection Institute. It is hoped that this research will provide the basis for more definitive spacing requirements in the near future.

An analytical method based on temperature rise data from the first phase of the Fire Detection Institute research is included in Appendix B. This method is very useful in aiding in detector spacing and placement for flaming fire scenarios and has become a very important tool for the fire protection systems designer. Furthermore, several computer models are now available, including FPETool and Hazard 1 to predict smoke detector activation. However, one must keep in mind that these computer models also use a temperature rise model to predict the activation of smoke detectors.

5-3.4.5.1.2* For smooth ceilings, all points on the ceiling shall have a detector within a distance equal to 0.7 times the selected spacing.

A-5-3.4.5.1.2 This is useful in calculating locations in corridors or irregular areas *(see A-5-2.4.1 and Figure A-5-2.4.1.1)*. For irregularly shaped areas, the spacing between detectors may be permitted to be greater than the selected spacing, provided the maximum spacing from a detector to the farthest point of a sidewall or corner within its zone of protection is not greater than 0.7 times the selected spacing (0.7S). *See Figure A-5-2.4.1.1.*

The concepts behind the spacing of smoke detectors follow directly from the concepts developed for heat detectors. Subsection A-5-2.4.1 develops the concepts that enable us to determine the area that will be covered by a detector. That area can vary in shape as long as the distance from the detector to the farthest point to be covered by the detector does not exceed 0.7 times the spacing factor. See A-5-2.4.1(a), (b), (d), (e) and (f) and Figure A-5-2.4.1.1.

5-3.4.5.2*   Projected Beam-Type Detectors. For location and spacing of projected beam-type detectors, the manufacturer's documented installation instructions shall be followed. *See Figure A-5-3.4.5.2.*

A-5-3.4.5.2 On smooth ceilings, a spacing of not more than 60 ft (18.3 m) between projected beams and not more than ¹/₂ that spacing between a projected beam and a side-

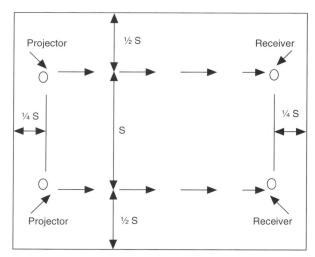

Figure A-5-3.4.5.2 *Maximum distance at which ceiling-suspended light projector and receiver may be permitted to be positioned from end wall is 1/4 selected spacing (S).*

wall (wall parallel to the beam travel) should be used as a guide. Other spacing should be determined based on ceiling height, airflow characteristics, and response requirements.

In some cases, the light beam projector is mounted on one end wall, with the light beam receiver mounted on the opposite wall. However, it is also permitted to suspend the projector and receiver from the ceiling at a distance from the end walls not exceeding ¹/₄ the selected spacing. *See Figure A-5-3.4.5.2.*

Notice the similarity between the installation and spacing concept developed for line-type heat detectors and line-type (projected beam) smoke detectors. The logic behind the design rules is consistent. Just as a line-type heat detector can be thought of as a row of spot-type heat detectors, it is often helpful to think of a projected beam detector as the equivalent to a row of spot-type detectors when developing a spacing strategy. The distance between the projected beams is analogous to the distance between rows of spot-type smoke detectors. Obviously, in high ceiling areas where stratification is probable and a serious concern, projected beams can be positioned at several levels.

5-3.4.6*   Solid Joist and Beam Construction. Solid joists shall be considered equivalent to beams for smoke detector spacing guidelines.

A-5-3.4.6 Detectors are placed at reduced spacings at right angles to joists or beams in an attempt to ensure that detection time is equivalent to that which would be experienced on a flat ceiling. It takes longer for the combustion products (smoke or heat) to travel at right angles to beams or joists because of the phenomenon wherein a plume from a relatively hot fire with significant thermal lift tends to fill the pocket between each beam or joist before moving to the next beam or joist.

Though it is true that this phenomenon might not be significant in a small smoldering fire where there is only enough thermal lift to cause stratification at the bottom of the joists, reduced spacing is still recommended to ensure that detection time is equivalent to that which would exist on a flat ceiling, even in the case of a hotter type of fire.

The logic behind these recommendations for the spacing of smoke detectors on ceilings with joists or beams is the same as is used in the requirements for heat detectors. In the case where the fire is a low-energy, smoldering fire the existence of joists and beams will slow the ambient air flow in the direction perpendicular to the joists or beams. In the case where the fire has developed a fire plume and a ceiling jet, the ceiling jet velocity across the beams and joists is slower than along the length of the beams or joists. In either case these projections from the ceiling retard the flow of smoke and hence necessitate reduced spacing if response time is to be held constant.

In the following spacing reduction recommendations keep in mind that joists are defined as solid projections of more than 4 inches in depth, placed on centers of less than 3 feet and beams are solid projections of more than 4 inches on centers greater than three feet.

5-3.4.6.1* Flat Ceilings.

A-5-3.4.6.1 The spacing guidelines in 5-3.4.6.1 are based on a detection design fire of 100 kW. For detection at a larger 1-MW fire and ceiling heights of 28 ft (8.53 m) or less, smooth ceiling spacings should be used and the detectors may be permitted to be located on the ceiling or the bottom of the beams.

(a) For ceiling heights of 12 ft (3.66 m) or lower and beam depths of 1 ft (0.3 m) or less, smooth ceiling spacings running in the direction parallel to the run of the beams shall be used and 1/2 the smooth ceiling spacing shall be in the direction perpendicular to the run of the beams. Spot-type detectors shall be permitted to be located either on the ceiling or on the bottom of the beams.

(b) For beam depths exceeding 1 ft (0.3 m) or for ceiling heights exceeding 12 ft (3.66 m), spot-type detectors shall be located on the ceiling in every beam pocket.

While the concept is the same, the dimension criteria regarding the depth of the joists is different for smoke detectors and heat detectors. When heat detectors are used, wherever the joists are more than 4 inches deep, a spacing reduction of 50 percent perpendicular to the

beams is required. When smoke detectors are used the spacing reduction is contingent upon both the ceiling height and the beam depth.

Paragraph 5-3.4.6.1(a) requires that when the beams or joists are less than 12 inches deep and the ceiling is less than 12 feet high, the spacing perpendicular to the beams must be reduced by 50% while the spacing running parallel to the beams is the same as smooth ceilings. The location criteria in this paragraph are consistent with the requirements of paragraph 5-2.2.1, pertaining to heat detectors. The air currents that convey smoke and heat, whether a ceiling jet or ambient air movement, ensure that the concentration of smoke at a detector mounted on the bottom of a beam less than 12 inches deep will be sufficient to achieve an alarm. As the smoke plume from a fire impinges upon the ceiling and begins expanding horizontally, a beam becomes a dam of sorts. Time and the expenditure of smoke energy forces the smoke downward far enough to spill over the dam (actually flow under the dam), encountering the smoke detector in the process. Energetic fires that produce large thermal outputs quickly force smoke across beams due to the large amounts of thermal energy available. Low-energy-output fires provide less energy, and the movement across beams is slower. When beams are less than 12 inches deep, detectors are permitted (not required) to be mounted on the beam bottoms.

When either the beam depths exceed 12 inches or when the ceiling height exceeds 12 feet smoke detectors must be placed in each beam pocket per 5-3.4.6.1(b), mounted on the ceiling surface between the beams.

These design criteria are the result of research conducted under auspices of the National Fire Protection Research Foundation and the Fire Detection Institute. Copies of the research reports are available from the National Fire Protection Research Foundation.

5-3.4.6.2* Sloped Ceilings.

(a) For beamed ceilings with beams running parallel to (up) the slope, the spacing for flat beamed ceilings shall be used. The ceiling height shall be taken as the average height over the slope. For slopes greater than 10 degrees, the detectors located at 1/2 the spacing from the low end shall not be required. Spacings shall be measured along a horizontal projection of the ceilings.

(b) For beamed ceilings with beams running perpendicular to (across) the slope, the spacing for flat beamed ceilings shall be used. The ceiling height shall be taken as the average height over the slope.

A-5-3.4.6.2 The spacing guidelines in 5-3.4.6.2 are based on a detection design fire of 100 kW. For detection at a larger 1-MW fire, the following spacings should be used:

Keep in mind that the "design fire " for a smoke detection system is usually orders of magnitude smaller than that for heat detectors.

(a) For beamed ceilings with beams running parallel to (up) the slope, with slopes 10 degrees or less, spacing for flat-beamed ceilings should be used. For ceilings with slopes greater that 10 degrees, twice the smooth ceiling spacing should be used in the direction parallel to (up) the slopes, and $\frac{1}{2}$ the spacing should be used in the direction perpendicular to (across) the slope. For slopes greater than 10 degrees, the detectors located at a distance of $\frac{1}{2}$ the spacing from the low end are not required. Spacing should be measured along the horizontal projection of the ceiling.

(b) For beamed ceilings with beams running perpendicular to (across) the slope for any slope, smooth ceiling spacing should be used in the direction parallel to the beams (across the slope), and $\frac{1}{2}$ the smooth ceiling spacing should be used in the direction perpendicular to the beams (up the slope).

These design criteria are the result of research conducted under auspices of the National Fire Protection Research Foundation and the Fire Detection Institute. Copies of the research reports are available from the National Fire Protection Research Foundation.

5-3.4.6.3 A projected beam-type smoke detector shall be considered equivalent to a row of spot-type smoke detectors for flat and sloped ceiling applications.

5-3.4.7 **Peaked.** Detectors shall first be spaced and located within 3 ft (0.9 m) of the peak, measured horizontally. The number and spacing of additional detectors, if any, shall be based on the horizontal projection of the ceiling. *See Figure A-5-2.4.4.1.*

5-3.4.8 **Shed.** Detectors shall first be spaced and located within 3 ft (0.9 m) of the high side of the ceiling, measured horizontally. The number and spacing of additional detectors, if any, shall be based on the horizontal projection of the ceiling. *See Figure A-5-2.4.4.2.*

5-3.4.9 **Raised Floors and Suspended Ceilings.** Spaces beneath raised floors and above suspended ceilings shall be considered separate rooms for smoke detector spacing. Detectors installed beneath raised floors or above suspended ceilings, or both, including raised floors and suspended ceilings used for environmental air, shall not be used in lieu of providing detection within the room.

When total coverage is required by the authority having jurisdiction or other codes, paragraph 5-1.4.2 requires detection in all accessible spaces (combustible or noncombustible and in inaccessible combustible spaces). The spaces beneath raised floors and above suspended ceilings usually fall into that category and, hence, require detection using the same location and spacing concepts as required for the occupied portion of a building.

5-3.4.9.1 **Raised Floors.** Detectors installed beneath raised floors shall be spaced in accordance with 5-3.4.1, 5-3.4.1.2, and 5-3.4.3.2. Where the area beneath the raised floor is also used for environmental air, detector spacing shall also conform to 5-3.5.1 and 5-3.5.2.

5-3.4.9.2 **Suspended Ceilings.** Detector spacing above suspended ceilings shall conform to the requirements of 5-3.4, as appropriate for the ceiling configuration. Where detectors are installed in ceilings used for environmental air, detector spacing shall also conform to 5-3.5.1 and 5-3.5.2.

5-3.4.10 **Partitions.** Where partitions extend upward to within 18 in. (460 mm) of the ceiling, they shall not influence the spacing. Where the partition extends to within less than 18 in. (460 mm) of the ceiling, the effect of smoke travel shall be considered in the reduction of spacing.

Research on fire plumes and ceiling jets indicates that the thickness of the ceiling jet under most conditions is approximately 10% of the distance from the floor to the ceiling in the fire compartment. However, the ceiling jet does not have an abrupt boundary, the dimension used for its thickness is dependent upon the velocity criterion used for the jet boundary. When considering partitions, the important factor is whether the partition impedes the flow of the ceiling jet across the ceiling in the case of a fire with an established plume or whether it impedes the flow of smoke entrained in the normal air currents in the case of a small, low-energy fire. In spaces where the ceiling is approximately 10 feet high and allowing a 50% margin of error on the impact on the ceiling jet and environmental air, the reader should see that a partition extending to within 18 inches of the ceiling will very likely affect the ceiling jet, restricting the horizontal flow of smoke across the ceiling. Research that further quantifies the effect has not yet been conducted. Consequently, the designer is instructed to consider it on a qualitative basis. The treatment of partitions in this Code is very different than the treatment of partitions in NFPA 13, Standard on Installation of Sprinklers where the principal concern is the discharge pattern of the sprinkler head and the impact of the partition on the control of the fire.

5-3.5 Heating, Ventilating, and Air Conditioning (HVAC).

5-3.5.1* In spaces served by air-handling systems, detectors shall not be located where air from supply diffusers could dilute smoke before it reaches the detectors. Detectors shall be located to intercept the airflow toward the return air opening(s) where the opening(s) is not adjacent to the supply. Any detectors needed to meet this requirement shall be in addition to, and not a substitute for, detectors required by 5-3.4 to protect the balance of the area when the air-handling system is shut down.

Exception: Where detector rearrangement complies with the requirements for protection under both airflow and static conditions.

A-5-3.5.1 Detectors should not be located in a direct airflow nor closer than 3 ft (1 m) from an air supply diffuser.

This paragraph recommends a separation of at least 3 feet between an air supply diffuser and the detector. There may be situations where 3 feet is not adequate, depending upon the air velocity, the "throw" characteristics of the diffuser and diffuser size. This is currently the subject of the International Fire Detection Research Project being conducted by the National Fire Protection Research Foundation.

5-3.5.2 Plenums.

5-3.5.2.1 In under-floor spaces and above-ceiling spaces that are used as HVAC plenums, detectors shall be listed for the anticipated environment (*see 5-3.6.1.1*). Detector spacings and locations shall be selected based upon anticipated airflow patterns and fire type.

In order to cool a room to 70°F, it may be necessary to introduce extremely frigid air into the room. Heating a room sometimes requires superheated air introduced into a room. Consequently, HVAC plenums usually have ambient conditions that are far more extreme than the spaces they support. Smoke detectors are electronic sensors whose operation is affected by the ambient temperature, the relative humidity, and, especially in the case of spot-type ionization detectors, the velocity of the air around the detector. Not all smoke detectors are listed for the range of conditions found in HVAC plenums.

5-3.5.2.2* Detectors placed in environmental air ducts or plenums shall not be used as a substitute for open area detectors (*see Section 5-10, Table A-5-3.6.1.1, A-5-10.1, and A-5-10.2*). Where open area protection is required, 5-3.4 shall apply.

A-5-3.5.2.2 Smoke might not be drawn into the duct or plenums when the ventilating system is shut down. Furthermore, when the ventilating system is operating, the detector(s) can be less responsive to a fire condition in the room of fire origin due to dilution by clean air.

5-3.6 Special Considerations.

It is important for the reader to recognize that there may be special considerations unique to a specific application or product that must be addressed if the system is to fulfill its design objective yet are not listed as minimum compliance criteria in this Code. While the Code makes every effort to establish minimum compliance criteria to address problems that have a documented history of affecting smoke detection systems, it cannot be assumed that this list is exhaustive and covers every conceivable contingency. The designer should be aware of any factor in the protected area that could contribute to unwanted alarms or could prevent the successful conveyance of smoke to the detector.

5-3.6.1 The selection and placement of smoke detectors shall take into consideration both the performance characteristics of the detector and the areas into which the detectors are to be installed to prevent nuisance alarms or improper operation after installation. Some of the considerations are provided in 5-3.6.1.1 through 5-3.6.1.3.

5-3.6.1.1* Smoke detectors shall be installed in areas where the normal ambient conditions are not likely to exceed the following range of environmental conditions:

(a) A temperature of 100°F (38°C), or a temperature 32°F (0°C); or

(b) A relative humidity of 93 percent; or

(c) An air velocity of 300 fpm (1.5 mps).

Exception: Detectors specifically designed for use in ambients exceeding the limits of 5-3.6.1.1(a) through (c) and listed for the temperature, humidity, and air velocity conditions expected.

A-5-3.6.1.1 Product-listing standards include tests for temporary excursions beyond normal limits. In addition to temperature, humidity, and velocity variations, smoke detectors should operate reliably under such common environmental conditions as mechanical vibration, electrical interference, and other environmental influences. Tests for these conditions are also conducted by the testing laboratories in their listing program. In those cases in which environmental conditions approach the limits shown in Table A-5-3.6.1.1, the detector manufacturer should be consulted for additional information and recommendations.

Different detection technologies are affected differently by these environmental extremes. Different makes and models within each group may be affected more or less than others. It is beyond the scope of this handbook to identify these effects beyond the generalities presented here. However, the reader must recognize that some detector designs are inherently more forgiving than others. The tests performed in the process of listing ascertain that a detector meets minimum performance criteria. There may be design features in specific devices that allow them to be effectively used in extreme environments that are beyond those considered in the listing evaluation. The manufacturer should be consulted when such an application is contemplated.

These environmental limits may require the designer to consider alternative detection modes. While smoke detection may be preferable from the early warning standpoint, heat or radiant energy detection may be a better choice when the hazard area is one which undergoes too broad a range of environmental conditions to allow the use of smoke detection.

5-3.6.1.2* To avoid nuisance alarms, the location of smoke detectors shall take into consideration normal sources of smoke, moisture, dust or fumes, and electrical or mechanical influences.

A-5-3.6.1.2 Smoke detectors can be affected by electrical and mechanical influences and by aerosols and particulate matter found in protected spaces. The location of detectors should be such that the influences of aerosols and particulate matter from sources such as those in Table A-5-3.6.1.2(a) are minimized. Similarly, the influences of electrical and mechanical factors shown in Table A-5-3.6.1.2(b) should be minimized. While it might not be possible to isolate environmental factors totally, an awareness of these factors during system layout and design favorably affects detector performance.

Table A-5-3.6.1.2(a)   Common Sources of Aerosols and Particulate Matter Moisture

Moisture	Combustion Products and Fumes (continued)
Live steam	Excessive tobacco smoke
Steam tables	Heat treating
Showers	Corrosive atmospheres
Humidifiers	Dust or lint
Slop sink	Linen/bedding handling
Humid outside air	Sawing, drilling, and grinding
Water spray	Pneumatic transport
	Textile and agricultural processing
Combustion Products and Fumes	
Cooking equipment	**Engine Exhaust**
Ovens	
Dryers	Gasoline forklift trucks
Fireplaces	Diesel trucks and locomotives
Exhaust hoods	Engines not vented to the outside
Cutting, welding, and brazing	**Heating Element with Abnormal Conditions**
Machining	
Paint spray	
Curing	Dust accumulations
Chemical fumes	Improper exhaust
Cleaning fluids	Incomplete combustion

Table A-5-3.6.1.2(b)   Sources of Electrical and Mechanical Influences on Smoke Detectors

Electrical Noise and Transients	Airflow
Vibration or shock	Gusts
Radiation	Excessive velocity
Radio frequency	
Intense light	
Lightning	
Electrostatic discharge	
Power supply	

Table A-5-3.6.1.1   Environmental Conditions that Influence Detector Response

Detection Protection	Air Velocity >300 ft (>91.44 m)/min	Altitude >3000 ft (>914.4 m)	Humidity >93% RH	Temp. <32°F >100°F (<0°C >37.8°C)	Color of Smoke
Ion	X	X	X	X	O
Photo	O	O	X	X	X
Beam	O	O	X	X	O
Air Sampling	O	O	X	X	O

X = Can affect detector response.
O = Generally does not affect detector response.

In applications where the factors outlined in Tables A-5-3.6.1.2(a) and Table A-5-3.6.1.2(b) cannot be sufficiently limited to allow reasonable stability and response times, alternate modes of fire detection should be considered.

5-3.6.1.3 Detectors shall not be installed until after the construction clean-up of all trades is complete and final.

Exception: Where required by the authority having jurisdiction for protection during construction.

Detectors that have been installed prior to final clean-up by all trades shall be cleaned or replaced in accordance with Chapter 7.

Many needless alarms have been caused by the early installation of smoke detectors. This subsection forbids that practice unless the authority having jurisdiction requires it. In this latter case the detectors must be either replaced or cleaned after all construction trades have finished their work. In all other cases, smoke detectors are not allowed to be installed until all finish work is complete.

The authority having jurisdiction may allow the installation of smoke detectors with protective covers. These covers may not keep the detector entirely free from contaminants therefore cleaning of the detectors after all construction trades have finished their work may still be necessary. If these covers are used, the contractor must ensure that they are all removed when the construction trades have completed their work. If the authority having jurisdiction requires the covers be removed at the end of each day, it is good practice to number the covers to ensure all have been removed and then replaced the next morning. Again, if the covers are removed during the con-

Figure 5.21 Smoke detector with protective plastic cover installed. Photograph courtesy of PDH System Co., Braintree, MA.

struction process, the detectors will need to be inspected closely and cleaned where necessary. See Figure 5.21.

5-3.6.2 Spot-Type Detectors.

5-3.6.2.1 Smoke detectors having a fixed temperature element as part of the unit shall be selected in accordance with Table 5-2.1.1.1 for the maximum ceiling temperature that can be expected in service.

5-3.6.2.2* Holes in the back of a detector shall be covered by a gasket, sealant, or equivalent, and the detector shall be mounted so that airflow from inside or around the housing does not prevent the entry of smoke during a fire or test condition.

A-5-3.6.2.2 Airflow through holes in the rear of a smoke detector can interfere with smoke entry to the sensing chamber. Similarly, air from the conduit system can flow around the outside edges of the detector and interfere with smoke reaching the sensing chamber. Additionally, holes in the rear of a detector provide a means for entry of dust, dirt, and insects, each of which can adversely affect the detector's performance.

The conditions listed in A-5-3.6.2.2 have been encountered frequently enough to warrant inclusion of the requirements in 5-3.6.2.2 into the code. However, the list of installation related problems in A-5-3.6.2.2 cannot be assumed to be exhaustive. Once again, the designer should be aware of any factor in the protected area that could contribute to unwanted alarms or could prevent the successful conveyance of smoke to the detector.

5-3.6.3 Projected Beam-Type Detectors.

5-3.6.3.1 Projected beam-type detectors and mirrors shall be firmly mounted on stable surfaces so as to prevent false or erratic operation due to movement. The beam shall be so designed that small angular movements of the light source or receiver do not prevent operation due to smoke and do not cause nuisance alarms.

Contrary to popular belief, buildings move. Portions of buildings vibrate due to passing traffic on nearby streets. They sway due to wind or uneven thermal expansion; even the ebb and flow of the tides can cause ocean-front buildings to flex. Modern curtain-wall/steel-frame buildings are designed to flex. This, however, places a demand on the fire detection systems, especially fire alarm systems utilizing projected beam smoke detection, as they must be able to accommodate the natural or designed movement of the building. The manufacturers of projected beam detection devices provide installation instructions that address the potential for this type of difficulty. Some manufacturers do not allow the use of mirrors due to the physical instability of mounting surfaces and building movement.

5-3.6.3.2* The light path of projected beam-type detectors shall be kept clear of opaque obstacles at all times.

A-5-3.6.3.2 Where the light path of a projected beam-type detector is abruptly interrupted or obscured, the unit should not initiate an alarm. It should give a trouble signal after verification of blockage.

Modern projected beam detectors use "obscuration algorithms" in their software that can distinguish the progressive obscuration that occurs during a fire with the step-wise obscuration that is usually indicative of interference in the beam by an opaque object. However, Christmas decorations, party balloons and hanging plants have been known to cause problems in spite of the most sophisticated software.

5-3.6.4 Air Sampling-Type Detectors.

In addition to the cloud chamber type of smoke detector, several varieties of aspirating-type air-sampling smoke detectors exist. These detectors are essentially photoelectric smoke detectors with aspirating fans and a control unit. The apparatus as a whole constitutes a smoke detector. These detectors are used in a variety of applications where the designer is concerned with the effects of high airflow on smoke detection. Because of their sensitivity ranges, air-sampling detectors are also used in areas housing very valuable equipment. See Figures 5.22, 5.23 and 5.24.

5-3.6.4.1* Sampling pipe networks shall be designed on the basis of and shall be supported by sound fluid dynamic principles to ensure proper performance. Network design details shall include calculations showing the flow characteristics of the pipe network and for each sample port.

A-5-3.6.4.1 **Air Sampling-Type Detectors.** A single-pipe network has a shorter transport time than a multiple-pipe network of similar length pipe; however, a multiple-pipe system provides a faster smoke transport time than a single-pipe system of the same total length. As the number of sampling holes in a pipe increases, the smoke transport time increases. Where practical, pipe run lengths in a multiple-pipe system should be nearly equal, or the system should be otherwise pneumatically balanced.

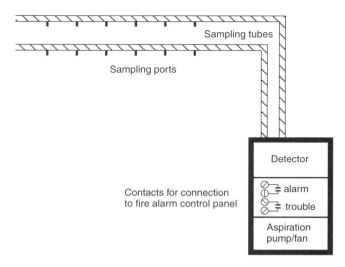

Figure 5.22 *The air sampling detector uses sampling tubes to convey smoke laden air to the central detection unit. Drawing courtesy of J.M.Cholin Consultants, Inc., Oakland, NJ.*

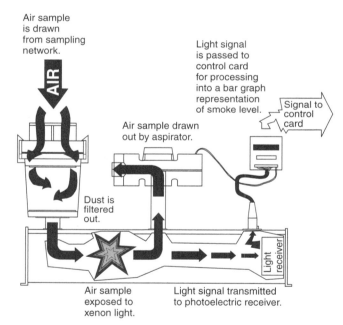

Figure 5.23 *How an optical air-sampling system works. Drawing courtesy of Vision Systems Inc., Hingham, MA.*

The manufacturers of this type of smoke detection unit provide engineering guidelines in their installation manuals that ensure that the products meet the criteria of 5-3.6.4.1. These guidelines are evaluated by testing laboratories as part of the listing evaluation procedure. The factors in A-5-3.6.4.1 are generalizations the designer can

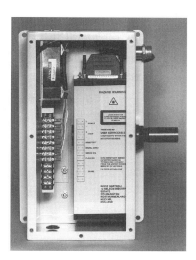

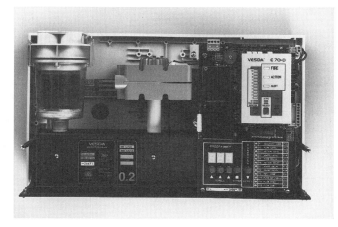

Figure 5.24 *Typical air sampling-type smoke detectors with cover in place [(a) and (c)] and with cover removed [(b) and (d)].* *Photographs courtesy of Top: Kidde-Fenwal Inc., Ashland, MA.; Bottom: Vision Systems Inc., Hingham, MA.*

use as guidance in deciding the type of piping network that best serves the application under consideration.

5-3.6.4.2* Air sampling detectors shall give a trouble signal where the airflow is outside the manufacturer's specified range. The sampling ports and in-line filter (where used) shall be kept clear in accordance with the manufacturer's documented instructions.

A-5-3.6.4.2 The air sampling-type detector system should be able to withstand dusty environments by either air filtering or electronic discrimination of particle size. The detector should be capable of providing optimal time delays of alarm outputs to eliminate nuisance alarms due to transient smoke conditions. The detector should also provide facilities for the connection of monitoring equipment for the recording of background smoke level information necessary in setting alert and alarm levels and delays.

5-3.6.4.3 Air sampling network piping and fittings shall be airtight and permanently fixed. Sampling system piping shall be conspicuously identified as "SMOKE DETECTOR SAMPLING TUBE. DO NOT DISTURB," as follows:

(a) At changes in direction or branches of piping;

(b) At each side of penetrations of walls, floors, or similar barriers;

(c) At intervals on piping sufficient to provide ready visibility within the space, but no greater than 20 ft (6 m).

5-3.6.5* **High Rack Storage.** Where smoke detectors are installed in high rack storage areas, consideration shall be given to installing detectors at several levels in the racks [*see Figures A-5-3.6.5(a) and (b).*] Where detectors are installed to actuate a suppression system, see NFPA 231C, *Standard for Rack Storage of Materials.*

Fire protection for high rack storage warehouses is a particularly difficult problem. The fuel load per unit of floor area is extremely high, the accessibility of the fuel is relatively low, and the combustibility of the materials in any given rack can vary from nominally noncombustible to flammable. The orientation of the fuel also creates vertical "flues" between the combustibles which produces ideal conditions for the propagation of the fire and the worst possible conditions for extinguishment. This makes early detection and rapid extinguishment of the fire in the incipient stages critical because once the fire becomes well established it is virtually impossible to extinguish. There have been a number of catastrophic total losses in high rack storage facilities in the past decade.

The location guidance provided strives to assure that any flue spaces created by the stored commodities are covered with a detector at some level. Care must also be used in installing such detectors as they are vulnerable to damage as commodities are moved in and out of the storage racks. While it may seem impossible to maintain accessibility for service and maintenance while both locating detectors for maximum speed of response and minimum exposure to damage from operations it is not! There are system designs that have satisfied all three of these apparently conflicting requirements.

Air sampling type smoke detectors, with the piping network extended throughout each rack have been used successfully in this application.

A-5-3.6.5 High Rack Storage.

For the most effective detection of fire in high rack storage areas, detectors should be located on the ceiling above each aisle and at intermediate levels in the racks. This is necessary to detect smoke that is trapped in the racks at an early stage of fire development, when insufficient thermal energy is released to carry the smoke to the ceiling. Earliest detection of smoke is achieved by locating the intermediate level detectors adjacent to alternate pallet sections as shown in Figures A-5-3.6.5(a) and (b). The detector manufacturer's recommendations and engineering judgment should be followed for specific installations.

A projected beam-type detector may be permitted to be used in lieu of a single row of individual spot-type smoke detectors.

Sampling ports of an air sampling-type detector may be permitted to be located above each aisle to provide coverage equivalent to the location of spot-type detectors. The manufacturer's recommendations and engineering judgment should be followed for the specific installation.

5-3.6.6 High Air Movement Areas.

5-3.6.6.1 General.

The purpose and scope of 5-3.6.6 are to provide location and spacing guidance for smoke detectors intended for early warning of fire in high air movement areas.

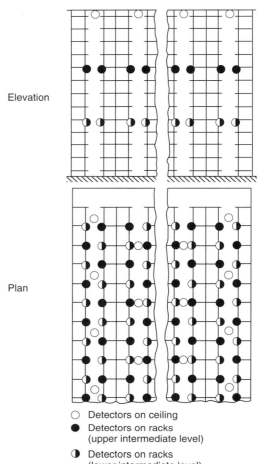

Elevation

Plan

○ Detectors on ceiling
● Detectors on racks
   (upper intermediate level)
◑ Detectors on racks
   (lower intermediate level)

Figure A-5-3.6.5(a) *Detector location for solid storage (closed rack) in which transverse and longitudinal flue spaces are irregular or nonexistent, as for slatted or solid shelved storage.*

Exception: Detectors provided for the control of smoke spread are covered by the requirements of Section 5-10.

The most regularly encountered example of a high-air-movement area is the data center (computer room), specifically its underfloor and above-ceiling plenums. This is by no means the only area that falls into this category. In general, areas where the air velocity across the detector exceeds 300 feet per minute (1.5 meters per second) are considered high-air-movement ambients. Table 5-3.6.6.3 and Figure 5-3.6.6.3 provide the detector spacing for high air movement ambients. Any time high airflow is encountered, consideration should be given to reducing the spacing of spot-type detectors or utilizing detectors that are not as affected by high airflow.

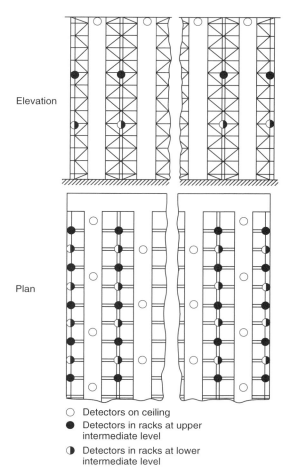

Figure A-5-3.6.5(b) *Detector location for pelletized storage (open rack) or no shelved storage in which regular transverse and longitudinal flue spaces are maintained.*

5-3.6.6.2 **Location.** Smoke detectors shall not be located directly in the airstream of supply registers.

5-3.6.6.3 **Spacing.** Smoke detector spacing depends upon the movement of air within the room (including both supplied and recirculated air), which can be designated as minutes per air change or air changes per hour. Spacing shall be in accordance with Table 5-3.6.6.3 and Figure 5-3.6.6.3.

Exception: Air sampling or projected beam smoke detectors installed in accordance with the manufacturer's documented instructions.

Because of the very high value of a data center, when a smoke detection system is being designed for such a site, it is common for spacing to be reduced. This spacing may be derived from Table 5-3.6.6.3 and Figure 5-3.6.6.3.

The AHJ ultimately determines how the reduced spacing provided in Table 5-3.6.6.3 and Figure 5-3.6.6.3 will be used for the underfloor and above ceiling plenums of the data center, especially when smoke detectors are used for the release of a special hazards extinguishing agent. Some AHJs will compute the rate of air change based upon the whole air volume, including the room, underfloor plenum and above ceiling plenum. In other circumstances the above ceiling space is not part of the working air volume of the hazard area and only the volume of the room and the underfloor are used to compute air changes per hour. Before the design process is begun the proprietary HVAC system must be well understood and the designer and the AHJ must agree on what air volume the calculations are to be based.

The reduced spacing for spot-type detectors is based upon the concerns over the dilution of smoke. Instead of forming a localized plume of relatively high smoke concentration the smoke is uniformly mixed throughout the entire proprietary air volume by the HVAC system, retarding the development of a detectable concentration until after the fire has matured. Data has not yet been presented to the Technical Committee that would enable

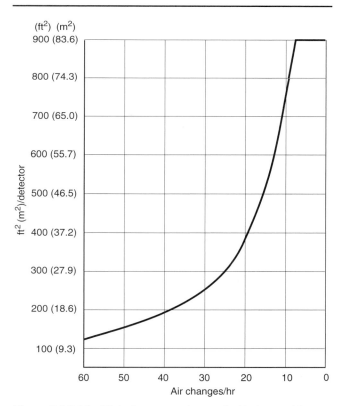

Figure 5-3.6.6.3 *High air movement areas (not to be used for under-floor or above-ceiling spaces).*

the committee to determine the effect of high air movement on projected beam or air sampling detectors. The designer must depend upon the data provided by the manufacturer.

Table 5-3.6.6.3   Smoke Detector Spacing Based on Air Movement

Min/Air Change	Air Changes/hr	ft^2 (m^2)/ Detector
1	60	125 (11.61)
2	30	250 (23.23)
3	20	375 (34.84)
4	15	500 (46.45)
5	12	625 (58.06)
6	10	750 (69.68)
7	8.6	875 (81.29)
8	7.5	900 (83.61)
9	6.7	900 (83.61)
10	6	900 (83.61)

5-4   Radiant Energy-Sensing Fire Detectors.

This section was entitled "Flame Detectors" until the 1990 edition of NFPA 72E, when very substantial revisions were made. The title was changed to "Radiant Energy-Sensing Fire Detectors" as this term was sufficiently inclusive to encompass both flame detectors and spark/ember detectors.

5-4.1   General.

5-4.1.1   The purpose and scope of Section 5-4 are to provide standards for the selection, location, and spacing of fire detectors that sense the radiant energy produced by burning substances. These detectors are categorized as flame detectors and spark/ember detectors.

5-4.1.1.1   Flame Detectors.   See Section 1-4, definition of Flame Detector.

5-4.1.1.2   Spark/Ember Detectors.   See Section 1-4, definition of Spark/Ember Detector.

Editions of NFPA 72E prior to 1990 did not address spark detection at all. Since the radiant emissions from an ember are very different from those of a flame and since spark/ember detectors are used in very different contexts than flame detectors, they require specific treatment.

5-4.1.2   Radiant Energy.   For the purpose of this code, radiant energy includes the electromagnetic radiation emitted as a by-product of the combustion reaction, which obeys the laws of optics. This includes radiation in the ultraviolet, visible, and infrared portions of the spectrum emitted by flames or glowing embers. These portions of the spectrum are distinguished by wavelengths as shown in Table 5-4.1.2.

Table 5-4.1.2   Spectrum Wavelength Ranges (μm)

Ultraviolet	**0.1 to 0.35**
Visible	0.36 to 0.75
Infrared	0.76 to 220

NOTE: 1.0 μm = 1000 nM = 10,000 Å.

This paragraph clarifies the distinction drawn in this Code between heat, which is commonly detected with heat detectors utilizing convective heat transfer, and radiant energy which is detected with either flame or spark/ember detectors using electro-optical methods to sense sparks, embers, and flames. See Section 1-4 for the definition of Wavelength and the associated Appendix material in A-1-4, Wavelength.

5-4.2*   Fire Characteristics and Detector Selection.

When using radiant energy sensing detectors the designer must match the detector to the fire to be detected with a degree of precision and attention to detail that is not generally required with other detection modes such as heat detection or smoke detection. In the effort to reduce unwanted alarms from non-fire radiant emission sources, detector designers have developed detectors which "look for" very specific radiant emission wavelengths that are uniquely associated with the combustion process. This has resulted in detectors which will detect one type of radiant emissions from one class of fuels while being virtually blind to fires involving other combustibles. A thorough understanding of how these detectors operate is necessary if they are to be properly applied. See Figures 5.25 through 5.30.

A-5-4.2   Operating Principles of Detectors.

(a)   Flame Detectors.   Ultraviolet flame detectors typically use a vacuum photodiode Geiger-Muller tube to detect the ultraviolet radiation that is produced by a flame. The photodiode allows a burst of current to flow for each ultraviolet photon that hits the active area of the tube. When the number of current bursts per unit time reaches a predetermined level, the detector initiates an alarm.

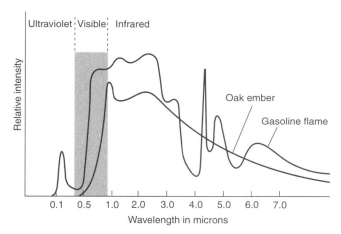

Figure 5.25 *Emission spectra of Class A and Class B combustibles. Drawing courtesy of J.M. Cholin Consultants, Inc., Oakland, NJ.*

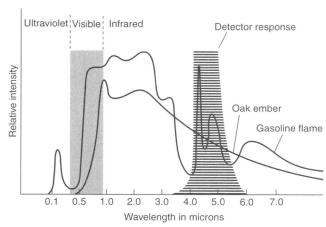

Figure 5.27 *The spectral response of a single wavelength infrared flame detector superimposed on the spectrum of "typical" radiators. Drawing courtesy of J.M. Cholin Consultants, Inc., Oakland, NJ.*

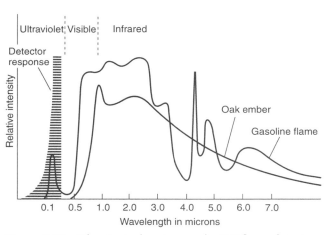

Figure 5.26 *The spectral response of a UV flame detector superimposed on the spectrum of "typical" radiators. Drawing courtesy of J.M. Cholin Consultants, Inc., Oakland, NJ.*

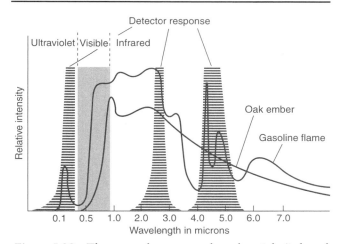

Figure 5.28 *The spectral response of an ultraviolet/infrared (UV/IR) flame detector superimposed on the spectrum of "typical" radiators. Drawing courtesy of J.M. Cholin Consultants, Inc., Oakland, NJ.*

A single wavelength infrared flame detector uses one of several different photocell types to detect the infrared emissions in a single wavelength band that are produced by a flame. These detectors generally include provisions to minimize alarms from commonly occurring infrared sources such as incandescent lighting or sunlight.

An ultraviolet/infrared (UV/IR) flame detector senses ultraviolet radiation with a vacuum photodiode tube and a selected wavelength of infrared radiation with a photocell and uses the combined signal to indicate a fire. These detectors need exposure to both types of radiation before an alarm signal can be initiated.

A multiple wavelength infrared (IR/IR) flame detector senses radiation at two or more narrow bands of wavelengths in the infrared spectrum. These detectors electronically compare the emissions between the band and initiate a signal where the relationship between the two bands indicates a fire.

NOTE: Some UV/IR flame detectors require radiant emissions at 0.2 microns (UV) and 2.5 microns (IR). Other UV/IR flame detectors require radiant emissions at 0.2 microns (UV) and nominal 4.7 microns (IR).

NOTE: Some IR/IR flame detectors compare radiant emissions at 4.3 microns (IR) to a reference at nominal 3.8 microns (IR). Other IR/IR flame detectors use a nominal 5.6 microns (IR) reference.

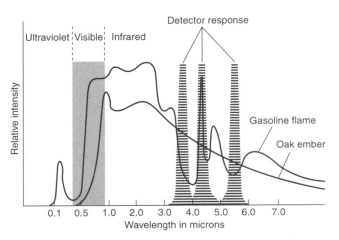

Figure 5.29 *The spectral response of a multiple wavelength infrared (IR/IR) flame detector superimposed on the spectrum of "typical" radiators. Drawing courtesy of J.M. Cholin Consultants, Inc., Oakland, NJ.*

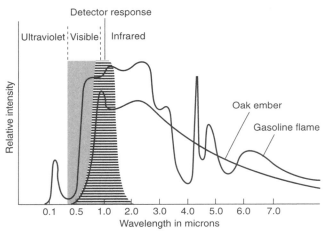

Figure 5.30 *The spectral response of an infrared spark/ ember detector superimposed on the spectrum of "typical" radiators. Drawing courtesy of J.M. Cholin Consultants, Inc., Oakland, NJ.*

(b) **Spark/Ember Detectors.** A spark/ember-sensing detector usually uses a solid state photodiode or phototransistor to sense the radiant energy emitted by embers, typically between 0.5 microns and 2.0 microns in normally dark environments. These detectors can be made extremely sensitive (microwatts), and their response times can be made very short (microseconds).

Product development in the field of radiant energy sensing fire detection has been very vigorous with new design concepts being introduced frequently. Consequently, it is impossible for the Code to provide an exhaustive list of the available technologies at the current rate of change. Recently, microcomputer based multispectrum flame detectors have become available which use a microcomputer to evaluate emissions from four, five and possibly six different bands in the UV, Visible and IR regions. However, the wavelength bands and operational architecture of these multispectrum devices have not been disclosed in sufficient detail to provide the Technical Committee with the requisite information for inclusion into this edition of the Code.

5-4.2.1* The type and quantity of radiant energy-sensing fire detectors shall be determined based upon the performance characteristics of the detector and an analysis of the hazard, including the burning characteristics of the fuel, the fire growth rate, the environment, the ambient conditions, and the capabilities of the extinguishing media and equipment.

The requirements of this paragraph effectively direct the system designer to work through a decision tree to arrive at the most appropriate detector for the fire hazard under consideration. The first decision is whether flame detection or spark/ember detection is the most appropriate type of radiant energy sensing fire detector.

The type of detector is often determined by the physical state of the material involved in the fire. Combustion occurs in the gas phase and in the solid phase. Flammable gases, flammable liquids, combustible liquids and many combustible solids will form a flame (See definition in Section 1-4). The combustion takes place in the gas phase, regardless of the physical state of the unburned fuel. The heat from the combustion gasifies the fuel allowing it to mix with air, supporting the flame. Since gas molecules are free to vibrate in free space the flame spectra show typical emission spikes that are indicative of flame intermediates and products. Many solids also burn in the solid phase. In solid phase combustion the molecules on the surface of the fuel particle are oxidized off the surface of the particle without the development of a layer of gasified fuel which could produce a true flame. Therefore, combustion intermediates (partially oxidized molecules), and, often combustion products are locked-up on the surface of the fuel particle and are not free to assume the diverse vibrational states of a gas phase molecule. Consequently the radiant emissions are profoundly different in solid phase combustion than they are in gas phase combustion. This difference in combustion radiant emissions necessitates different types of radiant energy sensing detectors for the different physical combustion states.

The paragraph lists the criteria that must be considered during the decision making process. The following material of A-5-4.2.1 provides additional insight in how the decision making process is driven by the detector performance criteria, the anticipated fire and hazard environment.

A-5-4.2.1   The radiant energy from a flame or spark/ember is comprised of emissions in various bands of the ultraviolet, visible, and infrared portions of the spectrum. The relative quantities of radiation emitted in each part of the spectrum are determined by the fuel chemistry, the temperature, and the rate of combustion. The detector should be matched to the characteristics of the fire.

Almost all materials that participate in flaming combustion emit ultraviolet radiation to some degree during flaming combustion, whereas only carbon-containing fuels emit significant radiation at the 4.35 micron (carbon dioxide) band used by many detector types to detect a flame. *See Figure A-5-4.2.1.*

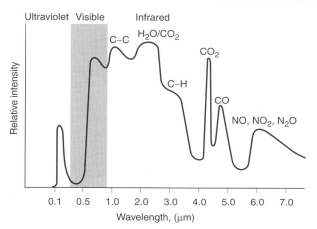

Figure A-5-4.2.1   *Spectrum of a "typical" flame (free-burning gasoline).*

The radiant energy emitted from an ember is determined primarily by the fuel temperature (Planck's Law Emissions) and the emissivity of the fuel. Radiant energy from an ember is primarily infrared and, to a lesser degree, visible in wavelength. In general, embers do not emit ultraviolet energy in significant quantities (0.1 percent of total emissions) until the ember achieves temperatures of 2000°K (1727°C or 3240°F). In most cases, the emissions are included in the band of 0.8 microns to 2.0 microns, corresponding to temperatures of approximately 750°F to 1830°F (398°C to 1000°C).

Most radiant energy detectors have some form of qualification circuitry within them that uses time to help distinguish between spurious, transient signals and legitimate fire alarms. These circuits become very important where the anticipated fire scenario and the ability of the detector to respond to that anticipated fire are considered. For example, a detector that utilizes an integration circuit or a timing circuit to respond to the flickering light from a fire might not respond well to a deflagration resulting from the ignition of accumulated combustible vapors and gases, or where the fire is a spark that is traveling

up to 328 ft/sec (100 m/sec) past the detector. Under these circumstances, a detector that has a high speed response capability is most appropriate. On the other hand, in applications where the development of the fire is slower, a detector that utilizes time for the confirmation of repetitive signals is appropriate. Consequently, the fire growth rate should be considered in selecting the detector. The detector performance should be selected to respond to the anticipated fire.

The radiant emissions are not the only criteria to be considered. The medium between the anticipated fire and the detector is also very important. Different wavelengths of radiant energy are absorbed with varying degrees of efficiency by materials that are suspended in the air or that accumulate on the optical surfaces of the detector. Generally, aerosols and surface deposits reduce the sensitivity of the detector. The detection technology utilized should take into account those normally occurring aerosols and surface deposits to minimize the reduction of system response between maintenance intervals. It should be noted that the smoke evolved from the combustion of middle and heavy fraction petroleum distillates is highly absorptive in the ultraviolet end of the spectrum. Where using this type of detection, the system should be designed to minimize the effect of smoke interference on the response of the detection system.

The environment and ambient conditions anticipated in the area to be protected impact the choice of detector. All detectors have limitations on the range of ambient temperatures over which they will respond, consistent with their tested or approved sensitivities. The designer should make certain that the detector is compatible with the range of ambient temperatures anticipated in the area in which it is installed. In addition, rain, snow, and ice attenuate both ultraviolet and infrared radiation to varying degrees. Where anticipated, provisions should be made to protect the detector from accumulations of these materials on its optical surfaces.

This Appendix material provides the salient criteria for radiant energy sensing fire detector selection in general terms. The hazard analysis must, therefore, begin with the determination as to whether the combustible will burn in the solid phase as an ember or in the gaseous phase as a flame. That determination then points the designer toward the spark/ember detector (for solid phase combustion) or the flame detector (for gas phase combustion). The engineering manuals provided by the manufacturers of the various detectors under consideration should be used to determine the usefulness of a particular device for the hazard under consideration.

5-4.2.2   The selection of the radiant energy-sensing detectors shall be based upon:

(a)   The matching of the spectral response of the detector to the spectral emissions of the fire or fires to be detected; and

(b)   Minimizing the possibility of spurious nuisance alarms from nonfire sources inherent to the hazard area. (*See A-5-4.2.1.*)

Once the type of combustion has been determined and the decision regarding type of detector to be used has been made, the decision regarding the most appropriate model or technology must be made. This is the second stage of the decision-tree.

The expected emission spectrum from the fuel is matched to the wavelength bands of the candidate detector to assure response to the fire, using the criteria listed and the detector manufacturer's engineering manual. The performance capabilities of the detector must be matched with the known radiant emissions of the fuel or the performance attributes that were verified by the nationally recognized testing laboratory during the listing evaluation to ascertain that the detector is appropriate for the fuels to be detected.

Then the candidate detector must be evaluated for its unwanted alarm immunity relative to the ambient or spurious alarm sources anticipated in the hazard area. The information provided in A-5-4.2.1 is also relevant to this paragraph and should be used for this stage of the decision process also.

Finally the impact on the full range of expected ambient conditions on both the detection capability as well as the stability of the candidate detector is considered. Both flame detectors and spark/ember detectors are routinely installed outdoors where they are exposed to the weather and fluctuations in temperature. Special attention must be given to the temperature range limits, and other limiting weather related conditions, provided by the manufacturer to ensure that the detector has been qualified for the anticipated extremes. The prudent designer will document his/her decision-making process in writing for future reference.

5-4.3   Spacing Considerations.

The spacing considerations for radiant energy sensing fire detectors are derived from the physics of light transmission rather than the fire plume dynamics and the fluid flow physics that govern the spacing of heat and smoke detectors. Consequently, when using radiant energy sensing fire detectors the spacing of the detectors is determined by the location and aiming of the devices which is in turn determined by two critical factors: the Field of View (see definition in Section 1-4) of the detector and the sensitivity (see definition in Section 1-4) of the detector.

5-4.3.1   General Rules.

5-4.3.1.1*   Radiant energy-sensing fire detectors shall be employed consistent with the listing or approval and the

inverse square law, which defines the fire size versus distance curve for the detector.

The inverse square law relates the size of the fire, the detector sensitivity, and the distance between the fire and the detector. It is applicable to all radiant energy sensing detectors. It enables the engineer to compute with considerable precision how large the fire must get before there is enough radiant energy hitting the detector to cause an alarm. This is critical as it will contribute to the determination of the number of detectors of given sensitivity, location and aiming that are necessary to detect a fire of given size. Remember that fire size is quantified in units of power output, either Btu/second or watts, regardless of the type of radiant energy sensing fire detector under consideration.

A-5-4.3.1.1   All optical detectors respond according to the following theoretical equation:

$$S = \frac{kpe\zeta d}{d^2}$$

where:

k = proportionality constant for the detector
p = radiant power emitted by the fire
e = Naperian logarithm base (2.7183)
ζ = extinction coefficient of air
d = distance between the fire and the detector
S = radiant power reaching the detector.

The sensitivity (S) typically is measured in nanowatts. This equation yields a family of curves similar to the one shown in Figure A-5-4.3.1.1.

The curve defines the maximum distance at which the detector consistently detects a fire of defined size and fuel. Detectors should be employed only in the shaded area beneath the curve.

Under the best of conditions, with no atmospheric absorption, the radiant power reaching the detector is reduced by a factor of 4 if the distance between the detector and the fire is doubled. For the consumption of the atmospheric extinction, the exponential term Zeta (ζ) is added to the equation. Zeta is a measure of the clarity of the air at the wavelength under consideration. Zeta is affected by humidity, dust, and any other contaminants in the air that are absorbent at the wavelength in question. Zeta generally has values between −0.001 and −0.1 for normal ambient air.

5-4.3.1.2   Detectors shall be used in sufficient quantity and positioned so that no point requiring detection in the hazard area is obstructed or outside the field of view of at least one detector.

A flame detector or spark/ember detector cannot detect what it cannot "see". Review the definition of the term "field of view" in Section 1-4. It has a sensitivity

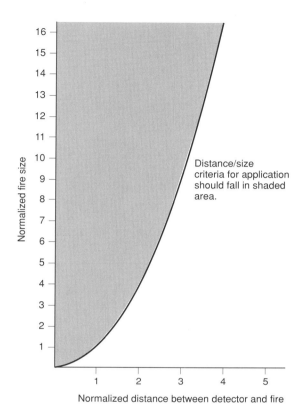

Figure A-5-4.3.1.1 Generalized fire size vs. distance.

other design considerations that are more specific to one type of detector or an other. These are addressed in sections 5-4.3.2 for flame detectors and 5-4.3.3 for spark/ember detectors, respectively.

5-4.3.2 Spacing Considerations for Flame Detectors.

5-4.3.2.1* The location and spacing of detectors shall be the result of an engineering evaluation that takes into consideration:

   (a) The size of the fire that is to be detected.

   (b) The fuel involved.

   (c) The sensitivity of the detector.

   (d) The field of view of the detector.

   (e) The distance between the fire and the detector.

   (f) The radiant energy absorption of the atmosphere.

   (g) The presence of extraneous sources of radiant emissions.

   (h) The purpose of the detection system.

   (i) The response time required.

Keep in mind that in the context of this section on radiant energy sensing fire detectors the term "spacing" includes the number, location and aiming of the detectors selected for the hazard area. In every system design using flame detectors, the spacing of each unit in the system must address the criteria listed in paragraph 5-4.3.2.1.

A-5-4.3.2.1 Flame Detector Applications and Stability.

   (a) The types of application for which flame detectors are suitable are:

     1. High-ceiling, open-spaced buildings such as warehouses and aircraft hangers.

     2. Outdoor or semioutdoor areas where winds or draughts can prevent smoke from reaching a heat or smoke detector.

     3. Areas where rapidly developing flaming fires can occur, such as aircraft hangers, petrochemical production, storage and transfer areas, natural gas installations, paint shops, or solvent areas.

     4. Areas needing high fire risk machinery or installations, often coupled with an automatic gas extinguishing system.

     5. Environments that are unsuitable for other types of detectors.

   (b) Some extraneous sources of radiant emissions that have been identified as interfering with the stability of flame detectors include:

     1. Sunlight

     2. Lightning

criterion attached to it. It is the angle off the optical axis of the detector where the effective sensitivity is 50% of the on-axis sensitivity. All points where a fire can exist in the hazard area must be within the field of view of at least one detector. This requirement also effectively demands that the manufacturer provide sensitivity versus angle of incidence data as part of the information in its engineering manual.

When flame detectors are used to release extinguishing agents such as Aqueous Film Forming Foam (AFFF), alarm signals from two or more detectors are usually required before the agent is released. Under those circumstances the designer should apply this paragraph in a manner that requires all points in the hazard area where a fire can exist to be within the fields of view of that number of detectors required to discharge the extinguishing agent. Otherwise, the fire could occur in a portion of the hazard area that is within the field of view of only one detector. The release of the extinguishing agent would be delayed until the fire grows to a size sufficient to alarm the additional confirmation detector(s). This would result in far more fire damage and a greater threat of loss of life.

The requirements in 5-4.3.1.1 and 5-4.3.1.2 pertain to both flame and spark/ember detectors. There are

3. X-rays

4. Gamma rays

5. Cosmic rays

6. Ultraviolet radiation from arc welding

7. Electromagnetic interference (EMI, RFI)

8. Hot objects

9. Artificial lighting.

Do not infer from this list that any one detector type or model is susceptible to all or even a majority of these unwanted alarm sources. Different types and models of flame detectors exhibit different degrees of susceptibility to some of these sources. Despite the best intentions and ardent efforts of the flame detector manufacturing community, the completely spurious alarm proof flame detector has not yet been invented.

5-4.3.2.2   The system design shall specify the size of the flaming fire of given fuel that is to be detected.

This is a performance-based code requirement. Because of the complexities inherent in the design of flame detection systems there must be a performance criterion driving the design. The performance criterion is the detection of a fire of specified size and fuel. Fire size is usually measured in kilowatts (kW) or British Thermal Units per second (Btu/second) but more information is necessary in this context because flames are optically dense radiators (meaning that the radiation from the back side of the flame does not travel through the flame toward the detector, it is reabsorbed). Consequently, the flame detector only "sees" the profile of the fire—its width and height. The flame height is proportional to the heat release rate (kW or Btu/sec). Consequently both fire width and heat release rate are necessary to quantify the size of a fire. Many designers have not yet made the conversion from simply stipulating a fire size criterion in terms of a pool fire of given fuel and area.

5-4.3.2.3*   In applications where the fire to be detected could occur in an area not on the optical axis of the detector, the distance shall be reduced or detectors added to compensate for the angular displacement of the fire in accordance with the manufacturer's documented instructions.

A-5-4.3.2.3   The greater the angular displacement of the fire from the optical axis of the detector, the larger the fire must become before it is detected. This phenomenon establishes the field of view of the detector. Figure A-5-4.3.2.3 shows an example of the effective sensitivity versus angular displacement of a flame detector.

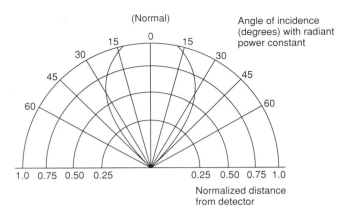

Figure A-5-4.3.2.3   *Normalized sensitivity vs. angular displacement.*

5-4.3.2.4*   In applications in which the fire to be detected is of a fuel that differs from the test fuel used in the process of listing or approval, the distance between the detector and the fire shall be adjusted consistent with the fuel specificity of the detector as established by the manufacturer.

In an effort to make flame detectors more sensitive yet more immune to unwanted alarms, designers began designing detectors that concentrated on very specific features of the flame spectrum, the emissions of the flame across the range of wavelengths from ultraviolet to infrared. In concept, such flame detectors infer that a flame exists if an emission of a specific wavelength or set of wavelengths is detected. However, one fuel emits a different radiant intensity at a given wavelength than another fuel. This gives rise to detectors that are fuel specific. There are cases where a flame detector may be several times more sensitive to one fuel than another. The designer is effectively required by this language to obtain flame spectra of potential fuels in the hazard area and response curves from the detector manufacturer to make certain the detector will respond to the fuel(s) involved. Furthermore, if the detector chosen for the system is less sensitive to one of the fuels in the hazard area the spacing (quantity, location and aiming) of the detectors must be adjusted commensurately.

A-5-4.3.2.4   Virtually all radiant energy-sensing detectors exhibit some kind of fuel specificity. Where burned at uniform rates [J/sec (W)], different fuels emit different levels of radiant power in the ultraviolet, visible, and infrared portions of the spectrum. Under free-burn conditions, a fire of given surface area but of different fuels burns at different rates [J/sec (W)] and emits varying levels of radiation in each of the major portions of the spectrum. Most radiant energy detectors designed to detect flame are qualified based upon a defined fire under specific conditions. Where employing these detectors for fuels other than the defined fire, the designer should make certain that the appropriate adjustments to the maximum distance between the detector and the fire are made consistent with the fuel specificity of the detector.

5-4.3.2.5 Since flame detectors are essentially line of sight devices, special care shall be taken to ensure that their ability to respond to the required area of fire in the zone that is to be protected is not compromised by the presence of intervening structural members or other opaque objects or materials.

Keep in mind that some atmospheric contaminants including vapors and gases may be opaque at the wavelengths used by some flame detectors. This can have significant effect on the performance of the system. See A-5-4.3.1.1

5-4.3.2.6* Provisions shall be made to sustain detector window clarity in applications where airborne particulates and aerosols coat the detector window between maintenance intervals and affect sensitivity.

A-5-4.3.2.6 The means by which this requirement has been satisfied include:

(a) Lens clarity monitoring and cleaning where a contaminated lens signal is rendered.

(b) Lens air purge.

The need to clean detector windows can be reduced by the provision of air purge devices. These devices are not foolproof, however, and are not a replacement for regular inspection and testing. Radiant energy-sensing detectors should not be placed in protective housings (e.g., behind glass) to keep them clean, unless such housings are listed for the purpose. Some optical materials are absorptive at the wavelengths used by the detector.

5-4.3.3 Spacing Considerations for Spark/Ember Detectors.

5-4.3.3.1* The location and spacing of detectors shall be the result of an engineering evaluation that takes into consideration:

(a) The size of the spark or ember that is to be detected.

(b) The fuel involved.

(c) The sensitivity of the detector.

(d) The field of view of the detector.

(e) The distance between the fire and the detector.

(f) The radiant energy absorption of the atmosphere.

(g) The presence of extraneous sources of radiant emissions.

(h) The purpose of the detection systems.

(i) The response time required.

A-5-4.3.3.1 Spark/ember detectors are installed primarily to detect sparks and embers that could, if allowed to continue to burn, precipitate a much larger fire or explosion. Spark/ember detectors are typically mounted on some form of duct or conveyor, monitoring the fuel as it passes by. Usually, it is necessary to enclose the portion of the conveyor where the detectors are located, as these devices generally require a dark environment. Extraneous sources of radiant emissions that have been identified as interfering with the stability of spark/ember detectors include:

(a) Ambient light.

(b) Electromagnetic interference (EMI, RFI).

(c) Electrostatic discharge in the fuel stream.

Figure 5.31 shows typical applications where spark/ember detectors are used. Note that the detectors are located at a point along the duct or conveyor, monitoring the cross-section of the duct or conveyor at that one point by essentially "looking across" the duct. Commercially available, listed spark/ember detectors are designed to monitor a fuel stream as it moves past the detector. They are NOT designed to "look down the duct". The "capacitive" nature of the circuitry of this type of detector generally makes them incapable of detecting a slowly growing radiator; the radiator must move past the detector rapidly if it is to be detected.

5-4.3.3.2* The system design shall specify the size of the spark or ember of the given fuel that the detection system is to detect.

The size of an ember is measured in terms of watts or milliwatts. The radiant energy from an ember and hence its size, cannot be accurately inferred from a description listing diameter and temperature only. See Section 1-4 Spark/Ember Detector Sensitivity. Furthermore, the equation for the inverse square law in A-5-4.3.1.1 cannot be used to calculate the ability of the detector to detect the ember in question unless both the detector sensitivity and the ember size are specified in the same terms of radiant power, watts, milliwatts or microwatts.

As with the corresponding paragraph regarding flame detectors, this is a performance based design criterion which drives the entire system design.

A-5-4.3.3.2 There is a minimum ignition power (watts) for all combustible dusts. Where the spark or ember is incapable of delivering that quantity of power to the adjacent combustible

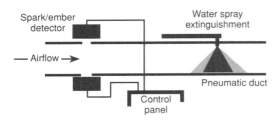

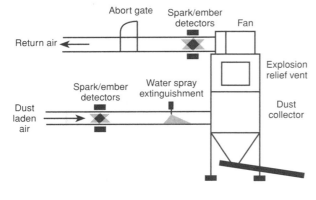

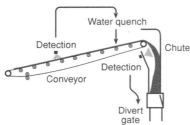

Figure 5.31 Spark/ember detectors are usually used on conveyance ducts and conveyors to detect embers in particulate solids as they are transported. The top drawing shows the general concept of spark/ember detectors. The middle drawing illustrates the application of spark/ember detectors to protect a dust collector. The bottom drawing illustrates the protection of a conveyor. Drawing courtesy of J.M. Cholin Consultants, Inc., Oakland, NJ.

material (dust), an expanding dust fire cannot occur. The minimum ignition power is determined by the fuel chemistry, fuel particle size, fuel concentration in air, and ambient conditions such as temperature and humidity.

5-4.3.3.3 Spark detectors shall be positioned so that all points within the cross section of the conveyance duct, conveyor, or chute where the detectors are located are within the field of view (as as defined in Section 1-4) of at least one detector.

5-4.3.3.4 The location and spacing of the detectors shall be adjusted using the inverse square law, modified for the atmospheric absorption and the absorption of nonburning fuel

suspended in the air in accordance with the manufacturer's documented instructions. *See A-5-4.3.1.1.*

5-4.3.3.5* In applications where the sparks to be detected could occur in an area not on the optical axis of the detector, the distance shall be reduced or detectors added to compensate for the angular displacement of the fire in accordance with the manufacturer's documented instructions.

A-5-4.3.3.5 The greater the angular displacement of the fire from the optical axis of the detector, the larger the fire must become before it is detected. This phenomenon establishes the field of view of the detector. Figure A-5-4.3.2.3 shows an example of the effective sensitivity versus angular displacement of a flame detector.

5-4.3.3.6* Provisions shall be made to sustain the detector window clarity in applications where airborne particulates and aerosols coat the detector window and affect sensitivity.

A-5-4.3.3.6 The means by which this requirement has been satisfied include:

(a) Lens clarity monitoring and cleaning where a contaminated lens signal is rendered.

(b) Lens air purge.

5-4.4   **Other Considerations.**

5-4.4.1 Radiant energy-sensing detectors shall be protected either by way of design or installation to ensure that optical performance is not compromised.

Since these types of detectors are usually installed in industrial environments where they must endure the rigors of difficult environments, the designer is cautioned to consider the long-term impact of the environment on the optical performance of the detectors.

5-4.4.2 Where necessary, radiant energy-sensing detectors shall be shielded or otherwise arranged to prevent action from unwanted radiant energy.

In some cases shielding a detector from radiant emissions coming from a portion of its field of view where the sole source of radiant emissions is a spurious source can be an effective way of dealing with the source of unwanted alarms. Many detectors are available with "scoops" or baffles to limit the field of view to a small portion of the total viewing area, providing the ability to operate in spite of the presence of a spurious alarm source. When considering such methods the manufacturer should be consulted. Also remember that reflected radiant emissions can also cause alarms.

5-4.4.3 Where used in outdoor applications, radiant energy-sensing detectors shall be shielded or otherwise arranged in a fashion to prevent diminishing sensitivity by conditions such as rain or snow and yet allow a clear field of vision of the hazard area.

5-4.4.4 A radiant energy-sensing fire detector shall not be installed in a location where the ambient conditions are known to exceed the extremes for which the detector has been listed.

5-5 Other Fire Detectors.

It is the intent of the Code to provide for the development of new technologies and to allow the use of such technologies consistent with sound principles of fire protection engineering. The requirements in this section provide for alternate methods not explicitly described in other sections of Chapter 5.

5-5.1 Detectors in the classification of "other fire detectors" are those that operate on principles different from detectors covered by Sections 5-2, 5-3, and 5-4. Such detectors shall be installed in all areas where they are required either by the appropriate NFPA standard or by the authority having jurisdiction.

5-5.2 Facilities for testing or metering or instrumentation to ensure adequate initial sensitivity and adequate retention thereof, relative to the protected hazard, shall be provided. These facilities shall be employed at regular intervals.

Chapter 7 of this Code outlines the required maintenance procedures and schedules for all components of a fire alarm system, including the initiating devices. Initiating devices covered by this section must be maintained pursuant to the criteria in Chapter 7 and the manufacturer's recommended practices.

5-5.3 These detectors shall operate where subjected to the abnormal concentration of combustion effects that occur during a fire, such as water vapor, ionized molecules, or other phenomena for which they are designed. Detection is dependent upon the size and intensity of fire to provide the necessary quantity of required products and related thermal lift, circulation, or diffusion for adequate operation.

5-5.4 Room sizes and contours, airflow patterns, obstructions, and other characteristics of the protected hazard shall be taken into account.

5-5.5 Location and Spacing.

5-5.5.1 The location and spacing of detectors shall be based on the principle of operation and an engineering survey of the conditions anticipated in service. The manufacturer's technical bulletin shall be consulted for recommended detector uses and locations.

5-5.5.2 Detectors shall not be spaced beyond their listed or approved maximums. Closer spacing shall be utilized where the structural or other characteristics of the protected hazard warrant.

5-5.5.3 Consideration shall be given to all factors that could affect the location and sensitivity of the detectors, including structural features such as the sizes and shapes of rooms and bays and their occupancies and uses, ceiling heights, ceiling and other obstructions, ventilation, ambient environment, stock piles, files, and fire hazard locations.

5-5.5.4 The overall situation shall be reviewed frequently to ensure that changes in structural or usage conditions that could interfere with fire detection are remedied.

5-5.6 Special Considerations. The selection and placement of detectors shall take into consideration both the performance characteristics of the detector and the areas into which the detectors are to be installed to prevent nuisance alarms or nonoperation after installation.

5-6 Sprinkler Waterflow Alarm-Initiating Devices.

5-6.1 The provisions of this section shall apply to devices that initiate an alarm indicating a flow of water in a sprinkler system.

5-6.2* Initiation of the alarm signal shall occur within 90 seconds of waterflow at the alarm-initiating device when flow equal to or greater than that from a single sprinkler of the smallest orifice size installed in the system occurs. Movement of water due to waste, surges, or variable pressure shall not be indicated.

A-5-6.2 The waterflow device should be field adjusted so that an alarm is initiated in no more than 90 seconds after a sustained flow of at least 10 gpm (40 L/min).

Features that should be investigated to minimize alarm response time include elimination of trapped air in the sprinkler system piping, use of an excess pressure pump, use of pressure drop alarm-initiating devices, or a combination thereof.

Care should be used where choosing waterflow alarm-initiating devices for hydraulically calculated looped systems and those systems using small orifice sprinklers. Such systems might incorporate a single point flow of significantly less than 10 gpm (40 L/min). In such cases, additional waterflow alarm-initiating devices or the use of pressure drop-type waterflow alarm-initiating devices might be necessary.

Care should be used where choosing waterflow alarm-initiating devices for sprinkler systems utilizing on-off sprinklers to ensure that an alarm is initiated in the event of a waterflow condition. On-off sprinklers open at a predetermined temperature and close when the temperature reaches a predetermined lower temperature. With certain types of fires, waterflow might occur in a series of short bursts of 10 seconds' to 30 seconds' duration each. An alarm-initiating device with retard might not detect waterflow under these conditions. An excess pressure system or a system that operates on pressure drop should be considered to facilitate waterflow detection on sprinkler systems utilizing on-off sprinklers.

Excess pressure systems can be used with or without alarm valves. The following is a description of one type of excess pressure system with an alarm valve.

An excess pressure system with an alarm valve consists of an excess pressure pump with pressure switches to control the operation of the pump. The inlet of the pump is connected to the supply side of the alarm valve, and the outlet is connected to the sprinkler system. The pump control pressure switch is of the differential type, maintaining the sprinkler system pressure above the main pressure by a constant amount. Another switch monitors low sprinkler system pressure to initiate a supervisory signal in the event of a failure of the pump or other malfunction. An additional pressure switch can be used to stop pump operation in the event of a deficiency in water supply. Another pressure switch is connected to the alarm outlet of the alarm valve to initiate a waterflow alarm signal when waterflow exists. This type of system also inherently prevents false alarms due to water surges. The sprinkler retard chamber should be eliminated to enhance the detection capability of the system for short duration flows.

In many facilities the sprinkler system is used as both a suppression system and a detection system. The flow of water is used to initiate the alarm. In a large system with large sprinkler risers, the flow from a single head has proven hard to detect. In addition, if there is any air in the piping it will act as a "gas cushion," allowing variations in water pressure from the street to cause water to slosh back and forth in the riser. This can cause unwanted alarms. Consequently, most flow switches are equipped with a retard feature that delays the transmis-

sion of a signal until after stable waterflow has been achieved.

The 90-second criterion can be a problem. If on/off systems are incorporated into the sprinkler system the retard adjustment on the waterflow alarm initiating device must be achieved in a manner that does not allow the failure to indicate on/off flow.

The 90 second criterion also can be a problem with large systems. Where the waterflow alarm initiating device is installed on a simple fire alarm control unit the retard adjustment is simple and straight forward. However, when addressable fire alarm systems are used to monitor waterflow alarm initiating devices, the worst-case polling delay of the addressable initiating device circuit must be added to the retard adjustment of the waterflow alarm initiating device in order to ensure that the total accumulated delay from the moment waterflow is begun, to the time that the fire alarm notification begins is not longer than 90 seconds. Care must be exercised when adjusting waterflow alarm initiating device retard delays on large sprinkler systems which are monitored by large fire alarm systems. The retard delay needed to prevent spurious waterflow alarm indications due to the extent of the sprinkler system added to the worst-case polling delay of the fire alarm system may result in an excessive delay in the activation of alarm notification.

Finally, it is clear the designer must be familiar with NFPA 13, *Standard for the Installation of Sprinkler Systems.*

5-6.3 Piping between the sprinkler system and a pressure actuated alarm-initiating device shall be galvanized or of nonferrous metal or other approved corrosion-resistant material of not less than ³⁄₈ in. (9.5 mm) nominal pipe size.

These requirements stem from the problems associated with piping corrosion and with the mechanical strength necessary to endure the environment of the sprinkler system.

5-7*   Detection of the Operation of Other Automatic Extinguishing Systems.

The operation of other fire extinguishing system(s) or suppression system(s) shall initiate an alarm signal by means appropriate to the system, such as agent flow or agent pressure, by alarm-initiating devices installed in accordance with their individual listings.

A-5-7 Appropriate means can include:

(a) Foam systems (flow of water).
(b) Pump activation.
(c) Differential pressure detectors.
(d) Halon (pressure detector).
(e) Carbon dioxide (pressure detector).

In any case, an alarm that activates the fire extinguishing system(s) or suppression system(s) may be permitted to be initiated from the detection system.

Many extinguishing systems include emergency mechanical manual release capability. This provides for the release of the extinguishing agent, bypassing the operation of the fire alarm system. See Figure 5.32. However, the fire alarm system usually serves as the nerve center for the all of the fire protection systems for the protected hazard area. The fire alarm system houses the electrical switching used to sound warnings, terminate fuel flow, release automatic door closers, activate emergency power interrupts, energize smoke management systems and similar functions that secure the hazard area. There must be a means to ensure that critical functions such as these are achieved if the extinguishing system discharge is caused by the mechanical release.

Discharge indication switches are the usual means of providing the extinguishing system operation signal to the fire detection control unit. Due to the critical function these switches perform, they must be listed for use with the specific make and model of extinguishing system. The connection of a listed extinguishing agent release initiating device to an appropriate initiating device circuit on the fire alarm control unit, consistent with the listings of both the unit and the initiating device, is critical for the successful operation of most special extinguishing systems. See Figure 5.33.

The reader should review:

NFPA 12, *Standard on Carbon Dioxide Extinguishing Systems*

NFPA 12A, *Standard on Halon 1301 Fire Extinguishing Systems*

NFPA 16, *Standard on the Installation of Deluge Foam-Water Sprinkler and Foam-Water Spray Systems*

NFPA 17, *Standard for Dry Chemical Extinguishing Systems*

NFPA 2001, *Standard on Clean Agent Fire Extinguishing Systems.*

Figure 5.32 Emergency manual cable release for extinguishing systems. Photograph courtesy of Kidde-Fenwal Protection Systems, Ashland, MA.

Figure 5.33 Typical extinguishing system pressure-actuated discharge switch. Photograph courtesy of Kidde-Fenwal Protection Systems, Ashland, MA.

5-8   Manually Actuated Alarm-Initiating Devices.

5-8.1   Manual fire alarm boxes shall be used only for fire alarm-initiating purposes. However, combination manual fire alarm boxes and guard's signaling stations shall be permitted.

This requirement stems from two concerns: credibility and reliability. If the manual fire alarm box is incorporated into some other non-fire-related assembly (with the single exception of guard's signaling stations), the probability of unwarranted operation is increased. This leads to false alarms and erodes the occupants' confidence in the system. Also, when manual fire alarms are combined with non-fire-related functions, there is an increased probability that a failure in the non-fire function will compromise the fire alarm system.

5-8.1.1   **Mounting.**   Each manual fire alarm box shall be securely mounted. The operable part of each manual fire alarm box shall be not less than 3½ ft (1.1 m) and not more than 4½ ft (1.37 m) above floor level.

The National Fire Alarm Code has addressed the fire alarm box "side reach" accessibility requirements that resulted from the adoption of the Americans with Disabilities Act (ADA). The "front reach" ADA requirement permits a maximum mounting height of 48 inches.

See Figure 5.34 for examples of manual fire alarm boxes.

5-8.1.2   **Distribution.**   Manual fire alarm boxes shall be distributed throughout the protected area so that they are unobstructed and readily accessible. They shall be located in the normal path of exit from the area with a manual fire alarm box at each exit on each floor. Additional manual fire alarm boxes shall be provided so that travel distance to the nearest fire alarm box will not be in excess of 200 ft (61 m) measured horizontally on the same floor.

5-8.1.3*   A coded manual fire alarm box shall produce at least three repetitions of the coded signal, with each repetition to consist of at least three impulses.

A-5-8.1.3   **Coded Signal Designations.**   The following recommended coded signal designations for buildings having four floors and multiple basements are provided in Table A-5-8.1.3:

Table A-5-8.1.3   Recommended Coded Signal Designations

Location	Coded Signal
4th floor	2-4
3rd floor	2-3
2nd floor	2-2
1st floor	2-1
Basement	3-1
Sub-basement	3-2

This is one of many possible code schemes. Others have been used for large buildings that divide the building up into a north and south end, for example, and use the first digit of the code to specify the floor (1, 2, 3, 4 etc.) and the second digit to specify either north (1) or south (2) wing. Coding of the signals allows the parties

Figure 5.34   *Typical manual fire alarm boxes. Photographs courtesy of: (Top) Simplex Time Recorder Company, Gardner, MA. (Bottom) Edwards Systems Technology/EST, Cheshire, CT.*

responsible for the site to proceed directly to the area where the alarm was activated. There are both benefits and disadvantages to coding manual fire alarm boxes. The decision to use coded signals must be part of the overall fire prevention and protection plan for the site.

5-8.2 Publicly Accessible Fire Service Boxes (Street Boxes).

5-8.2.1 Street boxes, when in an abnormal condition, shall leave the circuit usable.

Street boxes are exposed to all manner of possible damage, from intentional vandalism to automobile collisions. The operability of the remainder of the system must be ensured in the event of damage to the street box.

5-8.2.2 Street boxes shall be designed so that recycling does not occur if a box actuating device is held in the actuating position and so that they are ready to accept a new signal as soon as the actuating device is released.

Street boxes are available to the public. They must be able to operate in spite of the possible panicked actions of the person seeking to report a fire.

5-8.2.3 Street boxes, when actuated, shall give a visible or audible indication to the user that the box is operating or that the signal has been transmitted to the communications center.

NOTE: Where the operating mechanism of a box creates sufficient sound to be heard by the user, the requirements are satisfied.

5-8.2.4 The street box housing shall protect the internal components from the weather.

5-8.2.5 Doors on street boxes shall remain operable under adverse climatic conditions, including icing and salt spray.

5-8.2.6 Street boxes shall be recognizable as such. Street boxes shall have instructions for use plainly marked on their exterior surfaces.

Figure 5.35 shows a typical street box.

5-8.2.7 Street boxes shall be securely mounted on poles, pedestals, or structural surfaces as directed by the authority having jurisdiction.

5-8.2.8 Street boxes shall be as conspicuous as possible. Their color shall be distinctive, and they shall be visible from as many directions as possible. A wide band of distinctive

Figure 5.35 *Typical street box. Photograph courtesy of Mammoth Fire Alarms (Gamewell), Lowell, MA.*

colors visible over the tops of parked cars or adequate signs completely visible from all directions shall be applied to supporting poles.

5-8.2.9* Location-designating lights of distinctive color, visible for at least 1500 ft (460 m) in all directions, shall be installed over street boxes. The street light nearest the street box, where equipped with a distinctively colored light, shall be considered as meeting this requirement.

A-5-8.2.9 The current supply for location-designating lights at street boxes should be secured at lamp locations from the local electric utility company.

Alternating current power may be permitted to be superimposed on metallic fire alarm circuits for supplying designating lamps, or for control or actuation of equipment devices for fire alarm or other emergency signals, provided:

(a) Voltage between any wire and ground or between one wire and any other wire of the system does not exceed 150 volts, and the total resultant current in any line circuit does not exceed 1/4 ampere.

(b) Components such as coupling capacitors, transformers, chokes, or coils are rated for 600-volt working voltage and have a breakdown voltage of at least twice the working voltage plus 1000 volts.

(c) There is no interference with fire alarm service under any conditions.

5-8.2.10   Street box cases and parts that are, at any time, accessible to users shall be of insulating materials or permanently and effectively grounded. All ground connections to street boxes shall comply with the requirements of NFPA 70, *National Electrical Code*, Article 250.

5-8.2.11   Where a street box is installed inside a structure, it shall be placed as close as is practicable to the point of entrance of the circuit, and the exterior wire shall be installed in conduit or electrical metallic tubing in accordance with Chapter 3 of NFPA 70, *National Electrical Code*.

5-8.2.12   **Coded Radio Street Boxes.**

With the advent of modern microprocessor-based telecommunications systems, the features available in street boxes have expanded considerably. The requirements in 5-8.2.12 ensure that the transmission of a fire alarm remains the paramount priority.

5-8.2.12.1   Coded radio street boxes shall be designed and operated in compliance with all applicable rules and regulations of the FCC, as well as with the requirements established herein.

5-8.2.12.2   Coded radio street boxes shall provide no less than three specific and individually identifiable functions to the communications center, in addition to the street box number, as follows:

    (a)   Test;

    (b)   Tamper; and

    (c)   Fire.

5-8.2.12.3*   Coded radio street boxes shall transmit to the communications center no less than one repetition for "test," no less than one repetition for "tamper," and no less than three repetitions for "fire."

A-5-8.2.12.3   The following is an excerpt from the FCC Rules and Regulations, Vol. V, Part 90, March 1979:

"Except for test purposes, each transmission must be limited to a maximum of 2 seconds and may be automatically repeated not more than two times at spaced intervals within the following 30 seconds; thereafter, the authorized cycle may not be reactivated for 1 minute."

5-8.2.12.4   Where multifunction coded radio street boxes are used to transmit to the communications center a request(s) for emergency service or assistance in addition to those stipulated in 5-8.2.12.2, each such additional message function shall be individually identifiable.

5-8.2.12.5   Multifunction coded radio street boxes shall be so designed as to prevent the loss of supplemental or concurrently actuated messages.

5-8.2.12.6   An actuating device held or locked in the activating position shall not prevent the activation and transmission of other messages.

5-8.2.13   **Power Source.**

5-8.2.13.1   Box primary power shall be permitted to be from a utility distribution system, a photovoltaic power system, or user power, or shall be self-powered using either an integral battery or other stored energy source, as approved by the authority having jurisdiction.

5-8.2.13.2   Self-powered boxes shall have sufficient power for uninterrupted operation for a period of not less than 6 months. Self-powered boxes shall transmit a low power warning message to the communications center for at least 15 days prior to the time the power source will fail to operate the box. This message shall be part of all subsequent transmissions.

Use of a charger to extend the life of a self-powered box shall be permitted where the charger does not interfere with box operation. The box shall be capable of operation for not less than 6 months with the charger disconnected.

5-8.2.13.3   Boxes powered by a utility distribution system shall have an integral standby, sealed, rechargeable battery capable of powering box functions for at least 60 hours in the event of primary power failure. Transfer to standby battery power shall be automatic and without interruption to box operation. Where operating from primary power, the box shall be capable of operation with a dead or disconnected battery. A local trouble indication shall activate upon primary power failure. A battery charger shall be provided in compliance with 1-5.2.9.2, except as modified herein.

Where the primary power has failed, boxes shall transmit a power failure message to the communications center as part of subsequent test messages until primary power is restored. A low battery message shall be transmitted to the communications center where the remaining battery standby time is less than 54 hours.

5-8.2.13.4   Photovoltaic power systems shall provide box operation for not less than 6 months.

Photovoltaic power systems shall be supervised. The battery shall have power to sustain operation for a minimum period of 15 days without recharging. The box shall transmit a trouble message to the communications center when the charger has failed for more than 24 hours. This message shall be part of all subsequent transmissions. Where the remaining battery standby duration is less than 10 days, a low battery message shall be transmitted to the communications center.

5-8.2.13.5 User-powered boxes shall have an automatic self-test feature.

5-8.2.14 Design of Telephone Street Boxes (Series or Parallel).

5-8.2.14.1 If a handset is used, the caps on the transmitter and receiver shall be secured to reduce the probability of the telephone street box being disabled due to vandalism.

5-8.2.14.2 Telephone street boxes shall be designed to allow the communications center operator to determine whether or not the telephone street box has been restored to normal condition after use.

5-9 Supervisory Signal-Initiating Devices.

5-9.1 Control Valve Supervisory Signal-Initiating Device. Two separate and distinct signals shall be initiated: one indicating movement of the valve from its normal position, and the other indicating restoration of the valve to its normal position. The off-normal signal shall be initiated during the first two revolutions of the hand wheel or during 1/5 of the travel distance of the valve control apparatus from its normal position. The off-normal signal shall not be restored at any valve position except normal.

Control valve supervisory signal-initiating devices have traditionally been switches specifically designed, approved, and listed for service as valve-monitoring devices. See Figure 5.36. The requirement for two distinct signals does not necessarily mean two switches. A switch that transfers when the valve begins to close and stays transferred, and then returns to normal when the valve is reopened, satisfies the requirement. The initial transfer is the first signal. The return to normal is the second signal. For example, assume the switch on the valve is a normally open contact. As the operator begins to turn the valve, the switch closes, indicating an off-normal condition. The switch stays in the closed, off-normal position as the operator continues to close the valve. When the operator opens the valve, the closed contacts transfer back to the open state as the valve is

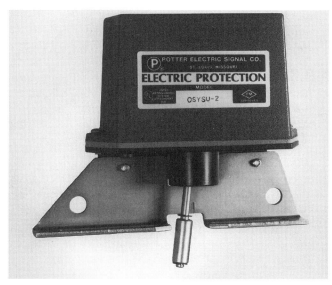

Figure 5.36 Outside screw and yoke (OS & Y) valve supervisory switch. Photograph courtesy of Potter Electric Signal Co., St. Louis, MO.

completely open. The opening of the contacts is the second distinct signal.

5-9.2 Pressure Supervisory Signal-Initiating Device. Two separate and distinct signals shall be initiated: one indicating that the required pressure has increased or decreased, and the other indicating restoration of the pressure to its normal value.

(a) A pressure tank supervisory signal-initiating device for a pressurized limited water supply, such as a pressure tank, shall indicate both high and low pressure conditions. A signal shall be initiated when the required pressure increases or decreases 10 psi (70 kPa) from the normal pressure.

(b) A pressure supervisory signal-initiating device for a dry-pipe sprinkler system shall indicate both high and low pressure conditions. A signal shall be initiated when the pressure increases or decreases 10 psi (70 kPa) from the normal pressure.

(c) A steam pressure supervisory signal-initiating device shall indicate a low pressure condition. A signal shall be initiated when the pressure reaches or exceeds 110 percent of the minimum operating pressure of the steam-operated equipment supplied.

(d) An initiating device for supervising the pressure of sources other than those specified in 5-9.2(a) through (c) shall be provided as required by the authority having jurisdiction.

As with supervisory initiating devices for valve operation, water pressure supervisory initiating devices can consist of a single switch. See Figure 5.37.

5-9.3 Water Level Supervisory Signal-Initiating Device.
Two separate and distinct signals shall be initiated: one indicating that the required water level has been lowered or raised, and the other indicating restoration to the normal level.

(a) A pressure tank signal-initiating device shall indicate both high and low level conditions. A signal shall be obtained when the water level falls 3 in. (76 mm) below or rises 3 in. (76 mm) above the normal level.

(b) A supervisory signal-initiating device for other than pressure tanks shall initiate a low level signal when the water level falls 12 in. (300 mm) below the normal level.

As with supervisory initiating devices for valve operation and water pressure, water level supervisory initiating devices can also consist of a single switch. See Figure 5.38.

5-9.4 Water Temperature Supervisory Signal-Initiating Device.
A temperature supervisory device for a water storage container exposed to freezing conditions shall initiate two separate and distinctive signals. One signal shall indicate that

Figure 5.38 *Tank water level supervisory switch. Photograph courtesy of Potter Electric Signal Co., St. Louis, MO.*

Figure 5.39 *Tank water temperature supervisory switch. Photograph courtesy of Potter Electric Signal Co., St. Louis, MO.*

the temperature of the water has dropped to 40°F (4.4°C), and the other shall indicate restoration to a proper temperature.

Water temperature supervisory initiating devices can consist of a single switch, as with supervisory initiating devices for valve operation, water pressure, and water level. See Figure 5.39.

5-9.5 Room Temperature Supervisory Signal-Initiating Device.
A room temperature supervisory device shall indicate a decrease in room temperature to 40°F (4.4°C) and its restoration to above 40°F (4.4°C).

As with other supervisory initiating devices mentioned in Section 5-10, room temperature supervisory initiating devices can also consist of a single switch. See Figure 5.40.

Figure 5.37 *Pressure supervisory switch. Photograph courtesy of Potter Electric Signal Co., St. Louis, MO.*

*Figure 5.40   Room temperature supervisory switch.
Photograph courtesy of Potter Electric Signal Co., St. Louis,
MO.*

5-10*   Smoke Detectors for Control of Smoke Spread.

A-5-10 See NFPA *101, Life Safety Code®*, for definition of smoke compartment; NFPA 90A, *Standard for the Installation of Air Conditioning and Ventilating Systems*, for definition of duct systems; and NFPA 92A, *Recommended Practice for Smoke-Control Systems*, for definition of smoke zone.

Between 1960 and late 1970 there were several fires in high-rise buildings that demonstrated the futility of trying to evacuate an entire building when a fire occurred. Not only were injuries incurred by occupants during the evacuation efforts but often the means of egress became untenable due to heavy smoke concentrations within them. As the improved building codes resulted in structures that could maintain their integrity in spite of the complete combustion of the interior fire load through passive fire-resistive construction and compartmentation, the option of defending occupants in place became viable. Strategies of establishing smoke compartments and refuge zones, and managing the flow of smoke by directing it away from the occupants were developed. Experiences with high-rise fires indicate that the proactive control of smoke with both automatic smoke detectors and HVAC systems is a viable strategy for occupant protection in high-rise buildings.

5-10.1* Smoke detectors installed and used to prevent smoke spread by initiating control of fans, dampers, doors, and other equipment shall be classified as:

(a) Area detectors that are installed in the related smoke compartments.

(b) Detectors that are installed in the air duct systems.

Section 5-10 does not require the installation of smoke detectors for smoke control. The purpose of the section is to describe the performance and installation requirements for smoke detectors when they are used for smoke control.

A-5-10.1 Smoke detectors located in the open area(s) should be used rather than duct-type detectors because of the dilution effect in air ducts. Active smoke management systems installed in accordance with NFPA 92A, *Recommended Practice for Smoke-Control Systems*, or NFPA 92B, *Guide for Smoke Management Systems in Malls, Atria, and Large Areas*, should be controlled by total coverage open area detection.

Paragraph 5-1.4.2 identifies all of the spaces that must have smoke detectors if total coverage is to be achieved.

5-10.2* Detectors that are installed in the air duct system per 5-10.1(b) shall not be used as a substitute for open area protection. Where open area protection is required, 5-3.4 shall apply.

A-5-10.2 Dilution of smoke-laden air by clean air from other parts of the building or dilution by outside air intakes can allow high densities of smoke in a single room with no appreciable smoke in the air duct at the detector location. Smoke might not be drawn from open areas where air conditioning systems or ventilating systems are shut down.

5-10.3 Smoke detectors in the related smoke compartment for open area protection shall be the preferred means to initiate control of smoke spread.

5-10.4* Purposes.

A-5-10.4 The purposes for which smoke detectors may be permitted to be applied in order to initiate control of smoke spread are:

(a) Prevention of the recirculation of dangerous quantities of smoke within a building.

(b) Selective operation of equipment to exhaust smoke from a building.

(c) Selective operation of equipment to pressurize smoke compartments.

(d) Operation of doors and dampers to close the openings in smoke compartments.

5-10.4.1 To prevent the recirculation of dangerous quantities of smoke, a detector approved for air duct use shall be installed on the supply side of air-handling systems as required by NFPA 90A, *Standard for the Installation of Air Conditioning and Ventilating Systems*, and 5-10.5.2.1.

5-10.4.2   Where smoke detectors are used to initiate selectively the operation of equipment to control smoke spread, the requirements of 5-10.5.2.2 shall apply.

5-10.4.3   Where detectors are used to initiate the operation of smoke doors, the requirements of 5-10.7 shall apply.

5-10.4.4   Where duct detectors are used to initiate the operation of smoke dampers within ducts, the requirements of 5-10.6 shall apply.

5-10.5   Application.

5-10.5.1   Area Detectors within Smoke Compartments. Area smoke detectors shall be permitted to be used to control the spread of smoke by initiating appropriate operation of doors, dampers, and other equipment.

This paragraph was added in this edition to allow specific area detectors to be used to control the spread of smoke. The prior wording implied that complete area coverage was required throughout the smoke compartment to initiate smoke control functions. From an engineering standpoint, smoke detectors are needed where they are intended to identify the presence of smoke at a particular location or movement of smoke past a particular location. The necessary locations for area smoke detectors are a function of building geometry, anticipated fire locations, and intended goals of smoke control functions. Complete area smoke detection is not necessary to provide for such control features, which can be accomplished by many possible detector locations for any given fire scenario. An example is those smoke detectors that are often placed at the perimeter of an atrium opening to detect smoke movement through the floor opening. Another example would be to release smoke doors only as their associated smoke detector is activated, thus avoiding premature release of all other doors which may impede evacuation during an emergency.

This paragraph also allows complete area coverage to be used if the designer wishes to do so. In this case, when a compartment detector activates in the smoke compartment, it signals the fire alarm control unit which in turn provides the signal to the HVAC control system or smoke door release system. The HVAC controller operates or controls fans and dampers to prevent the introduction of smoke into other smoke compartments and to vent the smoke from the fire compartment, facilitating occupant egress. The smoke door release system would either close all doors in the building or all doors in the smoke zone.

5-10.5.2   Smoke Detection for the Air Duct System.

5-10.5.2.1   Supply Air System.   Where the detection of smoke in the supply air system is required by other NFPA standards, a detector(s) listed for the air velocity present and located in the supply air duct downstream of both the fan and the filters shall be installed.

Exception: Additional smoke detectors shall not be required to be installed in ducts where the air duct system passes through other smoke compartments not served by the duct.

The NFPA standards relevant to 5-10.5.2.1 *are NFPA 90A, Standard for the Installation of Air Conditioning and Ventilating Systems*; NFPA 92A, *Recommended Practice for Smoke Control Systems*; and *NFPA 101, Life Safety Code*®. The purpose of supply-side smoke detection is to sense smoke that may be contaminating the area served by the duct. That smoke might be coming from the area via return air ducts, from outside via fresh air mixing ducts, or from a fire within the duct (such as in a filter or fan belt). If the source of smoke is not within the area served by the duct, but is from outside or from within the duct, a fire alarm response for ordinary area detection within the space would not normally be expected to produce the most appropriate set of responses to the fire within the duct. Different air-flow management programs are required for supply side smoke in-flow as opposed to smoke generated within the compartment. Furthermore, one could not rely upon compartment area detection to respond to a supply smoke in-flow because of the expected dilution of smoke-laden air as it enters the smoke compartment where the area detection is installed and mixes with fresh air. This necessitates the use of detectors downstream from the fan and filters in the supply air duct.

The exception is based upon the fire resistance of HVAC ducts and the unlikelihood of smoke escaping from the HVAC duct into a compartment not served by the duct. If this exception were not provided an additional detector would be necessary every time the duct passed from one smoke compartment into another. Figure A-5-10.5.2.2(c) shows an air duct passing through a smoke compartment without serving the compartment. The middle air supply duct serves the center smoke compartment. The top air supply duct serves only the left compartment and passes through the center and right compartments without serving them.

5-10.5.2.2*   **Return Air System.**   Where the detection of smoke in the return air system is required by other NFPA standards, a detector(s) listed for the air velocity present shall be located at every return air opening within the smoke compartment, where the air leaves each smoke compartment, or in the duct system before the air enters the return air system common to more than one smoke compartment. [*See Figures A-5-10.5.2.2(a), (b), and (c).*]

Exception No. 1: Where complete smoke detection is installed in the smoke compartment, installation of air duct detectors in the return air system shall not be required, provided their function can be accomplished by the design of the area detection system.

Exception No. 2: Additional smoke detectors shall not be required to be installed in ducts where the air duct system passes through other smoke compartments not served by the duct.

A-5-10.5.2.2 Detectors listed for the air velocity present may be permitted to be installed at the opening where the return air enters the common return air system. The detectors should be installed up to 12 in. (0.3 m) in front of or behind the opening and spaced according to the following opening dimensions [*see Figures A-5-10.5.2.2(a), (b), and (c)*]:

(a) *Width.*

1. Up to 36 in. (914 mm) — One detector centered in opening

2. Up to 72 in. (1829 mm) — Two detectors located at the ¼-points of the opening

3. Over 72 in. (1829 mm) — One additional detector for each full 24 in. (610 mm) of opening.

(b) *Depth.* The number and spacing of the detector(s) in the depth (vertical) of the opening should be the same as those given for the width (horizontal) above.

(c) *Orientation.* Detectors should be oriented in the most favorable position for smoke entry with respect to the direction of airflow. The path of a projected beam-type detector across the return air openings should be considered equivalent in coverage to a row of individual detectors.

Return air ducts must be addressed differently than the supply ducts. When duct detection is used for control of smoke spread, detectors must be installed at each and every return air opening from the smoke compartment. This is intended to minimize the effects of smoke dilution.

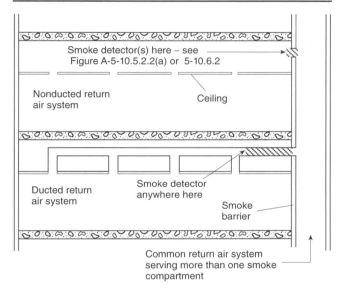

Figure A-5-10.5.2.2(b) *Location of a smoke detector(s) in return air systems for selective operation of equipment.*

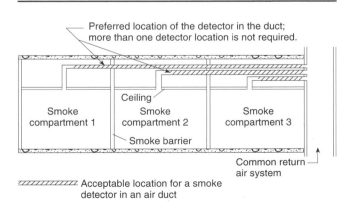

Figure A-5-10.5.2.2(c) *Detector location in a duct that passes through smoke compartments not served by the duct.*

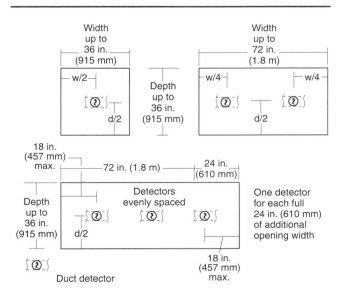

Figure A-5-10.5.2.2(a) *Location of a smoke detector(s) in return air systems for selective operation of equipment.*

The key phrase in Exception No. 1 is "provided their function can be accomplished." When an engineering analysis shows that the area smoke detection addresses all of the smoke ingress paths from the compartment into the return air duct, this exception is operative.

Exception No. 2. is based upon the same reasoning used in 5-10.5.2.1, Exception No. 1. Based upon this exception and referring to Figure A-5-10.5.2.2(c) the top duct does not need additional detectors and/or dampers where it passes through either the center compartment or the right compartment.

5-10.6   Location and Installation of Detectors in Air Duct Systems.

5-10.6.1   Detectors shall be listed for the purpose.

5-10.6.2*   Air duct detectors shall be securely installed in such a way as to obtain a representative sample of the airstream. This shall be permitted to be achieved by any of the following methods:

(a) Rigid mounting within the duct;

(b) Rigid mounting to the wall of the duct with the sensing element protruding into the duct;

(c) Installation outside the duct with rigidly mounted sampling tubes protruding into the duct;

(d) Installation through the duct with projected light beam.

See Figures 5.41 and 5.42.

A-5-10.6.2 Where duct detectors are used to initiate the operation of smoke dampers, they should be located so that the detector is between the last inlet or outlet upstream of the damper and the first inlet or outlet downstream of the damper.

In order to obtain a representative sample, stratification and dead air space should be avoided. Such conditions could be caused by return duct openings, sharp turns, or connections, as well as by long, uninterrupted straight runs. For this reason, duct smoke detectors should be located in the zone between 6 and 10 duct equivalent diameters of straight, uninterrupted run. In return air systems, the requirements of 5-10.5.2.2 take precedence over these considerations. [*See Figures A-5-10.6.2(a), (b), and (c).*]

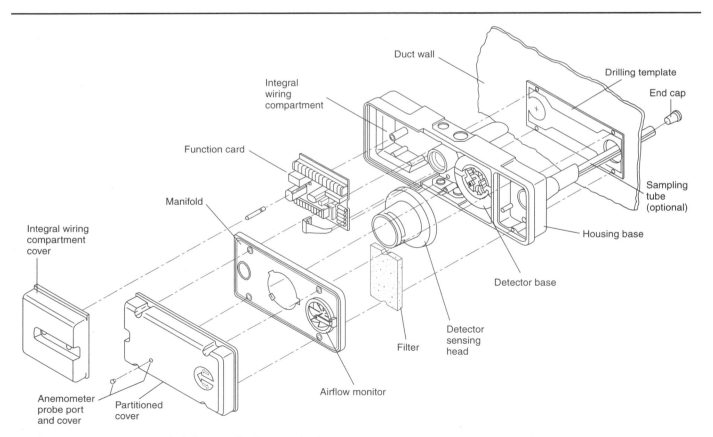

Figure 5.41 *Exploded view of a duct smoke detector. Drawing courtesy of ESL Sentrol, Inc., Portland, OR.*

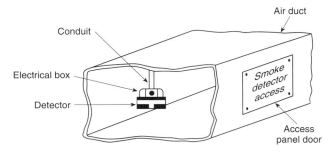

Figure A-5-10.6.2(a) *Pendant-mounted air duct installation.*

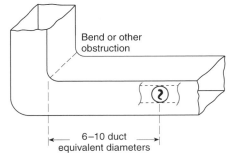

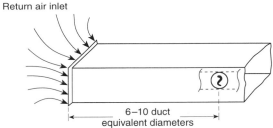

Figure A-5-10.6.2(b) *Typical duct detector placement.*

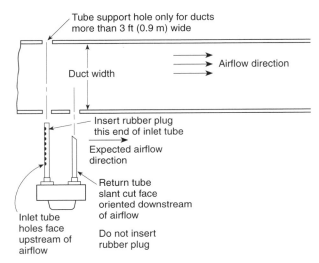

Figure A-5-10.6.2(c) *Inlet tube orientation.*

Figure 5.42 Duct-type smoke detector. Photograph courtesy of Kidde-Fenwal, Inc., Ashland, MA.

Note that the text is from Appendix A-5 and, therefore, is not a requirement of the Code. While it is good practice, it is recognized that the design of the air-handling ductwork may make it impossible to achieve a distance of 6 to 10 duct widths from a bend or opening.

Support of the detector by the conduit or raceway containing its wires is not permitted by NFPA 70, *National Electrical Code®*, unless the box is specifically listed for the purpose and installed in accordance with the listing.

The requirements in 5-10.6.2 and the guidance in A-5-10.6.2 are provided to ensure that the detectors in the air duct are suitably located to obtain an adequate sampling of air. In order to be certain that the detectors are placed where there is the highest probability that smoke will be evenly distributed throughout the duct cross-section, these location guidelines should be followed.

5-10.6.3 Detectors shall be accessible for cleaning and shall be mounted in accordance with the manufacturer's instructions. Access doors or panels shall be provided in accordance with NFPA 90A, *Standard for the Installation of Air Conditioning and Ventilating Systems.*

Chapter 7 provides recommended maintenance schedules for each type of detector. It is critical that the detectors be accessible in order to facilitate cleaning, which is, in turn, critical for reliable operation of the detector.

5-10.6.4 The location of all detectors in air duct systems shall be permanently and clearly identified and recorded.

It is advisable to place permanent placards outside the first point of access, indicating that a detector is accessible from that point. For example, the placard may be mounted on the wall beneath the ceiling tile that must be removed to access the duct. HVAC and fire alarm drawings should clearly show the actual as-built locations of the detectors. In most cases it is useful to generate one drawing that shows only the smoke detector locations.

5-10.6.5 Detectors mounted outside of a duct employing sampling tubes for transporting smoke from inside the duct to the detector shall be designed and installed to allow verification of airflow from the duct to the detector.

Sampling tubes are able to provide a flow of air through the detector enclosure due to a pressure differential that results from the flow of air across the tubes. Small errors in the orientation of the sampling tubes can reduce the pressure differential, rendering them ineffective in drawing air into the detector enclosure. In order for sampling tubes to take a representative sample of the air passing through the duct, they must be fabricated and installed in a manner consistent with their listing. Not all detectors are listed for use in a sampling tube enclosure. If the flow of air through the sampling tube and detector enclosure assembly cannot be verified there is no basis to presume that the air within the duct is being sampled by the detector.

5-10.6.6 Detectors shall be listed for proper operation over the complete range of air velocities, temperature, and humidity expected at the detector when the air-handling system is operating.

5-10.6.7 All penetrations of a return air duct in the vicinity of detectors installed on or in an air duct shall be sealed to prevent entrance of outside air and possible dilution or redirection of smoke within the duct.

5-10.6.8 Where in-duct smoke detectors are installed in concealed locations, more than 10 ft (3 m) above the finished floor, or in arrangements where the detector's alarm indicator is not readily visible to responding personnel, the detectors shall be provided with remote alarm indicators. Remote alarm indicators shall be installed in a readily accessible location and shall be clearly labeled to indicate both their function

(e.g., "In-Duct Smoke Detector Alarm") and the air-handling unit(s) associated with each detector.

Exception: Where the specific detector in alarm is indicated at the control unit.

Identification of which in-duct smoke detector is in alarm is a chronic problem, due to the difficulties typically involved in getting to the detectors. Because the fire causing the alarm may be located far from the in-duct smoke detector location, rapid identification of the individual detector in alarm and the associated air handling unit is critical.

5-10.7 **Smoke Detectors for Door Release Service.**

5-10.7.1 Smoke door release not initiated by a fire alarm system that includes smoke detectors protecting the areas on both sides of the door affected shall be accomplished by smoke detectors applied as specified in 5-10.7.

There are two general methods of controlling doors with smoke detectors. The first is to connect area smoke detectors to a selected circuit of a fire alarm control unit and use an output circuit in the control unit to operate magnetic door release devices. When one of the area smoke detectors renders an alarm, the control unit transfers to the alarm state and energizes the output circuit that controls the door holders. The requirements for such a system are addressed in Chapter 3, Protected Premises Fire Alarm Systems. The second method is to control the door holder mechanism directly with a dedicated smoke detector. The requirements in this section apply equally to both design concepts. See Figure 5.43.

5-10.7.2 Smoke detectors listed exclusively for door release service shall not be used for open area protection.

A smoke detector used concurrently for door release service and open area protection shall be permitted where listed for open area protection and installed in accordance with 5-3.4.

5-10.7.3 Smoke detectors shall be of the photoelectric, ionization, or other approved type.

5-10.7.4 **Number of Detectors Required.**

The placement requirements outlined in this section have been derived from the ceiling jet dynamics that have served as the physical principals from which the location and placement rules for area smoke detection

Figure 5.43 Typical magnetic door hold release. Photograph courtesy of Mammoth Fire Alarms (Edwards System Technology), Lowell, MA.

have been derived. As research continues, there may be additional insight developed for this application.

While it is not explicitly required, it is recommended that smoke detectors installed only for door release be connected to a fire alarm control unit and activate notification appliances when smoke is detected.

5-10.7.4.1 Where doors are to be closed in response to smoke flowing in either direction, the requirements of 5-10.7.4.1.1 through 5-10.7.4.1.3 shall apply.

5-10.7.4.1.1 Where the depth of wall section above the door is 24 in. (610 mm) or less, one ceiling-mounted detector shall be required on one side of the doorway only. *See Figure 5-10.7.4.1.1, parts B and D.*

5-10.7.4.1.2* Where the depth of wall section above the door is greater than 24 in. (610 mm), two ceiling-mounted detectors shall be required, one on each side of the doorway. *See Figure 5-10.7.4.1.1, part F.*

This requirement is similar to the requirements regarding smoke detectors and ceilings with deep beams (see Section 5-3.4.6). The difference in the depth requirements comes from the fact that doors for smoke control are typically located in corridors that are narrower than the bays encountered in deep beam ceilings.

A-5-10.7.4.1.2 Where the depth of wall section above the door is 60 in. (1520 mm) or greater, additional detectors might be required as indicated by an engineering evaluation.

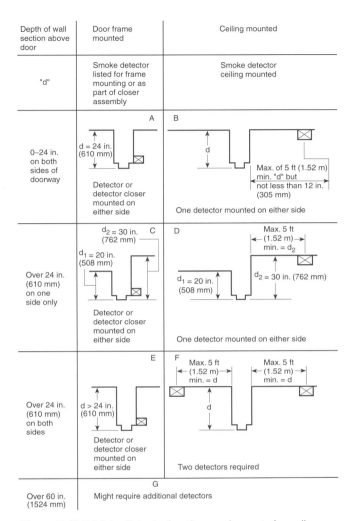

Figure 5-10.7.4.1.1 Detector location requirements for wall sections.

Since the average door height is a nominal 84 to 96 inches, the addition of 60 inches above the door results in a ceiling height greater than 148 inches (11.5 feet). This is above the height for which reduced spacing is required for heat detectors. The data in Appendix B indicate that when the ceiling height exceeds 10 feet, reduced spacing is required. An engineering evaluation should be performed to determine if reduced smoke detector spacing is appropriate for the specific application under consideration.

5-10.7.4.1.3 Where a detector is specifically listed for door frame mounting or where a listed combination or integral detector-door closer assembly is used, only one detector shall be required where installed in the manner recommended by the manufacturer.

Formal Interpretation 78-1
Reference: 5-10.7.4.1.3

Question: Does a door frame mounted combination automatic door closer incorporating a smoke detector listed by a testing laboratory for limited open area protection meet the requirements of 5-10.7.4.1.3?
Answer: Yes.

Issue Edition: NFPA 72E-1978
Reference: 9-2.2
Issue Date: February 1979 ∎

5-10.7.4.2 Where door release is intended to prevent smoke transmission from one space to another in one direction only, one detector located in the space to which smoke is to be confined shall be required, regardless of the depth of wall section above the door. Alternatively, a smoke detector conforming with 5-10.7.4.1.3 shall be permitted to be used.

5-10.7.4.3 Where there are multiple doorways, additional ceiling-mounted detectors shall be required as specified in 5-10.7.4.3.1 through 5-10.7.4.3.3.

5-10.7.4.3.1 Where the separation between doorways exceeds 24 in. (610 mm), each doorway shall be treated separately. *See Figure 5-10.7.4.3.1, part E.*

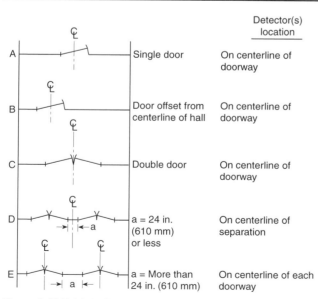

Figure 5-10.7.4.3.1   *Detector location requirements for single and double doors.*

5-10.7.4.3.2 Each group of three doorway openings shall be treated separately. *See Figure 5-10.7.4.3.2.*

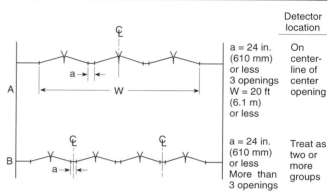

Figure 5-10.7.4.3.2   *Detector location requirements for group doorways.*

5-10.7.4.3.3 Each group of doorway openings that exceeds 20 ft (6.1 m) in width measured at its overall extremes shall be treated separately. *See Figure 5-10.7.4.3.3.*

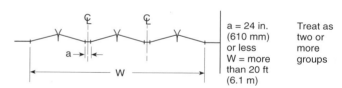

Figure 5-10.7.4.3.3   *Detector location requirements for group doorways over 20 ft (6.1 m) in width.*

5-10.7.4.4 Where there are multiple doorways and listed door frame-mounted detectors or where listed combination or integral detector-door closer assemblies are used, there shall be one detector for each single or double doorway.

5-10.7.5 **Location.**

5-10.7.5.1 Where ceiling-mounted smoke detectors are to be installed on a smooth ceiling for a single or double doorway, they shall be located as follows (*see Figure 5-10.7.4.3.1*):

(a) On the centerline of the doorway; and

(b) No more than 5 ft (1.5 m), measured along the ceiling and perpendicular to the doorway (*see Figure 5-10.7.4.1.1*); and

(c)   No closer than shown in Figure 5-10.7.4.1.1, parts B, D, and F.

5-10.7.5.2   Where ceiling-mounted detectors are to be installed in conditions other than those outlined in 5-10.7.5.1, an engineering evaluation shall be made.

6

Notification Appliances for Fire Alarm Systems

Contents, Chapter 6

6-1   Scope.

6-1.1   Minimum Requirements. This chapter covers minimum requirements for the performance, location, and mounting of notification appliances for fire alarm systems for the purpose of evacuation or relocation of the occupants.

This subsection recognizes that the fire emergency plan for a particular building may require evacuation of the building or relocation of occupants within protected premises.

6-1.2   Intended Use. These requirements are intended to be used with other NFPA standards that deal specifically with fire alarm, extinguishment, or control systems. Notification appliances for fire alarm systems add to fire protection by providing stimuli for initiating emergency action.

The audible or visible notification appliances referred to in other chapters of the Code are described in detail in this chapter.

6-1.3 All notification appliances or combinations thereof installed in conformity with this chapter shall be listed for the purpose for which they are used.

This requirement states that the listing of notification appliances be use-specific. This means that the listing of an appliance must relate to the exact manner in which it will be used. See Figure 6.1.

6-1.4 These requirements are intended to address the reception of a notification signal and not its information content.

The requirements in this chapter do not address the content of a textual message or a message contained in a coded audible or visible signal. Rather, the requirements address the ability of notification appliances to deliver a message. For example, Chapter 3 requires the use of the ANSI S3.41, *American National Standard Audible Emergency Evacuation Signal* where evacuation of an area is intended. Chapter 6 requires that signal to meet certain audibility and installation requirements.

6-1.5   Interconnection of Appliances. The interconnection of appliances, the control configurations, the power supplies, and the use of the information provided by notification appliances for fire alarm systems are described in Chapter 1 and Chapter 3.

Figure 6.1 *Typical listed audible notification appliances (bells). Photograph courtesy of Edwards Systems Technology/EST, Inc., Cheshire, CT.*

6-2   General.

6-2.1   Nameplates.

6-2.1.1 Notification appliances shall include on their nameplates reference to electrical requirements and rated audible or visible performance, or both, as defined by the listing authority.

6-2.1.2 The audible appliances shall include on their nameplates reference to their parameters or reference to installation documents (supplied with the appliance) that include the parameters in accordance with 6-3.2. The visible appliances shall include on their nameplates reference to their parameters or reference to installation documents (supplied with the appliance) that include the parameters in accordance with 6-4.2.1.

In order to guide designers and installers of fire alarm systems so that the system will deliver audible and visible information with appropriate intensity, the nameplate must state the capabilities of the appliance as determined through tests conducted by the listing organization.

6-2.2 **Physical Construction.** Appliances intended for use in special environments (e.g., outdoors versus indoors, high or low temperatures, high humidity, dusty conditions, hazardous locations) or where subject to tampering shall be listed for the intended application.

It is essential to maintain the operational integrity of audible and visible notification appliances despite their possible location in relatively hostile environments. Use of appliances not listed for use in the same type of environment in which the appliance is to be placed is a violation of the Code.

6-2.3* Where subject to obvious mechanical damage, appliances shall be suitably protected. Where guards or covers are employed, they shall be listed for use with the appliance. Their effect on the appliance's field performance shall be considered in accordance with the listing requirements.

This protection is usually provided by an enclosure that protects the actual audible or visible mechanism. In the case of speakers, a mechanical baffle protects the cone from being punctured by a sharp object. See Figure 6.2.

Any guards placed over an audible or visible appliance may affect the level of the audible signal or the light intensity of the appliance. For this reason, the guard or protective device must be tested with the specific appliance, and its effect must be measured and reported. Then, system designers, installers and inspectors can de-rate the appliance performance and make corrections in the application of the appliance with the guard. See Figure 6.3.

A-6-2.3 **Use of Guards with Notification Appliances.** Situations exist where supplemental enclosures are necessary to protect the physical integrity of a notification appliance. Protective enclosures should not interfere with the performance characteristics of the appliance. Where the enclosure degrades the performance, methods should be detailed in the installation instructions of the enclosure that clearly identify the degradation. For example, where the appliance signal is attenuated, it might be necessary to adjust the appliance spacings or appliance output.

6-2.4 In all cases, appliances shall be supported independently of their attachments to the circuit conductors.

It is not permitted to physically support the appliance by means of the wires that connect the appliance to the notification appliance circuit of the fire alarm system. See Figure 6.4.

Figure 6.3 Notification appliance showing external mechanical protection. Photograph courtesy of Safety Technology International, Inc. Waterford, MI.

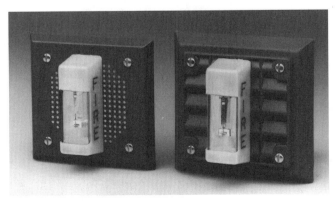

Figure 6.2 Notification appliances showing mechanical baffles. Photograph courtesy of Wheelock, Inc., Long Branch, NJ.

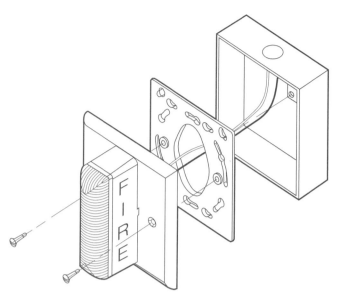

Figure 6.4   Notification appliance with independent support. (See 6-2.4.) Photograph courtesy of Gentex Corp., Zeeland, MI.

6-3   Audible Characteristics.

6-3.1   General Requirements.

6-3.1.1   Audibility.   The sound level of an installed audible signal shall be adequate to perform its intended function and shall meet the requirements of the authority having jurisdiction or other applicable standards.

Although additional requirements for textual audible appliances are covered in Section 6-7, this subsection can be used by an authority having jurisdiction to enforce the intelligibility of textual signals. If a textual audible notification appliance produced a signal of adequate sound level, but the message was not intelligible, then such a signal would not be adequate.

6-3.1.2   An average sound level greater than 105 dBA shall require the use of a visible signal appliance(s) in accordance with Section 6-4.

In some occupancies, the ambient sound level is so high that it would be impractical to rely solely on audible notification appliances. A drop forge shop, a large casino, a rock music dance hall, or a newspaper press room are all candidates for the addition of visible signal

appliances to help ensure that the signals will be perceived by the occupants. See Figures 6.5 and 6.6.

Visible signaling is required when the ambient sound pressure levels exceed 105 dBA because it is difficult, and possibly harmful, to try to overcome that level with audible fire alarm signals. In some occupancies, such as theaters, it may be possible to turn off the ambient noise when the fire alarm system is activated. The value of 105 dBA was chosen because an audible signal 15 dBA above the limit set here would produce 120 dBA. The limit of 120 dBA has been set as the upper allowable limit by the Occupational, Health and Safety Act (OSHA) and the Americans with Disabilities Act (ADA).

6-3.1.3   The total sound pressure level produced by combining the ambient sound pressure level with all audible signaling appliances operating shall not exceed 120 dBA anywhere in the occupied area.

6-3.1.4   Sound sources not normally found continuously in the occupied area shall not be required to be considered in measuring maximum ambient sound level.

Temporary sound sources, such as construction noise, are not required to be considered when measuring the maximum ambient sound pressure level. However, whether or not a sound source is "not normally found continuously" in a space must be determined with care. In a college dormitory room, for example, a student's radio may be portable, yet because it is usually part of the permanent equipment in the room, many authorities having jurisdiction do not believe it should be considered a "temporary" sound source.

Figure 6.5   Visible notification appliance. (See 6-3.1.2.) Photograph courtesy of Gentex Corp., Zeeland, MI.

*Figure 6.6 Visible notification appliance. (See 6-3.1.2.)
Photograph courtesy of Faraday Corp., Techumseh, MI.*

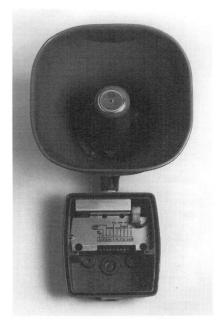

*Figure 6.7 Audible notification appliance for high ambient
noise areas. (See 6-3.1.5.) Photograph courtesy of Audiosone
Corp., Stratford, CT.*

6-3.1.5 **Mechanical Equipment Rooms.** Where audible appliances are installed in mechanical equipment rooms, the average ambient sound level used for design guidance shall be at least 85 dBA for all occupancies.

The required value of 85 dBA is only a minimum value and some mechanical equipment rooms may have an average ambient sound level that exceeds this value. Based on the average ambient sound level of 85 dBA, the audible notification appliances would need to deliver 100 dBA throughout the room. See Figure 6.7.

6-3.2* **Public Mode Audible Requirements.**

A-6-3.2 The typical average ambient sound level for the occupancies specified in Table A-6-3.2 are intended only for design guidance purposes.

The typical average ambient sound levels specified should not be used in lieu of actual sound level measurements.

These sound levels, previously known as "equivalent sound levels," are defined as the mean square A-weighted

sound pressure measured over a 24-hour period. This value can be experimentally determined for a particular occupancy by using a recording sound pressure level meter. The technician taking the measurement obtains the mean square result for the area under the curve by using integral calculus. The levels listed in A-6-3.2 were

Table A-6-3.2   Average Ambient Sound Level According toLocation

Locations	Average Ambient Sound Level
Business occupancies	55 dBA
Educational occupancies	45 dBA
Industrial occupancies	80 dBA
Institutional occupancies	50 dBA
Mercantile occupancies	40 dBA
Piers and water-surrounded structures	40 dBA
Places of assembly	55 dBA
Residential occupancies	35 dBA
Storage occupancies	30 dBA
Thoroughfares, high density urban	70 dBA
Thoroughfares, medium density urban	55 dBA
Thoroughfares, rural and suburban	40 dBA
Tower occupancies	35 dBA
Underground structures and windowless buildings	40 dBA
Vehicles and vessels	50 dBA

derived from acoustic reference literature and were originally based on this kind of calculation. See also 6-3.1.4 and 6-3.1.5.

It should be noted that measurements taken by a major manufacturer of fire alarm equipment in a large sampling of hotel rooms with through-the-wall air conditioning units determined that the average ambient sound level in those rooms, with the air conditioners operating, was 55 dBA.

6-3.2.1 Audible signal appliances intended for operation in the public mode shall have a sound level of not less than 75 dBA at 10 ft (3 m) or more than 120 dBA at the minimum hearing distance from the audible appliance.

Audible appliance ratings, as measured by the manufacturer and the nationally recognized testing laboratories, are given as a dB rating at a predetermined distance, usually 10 feet (3 m). The rule-of-thumb is that the output of an audible notification appliance is reduced by 6 dBA if the distance between the appliance and the listener is doubled. The accuracy of this rule-of-thumb depends on many intervening variables, particularly the acoustic properties of the materials in the listening space, such as ceiling materials and floor and wall coverings.

Using the appliance's rating along with this rule allows system designers to calculate audible levels in occupied spaces before a system is installed. See Figure 6.8 for an example of how this rule of thumb is applied.

More complex situations require calculating sound attenuation through doors and walls. See the SFPE *Handbook of Fire Protection Engineering*, Chapter 4-1, pp. 20-26) for appropriate calculation methods.

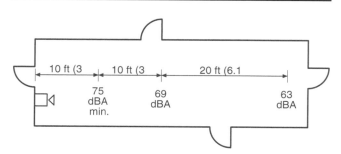

10 ft (3	10 ft (3	20 ft (6.1
75 dBA min.	69 dBA	63 dBA

Figure 6.8   Calculating audible levels using the 6 dBA "rule of thumb" method. Drawing courtesy of R.P. Schifiliti Associates, Inc., Reading, MA.

6-3.2.2* To ensure that audible public mode signals are clearly heard, they shall have a sound level at least 15 dBA above the average ambient sound level or 5 dBA above the maximum sound level having a duration of at least 60 seconds, whichever is greater, measured 5 ft (1.5 m) above the floor in the occupiable area.

Most authorities having jurisdiction will measure the sound level of audible notification appliances to ensure that it is at least 5 dBA above the maximum sound level. It is more difficult to determine the average ambient sound level than it is to determine the maximum sound level that lasts at least 60 seconds. Care must be exercised in selecting the source of the maximum sound level for each occupancy.

The measurement is made at 5 ft (1.5 m) above the floor to reduce the effects of walls and surfaces on the signal level.

Additional appliances may be required to ensure that the signal will be clearly heard throughout the occupiable area. These measurements are very important in especially when use of the building fire alarm signal to warn the occupants is planned.

A-6-3.2.2 The constantly changing nature of pressure waves, which are detected by ear can be measured by electronic sound meters, and the resulting electronic waveforms can be processed and presented in a number of meaningful ways.

Most simple sound level meters quickly average a sound signal and present a root mean square (RMS) level to the meter movement or display. However, this quick average of impressed sound results in fast movements of the meter's output that are best sent when talking into the microphone; the meter quickly rises and falls with speech. However, when surveying the ambient sound levels to establish the increased level at which a notificaton appliance will properly function, the sound source needs to be averaged over a longer period of time. Moderately priced sound level meters have such a function, usually called L_{eq} or "equivalent sound level." For example, an L_{eq} of speech in a quiet room would cause the meter movement to rise gradually to a peak reading and slowly fall well after the speech is over.

L_{eq} readings can be misapplied in situations where the background ambient noises vary greatly during a 24-hour period. L_{eq} measurements should be taken over the period of occupancy.

6-3.3 Private Mode Audible Requirements.

An example of private mode signaling is a hospital care area. The public occupants include patients who may not be able to respond to a fire alarm signal. In some cases, it may even be dangerous to alert them directly with audible (and possibly visible) signals. For this reason, the system is designed to alert trained staff.

Areas that use private mode signaling such as in a hospital, often have a less intense average ambient sound level and a lower maximum sound level, thus the reduced level cited in this section is appropriate. In delivering private mode signals it is important that the sound level of the audible notification appliance be adequate, but not unduly startle the occupants.

Lower audible levels are permitted because the part of the staff's job is to listen for, and respond appropriately to the fire alarm signals. In addition, they must communicate amongst themselves to be able to implement their emergency procedures and a louder alarm might interfere with this communication.

6-3.3.1 Private Mode.
Audible signals intended for operation in the private mode shall have a sound level of not less than 45 dBA at 10 ft (3 m) or more than 120 dBA at the minimum hearing distance from the audible appliance.

6-3.3.2
To ensure that audible private mode signals are clearly heard, they shall have a sound level at least 10 dBA above the average ambient sound level or 5 dBA above the maximum sound level having a duration of at least 60 seconds, whichever is greater, measured 5 ft (1.5 m) above the floor in the occupiable area.

6-3.4 Sleeping Areas.
Where audible appliances are installed to signal sleeping areas, they shall have a sound level of at least 15 dBA above the average ambient sound level or 5 dBA above the maximum sound level having a duration of at least 60 seconds or a sound level of at least 70 dBA, whichever is greater, measured at the pillow level in the occupiable area.

Subsection 6-3.4 indicates that, in rooms where people sleep, the sound level delivered by the audible notification appliance must be 15 dBA above the average ambient sound level, or 5 dBA above any peak sound level lasting 60 seconds or more, or at least 70 dBA. If the average ambient sound level in the sleeping area is 40 dBA, then the audible notification appliances must deliver at least 70 dBA. If the average ambient sound level in the sleeping area is 60 dBA, then the audible notification appliances must deliver at least 75 dBA.

6-3.5 Location of Audible Signal Appliances.

6-3.5.1
Where ceiling heights allow, wall-mounted appliances shall have their tops at heights above the finished floors of not less than 90 in. (2.30 m) and below the finished ceilings of not less than 6 in. (152 mm). This requirement shall not preclude ceiling-mounted or recessed appliances.

Exception: Combination audible/visible appliances installed in sleeping areas shall comply with 6-4.4.3.

In rooms with a ceiling height of at least 8 feet, the top of a wall-mounted audible notification appliance must be at least 7 feet 6 inches (2.30 m) above the floor and at least 6 inches (152 mm) below the ceiling. The exception points to 6-4.4.3 for the location of combination units within sleeping areas. If the linear dimension of a sleeping room exceeds 16 feet (4.87 m), 6-4.4.3.2 requires a combination audible/visible notification appliance to be located within 16 feet of the pillow.

6-3.5.2
Where combination audible/visible appliances are installed, the location of the installed appliance shall be determined by the requirements of 6-4.4.

Exception: Where the combination audible/visible appliance serves as an integral part of a smoke detector, the mounting location shall be in accordance with Chapter 2.

Subsection 6-3.5.1 requires that the location of a combination audible/visible notification appliance complies with the requirements of 6-4.4. The bottom of a wall-mounted appliance must be no less than 6 feet 8 inches above the floor, nor greater than 8 feet (2.44 m) above the floor. For additional information relating to the exception, see 6-4.4.3.1, which specifies that smoke detectors in sleeping areas be installed in accordance with Chapter 2 and Chapter 5.

6-4 Visible Characteristics, Public Mode.

Following passage of the Americans with Disabilities Act (ADA), there was a great deal of confusion concerning visible signaling requirements. In the past, the *National Fire Alarm Code* has differed from the ADA and from other accessibility standards such as ANSI 117.1. The fire alarm industry has worked with the various code and advocacy groups to develop reasonable, safe, and effective visible notification requirements.

The requirements contained in this (1996) edition of the *National Fire Alarm Code* have been accepted as "equivalent facilitation" (and in some cases superior) to the original ADA requirements. The ADA Accessibility Guidelines and ANSI 117.1 are being revised and it is expected that they will adopt visible signaling performance and location requirements as specified in this chapter.

6-4.1   There are two methods of visible signaling. These are methods in which the message of notification of an emergency condition is conveyed by direct viewing of the illuminating appliance or by means of illumination of the surrounding area.

NOTE: One method of determining compliance with Section 6-4 is that the product be listed in accordance with UL 1971, *Standard for Safety Signaling Devices for the Hearing Impaired.*

6-4.2   **Light Pulse Characteristics.**   The flash rate shall not exceed two flashes per second (2 Hz) nor be less than one flash every second (1 Hz) throughout the listed voltage range of the appliance.

6-4.2.1   A maximum pulse duration shall be 0.2 second with a maximum duty cycle of 40 percent. The pulse duration is defined as the time interval between initial and final points of 10 percent of maximum signal.

The light intensity of a pulsed source may be graphed as a bell-shaped curve. The duration of the pulse is measured beginning at the point where the upward side of the curve exceeds 10 percent of the maximum intensity to the point where the downward side of the curve drops below 10 percent of the maximum intensity.

6-4.2.2   The light source color shall be clear or nominal white and shall not exceed 1000 candela (cd) (effective intensity).

Source intensity is a measure of the light output of the appliance. The unit of measure is the candela (cd). (This unit was formerly called "candlepower." There is a one-to-one relationship between candela and candlepower.) As you move away from any light source, its illumination decreases. Illumination is measured in units of lumens (lm) per square meter (also called "lux"), or lumens per square foot. (Formerly, the unit used to describe illumination was the "foot-candle." One foot-candle equals one lumen per square foot. One lumen per square meter equals 0.926 foot-candles.)

Because strobe lights flash for a very brief period of time, the perceived brightness can vary depending on the actual peak source strength and duration of the flash. One appliance might reach a peak intensity of 1000 candelas in 0.1 seconds while another might reach 750 candelas in 0.2 seconds. Nevertheless, the human eye might perceive both as being equally bright. A mathematical relationship is used to relate the perceived brightness of a strobe light to that of a constantly burning light. The result is called the effective intensity (candela, effective or cd, eff.).

6-4.3   **Appliance Photometrics.**   Visible notification appliances used in the public mode shall be located so that the operating effect of the appliance can be seen by the intended viewers and shall be of a type, size, intensity, and number so that the viewer can discern when they have been illuminated, regardless of the viewer's orientation.

In the same manner that signals produced by audible notification appliances must be clearly heard, signals produced by visible notification appliances must also be clearly seen without regard to the viewer's position within the protected area.

6-4.4   **Appliance Location.**   Wall-mounted appliances shall have their bottoms at heights above the finished floor of not less than 80 in. (2 m) and no greater than 96 in. (2.4 m). Ceiling-mounted appliances shall be installed per Table 6-4.4.1(b).

Exception: Appliances installed in sleeping areas shall comply with 6-4.4.3.

In rooms with sufficient ceiling height, the bottom of a wall-mounted visible notification appliance must be at least 6 feet 8 inches (2 m) above the floor, but not more than 8 feet (2.4 m) above the floor. The exception refers to 6-4.4.3 for the location of combination units within sleeping areas. If the linear dimension of a sleeping room exceeds 16 feet, 6-4.4.3.2 requires a combination audible/visible notification appliance to be located within 16 feet (4.87 m) of the pillow.

The minimum mounting height is intended to locate appliances so they are not blocked by common furnishings or equipment. The maximum mounting is important because the illumination from a visible appliance reduces drastically with distance and angle from a horizontal plane through the appliance. For this reason, wall mounted appliances are limited to 96 inches (2.44 m) above the floor. Ceiling mounting is permitted, however, the appliances must be specifically listed for ceiling mounting.

6-4.4.1*   **Spacing in Rooms.**

A-6-4.4.1   Areas large enough to exceed the rectangular dimensions given in Figures A-6-4.4.1(a), (b), and (c) require additional appliances. Often, proper placement of appliances can be facilitated by breaking down the area into multiple squares and dimensions that fit most appropriately [*see* Figures A-6-4.4.1(a), (b), (c), and (d)]. An area 40 ft (12.2 m) wide and 74 ft (22.6 m) long can be covered with two 60-cd appliances. Irregular areas need more careful planning to make certain that at least one 15-cd appliance is installed for each 20-ft × 20-ft (6.09-m × 6.09-m) area.

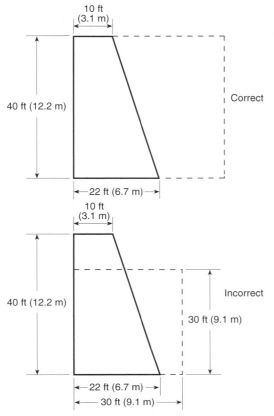

Figure A-6-4.4.1(a) *Irregular area spacing.*

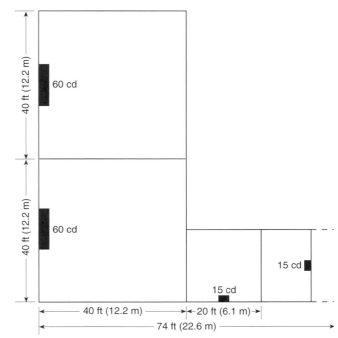

Note: Broken lines represent imaginary walls.

Figure A-6-4.4.1(b) *Spacing of wall-mounted visible appliances in rooms.*

These figures were added to avoid misinterpretation of the text. Figure A-6-4.4.1(a) demonstrates how a non-square or nonrectangular room can be fitted into the spacing allocation of Tables 6-4.4.1(a) and (b). Figure A-6-4.4.1(b) demonstrates how to divide a room or area into smaller areas to enable the use of lower intensity lights. Figures A-6-4.4.1(c) and (d) show the correct and incorrect placement of multiple visible notification appliances in a room.

6-4.4.1.1 Spacing shall be in accordance with Figure 6-4.4.1.1 and Tables 6-4.4.1.1(a) and (b). The separation between appliances shall not exceed 100 ft (30 m).

Visible notification appliances shall be installed in accordance with Table 6-4.4.1.1(a), using one of the following:

(a)* A single visible notification appliance; or

(b) Two visible notification appliances located on opposite walls; or

(c)* In rooms 80 ft × 80 ft (24.4 m × 24.4 m) or greater, where there are more than two appliances in any field of view, they shall be spaced a minimum of 55 ft (16.76 m) from each other; or

(d) More than two visible notification appliances that flash in synchronization.

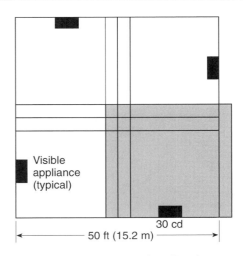

Figure A-6-4.4.1(c) *Room spacing allocation — correct.*

Table 6-4.4.1.1(a)   Room Spacing for Wall-Mounted Visible Appliances

		Minimum Required Light Output, Candela (cd) (Effective Intensity)		
Maximum Room Size		One Light Per Room	Two Lights per Room (Located on Opposite Walls)	Four Lights per Room One Light per Wall)
(ft)	(m)	(cd)	(cd)	(cd)
20 × 20	6.1 × 6.1	15	N/A	N/A
30 × 30	9.14 × 9.14	30	15	N/A
40 × 40	12.2 × 12.2	60	30	N/A
50 × 50	15.2 × 15.2	95	60	N/A
60 × 60	18.3 × 18.3	135	95	N/A
70 × 70	21.3 × 21.3	185	95	N/A
80 × 80	24.4 × 24.4	240	135	60
90 × 90	27.4 × 27.4	305	185	95
100 × 100	30.5 × 30.5	375	240	95
110 × 110	33.5 × 33.5	455	240	135
120 × 120	36.6 × 36.6	540	305	135
130 × 130	39.6 × 39.6	635	375	185

N/A: Not allowable.

A-6-4.4.1.1(a)

A design that delivers 0.0375 lumens/ft^2 (0.4037 lumens/m^2) effective intensity to all occupied spaces where visible notification is required is considered to meet the minimum light intensity requirements of this paragraph.

A-6-4.4.1.1(c)

The field of view is based on the focusing capability of the human eye, specified as 120 degrees in the IES *Handbook*. The apex of this angle is the viewer's eye. In order to ensure compliance with the requirements of 6-4.4.1.1, this angle should be increased to approximately 135 degrees.

Table 6-4.4.1.1(b)   Room Spacing for Ceiling-Mounted Visible Appliances

Maximum Room Size		Maximum Ceiling Height		Minimum Required Light Output, Candela (cd) (Effective Intensity) One Light
(ft)	(m)	(ft)	(m)	(cd)
20 × 20	6.1 × 6.1	10	3.05	15
30 × 30	9.14 × 9.14	10	3.05	30
40 × 40	12.2 × 12.2	10	3.05	60
50 × 50	15.2 × 15.2	10	3.05	95
20 × 20	6.1 × 6.1	20	6.1	30
30 × 30	9.14 × 9.14	20	6.1	45
40 × 40	12.2 × 12.2	20	6.1	80
50 × 50	15.2 × 15.2	20	6.1	115
20 × 20	6.1 × 6.1	30	9.14	55
30 × 30	9.14 × 9.14	30	9.14	75
40 × 40	12.2 × 12.2	30	9.14	115
50 × 50	15.2 × 15.2	30	9.14	150

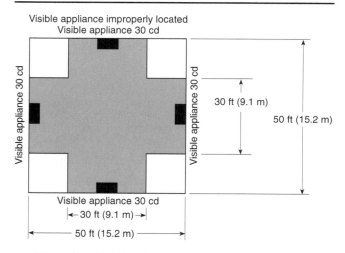

Figure A-6-4.4.1(d)   *Room spacing allocation — incorrect.*

NOTE 1:   Where ceiling heights exceed 30 ft (9.14 m), visible signaling appliances shall be suspended at or below 30 ft (9.14 m) or wall-mounted in accordance with Table 6-4.4.1.1(a).

NOTE 2:   Table 6-4.4.1.1(b) is based on locating the visible signaling appliance at the center of the room. Where it is not located at the center of the room, the effective intensity (cd) shall be determined by doubling the distance from the appliance to the farthest wall to obtain the maximum room size.

The 1996 edition of this Code has been modified to reduce the chances that strobe lights will induce epileptic seizures in persons with photosensitive epilepsy. The flash rate has been adjusted so that one, or even two appliances not flashing in unison can not produce a flash rate that is considered dangerous. If two or more appliances can be viewed at the same time, they must either be synchronized or located far enough apart so that their intensity at the viewer's location is low enough to be considered safe.

Visible signaling is a very complex topic. For this reason, the Code presents prescriptive requirements rather than the performance requirements, such as those for audible signaling. In essence, the Code provides preset designs that can be used for a variety of actual field conditions requiring these devices. The prescriptive requirements contained in the Code are based, in part, on extensive tests performed by Underwriters Laboratories, Inc., Northbrook, IL.

6-4.4.1.2 Where a room configuration is not square, the square room size that entirely encompasses the room or subdivides the room into multiple squares shall be used.

Figure 6-4.4.1.1 and Tables 6-4.4.1.1(a) and (b) help to ensure that a sufficient number of properly-sized visible notification appliances are installed in each protected space to provide complete coverage. The key to proper coverage in irregular spaces is to divide the space into a series of squares and provide proper coverage for each square as if it were an independent space.

6-4.4.2* Spacing in Corridors.

6-4.4.2.1 Table 6-4.4.2.1 shall apply to corridors not exceeding 20 ft (6.1 m) in width. For corridors greater than 20 ft (6.1 m) wide, refer to Figure 6-4.4.1.1 and Tables 6-4.4.1.1(a) and (b). In a corridor application, visible appliances shall be rated not less than 15 cd.

The intensity and spacing requirements for visible appliances located in corridors (less than 20 feet wide) are less stringent than for those in rooms. It is recognized that a person in a corridor is usually moving and alert. Because the occupants are usually alert, fewer appliances are required, which results in greater spacing in long corridors.

6-4.4.2.2 The visible appliances shall be located no more than 15 ft (4.57 m) from the end of the corridor with a separation no greater than 100 ft (30.4 m) between appliances. Where there is an interruption of the concentrated viewing path, such as a fire door, an elevation change, or any other obstruction, the area shall be considered as a separate corridor.

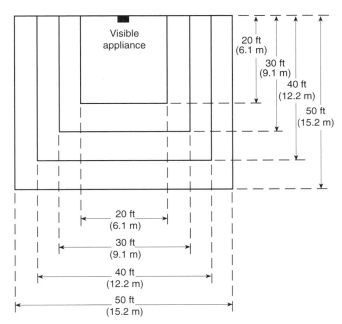

Note: Figure 6-4.4.1.1 is based on locating the visible signaling appliance at the halfway distance of the longest wall. In square rooms with appliances not centered or nonsquare rooms, the effective intensity (cd) from one visible signaling appliance shall be determined by maximum room size dimensions obtained either by measuring the distance to the farthest wall or by doubling the distance to the farthest adjacent wall, whichever is greater, as shown in Table 6-4.4.1.1(a).

Figure 6-4.4.1.1 *Room spacing for wall-mounted visible appliances.*

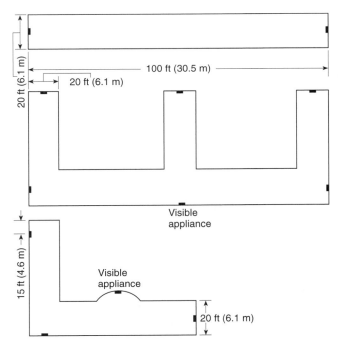

Figure A-6-4.4.2 *Corridor and elevator area spacing allocation.*

Table 6-4.4.2.1   Corridor Spacing for Visible Appliances

Corridor Length		Minimum Number of 15-cd Visible Appliances Required
(ft)	(m)	
0 – 30	0 – 9.14	1
31 – 130	9.45 – 39.6	2
131 – 230	39.93 – 70	3
231 – 330	70.4 – 100. 6	4
331 – 430	100.9 – 131.1	5
431 – 530	131.4 – 161.5	6

6-4.4.3*   Sleeping Areas.

A-6-4.4.3   Effective intensity is the conventional method of equating the brightness of a flashing light to that of a steady-burning light as seen by a human observer. The units of effective intensity are expressed in candelas. For example, a flashing light that has an effective intensity of 15 candelas has the same apparent brightness to an observer as a 15-candela steady-burning light source.

6-4.4.3.1   Combination smoke detectors and visible appliances shall be installed in accordance with the applicable requirements of Chapter 6, Chapter 2, and Chapter 5.

This requirement reinforces the detector coverage requirements of Chapters 2 and 5, as well as those of this Chapter. While not explicitly mentioned in 6-4.4.3.1, smoke alarms do have integral notification appliances, and these appliances should conform to the requirements of this chapter.

6-4.4.3.2   Table 6-4.4.3.2 shall apply to sleeping areas having no linear dimension greater than 16 ft (4.87 m). For larger rooms, the visible notification appliance shall be located within 16 ft (4.87 m) of the pillow.

In order to offset the obscuration effect of smoke near the ceiling, it was determined that a value of 177 cd is required when the notification appliance is located within 24 inches (610 mm) of the ceiling, and a value of 110 cd is sufficient when the distance is greater than 24 inches (610 mm).

6-4.4.4   Where visible appliances are required, a minimum of one appliance shall be installed in the concentrated viewing path such as might be experienced in such areas as classrooms or theater stages.

Table 6-4.4.3.2   Effective Intensity Requirements for Sleeping Area

Visible Notification Appliance Distance from Ceiling to Top of Lens	Intensity
Greater than or equal to 24 in. (610 mm)	110 cd
Less than 24 in. (610 mm)	177 cd

6-5   Visible Characteristics, Private Mode.   Visible signals used in the private mode shall be adequate for their intended purpose.

Visible notification appliances in the private mode are almost always used in conjunction with an audible notification appliance to call the viewer's attention to the visible appliance. Many visible appliances in the private mode provide annunciated information that helps the viewer to locate the source of an alarm, supervisory, or trouble signal.

6-6   Supplementary Visible Signaling Method.   A supplementary visible appliance is intended to augment an audible or visible signal.

A supplementary visible notification appliance is not intended to serve as one of the required visible notification appliances. Examples include non-required remote annunciators, or non-required flashing lights located in a guard's or office maintenance office.

6-6.1   A supplementary visible appliance shall comply with its marked rated performance.

Recognizing that this appliance is not satisfying a requirement, but is providing a supplemental function, this subsection makes it mandatory that the appliance function as marked and rated. This requirement discourages manufacturers from overrating the marking, which might not be detected since the appliances are supplementary, and gives the authority having jurisdiction a basis for verifying the performance of such appliances.

6-6.2   Supplementary visible notification appliances shall be permitted to be located less than 80 in. (2 m) above the floor.

Because such an appliance is supplementary, it need not meet the mandatory height requirement for visible appliances.

6-7 Textual Audible Appliances.

6-7.1 **Performance.** The textual appliance shall reproduce normal voice frequencies.

Textual audible appliances convey voice information. While the actual range of normal voice frequencies is not specified here, a textual appliance having a bandpass similar to telephonic communications would normally serve as an appropriate measure of the suitability of the frequency response.

6-7.2 **Loudspeaker Appliance.** The sound level in dBA of the loudspeaker appliance evacuation tone signals of the particular mode installed shall comply with all the requirements in 6-3.2.

In addition to conveying textual information, textual audible appliances are also used to produce tones used to warn occupants to evacuate the protected premises. This requirement ensures that the sound level requirements of 6-3.2 will be met by the textual audible appliances. See Figure 6.9.

6-7.3 **Location of Loudspeaker Appliances.**

This subsection requires textual audible appliances to be mounted at least 6 inches (152 mm) below the finished ceiling, and at least 7 feet 6 inches (2.30 m) above the finished floor when the ceiling height of the protected space permits. This required location helps ensure that the output of the appliances will be appropriately distributed throughout the protected space.

6-7.3.1 Where ceiling heights allow, wall-mounted loudspeaker appliances shall have their tops at heights above the finished floors of not less than 90 in. (2.30 m) and below the finished ceilings of not less than 6 in. (152 mm). This requirement shall not preclude ceiling-mounted or recessed appliances.

6-7.3.2 Where loudspeaker/visible appliances are installed, the height of the installed appliance shall comply with 6-4.4.

Exception: Combination loudspeaker/visible appliances installed in sleeping areas shall comply with 6-4.4.3.

As with other combination audible/visible appliances, 6-7.3.2 and its exception give preference to the visible appliance with respect to location.

6-7.4 **Telephone Appliance.** The telephone appliance shall be in accordance with EIA Tr 41.3, *Telephones.*

The referenced Electronic Industries Association standard helps to ensure the quality and technical suitability of a telephone handset.

6-7.5 **Location of Telephone Appliances.** Wall-mounted telephone appliances or related jacks shall be of convenient heights not to exceed 66 in. (1.7 m).

Exception: Where accessible to the general public, one telephone appliance per location shall be no higher than 54 in. (1.37 m) with clear access to the wall of at least 30 in. (0.76 m) wide.

The word "accessible" in this context means "available to and intended to be used by the general public." This includes use by floor or section fire wardens who might be required to communicate with the building emergency communication center by means of the fire alarm system telephones. However, fire alarm system telephones reserved exclusively for the use of fire fighters or other emergency services personnel would not need to meet the mounting height requirement of 6-7.5.

Figure 6.9 Typical loudspeakers used as textual notification appliances. Photograph courtesy of Wheelock, Inc., Long Branch, NJ.

6-8*   Textual Visible Appliances.

A-6-8   Textual visible appliances are selected and installed to provide temporary text, permanent text, or symbols. Textual visible information should be of a size and visual quality that is easily read and is distinguishable from the distance and lighting conditions anticipated during a fire. Appliances that provide temporary text or symbols should be arranged to display, or should be easily retrievable to display on demand, current alarm information for the entire duration of the alarms.

Paragraph 6-8.1.2 references cathode ray tube (CRT) displays having 250 lines per frame and 250 points per line scan, which reflect the technology of the 1980s. This can be compared to 640 pixels × 480 pixels up to 1280 pixels × 1224 pixels in technologies found in the 1990s. The code requires the screen to refresh at a rate of 30 times per second as compared to 1990s'-style monitors that refresh at 60 to 72 times per second. Ten shades of grey represented the minimum technology of the 1980s. Sixty-four shades of grey are common today. The aspect ratio is the ratio of pixels in the x and y planes. The dot size at the time this code was written was 0.5 mm, as opposed to the current screen technology of 0.28 mm.

6-8.1   The temporary textual visible appliance shall be a nonstorage display that produces either visible alphanumerics subtending a character angle to the observing eye of not less than 10 minutes of arc or visible pictorial images.

A temporary textual visible appliance has no memory that would permit scrolling backward and forward to display information that has been received over time. The viewing angle helps define the types of display equipment that would be suitable for fire alarm use, ensuring that a reasonable viewing angle is provided. See Figure 6.10.

6-8.1.1   The alphanumeric display shall have an equivalent minimum 7 × 5 matrix character definition, a minimum grey scale contrast as defined by 10 shades of grey, and a character retentivity from $1/2$ minute to 5 minutes.

6-8.1.2   The pictorial display shall have a minimum 250-line scan per frame and a minimum of 250 points per line scan,

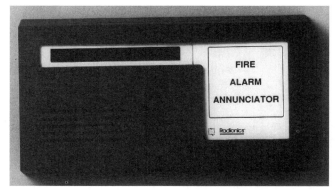

Figure 6.10   *Temporary textual visible appliance (annunciator). Photograph courtesy of Radionics, Inc., Salinas, CA*

each arranged on a scale of 10 shades of grey, and shall have 30 frames per second. The display shall have an aspect ratio of 1:1.33.

Subsections 6-8.1.1 and 6-8.1.2 help to ensure the overall quality of the display and its ability to accurately reproduce the information it is intended to convey.

6-8.2   **Location.**   All textual visible appliances in the private mode shall be located in rooms accessible only to those persons directly concerned with the implementation and direction of emergency action initiation and procedure in the areas protected by the fire alarm system.

Exception: In the lobby of a building where required by the authority having jurisdiction.

This subsection intends to limit access to private mode textual visible displays to only those persons authorized to obtain such information. The exception permits the authority having jurisdiction to specify a more public location of the textual visible appliance, presumably for use by responding emergency personnel.

7

Inspection, Testing, and Maintenance

Contents, Chapter 7

1. Control Equipment: Fire Alarm Systems Monitored for Alarm, Supervisory, Trouble Signals

2. Control Equipment: Fire Alarm Systems Unmonitored for Alarm, Supervisory, Trouble Signals

3. Batteries

4 Transient Suppressors

5. Control Panel Trouble Signals

6. Fiber Optic Cable Connections

7. Emergency Voice/Alarm Communications Equipment

8. Remote Annunciators

9. Initiating Devices

10. Guard's Tour Equipment

11. Interface Equipment

12. Alarm Notification Appliances - Supervised

13. Supervising Station Fire Alarm Systems - Transmitters

14. Special Procedures

15. Supervising Station Fire Alarm Systems - Receivers

7-3.2 Testing

Table 7-3.2 Testing Frequencies

1. Control Equipment: Fire Alarm Systems Monitored for Alarm, Supervisory, Trouble Signals

2. Control Equipment: Fire Alarm Systems Unmonitored for Alarm, Supervisory, Trouble Signals

3. Engine-Driven Generator

4. Batteries - Central Station Facilities

5. Batteries - Fire Alarm Systems

6. Batteries - Public Fire Alarm Reporting Systems

7. Fiber-Optic Cable Power

8. Control Unit Trouble Signals

9. Conductors/Metallic

10. Conductors/Nonmetallic

11. Emergency Voice/Alarm Communications Equipment

12. Retransmission Equipment

13. Remote Annunciators

14. Initiating Devices

15. Guard's Tour Equipment

16. Interface Equipment

17. Special Hazard Equipment

18. Alarm Notification Appliances

19. Off-Premises Transmission Equipment

20. Supervising Station Fire Alarm Systems - Transmitters

21. Special Procedures

22. Supervising Station Fire Alarm Systems - Receivers

7-4 Maintenance

7-5 Records

7-5.1 Permanent Records

7-5.2 Maintenance, Inspection, and Testing Records

Figure 7-5.2.2 Inspection and Testing Form

7-1   General.

7-1.1   **Scope.**   This chapter covers the minimum requirements for the inspection, testing, and maintenance of the fire alarm systems described in Chapters 1, 3, and 4 and for their initiation and notification components described in Chapters 5 and 6. The testing and maintenance requirements for one- and two-family dwelling units are located in Chapter 2. Single station detectors used for other than one- and two-family dwelling units shall be tested and maintained in accordance with this chapter. More stringent inspection, testing, or maintenance procedures that are required by other parties shall be permitted.

This chapter addresses testing and maintenance requirements for protected premises fire alarm systems and for supervising station fire alarm systems. It does not include the testing and maintenance requirements for household fire warning equipment, which remain in Chapter 2.

Single station smoke detectors used for other than one- and two-family dwelling units are often found in a variety of residential occupancies such as apartments, hotel and motel rooms, and dormitory living units. Non-system connected detection devices (sometimes called "stand alone" detectors) are sometimes found in HVAC systems, door releasing applications, and special hazard releasing devices. The requirements in Chapter 7, including sensitivity testing, apply to these types of detectors.

7-1.1.1 Inspection, testing, and maintenance programs shall satisfy the requirements of this code, conform to the equipment manufacturer's recommendations, and verify proper operation of the fire alarm system.

The requirements of Chapter 7 apply to both new and existing systems and are, therefore, retroactive. The intent of the Technical Committee on Testing and Maintenance was made clear by Proposal 72-601 in the NFPA Annual 1996 Report On Proposals. This proposal would have required systems to be tested to the requirements of the Code adopted at the time of installation. The proposal would have caused documentation problems and was rejected by the Technical Committee. By rejecting this proposal, the Committee intends that the most current edition of the Code be used for testing, inspection and maintenance of both new and existing fire alarm systems. The requirements of the other chapters are not retroactive and only apply to new installations. See Section 1-2.3 for further details on these requirements.

This subsection incorporates any manufacturer's instructions into the requirements of this Code. Therefore, these instructions should be enforced as Code.

7-1.1.2 Nothing in this chapter is intended to prevent the use of alternate test methods or testing devices, provided such methods or devices are equivalent in effectiveness and safety and meet the intent of the requirements of this chapter.

The authority having jurisdiction has the responsibility of ensuring that the alternative methods are equivalent.

7-1.2 The owner or owner's designated representative shall be responsible for inspection, testing, and maintenance of the system and alterations or additions to this system. The delegation of responsibility shall be in writing, with a copy of such delegation made available to the authority having jurisdiction.

The owner of the system, or his or her designated representative, is responsible for testing and maintaining the system. If the owner appoints a representative for this responsibility, the delegation must be confirmed in writing. This may take the form of a testing and maintenance contract with a qualified contractor or delegation to a qualified staff specialist. See 7-1.2.1.

7-1.2.1 Inspection, testing, or maintenance shall be permitted to be done by a person or organization other than the owner where conducted under a written contract. Testing and maintenance of central station service systems shall be performed under the contractual arrangements specified in 4-2.2.2.

Figure 7.1 *Technician testing fire alarm system. Photograph courtesy of Simplex Time Recorder Co., Gardner, MA.*

New language was added to the 1996 edition of the Code to clarify the requirement that a contractual agreement is required for central station systems.

7-1.2.2 Service personnel shall be qualified and experienced in the inspection, testing, and maintenance of fire alarm systems. Examples of qualified personnel shall be permitted to include, but shall not be limited to, individuals who are:

(a) Factory trained and certified.

(b) National Institute for Certification in Engineering Technologies fire alarm certified.

(c) International Municipal Signal Association fire alarm certified.

(d) Certified by a state or local authority.

(e) Trained and qualified personnel employed by an organization listed by a national testing laboratory for the servicing of fire alarm systems.

The qualification requirement stated in 7-1.2.2 will help ensure that the persons testing and maintaining fire alarm systems have an appropriate level of knowledge, skill, and understanding of the systems and equipment.

Because each manufacturer's equipment is different, personnel who are factory trained and certified should receive training from all manufacturers of the equipment encountered on the job. Fire alarm installation and testing personnel should also be trained to know and understand the concepts of this Code.

The state or local authority having jurisdiction may have specific certification or licensing tests or other requirements that must be met.

The International Municipal Signal Association (IMSA) is the professional association of those individuals who oversee public fire service communication systems and traffic signals. This organization offers educational programs as well as fire alarm certification programs. See Section 4-6.8.1.3.

7-1.3*   Notification.

A-7-1.3   Advance Notification.   Prior to any scheduled inspection or testing, the service company should consult with the building owner or owner's designated representative. Issues of advance notification in certain occupancies, including advance notification time, building posting, systems interruption and restoral, evacuation procedures, accommodation for evacuees, and other related issues, should be agreed upon by all parties prior to any inspection or testing.

7-1.3.1   Before proceeding with any testing, all persons who receive and facilities that receive alarm, supervisory, or trouble signals, and all building occupants, shall be notified to prevent unnecessary response. At the conclusion of testing, those previously notified (and others, as necessary) shall be notified that testing has been concluded.

Everyone who may be affected by fire alarm testing in a protected premises must be notified that testing will take place. This notification should include, but not be limited to, the building owner, building manager, switchboard operator, building engineer, building or floor fire wardens and building maintenance personnel. In addition, the staff of the public fire service communication center, alarm company supervising station, and building occupants should also be notified prior to testing. Effective methods of notification include bulletin board postings, electronic mail, public address announcements and lobby signs.

A fire emergency plan that makes provision for notifying occupants, the public fire service communication center, and the supervising station of the alarm company in case a fire occurs while testing is being conducted should also be established for each protected premises.

Section 4-2.7.3 of the Code now requires the prime contractor (or designated representative) to provide a unique identification code to the central station before placing the central station fire alarm system into test status. This requirement is intended to prevent unauthorized tampering.

7-1.3.2   The owner or owner's designated representative and service personnel shall coordinate system testing to prevent interruption of critical building systems or equipment.

If the fire alarm system has interface connections to other building systems, such as elevators and HVAC systems, the interfaces must be managed so that testing does not disrupt building systems or equipment that may be critical to the continuity of building operations.

7-1.4   Prior to system maintenance or testing, the system certificate and the information regarding the system and system alterations, including specifications, wiring diagrams, and floor plans, shall be made available by the owner or a designated representative to the service personnel.

Service personnel cannot effectively maintain or test a system without full access to the fire alarm system documentation.

7-1.5   Releasing Systems.   This subsection covers requirements pertinent to testing fire alarm systems initiating fire suppression system releasing functions.

Testing of special hazards fire protection systems that are equipped with their own fire alarm control unit should be conducted as a separate series of tests. Only the interface functions between the separate control unit and the building fire alarm system should be tested as part of the building fire alarm system testing.

7-1.5.1   Testing personnel shall be familiar with the specific arrangement and operation of the suppression system(s) and releasing function(s) and cognizant of the hazards associated with inadvertent system discharge of suppression systems.

It is necessary to prevent unwanted discharges that can cause significant property damage or cause accidental injury or death to the building occupants.

7-1.5.2   Occupant notification shall be required whenever a fire alarm system configured for releasing service is being serviced or tested.

Notification will allow the affected occupant(s) to secure the releasing system(s) prior to testing the related fire alarm system in order to avoid accidental discharges.

7-1.5.3   Discharge testing of suppression systems shall not be required by this code. Suppression systems shall be secured from inadvertent actuation, including disconnection of releasing solenoids/electric actuators, closing of valves, other actions, or combinations thereof, as appropriate for the specific system, for the duration of the fire alarm system testing.

7-1.5.4 Testing shall include verification that the releasing circuits and components energized or actuated by the fire alarm system are electrically supervised and operate as intended on alarm.

If the building fire alarm system also controls the release of a special hazard fire protection system, the special hazard system operation (without the discharge of suppression agent) must be tested as part of the building fire alarm system testing procedures. Care must be taken to ensure that the special hazards system is not inadvertently actuated.

7-1.5.5 Suppression systems and releasing components shall be returned to their normal condition upon completion of system testing.

7-1.6 System Testing.

7-1.6.1 Initial Acceptance Testing. All new systems shall be inspected and tested in accordance with the requirements of this chapter.

7-1.6.2* Reacceptance Testing.

A-7-1.6.2 The additional devices to be tested should be a sample representation of the types of devices and locations on the system.

7-1.6.2.1 Reacceptance testing shall be performed after system components are added or deleted; after any modification, repair, or adjustment to system hardware or wiring; or after any change to software. All components, circuits, systems operations, or site-specific software functions known to be affected by the change or identified by a means that indicates the system operational changes shall be 100 percent tested. In addition, 10 percent of initiating devices that are not directly affected by the change, up to a maximum of 50 devices, also shall be tested and proper system operation shall be verified. A revised record of completion in accordance with 1-7.2.1 shall be prepared to reflect any changes.

Whenever a fire alarm system is modified, the modification, no matter how small it may seem, may affect the overall operation of the modified portion of the system. Seemingly harmless changes in software have caused tremendous changes in operation, and have sometimes resulted in disastrous events. This Code requirement

ensures that the affected portion of the system will be completely reacceptance tested. It further requires that a random sampling of other portions of the fire alarm system be tested to help ensure that other, seemingly unrelated, portion(s) of the system have not been adversely affected by the modification. See Section 1-4 of this Code for definitions of software types.

The 10 percent sample should be random, and should include at least one device per initiating device circuit to ensure correct operation. Use of software comparison algorithms can also assist in determining where changes may have occurred.

Formal Interpretation 93-1
Reference: 7-1.6.2

Question: Is it the intent of the committee to mandate a complete retesting of an entire system including all devices, circuits, and connections when only a single device or circuit has been modified?
Answer: No.

Issue Edition: NFPA 72-1993
Reference: 7-1.6
Issue Date: September 9, 1993 ■

7-1.6.2.2 Changes to all control units connected or controlled by the system executive software shall require a 10-percent functional test of the system, including a test of at least one device on each input and output circuit to verify critical system functions such as notification appliances, control functions, and off-premises reporting.

See Figure 7.2.

Figure 7.2 Technician using a laptop computer to modify software. Photograph courtesy Mammoth Fire Alarms, Lowell, MA.

7-2   Test Methods.

7-2.1*   Central Stations. The installation shall be inspected at the request of the authority having jurisdiction for complete information regarding the system, including specifications, wiring diagrams, and floor plans that have been submitted for approval prior to installation of equipment and wiring.

A-7-2.1 Where the authority having jurisdiction strongly suspects significant deterioration or otherwise improper operation by a central station, a surprise inspection to test the operation of the central station should be made, but extreme caution should be exercised. This test is to be conducted without advising the central station, but the public fire service communications center is definitely to be contacted when manual alarms, waterflow alarms, or automatic fire detection systems are tested so that the fire department will not respond. In addition, persons normally receiving calls for supervisory alarms should be notified when items such as gate valves and functions such as pump power are tested. Confirmation of the authenticity of the test procedure should be obtained and should be a matter for resolution between plant management and the central station.

7-2.2* Fire alarm systems and other systems and equipment that are associated with fire alarm systems and accessory equipment shall be tested according to Table 7-2.2.

Table 7-2.2   Test Methods

Device	Method
1. Control Equipment a. Functions	At a minimum, control equipment shall be tested to verify proper receipt of alarm, supervisory, and trouble signals (inputs), operation of evacuation signals and auxiliary functions (outputs), circuit supervision including detection of open circuits and ground faults, and power supply supervision for detection of loss of ac power and disconnection of secondary batteries.
	The input/output control equipment functions must be tested to ensure proper operation. It is not intended to check functions internal to the equipment, such as software algorithms and communications protocols (sometimes called firmware). Another example of a functional test would be to verify that Class A circuits will transmit an alarm in either direction under a single fault condition.
b. Fuses	Remove fuse and verify rating and supervision.
	Incorrect fuse rating can lead to equipment damage or unnecessary loss of power.
c. Interfaced Equipment	Integrity of single or multiple circuits providing interface between two or more control panels shall be verified. Interfaced equipment connections shall be tested by operating or simulating operation of the equipment being supervised. Signals required to be transmitted shall be verified at the control panel.
	The wiring connections must be tested by simulating a single open and a single ground to verify proper indications for the monitoring of the interfaced equipment wiring integrity. In addition, the interfaced equipment must be placed in a simulated trouble condition to test for proper supervisory signal receipt and reaction at the main control unit. See Figure 7.3.
d. Lamps and LEDs	Lamps and LEDs shall be illuminated.
e. Primary (Main) Power Supply	All secondary (standby) power shall be disconnected and tested under maximum load, including all alarm appliances requiring simultaneous operation. All secondary (standby) power shall be reconnected at end of test. For redundant power supplies, each shall be tested separately.
2. Engine-Driven Generator	Where an engine-driven generator dedicated to the fire alarm system is used as a required power source, operation of the generator shall be verified in accordance with NFPA 110, *Standard for Emergency and Standby Power Systems*, by the building owner.
3. Secondary (Standby) Power Supply	Disconnect all primary (main) power supplies and verify that required trouble indication for loss of primary power occurs. Measure or verify system's standby and alarm current demand and, using manufacturer's data, verify whether batteries are adequate to meet standby and alarm requirements. Operate general alarm systems for a minimum of 5 minutes and emergency voice communications systems for a minimum of 15 minutes. Reconnect primary (main) power supply at end of test.
4. Uninterrupted Power Supply (UPS)	Where a UPS system dedicated to the fire alarm system is used as a required power source, operation of the UPS system shall be verified by the building owner in accordance with NFPA 111, *Standard on Stored Electrical Energy Emergency and Standby Power Systems*.

Table 7-2.2 Test Methods (cont.)

Device	Method
5. Batteries — General Tests	
a. Visual Inspection	Inspect batteries for corrosion or leakage. Check and ensure tightness of connections. Where necessary, clean and coat the battery terminals or connections. Visually inspect electrolyte level in lead-acid batteries.
b. Battery Replacement	Batteries shall be replaced in accordance with the recommendations of the alarm equipment manufacturer, or when the recharged battery voltage or current falls below the manufacturer's recommendations.
c. Charger Test	Check operation of battery charger in accordance with charger test for the specific type of battery.
d. Discharge Test	With the battery charger disconnected, load test the batteries following the manufacturer's recommendations. The voltage level shall not fall below the levels specified. Exception: An artificial load equal to the full fire alarm load connected to the battery shall be permitted to be utilized in conducting this test.
e. Load Voltage Test	With the battery charger disconnected, measure the terminal voltage while supplying the maximum load required by its application. The voltage level shall not fall below the levels specified for the specific type of battery. Where the voltage falls below the level specified, corrective action shall be taken and the batteries shall be retested. Exception: An artificial load equal to the full fire alarm load connected to the battery shall be permitted to be utilized in conducting this test.
f. Open Circuit Voltage	With the battery charger disconnected, measure the open circuit voltage of the battery.
6. Battery Tests (Specific Types)	
a. Primary Battery Load Voltage Test	The maximum load for a No. 6 primary battery shall not be more than 2 amperes per cell. An individual (1.5-volt) cell shall be replaced when a load of 1 ohm reduces the voltage below 1 volt. A 6-volt assembly shall be replaced where a test load of 4 ohms reduces the voltage below 4 volts. It is not desirable to completely drain a battery during the discharge test because, in the event of a power failure shortly after the test, the system could be left without a power supply. A typical test would place the battery under load for a shorter period (1 - 2 hours). However, the battery should be tested to ensure that it can deliver the required current at rated voltage under maximum expected load. Battery calculations must be relied upon to ensure capacity.
b. Lead-Acid Type:	
1. Charger Test	With the batteries fully charged and connected to the charger, measure the voltage across the batteries with a voltmeter. The voltage shall be 2.30 volts per cell (0.02 volt at 25°C (77°F) or as specified by the equipment manufacturer.
2. Load Voltage Test	Under load, the battery shall not fall below 2.05 volts per cell.
3. Specific Gravity	The specific gravity of the liquid in the pilot cell or all of the cells shall be measured as required. The specific gravity shall be within the range specified by the manufacturer. Although the specified specific gravity can vary from manufacturer to manufacturer, a range of 1.205 – 1.220 is typical for regular lead-acid batteries, while 1.240 – 1.260 is typical for high performance batteries. A hydrometer that shows only a pass or fail condition of the battery and does not indicate the specific gravity shall not be used, since such a reading does not give a true indication of the battery condition. Caution: Adding acid to a lead-acid type battery cell may destroy the plates, affecting battery capacity.
c. Nickel-Cadmium Type:	
1. Charger Test	With the batteries fully charged and connected to the charger, place an amp meter in series with the battery under charge. The charging current shall be in accordance with the manufacturer's recommendations for the type of battery used. In the absence of specific information, this usually is $1/30$ to $1/25$ of the battery rating. [Example: 4000 mAh $\times$ $1/25$ = 160 mA charging current at 25°C (77°F).]
2. Load Voltage Test	Under load, the float voltage for the entire battery shall be 1.42 volts per cell, nominal. If possible, cells shall be measured individually.
d. Sealed Lead-Acid Type:	
1. Charger Test	With the batteries fully charged and connected to the charger, measure the voltage across the batteries with a voltmeter. The voltage should be 2.30 volts per cell ±0.02 volt at 25°C (77°F) or as specified by the equipment manufacturer.
2. Load Voltage Test	Under load, the battery shall perform in accordance with the battery manufacturers' specifications.
7. Public Reporting System Tests	In addition to the tests and inspection required above, the following requirements shall apply. Manual tests of the power supply for public reporting circuits shall be made and recorded at least once during each 24-hour period. Such tests shall include: (a) Current strength of each circuit. Changes in current of any circuit, amounting to 10 percent of normal current, shall be investigated immediately.

Table 7-2.2   Test Methods   (cont.)

Device	Method
	(b) Voltage across terminals of each circuit, inside of terminals of protective devices. Changes in voltage of any circuit, amounting to 10 percent of normal voltage, shall be investigated immediately. (c) Voltage between ground and circuits. Where this test shows a reading in excess of 50 percent of that shown in the test specified in (b) above, the trouble shall be immediately located and cleared; readings in excess of 25 percent shall be given early attention. These readings shall be taken with a voltmeter of not more than 100-ohms resistance per volt. NOTE 1: The voltmeter sensitivity has been changed from 1000 ohms per volt to 100 ohms per volt so that false ground readings (caused by induced voltages) are minimized. NOTE 2: Systems in which each circuit is supplied by an independent current source (Forms 3 and 4) require tests between ground and each side of each circuit. Common current source systems (Form 2) require voltage tests between ground and each terminal of each battery and other current source. (d) A ground current reading shall be permitted in lieu of (c) above. Where this method of testing is used, all grounds showing a current reading in excess of 5 percent of the normal line current shall be given immediate attention. (e) Voltage across terminals of common battery, on switchboard side of fuses. (f) Voltage between common battery terminals and ground. Abnormal ground readings shall be investigated immediately. NOTE: Tests specified in (e) and (f) above apply only to those systems using a common battery. Where more than one common battery is used, each common battery is to be tested.
8. Transient Suppressors	Lightning protection equipment shall be inspected and maintained per manufacturer's specifications. Additional inspections shall be required after any lightning strikes. Equipment located in moderate to severe areas outlined in NFPA 780, *Standard for the Installation of Lightning Protection Systems*, Appendix I, shall be inspected semiannually and after any lightning strikes. In areas prone to lightning storms, the owner should be advised to notify the service company when a storm has occurred so that all original lightning protection can be checked.
9. Control Panel Trouble Signals a. Audible and Visual	Verify operation of panel trouble signals and ring-back feature for systems using a trouble-silencing switch that requires resetting.
b. Disconnect Switches	Where control unit (panel) has disconnect or isolating switches, verify that each switch performs its intended function and a trouble signal is received when a supervised function is disconnected.
c. Ground-Fault Monitoring Circuit	Where system has ground detection feature, verify that a ground-fault indication is given whenever any installation conductor is grounded. Each conductor should be grounded temporarily to ensure proper ground detection circuit operation. This information should be recorded on the acceptance test report for future troubleshooting information.
d. Transmission of Signals to Off-Premises Location	Actuate an appropriate initiating device and verify that alarm signal is received at the off-premises location. Create a trouble condition and verify that a trouble signal is received at the off-premises location. Actuate a supervisory device and verify that a supervisory signal is received at the off-premises location. Where a transmission carrier is capable of operation under a single or multiple fault condition, activate an initiating device during such fault condition and verify that a trouble signal is received at the off-premises location in addition to the alarm signal.
10. Remote Annunciators	Verify for proper operation and confirm proper identification. Where provided, verify proper operation under a fault condition. Remote annunciation is very important to the fire department personnel responding to the alarm. Its intent is to reduce the time in finding the source of the alarm by providing clear and accurate information to the responding fire service.
11. Conductors/Metallic a. Stray Voltage	All installation conductors shall be tested with a volt/ohm meter to verify that there are no stray (unwanted) voltages between installation conductors or between installation conductors and ground. Unless a different threshold is specified in the system installed equipment manufacturer's specifications, the maximum allowable stray voltage shall not exceed 1 volt ac/dc.

Table 7-2.2 Test Methods (cont.)

Device	Method
b. Ground Faults	All installation conductors other than those intentionally and permanently grounded shall be tested for isolation from ground per the installed equipment manufacturer's specifications.
c. Short Circuit Faults	All installation conductors other than those intentionally connected together shall be tested for conductor-to-conductor isolation per the installed equipment manufacturer's specifications. These same circuits also shall be tested conductor-to-ground.
d. Loop Resistance	With each initiating and indicating circuit installation conductor pair short-circuited at the far end, measure and record the resistance of each circuit. Verify that the loop resistance does not exceed the installed equipment manufacturer's specified limits.

The assumption here is that if the loop resistance exceeds the installed equipment manufacturer's specified limits, the wiring will be changed as it should be. This item was revised in the 1996 edition of this Code to specify the equipment manufacturer, rather than the conductor manufacturer, because the equipment manufacturer generally has more stringent operational requirements.

Device	Method
12. Conductors/Nonmetallic	
a. Circuit Integrity	Test each initiating device, indicating appliance, and signaling line circuit to confirm that the integrity of installation conductors is being properly supervised.
b. Fiber Optics	The fiber-optic transmission line shall be tested in accordance with the manufacturer's instructions, by the use of an optical power meter, or by an optical time domain reflectometer to measure the relative power loss of the line. This relative figure for each fiber-optic line shall be recorded in the fire alarm control panel. Where the power level drops 2 percent or more from the value recorded during the initial acceptance test, the transmission line, section thereof, or connectors shall be repaired or replaced by a qualified technician to bring the line back into compliance with the accepted transmission level per manufacturer's recommendations.
c. Supervision	Introduction of a fault in any supervised circuit shall result in a suitable trouble indication at the control unit. One connection shall be opened at not less than 10 percent of the initiating device, indicating appliance, and signaling line circuits.

The word "supervision" here means the monitoring of the circuit conductors integrity.

Test each initiating device, indicating appliance, and signaling line circuit for proper alarm response.

Device	Method
13. Initiating Devices	NOTE: See Table 3-6 for description of circuit performance and capacity.
a. Electromechanical Releasing Device:	
1. Nonrestorable-Type Link	Remove the fusible link and operate the associated device to ensure proper operation. Lubricate any moving parts as necessary.
2. Restorable-Type Link	Remove the fusible link and operate the associated device to ensure proper operation. Lubricate any moving parts as necessary. NOTE: Fusible thermal link detectors are commonly used to close fire doors and fire dampers. They can be actuated by the presence of external heat, which causes a solder element in the link to fuse, or by an electric thermal device, which, when energized, generates heat within the body of the link, causing the link to fuse and separate.
b. Fire Extinguishing System(s) or Suppression System(s) Alarm Switch	Mechanically or electrically operate the switch and verify receipt of signal by the control panel.
c. Fire-Gas and Other Detectors	Fire-gas detectors and other fire detectors shall be tested as prescribed by the manufacturer and as necessary for the application.
d. Heat Detectors:	
1. Fixed-Temperature, Rate-of-Rise, Rate-of-Compensation, Restorable Line, Spot Type (excluding Pneumatic Tube Type)	Heat test with a heat source per manufacturer's recommendations for response within 1 minute. Precaution should be taken to avoid damage to the nonrestorable fixed-temperature element of a combination rate-of-rise/fixed-temperature element. *Use extreme caution in hazardous locations (those containing explosive vapors or dusts) when heat testing these types of detectors. Use of a bucket of hot water or hot towels is recommended.*
2. Fixed-Temperature, Nonrestorable Line Type	Do not heat test. Test mechanically and electrically for function. Measure and record loop resistance. Investigate changes from acceptance test.

Table 7-2.2   Test Methods   (cont.)

Device	Method
3. Fixed-Temperature, Nonrestorable Spot Type	After 15 years, replace all devices or laboratory test two detectors per 100. Replace the two detectors with new devices. Where a failure occurs on any of the detectors removed, additional detectors shall be removed and tested to determine either a general problem involving faulty detectors or a localized problem involving one or two defective detectors. The laboratory test referenced here is conducted by an independent testing laboratory engaged in the listing or approval of heat detectors.
4. Nonrestorable (General)	Do not heat test. Test mechanically and electrically for function. Contacts may be operated by hand or electrically jumpered to ensure alarm response.
5. Restorable Line Type, Pneumatic Tube Only	Heat source (where test chambers are in circuit) or pressure pump.
e. Fire Alarm Boxes	Operate per manufacturer's instruction. For key-operated presignal fire alarm boxes, test both presignal and general alarm circuits.
f. Radiant Energy Fire Detectors	Flame detectors and spark/ember detectors shall be tested in accordance with the manufacturer's instructions to determine that each detector is operative. Flame detector and spark/ember detector sensitivity shall be determined using either: (a) A calibrated test method; or (b) The manufacturer's calibrated sensitivity test instrument; or (c) Listed control panel arranged for the purpose; or (d) Other calibrated sensitivity test method acceptable to the authority having jurisdiction that is directly proportional to the input signal from a fire consistent with the detector listing or approval. Detectors found to be outside of the approved range of sensitivity shall be replaced or adjusted to bring them into the approved range where designed to be field adjustable. Flame detector and spark/ember detector sensitivity shall not be determined using a light source that administers an unmeasured quantity of radiation at an undefined distance from the detector.
g. Smoke Detectors: 1. Systems Detectors	The detectors shall be tested in place to ensure smoke entry into the sensing chamber and an alarm response. Testing with smoke or listed aerosol acceptable to the manufacturer, or other means acceptable to the detector manufacturer, shall be permitted as one acceptable test method. This is a "go, no-go" type of test to ensure smoke entry into the chamber and alarm response. It does not test the detector's sensitivity. This section requires more than a visual test. See Figures 7-4 and 7-5. Ensure that each smoke detector is within its listed and marked sensitivity range by testing using either: (a) A calibrated test method; or (b) The manufacturer's calibrated sensitivity test instrument; or (c) Listed control equipment arranged for the purpose; or (d) A smoke detector/control unit arrangement whereby the detector causes a signal at the control unit when its sensitivity is outside its acceptable sensitivity range; or (e) Other calibrated sensitivity test method acceptable to the authority having jurisdiction. NOTE: The detector sensitivity cannot be tested or measured using any spray device that administers an unmeasured concentration of aerosol into the detector.
2. Single Station Detectors	The detectors shall be tested in place to ensure smoke entry into the sensing chamber and an alarm response. Testing with smoke or listed aerosol acceptable to the manufacturer, or other means acceptable to the detector manufacturer, shall be permitted as one acceptable test method. Single-station detectors other than those used in one- and two-family dwellings are often found in residential occupancies, such as apartments, hotel and motel rooms, and dormitory living units. Other non-system connected detection devices (sometimes called "stand alone" detectors) are sometimes found in HVAC systems, door releasing applications, and special hazards releasing devices. The requirements in Chapter 7 of this Code, including sensitivity testing, apply to these types of detectors. Single-station detectors used in one- and two-family dwellings, are required to be functionally tested, but are not required to be sensitivity tested. See Section 7-3.2.1. for additional information on this requirement.
3. Air Sampling	Per manufacturer's recommended test methods, and verify detector alarm response through the end sampling port on each pipe run, as well as verifying airflow through all other ports.

Table 7-2.2 Test Methods (cont.)

Device	Method
4. Duct Type	Air duct detectors shall be tested or inspected to ensure that the device will sample the airstream. The test shall be made in accordance with the manufacturer's instructions.
5. Projected Beam Type	The detector shall be tested by introducing smoke, other aerosol, or an optical filter into the beam path.
6. Smoke Detector with Built-in Thermal Element	Operate both portions of the detector independently as described for the respective devices.

This requires a test of both portions of a combination unit if possible. The Code does not explicitly address the issue of one feature failing, but it is assumed that if a combination smoke/heat detector is being used, and one feature fails a test, the entire unit should be removed and replaced.

Device	Method
7. Smoke Detectors with Control Output Functions	Where individual fire detectors are used to control the operation of equipment as permitted by 3-7.1, the control capability shall remain operable even where all of the initiating devices connected to the same initiating circuit are in an alarm state.

This requirement forbids the use of two-wire smoke detectors for controlling operations on an initiating device circuit (e.g., fan shutdown) when other devices are installed on the same circuit. If, for instance, the smoke detector tries to activate after a manual fire alarm box has been actuated, the smoke detector may not alarm. See Section 3-8.2.4 of this Code and the related section commentary for more detailed information on this issue.

Device	Method
h. Initiating Devices, Supervisory:	
1. Control Valve Switch	Operate valve and verify signal receipt within the first two revolutions of the hand wheel or within $^{1}/_{5}$ of the travel distance, or per the manufacturer's specifications.
2. High- or Low-Air Pressure Switch	Operate switch and verify that receipt of signal is obtained where the required pressure is increased or decreased 10 psi (70 kPa) from the required pressure level.
3. Room Temperature Switch	Operate switch and verify receipt of signal to indicate the decrease in room temperature to 40°F (4.4°C) and its restoration to above 40°F (4.4°C).
4. Water Level Switch	Operate switch and verify the receipt of signal indicating the water level raised or lowered 3 in. (76.2 mm) from the required level within a pressure tank, or 12 in. (305 mm) from the required level of a non-pressure tank, and its restoral to required level.
5. Water Temperature Switch	Operate switch and verify receipt of signal to indicate the decrease in water temperature to 40°F (4.4°C) and its restoration to above 40°F (4.4°C).
i. Mechanical, Electrosonic, or Pressure-Type Waterflow Device	Flow water through an inspector's test connection indicating the flow of water equal to that from a single sprinkler of the smallest orifice size installed in the system for wet-pipe systems, or an alarm test bypass connection for dry-pipe, pre-action, or deluge systems in accordance with NFPA 25, *Standard for the Inspection, Testing, and Maintenance of Water-Based Fire Protection Systems*.
14. Alarm Notification Appliances	
a. Audible	Measure sound pressure level with sound level meter meeting ANSI S-1.4a, *Specifications for Sound Level Meters*, Type 2 requirements. Measure and record levels throughout protected area.

Choose areas that are physically remote from the audible notification appliances to measure and record sound pressure levels. If these areas comply with Code requirements, then further measurements may be deemed unnecessary by the authority having jurisdiction. However, areas that fail to meet the requirements of Section 6-3 of this Code may require additional appliances.

Device	Method
b. Speakers	Measure sound pressure level with sound level meter meeting ANSI S-1.4a, *Specifications for Sound Level Meters*, Type 2 requirements. Measure and record levels throughout protected area.

In order to comply with the sound level requirements of the Code, many installers will attempt to tap speakers at a higher wattage, rather than increase the number of speakers in an area. This incorrect approach to sound level compliance leads to distortion of voice messages through the speakers. Therefore, in addition to meeting the sound level requirements of the Code, the clarity or intelligibility of the voice message must be checked when speakers are used for voice communications.

Device	Method
	Verify voice clarity.
c. Visible	Test in accordance with manufacturer's instructions. Verify device locations are per approved layout and confirm that no floor plan changes affect the approved layout.

Ensure that none of the visible notification appliances are blocked by shelving, furniture, ceiling-mounted light fixtures, or movable partitions. The owner also should be advised to keep all viewing paths to visible notification appliances clear.

Figure 7.3 *Technician removing wire from device to check electrical supervision. Photograph courtesy Simplex Time Recorder Co., Gardner, MA.*

Figure 7.4 *Functional test of a smoke detector. Photograph courtesy No Climb Products Ltd., Hertfordshire, UK.*

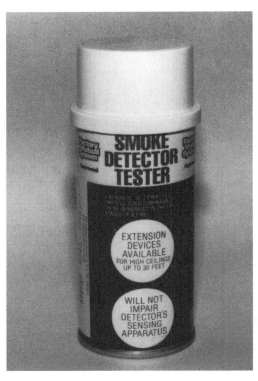

Figure 7.5 *Aerosol smoke product for functional test of a smoke detector. Photograph courtesy of R.P. Schifiliti & Associates (Home Safeguard Industries, Inc.) Reading, MA.*

Table 7-2.2   Test Methods   (cont.)

Device	Method
15. Special Hazard Equipment	Use caution when testing the interfaced special hazard equipment to avoid unnecessary actuation. Never assume that the previous tests were conducted or conducted properly when testing these interconnections. After all equipment has been tested independently, ensure that all previous connections or test switches are placed in their normal positions.
a. Abort Switch (IRI Type)	Operate abort switch. Verify correct sequence and operation.
b. Abort Switch (Recycle Type)	Operate abort switch. Verify correct matrix develops with each sensor operated.
c. Abort Switch (Special Type)	Operate abort switch. Verify correct sequence and operation in accordance with authority having jurisdiction. Note sequence on as-built drawings or in owner's manual.
d. Cross Zone Detection Circuit	Operate one sensor or detector on each zone. Verify that correct sequence occurs with operation of first zone and then with operation of second zone.
e. Matrix-Type Circuit	Operate all sensors in system. Verify correct matrix develops with each sensor operated.
f. Release Solenoid Circuit	Use solenoid with equal current requirements. Verify operation of solenoid.
g. Squibb Release Circuit	Use AGI flashbulb or other test light acceptable to the manufacturer. Verify operation of flashbulb or light.
h. Verified, Sequential, or Counting Zone Circuit	Operate required sensors at a minimum of four locations in circuit. Verify correct sequence with both the first and second detector in alarm.
i. All Above Devices or Circuits or Combinations Thereof	Verify supervision of circuits by creating an open circuit. Note specific trouble indications.
16. Supervising Station Fire Alarm Systems — Transmission Equipment	Items 16 and 17 have been totally revised and expanded to provide more comprehensive requirements for testing supervising station transmission and receiving equipment.
a. All Equipment	Test all system functions and features in accordance with the equipment manufacturer's instructions for proper operation in conformance with the applicable sections of Chapter 4.
	Actuate an initiating device, and verify that the appropriate initiating device signal is received at the supervising station within 90 seconds. Upon completion of the test, restore the system to its normal condition.
b. Digital Alarm Communicator Transmitter (DACT)	Where test jacks are used, the first and last tests shall be made without the use of the test jack. Ensure that the DACT is connected to two separate means of transmission.
	Exception: DACTs that are connected to a telephone line (number) that is also supervised for adverse conditions by a derived local channel.
	Test the DACT for line seizure capability by initiating a signal while utilizing the primary line for a telephone call. Verify that the appropriate signal is received at the supervising station. Verify that completion of the transmission attempt was completed within 90 seconds from going off-hook to on-hook.
	Disconnect the primary line from the DACT. Verify that the DACT trouble signal is indicated at the premises and is transmitted to the supervising station within 4 minutes of detection of the fault.
	Check that all timing devices are synchronized to ensure the proper compliance to the time limits for reporting. The primary line must be a loop start telephone line (number). See Sections 4-5.3.2.1.1 and 4-5.3.2.1.6. for more information on this requirement.
	Disconnect the secondary means of transmission from the DACT. Verify that the DACT trouble signal is indicated at the premises and is transmitted to the supervising station within 4 minutes of detection of the fault.
	Cause the DACT to transmit a signal to the DACR while a fault in the primary telephone number is simulated. Verify that the DACT utilized the secondary telephone number to complete the transmission to the DACR.
c. Digital Alarm Radio Transmitter (DART)	Disconnect the primary telephone line. Verify that the DART transmits a trouble signal to the supervising station within 4 minutes.
d. McCulloh Transmitter	Actuate an initiating device. Verify that the McCulloh transmitter produces not less than three complete rounds of not less than three signal impulses each.
	Where end-to-end metallic continuity is present and with a properly balanced circuit, cause each of the following four transmission channel fault conditions in turn and verify receipt of proper signals at the supervising station:
	(a) Open;
	(b) Ground;

Table 7-2.2   Test Methods   (cont.)

Device	Method
d. McCulloh Transmitter *(cont.)*	(c) Wire-to-wire short; (d) Open and ground. Where end-to-end metallic continuity is not present and with a properly balanced circuit, cause each of the following three transmission channel fault conditions in turn and verify receipt of proper signals at the supervising station: (a) Open; (b) Ground; (c) Wire-to-wire short.
e. Radio Alarm Transmitter (RAT)	Cause a fault between elements of the transmitting equipment. Verify that an indication of the fault is indicated at the protected premises or that a trouble signal is transmitted to the supervising station.
17. Supervising Station Fire Alarm Systems — Receiving Equipment a. All Equipment	Test all system functions and features in accordance with the equipment manufacturer's instructions for proper operation in conformance with the applicable sections of Chapter 4. Actuate an initiating device, and verify that the appropriate initiating device signal is received at the supervising station within 90 seconds. Upon completion of the test, restore the system to its normal condition.
b. Digital Alarm Communicator Receiver (DACR)	Where test jacks are used, the first and last tests shall be made without the use of the test jack. Disconnect in turn each telephone line (number) from the DACR and verify that a signal of trouble is audibly and visually annunciated in the supervising station. Cause a signal to be transmitted on each individual incoming DACR line at least once every 24 hours. Verify receipt of these signals.
c. Digital Alarm Radio Receiver (DARR)	Cause the following conditions of all DARRs on all subsidiary and repeater station receiving equipment. Verify that the supervising station receives appropriate signals for each of the following conditions: (a) AC power failure of the radio equipment; (b) Receiver malfunction; (c) Antenna and interconnecting cable failure; (d) Indication of automatic switchover of the DARR; (e) Data transmission line failure between the DARR and the supervising or subsidiary station.
d. McCulloh Systems	Test and record the current on each circuit at each supervising and subsidiary station under the following conditions: (a) Normal; (b) On each side of the circuit with the receiving equipment conditioned for an open circuit. Cause a single break or ground condition on each transmission channel. Where such a fault prevents the normal functioning of the circuit, verify that a trouble signal has been received. Cause each of the following conditions at each of the supervising or subsidiary stations and all repeater station radio transmitting and receiving equipment. Verify receipt of appropriate signals at the supervising station: (a) RF transmitter in use (radiating); (b) AC power failure supplying the radio equipment; (c) RF receiver malfunction; (d) Indication of automatic switchover.
e. Radio Alarm Supervising Station Receiver (RASSR) and Radio Alarm Repeater Station Receiver (RARSR)	Cause each of the following conditions at each of the supervising or subsidiary stations and all repeater station radio transmitting and receiving equipment. Verify receipt of appropriate signals at the supervising station: (a) AC power failure supplying the radio equipment; (b) RF receiver malfunction; (c) Indication of automatic switchover (where applicable).
f. Private Microwave Radio Systems	Cause each of the following conditions at each of the supervising or subsidiary stations and all repeater station radio transmitting and receiving equipment. Verify receipt of appropriate signals at the supervising station: (a) RF transmitter in use (radiating); (b) AC power failure supplying the radio equipment;

Table 7-2.2 Test Methods (cont.)

Device	Method
f. Private Microwave Radio Systems *(cont.)*	(c) RF receiver malfunction; (d) Indication of automatic switchover.
18. Emergency Communications Equipment a. Amplifier/Tone Generators b. Call-in Signal Silence c. Off-hook Indicator (Ring Down) d. Phone Jacks	 Verification of proper switching and operation of backup equipment. Operate function and verify receipt of proper visual and audible signals at control panel. Install phone set or remove phone from hook and verify receipt of signal at control panel. Visual inspection and initiate communications path through jack. *During an acceptance test, all phone jacks on each floor or zone must be checked for proper operation.*
e. Phone Set f. System Performance	Activate each phone set and verify proper operation. Operate system with a minimum of any five handsets simultaneously. Verify acceptable voice quality and clarity.
19. Interface Equipment	Interface equipment connections shall be tested by operating or simulating the equipment being supervised. Signals required to be transmitted shall be verified at the control panel. Test frequency for interface equipment shall be the same as the frequency required by the applicable NFPA standard(s) for the equipment being supervised. *The signals being verified include the status (alarm, trouble, supervisory conditions) of the interfaced equipment. The main fire alarm control unit will indicate a supervisory signal for any trouble or supervisory conditions at the interfaced equipment.*
20. Guard's Tour Equipment	Test the device in accordance with manufacturer's specifications.
21. Special Procedures a. Alarm Verification b. Multiplex Systems	 Verify time delay and alarm response for smoke detector circuits identified as having alarm verification. Verify communications between sending and receiving units under both normal and standby power. *System functions and features should be verified in accordance with the circuit styles as designed, as well as the manufacturer's specifications.* Verify communications between sending and receiving units under open circuit and short circuit trouble conditions. Verify communications between sending and receiving units in all directions where multiple communications pathways are provided. Where redundant central control equipment is provided, verify switchover and all required functions and operations of secondary control equipment. Verify all system functions and features in accordance with manufacturer's instructions.
22. Low Power Radio (Wireless Systems)	The following procedures describe additional acceptance and reacceptance test methods to verify wireless protection system operation: (a) The manufacturer's manual and the as-built drawings provided by the system supplier shall be used to verify proper operation after the initial testing phase has been performed by the supplier or by the supplier's designated representative. (b) Starting from the normal condition, initialize the system in accordance with the manufacturer's manual. A test shall be conducted to verify the alternative path, or paths, by turning off or disconnecting the primary wireless repeater. The alternative communications path shall exist between the wireless control panel and peripheral devices used to establish initiation, indicating, control, and annunciation. The system shall be tested for both alarm and trouble conditions. (c) Batteries for all components in the system shall be checked monthly. Where the control panel checks all batteries and all components daily, the system shall not require monthly testing of the batteries. *This section was added to Table 7-2.2 in this edition of the Code to address wireless technology introduced in the 1993 edition. This section applies to low power wireless systems covered by Section 3-13.*

A-7-2.2 **Test Methods.** The following wiring diagrams [see Figures A-7-2.2(a) through (oo)] are representative of typical circuits encountered in the field and are not intended to be all-inclusive.

The noted styles are as indicated in Tables 3-5, 3-6, 3-7.1, and 4-5.3.2.2.2.3.

The noted systems are as indicated in NFPA 170, *Standard for Fire Safety Symbols*.

Since ground-fault detection is not required for all circuits, tests for ground-fault detection should be limited to those circuits equipped with ground-fault detection.

An individual point-identifying (addressable) initiating device operates on a signaling line circuit and not on a Style A, B, C, D, or E (Class B and Class A) initiating device circuit.

All of the following initiating device circuits are illustrative of either alarm or supervisory signaling. Alarm-initiating devices and supervisory initiating devices are not permitted on the same initiating device circuit.

In addition to losing its ability to receive an alarm from an initiating device located beyond an open fault, a Style A (Class B) initiating device circuit also loses its ability to receive an alarm when a single ground fault is present.

Style C and Style E (Class B and Class A) initiating device circuits can discriminate between an alarm condition and a wire-to-wire short. In these circuits, a wire-to-wire short provides a trouble indication. However, a wire-to-wire short prevents alarm operation. Shorting-type initiating devices cannot be used without an additional current or voltage limiting element.

Directly-connected system smoke detectors, commonly referred to as two-wire detectors, should be listed as being electrically and functionally compatible with the control unit and the specific subunit or module to which they are connected. Where the detectors and the units or modules are not compatible, it is possible that, during an alarm condition, the detector's visible indicator will illuminate, but no change of state to the alarm condition will occur at the control unit. Incompatibility can also prevent proper system operation at extremes of operating voltage, temperature, and other environmental conditions.

Where two or more two-wire detectors with integral relays are connected to a single initiating device circuit and their relay contacts are used to control essential building functions (e.g., fan shutdown, elevator recall), it should be clearly noted that the circuit might be capable of supplying only enough energy to support one detector/relay combination in an alarm mode. Where control of more than one building function is required, each detector/relay combination used to control separate functions should be connected to separate initiating device circuits, or they should be connected to an initiating device circuit that provides adequate power to allow all the detectors connected to the circuit to be in the alarm mode simultaneously. During acceptance and reacceptance testing, this feature should always be tested and verified.

A speaker is an alarm indicating appliance, and, where used in the following diagrams, the principle of operation and supervision is the same as for other audible alarm indicating appliances (e.g., bells, horns).

Testing of supervised remote relays is to be conducted in same manner as for indicating appliances.

(a) **Wiring Diagrams.**

NOTE: Where testing circuits, the correct wiring size, insulation type, and conductor fill should be verified in accordance with the requirements of NFPA 70, *National Electrical Code*.

A-7-2.2(a) **Diagram of Nonpowered Alarm-Initiating or Supervisory-Initiating Devices (e.g., Manual Station or Valve Supervisory Switch) Connected to Style A, B, or C Initiating Device Circuits.** Disconnect conductor at device or control unit, then reconnect. Temporarily connect a ground to either leg of conductors, then remove ground. Both operations should indicate audible and visual trouble with subsequent restoral at control unit. Conductor-to-conductor short should initiate alarm, Style A and Style B (Class B) indicate trouble Style C (Class B). Style A (Class B) does not initiate alarm while in trouble condition.

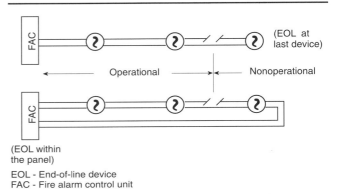

EOL - End-of-line device
FAC - Fire alarm control unit

Figure A-7-2.2(a).

A-7-2.2(b) **Diagram of Nonpowered Alarm-Initiating or Supervisory-Initiating Devices Connected to Style D or E Initiating Device Circuits.** Disconnect a conductor at a device at midpoint in the circuit. Operate a device on either side of the device with the disconnected conductor. Reset control unit and reconnect conductor. Repeat test with a ground applied to either conductor in place of the disconnected conductor. Both operations should indicate audible and visual trouble, then alarm or supervisory indication with subsequent restoral.

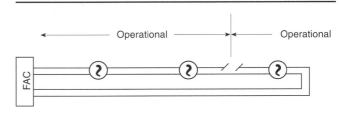

Figure A-7-2.2(b).

A-7-2.2(c) Diagram of Circuit-Powered (Two-Wire) Smoke Detectors for Style A, B, or C Initiating Device Circuits. Remove smoke detector where installed with plug-in base or disconnect conductor beyond first device from control unit. Activate smoke detector per manufacturer's recommendations between control unit and circuit break. Restore detector or circuit, or both. Control unit should indicate trouble where fault occurs and alarm where detectors are activated between the break and the control unit.

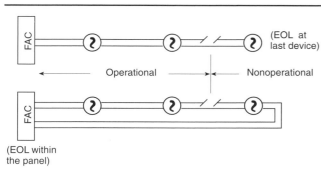

Figure A-7-2.2(c).

A-7-2.2(d) Diagram of Circuit-Powered (Two-Wire) Smoke Detectors for Style D or E Initiating Device Circuits. Disconnect conductor at a smoke detector or remove where installed with a plug-in base at midpoint in the circuit. Operate a device on either side of the device with the fault. Reset control unit and reconnect conductor or detector. Repeat test with a ground applied to either conductor in place of the disconnected conductor or removed device. Both operations should indicate audible and visual trouble, then alarm indication with subsequent restoral.

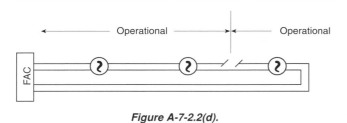

Figure A-7-2.2(d).

A-7-2.2(e) Diagram of Combination Alarm-Initiating Device and Indicating Appliance Circuits. Disconnect a conductor either at indicating or initiating device. Activate initiating device between the fault and the control unit. Activate additional smoke detectors between the device first activated and the control unit. Restore circuit, initiating devices, and control unit. Confirm that all indicating appliances on the cir-

cuit operate from the control unit up to the fault and that all smoke detectors tested and their associated ancillary functions, if any, operate.

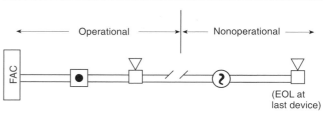

Figure A-7-2.2(e).

A-7-2.2(f) Diagram of Combination Alarm-Initiating Device and Indicating Appliance Circuits Arranged for Operation with a Single Open or Ground Fault. Testing of the circuit is similar to that described above. Confirm all indicating appliances operate on either side of fault.

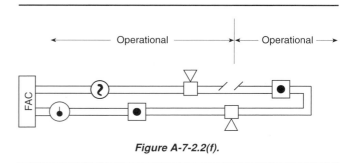

Figure A-7-2.2(f).

A-7-2.2(g) Diagram of Style A, B, or C Circuits with Four-Wire Smoke Detectors and an End-of-Line Power Supervision Relay. Testing of the circuit is similar to that described in A-7-2.2(c) and A-7-2.2(d). Disconnect a leg of the power supply circuit beyond the first device on the circuit. Activate initiating device between the fault and the control unit. Restore circuits, initiating devices, and control unit. Audible and visual trouble should indicate at the control unit where either the initiating or power circuit is faulted. All initiating devices between the circuit fault and the control unit should activate. In addition, removal of a smoke detector from a plug-in-type base can also break the power supply circuit. Where circuits contain various powered and nonpowered devices on the same initiating circuit, verify that the nonpowered devices beyond the power circuit fault can still initiate an alarm. A return loop should be brought back to the last powered device and the power supervisory relay to incorporate into the end-of-line device.

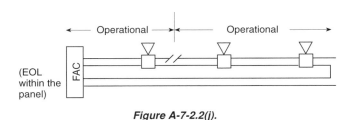

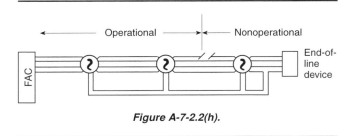

Figure A-7-2.2(g).

Figure A-7-2.2(j).

A-7-2.2(h)   Diagram of Style A, B, or C Initiating Device Circuits with Four-Wire Smoke Detectors that Include Integral Individual Supervision Relays.   Testing of the circuit is similar to that described in A-7-2.2(c) with the addition of a power circuit.

A-7-2.2(k)   Diagram of System with a Supervised Audible Indicating Appliance Circuit and an Unsupervised Visible Indicating Appliance Circuit.   Testing of the indicating appliances connected to Style X and Style Z (Class B and Class A) is similar to that described in A-7-2.2(d).

Figure A-7-2.2(h).

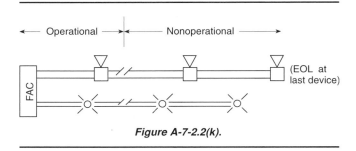

Figure A-7-2.2(k).

A-7-2.2(i)   Diagram of Alarm Indicating Appliances Connected to Styles W and Y (Two-Wire) Circuits.   Testing of the indicating appliances connected to Style W and Style Y (Class B) is similar to that described in A-7-2.2(c).

A-7-2.2(l)   Diagram of System with Supervised Audible and Visible Indicating Appliance Circuits.   Testing of the indicating appliances connected to Style X and Style Z (Class B and Class A) is similar to that described in A-7-2.2(d).

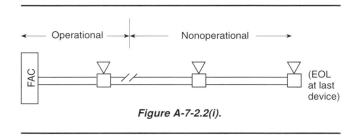

Figure A-7-2.2(i).

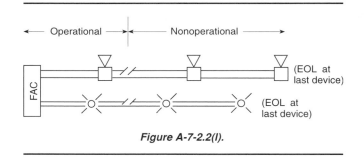

Figure A-7-2.2(l).

A-7-2.2(j)   Diagram of Alarm Indicating Appliances Connected to Styles X and Z (Four-Wire) Circuits.   Testing of the indicating appliances connected to Style X and Style Z (Class B and Class A) is similar to that described in A-7-2.2(d).

A-7-2.2(m)   Diagram of Series Indicating Appliance Circuit, which No Longer Meets the Requirements of NFPA 72.   An open fault in the circuit wiring should cause a trouble condition.

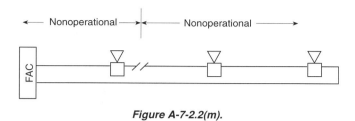

Figure A-7-2.2(m).

A-7-2.2(n) **Diagram of a Supervised Series Supervisory-Initiating Circuit with Sprinkler Supervisory Valve Switches Connected, which No Longer Meets the Requirements of NFPA 72.** An open fault in the circuit wiring of operation of the valve switch (or any supervisory signal device) should cause a trouble condition.

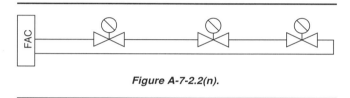

Figure A-7-2.2(n).

A-7-2.2(o) **Diagram of Initiating Device Circuit with Parallel Waterflow Alarm Switches and a Series Supervisory Valve Switch, which No Longer Meets the Requirements of NFPA 72.** An open fault in the circuit wiring or operation of the valve switch should cause a trouble signal.

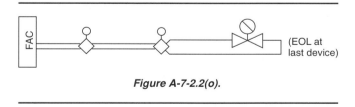

Figure A-7-2.2(o).

A-7-2.2(p) **Diagram of a System Connected to a Municipal Fire Alarm Master Box Circuit.** Disconnect a leg of municipal circuit at master box. Verify alarm sent to public communications center. Disconnect leg of auxiliary circuit. Verify trouble condition on control unit. Restore circuits. Activate control unit and send alarm signal to communications center. Verify control unit in trouble condition until master box reset.

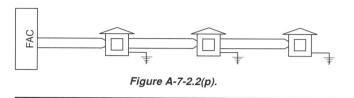

Figure A-7-2.2(p).

A-7-2.2(q) **Diagram of an Auxiliary Circuit Connected to a Municipal Fire Alarm Master Box.** For operation with a master box, an open or ground fault (where ground detection is provided) on the circuit should result in a trouble condition at the control unit. A trouble signal at the control unit should persist until the master box is reset. For operation with a shunt trip master box, an open fault in the auxiliary circuit should cause an alarm on the municipal system.

(b) **Circuit Styles.**

NOTE: Some testing laboratories and authorities having jurisdiction permit systems to be classified as a Style 7 (Class A) by the application of two circuits of the same style operating in parallel. An example of this is to take two series circuits, either Style 0.5 or Style 1.0 (Class B), and operate them in parallel. The logic is that if a condition occurs on one of the circuits, the other parallel circuit remains operative.

In order to understand the principles of the circuit, alarm receipt capability should be performed on a single circuit, and the style type, based on the performance, should be indicated on the record of completion.

Style 0.5. This signaling circuit operates as a series circuit in performance. This is identical to the historical series audible signaling circuits. Any type of break or ground in one of the conductors, or the internal of the multiple interface device, and the total circuit is rendered inoperative.

To test and verify this type of circuit, either a conductor should be lifted or an earth ground should be placed on a conductor or a terminal point where the signaling circuit attaches to the multiplex interface device.

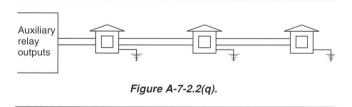

Figure A-7-2.2(q).

A-7-2.2(r) **Style 0.5(a) (Class B) Series.** Style 0.5(a) functions so that, when a box is operated, the supervisory contacts open, making the succeeding devices nonoperative while the operating box sends a coded signal. Any alarms occurring in any successive devices will not be received at the receiving station during this period.

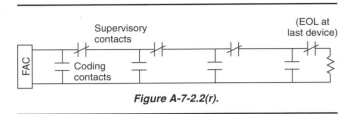

Figure A-7-2.2(r).

A-7-2.2(s) **Style 0.5(b) (Class B) Shunt.** The contact closes when the device is operated and remains closed to shunt out the remainder of the system until the code is complete.

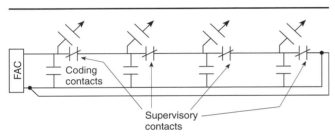

Figure A-7-2.2(s).

A-7-2.2(t) **Style 0.5(c) (Class B) Positive Supervised Successive.** An open or ground fault on the circuit should cause a trouble condition at the control unit.

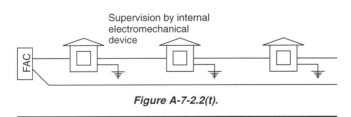

Figure A-7-2.2(t).

A-7-2.2(u) **Style 1.0 (Class B).** This is a series circuit identical to diagram for Style 0.5, except that the fire alarm system hardware has enhanced performance. A single earth ground can be placed on a conductor or multiplex interface device, and the circuit and hardware still have alarm operability.

If a conductor break or an internal fault occurs in the pathway of the circuit conductors, the entire circuit becomes inoperative.

To verify alarm receipt capability and the resulting trouble signal, place an earth ground on one of the conductors or at the point where the signaling circuit attaches to the multiplex interface device. One of the transmitters or an initiating devices should then be placed into alarm.

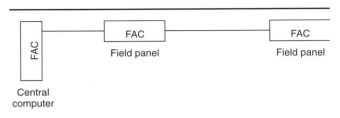

Figure A-7-2.2(u).

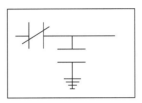

Figure A-7-2.2(v)   Typical transmitter layout.

A-7-2.2(w) **Typical McCulloh Loop.** This is the central station McCulloh redundant-type circuit and has alarm receipt capability on either side of a single break.

(a) To test, lift one of the conductors and operate a transmitter or initiating device on each side of the break. This activity should be repeated for each conductor.

(b) Place an earth ground on a conductor and operate a single transmitter or initiating device to verify alarm receipt capability and trouble condition for each conductor.

(c) Repeat the instructions of (a) and (b) at the same time and verify alarm receipt capability, and verify that a trouble condition results.

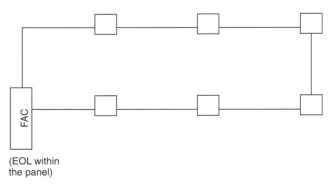

Figure A-7-2.2(w).

A-7-2.2(x) Style 3.0 (Class B). This is a parallel circuit whose multiplex interface devices transmit signal and operating power over the same conductors. The multiplex interface devices might be operable up to the point of a single break. Verify by lifting a conductor and causing an alarm condition on one of the units between the central alarm unit and the break. Either lift a conductor to verify the trouble condition or place an earth ground on the conductors. Test for all the valuations shown on the signaling table.

On ground-fault testing, verify alarm receipt capability by actuating a multiplex interface initiating device or a transmitter.

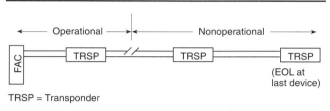

TRSP = Transponder

Figure A-7-2.2(x).

A-7-2.2(y) Style 3.5 (Class B). Repeat the instructions for Style 3.0 (Class B) and verify the trouble conditions by either lifting a conductor or placing a ground on the conductor.

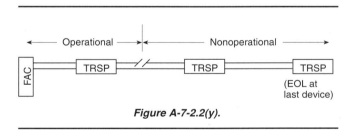

Figure A-7-2.2(y).

A-7-2.2(z) Style 4.0 (Class B). Repeat the instructions for Style 3.0 (Class B) and include a loss of carrier where the signal is being used.

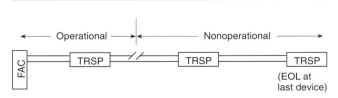

Figure A-7-2.2(z).

A-7-2.2(aa) Style 4.5 (Class B). Repeat the instructions for Style 3.5 (Class B). Verify alarm receipt capability while lifting a conductor by actuating a multiple interface device or transmitter on each side of the break.

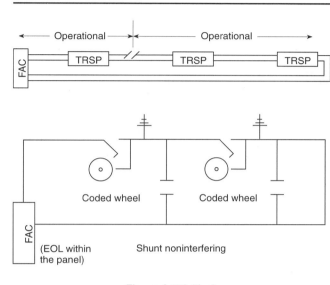

Figure A-7-2.2(aa).

A-7-2.2(bb) Style 5.0 (Class A). Verify the alarm receipt capability and trouble annunciation by lifting a conductor and actuating a multiplex interfacing device or a transmitter on each side of the break. For the earth ground verification, place an earth ground and certify alarm receipt capability and trouble annunciation by actuating a single multiplex interfacing device or a transmitter.

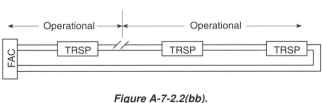

Figure A-7-2.2(bb).

A-7-2.2(cc) Style 6.0 (Class A). Repeat the instructions for Style 2.0 [(Class A (a) through (c)]. Verify the remaining steps for trouble annunciation for the various combinations.

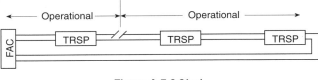

Figure A-7-2.2(cc).

A-7-2.2(dd) Style 6.0 (with Circuit Isolators) (Class A).

For the portions of the circuits electrically located between the monitoring points of circuit isolators, follow the instructions for a Style 7.0 (Class A) circuit. It should be clearly noted that the alarm receipt capability for remaining portions of the circuit protection isolators is not the capability of the circuit, but is permitted with enhanced system capabilities.

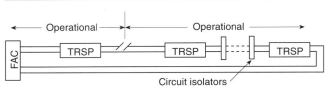

Figure A-7-2.2(dd).

A-7-2.2(ee) Style 7.0 (Class A).

Repeat the instructions for testing of Style 6.0 (Class A) for alarm receipt capability and trouble annunciation.

NOTE 1:   A portion of the circuit between the alarm processor or central supervising station and the first circuit isolator does not have alarm receipt capability in the presence of a wire-to-wire short. The same is true for the portion of the circuit from the last isolator to the alarm processor or the central supervising station.

NOTE 2:   Some manufacturers of this type of equipment have isolators as part of the base assembly. Therefore, in the field, this component might not be readily observable without the assistance of the manufacturer's representative.

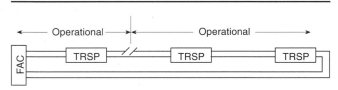

Figure A-7-2.2(ee).

Figures A-7-2.2(ff) through A-7-2.2(ii) depict block diagrams of radio fire alarm systems.

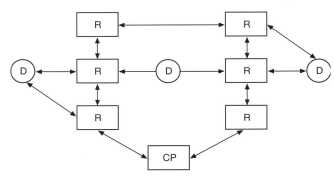

Figure A-7-2.2(ff)   Low power radio (wireless) fire alarm system.

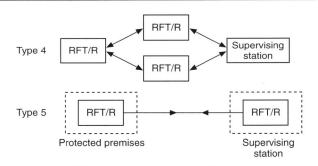

RFT/R = Radio frequency transmitter/receiver

Figure A-7-2.2(gg)   Two-way RF multiplex systems.

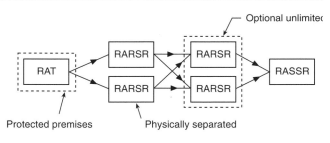

RAT   = Radio alarm transmitter
RARSR = Radio alarm repeater station receiver
RASSR = Radio alarm supervising station receiver

Figure A-7-2.2(hh)   One-way radio alarm system.

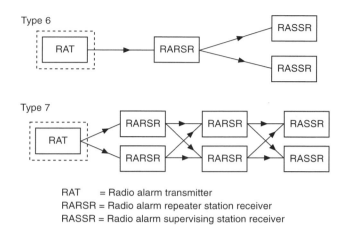

RAT = Radio alarm transmitter
RARSR = Radio alarm repeater station receiver
RASSR = Radio alarm supervising station receiver

Figure A-7-2.2(ii) *One-way radio alarm system.*

Figures A-7-2.2(jj) through A-7-2.2(oo) show fiber optic network operation under various fault conditions.

CC Control center

FACP Fire alarm control panel

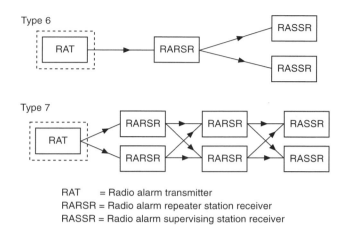

Style 4 fiber network where the panel has a two-way path communications capability. Using MultiMode Fiber for short distances, SingleMode Fiber if for long distances. Repeaters used to increase distances as needed.

Figure A-7-2.2(jj) *Style 4 fiber network.*

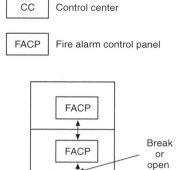

Style 4 fiber network where the panel has a two-way path communications capability. A single break separates the system into two LANs both with Style 4 capabilities.

Figure A-7-2.2(kk) *Style 4 fiber network.*

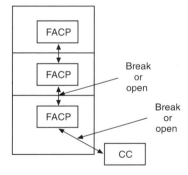

Style 4 fiber network where the panel has a two-way path communications capability. A double break isolates the panels and the control center in this case. In this case, there is one LAN and one isolated panel operating on its own. Control center is isolated completely with no communications with the network.

Figure A-7-2.2(ll) *Style 4 fiber network.*

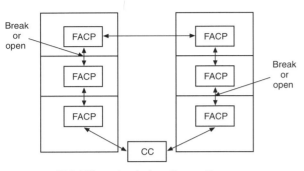

Style 7 fiber network where the panel has a two-way path communications capability with the two breaks now breaking into two LANs, both functioning as independent networks with the same Style 7 capabilities.

Figure A-7-2.2(mm)   *Style 7 fiber network.*

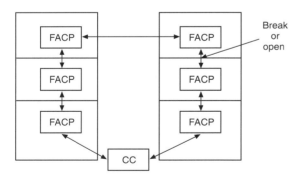

Style 7 fiber network where the panel has a two-way path communications capability, with one break. System remains as one LAN and meets Style 7.

Figure A-7-2.2(nn)   *Style 7 fiber network.*

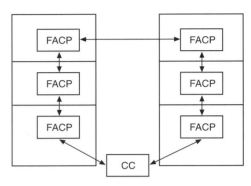

Style 7 fiber network where the panel has a two-way path communications capability.

Figure A-7-2.2(oo)   *Style 7 fiber network.*

(c) Batteries.   To maximize battery life, nickel-cadmium batteries should be charged as follows.

Table A-7-2.2(c)1	Voltage for Nickel-Cadmium Batteries
Float voltage	1.42 volts/cell + 0.01 volt
High rate voltage	1.58 volts/cell + 0.07 volt − 0.00 volt

NOTE: High and low gravity voltages are (+) 0.07 volt and (−) 0.03 volt, respectively.

To maximize battery life, the battery voltage for lead-acid cells should be maintained within the limits shown in the following table:

Table A-7-2.2(c)2 Voltage for Lead-Acid Batteries		
Float Voltage	**High Gravity Battery (Lead Calcium)**	**Low Gravity Battery (Lead Antimony)**
Maximum	2.25 volts/cell	2.17 volts/cell
Minimum	2.20 volts/cell	2.13 volts/cell
High rate voltage		2.33 volts/cell

The following procedure is recommended for checking the state of charge for nickel-cadmium batteries:

(a) The battery charger is switched from float to high-rate mode.

(b) The current, as indicated on the charger ammeter, will immediately rise to the maximum output of the charger, and the battery voltage, as shown on the charger voltmeter, will start to rise at the same time.

(c) The actual value of the voltage rise is unimportant, since it depends on many variables; the length of time it takes for the voltage to rise is the important factor.

(d) If, for example, the voltage rises rapidly in a few minutes, then holds steady at the new value, the battery is fully charged. At the same time, the current will drop to slightly above its original value.

(e) In contrast, if the voltage rises slowly and the output current remains high, the high-rate charge should be continued until the voltage remains constant. Such a condition is an indication that the battery is not fully charged, and the float voltage should be increased slightly.

7-3 Inspection and Testing Frequency.

7-3.1* **Visual Inspection.** Visual inspection shall be performed in accordance with the schedules in this chapter or more frequently where required by the authority having jurisdiction. The visual inspection shall be made to ensure that there are no changes that can affect equipment performance.

Where the authority having jurisdiction suspects that building conditions are changing more rapidly than normal and that these changes are likely to affect the performance of the fire alarm system, the authority having jurisdiction may require more frequent visual inspections.

Some fire alarm system control units monitor analog information for the alarm initiating devices and, thus,

are able to detect changes in ambient conditions at or internal to the initiating device. Extending the visual inspection frequency to annually when such control units are employed recognizes that these units provide more complete monitoring for integrity.

Table 7-3.1 indicates how often visual inspections are to be performed on various components, portions, and subsystems of the fire alarm system.

A visual inspection should always be conducted prior to any testing. A copy of the as-built drawings and system documentation will provide quantities and locations of devices. Improperly located, damaged, or nonfunctional equipment should be identified and corrected before tests begin.

Exception No. 1: Devices or equipment that is inaccessible for safety considerations (e.g., continuous process operations, energized electrical equipment, radiation, excessive height) shall be inspected during scheduled shutdowns where approved by the authority having jurisdiction. Extended intervals shall not exceed 18 months.

Exception No. 2: Where automatic inspection is performed at a frequency of not less than weekly by a remotely monitored fire alarm control unit specifically listed for such application, the visual inspection frequency shall be permitted to be annual. (See Table 7-3.1.*)*

Exception No. 2 was revised to allow new technology to be developed that will be able to remotely inspect equipment. This technology is not yet available for use in the field but is in development and may, during the course of this Code cycle, become available.

A-7-3.1 Equipment performance can be affected by building modifications, occupancy changes, changes in environmental conditions, device location, physical obstructions, device orientation, physical damage, improper installation, degree of cleanliness, or other obvious problems that might not be indicated through electrical supervision.

Table 7-3.1 Visual Inspection Frequencies

Component	Init./Reaccpt.	Monthly	Quarterly	Semiann.	Ann.
1. **Control Equipment: Fire Alarm Systems Monitored for Alarm, Supervisory, Trouble Signals**					
a. Fuses	X				X
b. Interfaced Equipment	X				X
c. Lamps and LEDs	X				X
d. Primary (Main) Power Supply	X				X

Table 7-3.1   Visual Inspection Frequencies   (cont.)

Component	Init./Reaccpt.	Monthly	Quarterly	Semiann.	Ann.
2. Control Equipment: Fire Alarm Systems Unmonitored for Alarm, Supervisory, Trouble Signals					
a. Fuses	X (weekly)				
b. Interfaced Equipment	X (weekly)				
c. Lamps and LEDs	X (weekly)				
d. Primary (Main) Power Supply	X (weekly)				

The term "monitored" refers to systems connected to a supervising station that receives all three signals. In unmonitored systems, signals are not transmitted to a supervising station for appropriate action to be taken. Therefore, weekly inspection of these items to ensure reliability is necessary.

Component	Init./Reaccpt.	Monthly	Quarterly	Semiann.	Ann.
3. Batteries					
a. Lead-Acid	X	X			
b. Nickel-Cadmium	X			X	
c. Primary (Dry Cell)	X	X			
d. Sealed Lead-Acid	X			X	
4. Transient Suppressors	X			X	
5. Control Panel Trouble Signals	X			X	
6. Fiber-Optic Cable Connections	X				X
7. Emergency Voice/Alarm Communications Equipment	X			X	
8. Remote Annunciators	X			X	
9. Initiating Devices					
a. Air Sampling	X			X	
b. Duct Detectors	X			X	
c. Electromechanical Releasing Devices	X			X	
d. Fire Extinguishing System(s) or Suppression System(s) Switches	X			X	
e. Fire Alarm Boxes	X			X	
f. Heat Detectors	X			X	

Heat and smoke detectors should be inspected to ensure that there is no mechanical damage, that they are properly located, and that building conditions have not changed, possibly reducing the effectiveness of the devices.

Component	Init./Reaccpt.	Monthly	Quarterly	Semiann.	Ann.
g. Radiant Energy Fire Detectors	X		X		

Radiant energy fire detectors should be inspected to ensure that there are no obstructions, that the lenses are clear and free of contaminates, that there is no mechanical damage, and that the unit is directed toward the intended hazard.

Component	Init./Reaccpt.	Monthly	Quarterly	Semiann.	Ann.
h. Smoke Detectors	X			X	
i. Supervisory Signal Devices	X		X		
j. Waterflow Devices	X		X		
10. Guard's Tour Equipment	X			X	
11. Interface Equipment	X			X	
12. Alarm Notification Appliances — Supervised	X			X	

Notification appliances should be inspected to ensure that there are no obstructions that impair effectiveness, that there is no mechanical damage, and that building conditions have not rendered the device ineffective.

Table 7-3.1   Visual Inspection Frequencies   (cont.)

Component	Init./Reaccpt.	Monthly	Quarterly	Semiann.	Ann.
13. Supervising Station Fire Alarm Systems — Transmitters					
a. DACT	X			X	
b. DART	X			X	
c. McCulloh	X			X	
d. RAT	X			X	
14. Special Procedures	X			X	
15. Supervising Station Fire Alarm Systems — Receivers					
a. DACR[1]	X	X			
b. DARR[1]	X			X	
c. McCulloh Systems[1]	X			X	
d. Two-Way RF Multiplex[1]	X			X	
e. RASSR[1]	X			X	
f. RARS[1]	X			X	
g. Private Microwave[1]	X			X	

[1]Reports of automatic signal receipt shall be verified daily.

7-3.2*   Testing. Testing shall be performed in accordance with the schedules in this chapter or more frequently where required by the authority having jurisdiction. Where automatic testing is performed at least weekly by a remotely monitored fire alarm control unit specifically listed for the application, the manual testing frequency shall be permitted to be extended to annual. *See* Table 7-3.2.

Exception: Devices or equipment that is inaccessible for safety considerations (e.g., continuous process operations, energized electrical equipment, radiation, excessive height) shall be tested during scheduled shutdowns where approved by the authority having jurisdiction but not more than every 18 months.

The exception was revised in the 1996 edition of this Code to include equipment (such as controls) and clearly define the intended safety considerations. Table 7-3.2 indicates the frequencies for tests on the various components, portions, and subsystems of the fire alarm system.

Table 7-3.2   Testing Frequencies

Component	Init./Reaccpt.	Monthly	Quarterly	Semiann.	Ann.	Table 7-2.2 Reference
1. Control Equipment: Fire Alarm Systems Monitored for Alarm, Supervisory, Trouble Signals						1, 7, 16, 17
a. Functions	X				X	
b. Fuses	X				X	
c. Interfaced Equipment	X				X	
d. Lamps and LEDs	X				X	
e. Primary (Main) Power Supply	X				X	
f. Transponders	X				X	
2. Control Equipment: Fire Alarm Systems Unmonitored for Alarm, Supervisory, Trouble Signals						1
a. Functions	X		X			
b. Fuses	X		X			
c. Interfaced Equipment	X		X			
d. Lamps and LEDs	X		X			
e. Primary (Main) Power Supply	X		X			
f. Transponders	X		X			

See commentary following Table 7-3.1, item 2.d.

| **3. Engine-Driven Generator** | X (weekly) | | | | | |

Table 7-3.2 Testing Frequencies (cont.)

Component	Init./Reaccpt.	Monthly	Quarterly	Semiann.	Ann.	Table 7-2.2 Reference
4. Batteries — Central Station Facilities						
a. Lead-Acid Type						6b
1. Charger Test	X				X	
(Replace battery as needed.)						
2. Discharge Test (30 min)	X	X				
3. Load Voltage Test	X	X				
4. Specific Gravity	X			X		
b. Nickel-Cadmium Type						6c
1. Charger Test	X		X			
(Replace battery as needed.)						
2. Discharge Test (30 min)	X				X	
3. Load Voltage Test	X				X	
c. Sealed Lead-Acid Type	X	X				6d
1. Charger Test		X	X			
(Replace battery as needed.)						
2. Discharge Test (30 min)	X	X				
3. Load Voltage Test	X	X				
5. Batteries — Fire Alarm Systems						
a. Lead-Acid Type						6b
1. Charger Test	X				X	
(Replace battery as needed.)						
2. Discharge Test (30 min)	X			X		
3. Load Voltage Test	X			X		
4. Specific Gravity	X			X		
b. Nickel-Cadmium Type						6c
1. Charger Test	X				X	
(Replace battery as needed.)						
2. Discharge Test (30 min)	X				X	
3. Load Voltage Test	X			X		
c. Primary Type (Dry Cell)						6a
1. Load Voltage Test	X	X				
d. Sealed Lead-Acid Type						6d
1. Charger Test	X				X	
(Replace battery every 4 years.)						
2. Discharge Test (30 min)	X				X	
3. Load Voltage Test	X			X		
6. Batteries — Public Fire Alarm Reporting Systems	X (daily)					

Voltage tests in accordance with Table 7-2.2, item 7, Public Reporting System Tests, paragraphs (a) through (f).

Component	Init./Reaccpt.	Monthly	Quarterly	Semiann.	Ann.	Table 7-2.2 Reference
a. Lead-Acid Type						6b
1. Charger Test	X				X	
(Replace battery as needed.)						
2. Discharge Test (2 hr)	X		X			
3. Load Voltage Test	X		X			
4. Specific Gravity	X			X		
b. Nickel-Cadmium Type						6c
1. Charger Test	X				X	
(Replace battery as needed.)	R					
2. Discharge Test (2 hr)	X				X	
3. Load Voltage Test	X		X			
c. Sealed Lead-Acid Type						6d
1. Charger Test	X				X	
(Replace battery as needed.)						
2. Discharge Test (2 hr)	X				X	
3. Load Voltage Test	X		X			
7. Fiber-Optic Cable Power	X				X	12b
8. Control Unit Trouble Signals	X				X	9

Table 7-3.2 Testing Frequencies

	Component	Init./Reaccpt.	Monthly	Quarterly	Semiann.	Ann.	Table 7-2.2 Reference
9.	Conductors/Metallic	X					11
10.	Conductors/Nonmetallic	X					12
11.	Emergency Voice/Alarm Communications Equipment	X				X	18
12.	Retransmission Equipment	X (*See* 7-3.4.)					
13.	Remote Annunciators	X				X	10
14.	**Initiating Devices**						13
	a. Duct Detectors	X				X	
	b. Electromechanical Releasing Device	X				X	
	c. Fire Extinguishing System(s) or Suppression System(s) Switches	X				X	
	d. Fire-Gas and Other Detectors	X				X	
	e. Heat Detectors	X				X	
	f. Fire Alarm Boxes	X				X	
	g. Radiant Energy Fire Detectors	X			X		
	h. All Smoke Detectors — Functional	X				X	
	i. Smoke Detectors — Sensitivity (*See* 7-3.2.1.)						
	j. Supervisory Signal Devices (except valve tamper switches)	X		X			
	1. Valve Tamper Switches				X		
	k. Waterflow Devices	X			X		
15.	**Guard's Tour Equipment**	X				X	
16.	**Interface Equipment**	X				X	19
17.	**Special Hazard Equipment**	X				X	15
18.	**Alarm Notification Appliances**						14
	a. Audible Devices	X				X	
	b. Speakers	X				X	
	c. Visible Devices	X				X	
19.	**Off-Premises Transmission Equipment**	X		X			
20.	**Supervising Station Fire Alarm Systems — Transmitters**						16
	a. DACT	X				X	
	b. DART	X				X	
	c. McCulloh	X				X	
	d. RAT	X				X	
21.	**Special Procedures**	X				X	21
22.	**Supervising Station Fire Alarm Systems — Receivers**						17
	a. DACR	X	X				
	b. DARR	X	X				
	c. McCulloh Systems	X	X				
	d. Two-Way RF Multiplex	X	X				
	e. RASSR	X	X				
	f. RARSR	X	X				
	g. Private Microwave	X	X				

NOTE: For testing addressable and analog-described devices, which are normally affixed to either a single, molded assembly or are a twist-lock type affixed to a base, TESTING SHALL BE DONE UTILIZING THE SIGNALING STYLE CIRCUITS (Styles 0.5 through 7). The addressable term was determined by the Technical Committee in Formal Interpretation 79-8 on NFPA 72D and Formal Interpretation 87-1 on NFPA 72A. Analog-type detectors shall be tested with the same criteria.

A-7-3.2 Detector-Caused Unwanted Alarms.

Detectors that cause unwanted alarms should be tested at their lower listed range (or at 0.5 percent obscuration if unmarked or unknown). Detectors that activate at less than this level should be replaced.

7-3.2.1

Detector sensitivity shall be checked within 1 year after installation and every alternate year thereafter. After the second required calibration test, where sensitivity tests indicate that the detector has remained within its listed and marked sensitivity range (or 4 percent obscuration light grey smoke, if not marked), the length of time between calibration tests shall be permitted to be extended to a maximum of 5 years. Where the frequency is extended, records of detector-caused nuisance alarms and subsequent trends of these alarms shall be maintained. In zones or in areas where nuisance alarms show any increase over the previous year, calibration tests shall be performed.

To ensure that each smoke detector is within its listed and marked sensitivity range, it shall be tested using either:

(a) A calibrated test method; or

(b) The manufacturer's calibrated sensitivity test instrument; or

(c) Listed control equipment arranged for the purpose; or

(d) A smoke detector/control unit arrangement whereby the detector causes a signal at the control unit where its sensitivity is outside its acceptable sensitivity range; or

(e) Other calibrated sensitivity test method acceptable to the authority having jurisdiction.

Detectors found to have a sensitivity outside the listed and marked sensitivity range shall be cleaned and recalibrated or replaced.

Exception No. 1:   Detectors listed as field adjustable shall be permitted to be either adjusted within the listed and marked sensitivity range and cleaned and recalibrated, or they shall be replaced.

Exception No. 2:   This requirement shall not apply to single station detectors referenced in 7-3.3 and Table 7-2.2.

Single station detectors used in one- and two- family dwelling units are not required to undergo sensitivity testing.

The detector sensitivity shall not be tested or measured using any device that administers an unmeasured concentration of smoke or other aerosol into the detector.

Older detectors that are not marked with a sensitivity range were manufactured prior to current standards that require a sensitivity range to be marked on the product. These detectors should have a sensitivity of between 0.5 and 4 percent per foot obscuration (light gray smoke). Sensitivities less than 0.5 percent obscuration per foot may lead to unwanted alarms, and sensitivities over 4 percent per foot may cause delays in, or failure of, the alarm signal.

The requirements of this subsection give the most possible options in testing the sensitivity of smoke detectors. It is important to note that each of the five options provides a measured means of ensuring sensitivity. Furthermore, after two successful tests in which sensitivity has remained stable, extending sensitivity testing to 5-year intervals recognizes the apparent stability of the environment in which the detector is installed, as well as the apparent stability of the detector itself. When frequency of sensitivity testing is extended, the required records of detector operation will help warn of changes in the environment or changes in the stability of the detector. Any such changes may warrant more frequent testing.

Formal Interpretation 87-1
Reference: 7-2.2

Background: There is a large population of installed smoke detectors which were manufactured in compliance with testing laboratory standards which are no longer used. It is not possible to test the listed sensitivity range of some of these detectors. However, detectors that are not marked with a sensitivity range are now required to be tested.

Question 1: Is it the code's intent that these detectors need not be tested for sensitivity?
Answer: No.

Question 2: It is the code's intent that these detectors be replaced with new detectors whose sensitivity can be tested?
Answer: No.

Issue Edition: NFPA 72E-1987
Reference: 8-2.4.2
Issue Date: July 11, 1989 ∎

A-7-3.2.1
It is suggested that the annual test can be conducted in segments so that all devices are tested annually.

The acceptable test methods include:

(a) The use of a calibrated test method, such as the operation of the test button where a calibrated wire is raised into the reflective path of a photoelectric detector. The wire provides a test of the outside sensitivity limit of the detector.

(b) The use of a manufacturer's sensitivity instrument to check the sensitivity window of the detector.

(c) A system control unit designed to connect internal sensitivity test instruments to each detector upon manual command from the control unit. This test method in effect provides remote sensitivity testing.

(d) A system control unit/detector combination which sends the control unit analog information from the detector and stores it in memory. Thus, the control unit contains a history of both the stability of the environment in which the detector is installed and the stability of the detector itself. By comparing the information stored in memory, the control unit can detect changes in the environment or in the detector's stability that might affect system operation.

(e) Other calibrated test methods. For example, the test in which a precisely measured amount of a test aerosol product acceptable to the detector manufacturer is used to activate a device. By comparing the minimum and maximum quantities of aerosol product delivered to the detector, the window of sensitivity for the device can

be determined. This type of test is generally conducted with a smoke box or a calibrated testing instrument. Most types of smoke detectors can be tested using this method. See Figure 7.6.

7-3.2.2 Test frequency of interfaced equipment shall be the same as specified by the applicable NFPA standards for the equipment being supervised.

For example, the test frequency for a carbon dioxide special hazard fire extinguishing system is specified in NFPA 12, Standard on Carbon Dioxide Extinguishing Systems.

7-3.2.3 For restorable fixed-temperature spot-type heat detectors, two or more detectors shall be tested on each initiating circuit annually. Different detectors shall be tested each year, with records kept by the building owner specifying which detectors have been tested. Within 5 years, each detector shall have been tested.

This is actually a reduction in the testing frequency requirements from the 1993 edition of the Code. It is imperative that accurate records are kept so that the same detectors are not tested each year. This section only applies to restorable fixed-temperature type detectors. All other heat detectors still require annual tests. See Figure 7.7.

7-3.3 Single-station smoke detectors installed in one- and two-family living units shall be inspected, tested, and maintained as specified in Chapter 2. Single station detectors installed in other than one- and two-family dwelling units shall be tested and maintained in accordance with this chapter.

Figure 7.6   Calibrated test instrument. Photograph courtesy of Gemini Scientific, Sunnyvale, CA.

Figure 7.7   Technician using a heat detector tester. Photograph courtesy Home Safeguard Industries, Inc. Malibu, CA.

7-3.4 Test of all circuits extending from the central station shall be made at intervals of not more than 24 hours.

Operators at the central station initiate these tests to verify that all circuits are operational.

7-4   Maintenance.

See Section 1-4 for definition of maintenance.

7-4.1 Fire alarm system equipment shall be periodically maintained in accordance with the manufacturer's instructions. The frequency of maintenance depends on the type of equipment and the local ambient conditions.

7-4.2 Any accumulation of dust and dirt can adversely affect device and appliance performance. The frequency of cleaning depends on the type of equipment and the local ambient conditions.

Examples of areas subject to accumulations of dust and dirt are elevator hoistways and machine rooms, HVAC ducts, and boiler rooms.

7-4.3 All apparatus requiring rewinding or resetting to maintain normal operation shall be restored to normal as promptly as possible after each test and alarm and kept in normal condition for operation. All test signals received shall be recorded to indicate date, time, and type.

Subsections 7-4.2 and 7-4.3 require that periodic maintenance be performed. The emphasis is on cleaning, which should be done in strict accordance with the manufacturer's instructions and as frequently as the ambient conditions of the placement area necessitate. Subsection 7-4.4 of this Code offers the testing frequency of retransmission means between the supervising station and the public fire service communication center.

7-4.4 The retransmission means as defined in Section 4-2 shall be tested at intervals of not more than 12 hours. The retransmission signal and the time and date of the retransmission shall be recorded in the central station.

Exception: Where the retransmission means is the public switched telephone network, it shall be permitted to be tested weekly to confirm its operation to each public fire service communications center.

7-5   Records.

7-5.1* **Permanent Records.** After successful completion of acceptance tests satisfactory to the authority having jurisdiction, a set of reproducible as-built installation drawings, operation and maintenance manuals, and a written sequence of operation shall be provided to the building owner or the owner's designated representative. It shall be the responsibility of the owner to maintain these records for the life of system and to keep them available for examination by any authority having jurisdiction. Paper or electronic media shall be permitted.

A historic record of the system installation that includes the information required by 7-5.1 gives the troubleshooters valuable assistance in promptly diagnosing and repairing system faults.

A-7-5.1 For final determination of record retention, refer to 7-3.2.1 for sensitivity options.

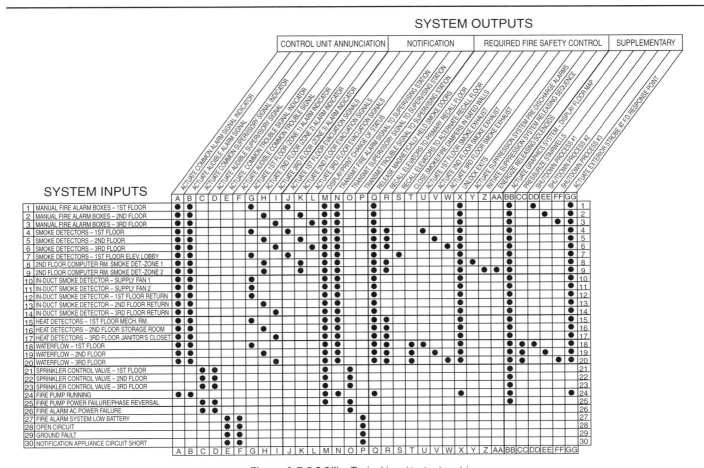

7-5.2 Maintenance, Inspection, and Testing Records.

7-5.2.1 Records shall be retained until the next test and for 1 year thereafter.

7-5.2.2 A permanent record of all inspections, testing, and maintenance shall be provided that includes the following information of periodic tests and all the applicable information requested in Figure 7-5.2.2.

(a) Date;

(b) Test frequency;

(c) Name of property;

(d) Address;

(e) Name of person performing inspection, maintenance, tests, or combination thereof, and affiliation, business address, and telephone number;

(f) Name, address, and representative of approving agency(ies);

(g) Designation of the detector(s) tested ("Tests performed in accordance with Section _____.");

(h) Functional test of detectors;

(i)* Functional test of required sequence of operations;

(j) Check of all smoke detectors;

(k) Loop resistance for all fixed-temperature line-type heat detectors;

(l) Other tests as required by equipment manufacturers;

(m) Other tests as required by the authority having jurisdiction;

(n) Signatures of tester and approved authority representative;

(o) Disposition of problems identified during test (e.g., owner notified, problem corrected/successfully retested, device abandoned in place).

A-7-5.2.2(i) One method used to define the required sequence of operations and document the actual sequence of operations is an input/output matrix. [See Figure A-7-5.2.2(i).]

Figure A-7-5.2.2(i) Typical input/output matrix.

INSPECTION AND TESTING FORM

DATE: _____

TIME: _____

SERVICE ORGANIZATION

NAME: _____

ADDRESS: _____

REPRESENTATIVE: _____

LICENSE NO.: _____

TELEPHONE: _____

PROPERTY NAME (USER)

NAME: _____

ADDRESS: _____

OWNER CONTACT: _____

TELEPHONE: _____

MONITORING ENTITY

CONTACT: _____

TELEPHONE: _____

MONITORING ACCOUNT REF. NO.: _____

APPROVING AGENCY

CONTACT: _____

TELEPHONE: _____

TYPE TRANSMISSION

[   ]- McCulloh

[   ]- Multiplex

[   ]- Digital

[   ]- Reverse Priority

[   ]- RF

[   ]- Other (Specify)

SERVICE

[   ]- Weekly

[   ]- Monthly

[   ]- Quarterly

[   ]- Semiannually

[   ]- Annually

[   ]- Other (Specify)

_____

_____

_____

PANEL MANUFACTURER: _____   MODEL NO.: _____

CIRCUIT STYLES: _____

NO. OF CIRCUITS: _____

SOFTWARE REV.: _____

LAST DATE SYSTEM HAD ANY SERVICE PERFORMED: _____

LAST DATE THAT ANY SOFTWARE OR CONFIGURATION WAS REVISED: _____

Figure 7-5.2.2   *Inspection and Testing Form.*

ALARM-INITIATING DEVICES AND CIRCUIT INFORMATION

QTY OF   CIRCUIT STYLE

_____   _____   MANUAL STATIONS

_____   _____   ION DETECTORS

_____   _____   PHOTO DETECTORS

_____   _____   DUCT DETECTORS

_____   _____   HEAT DETECTORS

_____   _____   WATERFLOW SWITCHES

_____   _____   SUPERVISORY SWITCHES

_____   _____   OTHER (SPECIFY): _____

_____

ALARM INDICATING APPLIANCES AND CIRCUIT INFORMATION

QTY OF   CIRCUIT STYLE

_____   _____   BELLS

_____   _____   HORNS

_____   _____   CHIMES

_____   _____   STROBES

_____   _____   SPEAKERS

_____   _____   OTHER (SPECIFY): _____

_____

NO. OF ALARM INDICATING CIRCUITS:

ARE CIRCUITS SUPERVISED?   [ ]YES   [ ]NO

SUPERVISORY SIGNAL INITIATING DEVICES AND CIRCUIT INFORMATION

QTY OF   CIRCUIT STYLE

_____   _____   BUILDING TEMP.

_____   _____   SITE WATER TEMP.

_____   _____   SITE WATER LEVEL

_____   _____   FIRE PUMP POWER

_____   _____   FIRE PUMP RUNNING

Figure 7-5.2.2 Continued.

QTY OF	CIRCUIT STYLE	
_____	_____	FIRE PUMP AUTO POSITION
_____	_____	FIRE PUMP OR PUMP CONTROLLER
_____	_____	FIRE PUMP RUNNING
_____	_____	GENERATOR IN AUTO POSITION
_____	_____	GENERATOR OR CONTROLLER TROUBLE
_____	_____	SWITCH TRANSFER
_____	_____	GENERATOR ENGINE RUNNING
_____	_____	OTHER: _____

_____

SIGNALING LINE CIRCUITS

Quantity and style (See NFPA 72, Table 3-6) of signaling line circuits connected to system:

Quantity _____   Style(s) _____

SYSTEM POWER SUPPLIES

a. Primary (Main): Nominal Voltage _____, Amps _____

Overcurrent Protection: Type _____, Amps _____

Location (Panel Number): _____

Disconnecting Means Location: _____

b. Secondary (Standby):

_____ Storage Battery: Amp-Hr. Rating _____

Calculated capacity to operate system, in hours: _____ 24 _____ 60 _____

_____ Engine-driven generator dedicated to fire alarm system:

Location of fuel storage: _____

TYPE BATTERY

[ ]Dry Cell

[ ]Nickel-Cadmium

[ ]Sealed Lead-Acid

[ ]Lead-Acid

[ ]Other (Specify)

c. Emergency or standby system used as a backup to primary power supply, instead of using a secondary power supply:

_____ Emergency system described in NFPA 70, Article 700

_____ Legally required standby described in NFPA 70, Article 701

_____ Optional standby system described in NFPA 70, Article 702, which also meets the performance requirements of Article 700 or 701.

Figure 7-5.2.2 *Continued.*

PRIOR TO ANY TESTING

NOTIFICATIONS ARE MADE:	YES	NO	WHO	TIME
MONITORING ENTITY	[ ]	[ ]	_____	_____
BUILDING OCCUPANTS	[ ]	[ ]	_____	_____
BUILDING MANAGEMENT	[ ]	[ ]	_____	_____
OTHER (SPECIFY)	[ ]	[ ]	_____	_____
AHJ (NOTIFIED) OF ANY IMPAIRMENTS	[ ]	[ ]	_____	_____

SYSTEM TESTS AND INSPECTIONS

TYPE	VISUAL	FUNCTIONAL	COMMENTS
CONTROL PANEL	[ ]	[ ]	_____
INTERFACE EQ.	[ ]	[ ]	_____
LAMPS/LEDS	[ ]	[ ]	_____
FUSES	[ ]	[ ]	_____
PRIMARY POWER SUPPLY	[ ]	[ ]	_____
TROUBLE SIGNALS	[ ]	[ ]	_____
DISCONNECT SWITCHES	[ ]	[ ]	_____
GROUND FAULT MONITORING	[ ]	[ ]	_____

SECONDARY POWER

TYPE	VISUAL	FUNCTIONAL	COMMENTS
BATTERY CONDITION	[ ]		_____
LOAD VOLTAGE		[ ]	_____
DISCHARGE TEST		[ ]	_____
CHARGER TEST		[ ]	_____
SPECIFIC GRAVITY		[ ]	_____

TRANSIENT SUPPRESSORS [ ]

REMOTE ANNUNCIATORS [ ] [ ] _____

NOTIFICATION APPLIANCES

	VISUAL	FUNCTIONAL	COMMENTS
AUDIBLE	[ ]	[ ]	_____
VISUAL	[ ]	[ ]	_____
SPEAKERS	[ ]	[ ]	_____
VOICE CLARITY	[ ]		_____

Figure 7-5.2.2 Continued.

INITIATING AND SUPERVISORY DEVICE TESTS AND INSPECTIONS

LOC. & S/N	DEVICE TYPE	VISUAL CHECK	FUNCTIONAL TEST	FACTORY SETTING	MEAS. SETTING	PASS	FAIL
_____	_____	[ ]	[ ]	_____	_____	[ ]	[ ]
_____	_____	[ ]	[ ]	_____	_____	[ ]	[ ]
_____	_____	[ ]	[ ]	_____	_____	[ ]	[ ]
_____	_____	[ ]	[ ]	_____	_____	[ ]	[ ]
_____	_____	[ ]	[ ]	_____	_____	[ ]	[ ]
_____	_____	[ ]	[ ]	_____	_____	[ ]	[ ]

COMMENTS: _____

_____

_____

_____

EMERGENCY COMMUNICATIONS EQUIPMENT	VISUAL	FUNCTIONAL	COMMENTS
PHONE SET	[ ]	[ ]	_____
PHONE JACKS	[ ]	[ ]	_____
OFF-HOOK INDICATOR	[ ]	[ ]	_____
AMPLIFIER(S)	[ ]	[ ]	_____
TONE GENERATOR(S)	[ ]	[ ]	_____
CALL IN SIGNAL	[ ]	[ ]	_____
SYSTEM PERFORMANCE	[ ]	[ ]	_____

INTERFACE EQUIPMENT	VISUAL	DEVICE OPERATION	SIMULATED OPERATION
(SPECIFY) _____	[ ]	[ ]	[ ]
(SPECIFY) _____	[ ]	[ ]	[ ]
(SPECIFY) _____	[ ]	[ ]	[ ]

SPECIAL HAZARD SYSTEMS			
(SPECIFY) _____	[ ]	[ ]	[ ]
(SPECIFY) _____	[ ]	[ ]	[ ]
(SPECIFY) _____	[ ]	[ ]	[ ]

Figure 7-5.2.2   Continued.

SPECIAL PROCEDURES: _____

_____

_____

_____

COMMENTS: _____

_____

_____

_____

ON/OFF PREMISES MONITORING:	YES	NO	TIME	COMMENTS
ALARM SIGNAL	[ ]	[ ]	_____	_____
ALARM RESTORAL	[ ]	[ ]	_____	_____
TROUBLE SIGNAL	[ ]	[ ]	_____	_____
SUPERVISORY SIGNAL	[ ]	[ ]	_____	_____
SUPERVISORY RESTORAL	[ ]	[ ]	_____	_____

NOTIFICATIONS THAT TESTING IS COMPLETE:	YES	NO	WHO	TIME
BUILDING MANAGEMENT	[ ]	[ ]	_____	_____
MONITORING AGENCY	[ ]	[ ]	_____	_____
BUILDING OCCUPANTS	[ ]	[ ]	_____	_____
OTHER (SPECIFY)	[ ]	[ ]	_____	_____

THE FOLLOWING DID NOT OPERATE CORRECTLY: _____

_____

_____

_____

SYSTEM RESTORED TO NORMAL OPERATION: DATE _____ TIME _____

THIS TESTING WAS PERFORMED IN ACCORDANCE WITH APPLICABLE NFPA STANDARDS.

NAME OF INSPECTOR: _____

DATE: _____   TIME: _____

SIGNATURE: _____

NAME OF OWNER OR REPRESENTATIVE: _____

DATE: _____   TIME: _____

SIGNATURE: _____

Figure 7-5.2.2 Continued.

7-5.3 Where off-premises monitoring is provided, records of signals, tests, and operations recorded at the monitoring center shall be maintained for not less than 12 months. Upon request, a hard copy record shall be available for examination by the authority having jurisdiction. Paper or electronic media shall be permitted.

7-5.4 Where the operation of a device, circuit, control panel function, or special hazard system interface is simulated, it shall be noted on the certificate that the operation was simulated, and the certificate shall indicate by whom it was simulated.

For future reference in determining overall system reliability, it is important to document if the interfaced system operation was fully tested by actual operation of the interfaced system or if its operation was simulated.

8

Referenced Publications

The following documents or portions thereof are referenced within this code and shall be considered part of the requirements of this document. The edition indicated for each reference is the current edition as of the date of the NFPA issuance of this document.

8-1.1 **NFPA Publications.** National Fire Protection Association, 1 Batterymarch Park, P.O. Box 9101, Quincy, MA 02269-9101.

NFPA 10, *Standard for Portable Fire Extinguishers,* 1990 edition.

NFPA 13, *Standard for the Installation of Sprinkler Systems*, 1996 edition.

NFPA 13D, *Standard for the Installation of Sprinkler Systems in One- and Two-Family Dwellings and Manufactured Homes,* 1996 edition.

NFPA 13R, *Standard for the Installation of Sprinkler Systems in Residential Occupancies up to and Including Four Stories in Height,* 1996 edition.

NFPA 20, *Standard for the Installation of Centrifugal Fire Pumps*, 1996 edition.

NFPA 25, *Standard for the Inspection, Testing, and Maintenance of Water-Based Fire Protection Systems,* 1995 edition.

NFPA 37, *Standard for the Installation and Use of Stationary Combustion Engines and Gas Turbines*, 1994 edition.

NFPA 54, *National Fuel Gas Code*, 1996 edition.

NFPA 58, *Standard for the Storage and Handling of Liquefied Petroleum Gases*, 1995 edition.

NFPA 70, *National Electrical Code®*, 1996 edition.

NFPA 90A, *Standard for the Installation of Air Conditioning and Ventilating Systems*, 1996 edition.

NFPA 110, *Standard for Emergency and Standby Power Systems*, 1993 edition.

NFPA 111, *Standard on Stored Electrical Energy Emergency and Standby Power Systems*, 1996 edition.

NFPA 601, *Standard for Security Services in Fire Loss Prevention*, 1996 edition.

NFPA 780, *Standard for the Installation of Lightning Protection Systems,* 1995 edition.

NFPA 1221, *Standard for the Installation, Maintenance, and Use of Public Fire Service Communication Systems,* 1994 edition.

8-1.2 Other Publications.

8-1.2.1 ANSI Publications. American National Standards Institute, 1430 Broadway, New York, NY 10036.

ANSI A-58.1, Building Code Requirements for Minimum Design Loads in Buildings and Other Structures, 1982.

ANSI S-1.4a, Specifications for Sound Level Meters, 1985.

ANSI S3.41, Audible Emergency Evacuation Signal, 1990.

ANSI/ASME A17.1, Safety Code for Elevators and Escalators, 1993.

ANSI/IEEE C2, National Electrical Safety Code, 1993.

ANSI/UL 217, Standard for Safety Single and Multiple Station Smoke Detectors, 1993.

ANSI/UL 268, Standard for Safety Smoke Detectors for Fire Protective Signaling Systems, 1989.

ANSI/UL 827, Standard for Safety Central-Station for Watchman, Fire-Alarm and Supervisory Services, 1993.

8-1.2.2 EIA Publication. Electronic Industries Association, 2500 Wilson Boulevard, Arlington, VA 22201-3834.

EIA Tr 41.3, Telephones.

8-1.3 Additional References.

International Municipal Signal Association, P.O. Box 539, Newark, NY 14513.

National Institute for Certification in Engineering Technologies, 1420 King Street, Alexandria, VA 22314-2794.

B

Engineering Guide for Automatic Fire Detector Spacing

The changes to Appendix B in the 1996 Edition of the Code include updates that provide a broader overview of performance based design work (i.e., sections regarding establishing design objectives, and the factors affecting achieving these objectives) as well as an attempt to make this a more "user friendly" document. This includes additions of a Fire Detection and Analysis Worksheet that provides a "roadmap" for the Appendix, as well as information regarding stratification, flame height versus heat release rate, and an overview of detection models.

This Appendix is not a part of the requirements of this NFPA document but is included for informational purposes only.

B-1   Introduction.

B-1.1   **Scope.**   This appendix provides information intended to supplement Chapter 5 and includes a procedure for determining heat detector spacing based on the size and rate of growth of fire to be detected, various ceiling heights, and ambient temperature. The effects of ceiling height and the size and rate of growth of a flaming fire on smoke detector spacing are also treated. A procedure for analyzing the

response of existing heat detection systems is also presented.

B-1.1.1   This appendix utilizes the results of fire research funded by the Fire Detection Institute to provide test data and analysis to the NFPA Technical Committee on Detection Devices. (*See reference 10 in Appendix C.*)

B-1.1.2   This appendix is based on full-scale fire tests in which all fires were geometrically growing flaming fires.

B-1.1.3   The tables and graphs in this guide were produced using test data and data correlations for wood fuels having a total heat of combustion of about 20,900 kJ/kg and a convective heat release rate fraction equal to 75 percent of the total heat release rate. Users should refer to references 12 and 13 in Appendix C for fuels or burning conditions substantially different from these conditions.

B-1.1.4   The guidance applicable to smoke detectors is limited to a theoretical analysis based on the flaming fire test data and is not intended to address the detection of smoldering fires.

B-1.2   **Purpose.**   The purpose of this guide is to provide a performance basis for the location and spacing of heat or smoke detectors. A performance based approach differs from a prescriptive approach in that listed spacing is used as a starting point for the design of a fire detection system that meets specific performance objectives, considering the individual building/room characteristics, potential fire growth rates, and damageability characteristics of the targets (e.g., building occupants, equipment and contents, or structures).

Appendix B provides a performance-based approach as an alternative to the prescriptive-based approach of Chapter 5. Under the prescriptive approach, heat detectors are typically installed according to their listed spacing. It is important to note that listed spacing is determined by the spacing which will give approximately the same response (approximately 2 minutes, plus or minus 10 seconds) as that of a specific, 165°F sprinkler installed on a 10 foot spacing, in a room with a ceiling height of 15 feet 9 inches, using a steady state, flammable liquid fire located three feet above the floor. Thus, if the room dimensions, ambient conditions, fire, and response characteristics of the detector are different from the test room, then response of the heat detector would be expected to change as well. In addition, if your design objectives are different, your results will also vary.

B-1.2.1 Establishing Design Objectives.

B-1.2.1.1 Fire detection system design should be based upon established design objectives that can include life safety, property protection, business interruption, or protection of the environment. The owner/occupant might define these in terms of maximum allowable loss (e.g., all occupants shall have sufficient egress time, 10-minute maximum downtime of equipment, or waste-basket maximum fire size.) These objectives are then put into engineering terms by the designer (e.g., detection of a 100 kW-fire).

B-1.2.1.2 There is a difference between maximum fire size and detected fire size. Although a fire has been detected, this does not mean that it stops growing or is extinguished. Fires typically grow exponentially until they either become ventilation controlled, until the fuel is consumed, or until manual or automatic extinguishment, or both, begins. Figure B-2.2.3.2 shows that there can be a significant increase in the heat release rate with only a small change in time due to the exponential growth rate of the fire.

It is important to note that neither the occupants nor the Fire Department *will necessarily* appear immediately on the scene at the first signs of a fire. Response to a fire incident involves several different sequential actions that must occur before containment and extinguishment efforts can even begin. These elements must to be taken into account to properly design detection systems that meet design goals. These actions typically include:

Detection
Notification to the monitoring station
Notification of the Fire Department
Alarm handling time at the Fire Department
Turnout time at the station
Travel time to incident
Access to site
Access to building
Access to fire floor
Access to area of involvement
Application of extinguishant on fire

Neither the clock, the growth of the fire, nor the resultant damage stop during these actions. Thus, when designing a detection system, these times need to be calculated (t_{delay}) and then subtracted from the time to the maximum allowed fire size in order to indicate a maximum time and fire size at detection. See Figure B.1.

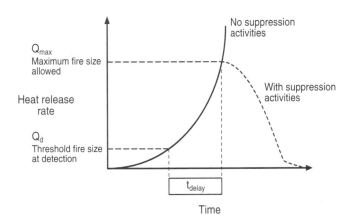

Figure B.1 *Heat release rate versus time. Drawing courtesy of C. Marrion, FireTech, Meilen, Switzerland.*

B-1.2.2 Factors Affecting Attainment of Objectives.

B-1.2.2.1 A performance based analysis should consider the factors that delay manual extinguishment or suppression activities once appropriate personnel are notified. These factors can include fire department dispatch time, response time to fire site, building access, incident location, equipment set-up, and suppressant discharge. Delays are also possible with automatic fire extinguishing system(s) or suppression system(s). Delay can be introduced by alarm verification or crossed zone detection systems, filling and discharge times of pre-action systems, delays in agent release required for occupant evacuation (i.e., CO_2 systems), and the time required to achieve extinguishment.

B-1.2.2.2 A performance based analysis also should consider the interaction between the expected fire, the space in which it is located, and the response of the specific detector to the expected fire in that space. Important factors to be considered regarding the fire include the type and quantity of fuel, fuel configuration, and fuel location within the space in relation to the walls, corners, and ceilings. These factors together allow for estimates of fire growth and smoke production rates. Paragraph B-2.2 provides guidance in evaluating fire development potentials.

B-1.2.2.3 Factors to be considered in the evaluation of the room or space to be protected include the physical configurations of the space, environmental factors present, and mechanical (HVAC) systems that could affect detector performance. Physical configuration parameters include floor area, ceiling height, ceiling configuration (e.g., flat vs. sloped, beamed vs. smooth), and thermodynamic properties of walls, floors, and ceilings. Environmental factors to consider include electromagnetic or radio frequency interference, ambient temperature and humidity conditions, and availability of free oxygen to support combustion.

B-1.2.2.4   Once the design objectives, the room, and the potential fire characteristics are well understood, the designer can select an appropriate detection strategy.   Important factors to consider include the type of detector, sensitivity to expected fire signatures, and freedom from nuisance response to environmental factors.

B-1.2.3   Design.   This guide provides a method for modifying the listed spacing of both rate-of-rise and fixed-temperature heat detectors required to achieve detector response to a geometrically growing flaming fire at a specific fire size, taking into account the height of the ceiling on which the detectors are mounted and the fire safety objectives for the space. This procedure also allows modification of listed spacing of fixed-temperature heat detectors to account for variation of ambient temperature (T_o) from standard test conditions.

B-1.2.4   Analysis.   This guide can be used to estimate the fire size that can be detected by an existing array of listed heat detectors installed at a given spacing for a given ceiling height in known ambient conditions.

B-1.2.5   This guide is also intended to explain the effect of rate of fire growth and fire size of a flaming fire, as well as the effect of ceiling height on the spacing and response of smoke detectors.

B-1.2.6   This methodology utilizes theories of fire development, fire plume dynamics, and detector performance, which are the major factors influencing detector response. However, it does not consider several lesser phenomena that, in general, are unlikely to have significant influence. A discussion of ceiling drag, heat loss to the ceiling, radiation to the detector from a fire, re-radiation of heat from a detector to its surroundings, and the heat of fusion of eutectic materials in fusible elements of heat detectors and their possible limitations on the design method are provided in references 4, 11, and 16 in Appendix C.

B-1.3   Relationship to Listed Spacings.   Listed spacings for heat detectors are based on relatively large fires (approximately 1200 Btu/sec), burning at a constant rate. [The listed spacing is based on the distance from a fire at which an ordi-

nary degree heat detector actuates prior to operation of a 160°F (71°C) sprinkler installed at a 10-ft (3-m) spacing.] [*See Figure A-5-2.4.1(c).*]

Design spacing for this type of fire can be determined using the material in Chapter 5.

Where smaller or larger fires and varying growth rates are to be considered, the designer can use the material presented in this guide.

B-1.4   Required Data.   The following data are necessary in order to use the methods in this guide for either analysis or design.

B-1.4.1   Analysis.

T_o	Ambient temperature
H	Ceiling height or clearance above fuel
T_s	Detector operating temperature (heat detectors only)
ΔT_s/min	Rate of temperature change set point for rate-of-rise heat detectors
RTI	Response time index for the detector (heat detectors only) or its listed spacing
α or t_g	Fuel fire intensity coefficient or t_g, the fire growth time
S	The actual installed spacing of the existing detectors

B-1.4.2   Design.

T_o	Ambient temperature
H	Ceiling height or clearance above fuel
T_s	Detector operating temperature (heat detectors only)
ΔT_s/min	Rate of temperature change set point for rate-of-rise heat detectors
RTI	Response time index for the detector (heat detectors only) or its listed spacing
α or t_g	Fuel fire intensity coefficient or t_g, the fire growth time
Q_d or t_d	The threshold fire size at which response must occur or the time to detector response

B-1.4.3   The terms and data in B-1.4.1 are defined in more detail in the following sections.

B-1.4.4   Fire Detection and Analysis Worksheet.   Figure B-1.4.4 is a fire detection and analysis worksheet, which is provided to assist in using this guide.

This worksheet has been developed in order to provide a "roadmap" to guide the user through the Appendix. Examples of the use of this worksheet are shown throughout the Appendix.

I. Input Data for Design and Analysis (*B-2*)

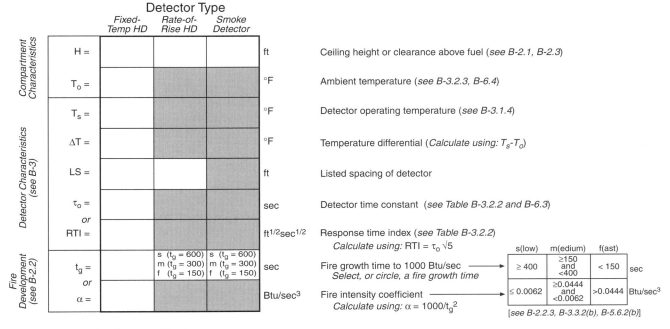

		Detector Type			
		Fixed-Temp HD	*Rate-of-Rise HD*	*Smoke Detector*	
Compartment Characteristics	H =				ft — Ceiling height or clearance above fuel (*see B-2.1, B-2.3*)
	T_o =				°F — Ambient temperature (*see B-3.2.3, B-6.4*)
Detector Characteristics (*see B-3*)	T_s =				°F — Detector operating temperature (*see B-3.1.4*)
	ΔT =				°F — Temperature differential (*Calculate using: T_s-T_o*)
	LS =				ft — Listed spacing of detector
	τ_o = *or* RTI =				sec — Detector time constant (*see Table B-3.2.2 and B-6.3*); $ft^{1/2}sec^{1/2}$ — Response time index (*see Table B-3.2.2*) *Calculate using: RTI = $\tau_o \sqrt{5}$*
Fire Development (*see B-2.2*)	t_g = *or* α =		s (t_g = 600) m (t_g = 300) f (t_g = 150)	s (t_g = 600) m (t_g = 300) f (t_g = 150)	sec — Fire growth time to 1000 Btu/sec *Select, or circle, a fire growth time*; Btu/sec^3 — Fire intensity coefficient *Calculate using: $\alpha = 1000/t_g^2$*

	s(low)	m(edium)	f(ast)	
	≥ 400	≥150 and <400	< 150	sec
	≤ 0.0062	≥0.0444 and <0.0062	>0.0444	Btu/sec^3

[*see B-2.2.3, B-3.3.2(b), B-5.6.2(b)*]

IIa. Input Data for *Design* (B-3)

IIa.1 Establish Design Objectives

> Determine the size of the fire at which detector response is desired using either 1 or 2:
>
> 1. *Select* Q_d = _____ Btu/sec [see B-2.2.4, Table B-2.2.2.1(a), (b), or B-2.2.2.3], or
>
> 2. *Calculate* Q_d using the time after ignition at which detector response is desired, t_d, using: $Q_d = \alpha t_d^2$ = _____ Btu/sec

IIa.2 Calculate Detector Response

Fixed-Temperature HD (B-3.2)	*Rate-of-Rise HD (B-3.3)*	*Smoke Detector (B-5)*
1. Fill in the variables: Q_d = _____ Btu/sec t_g = _____ sec, or α = _____ Btu/sec^3 τ_o = _____ sec, or RTI = _____ $ft^{1/2} sec^{1/2}$ ΔT = _____ °F and H = _____ ft 2. Using Q_d and t_g (or α), select the appropriate design table (a) through (y) from Table B-3.2.4: _____ 3. Using τ_o (or RTI), ΔT, and H, determine the installed spacing: _____ ft	1. Fill in the variables: Fire growth (s,m,f) = _____ H = _____ ft Q_d = _____ Btu/sec 2. Select installed spacing from Table B-3.3.2(a): _____ ft 3. Select spacing modifier from Table B-3.3.2(b): x _____ 4. Calculate installed spacing: _____ ft	1. Fill in the variables: Fire growth (s,m,f) = _____ H = _____ ft Q_d = _____ Btu/sec 2. Using fire growth (s, m, or f), select Figure B-5.5.1(a), (b), or (c): _____ 3. Using H and Q_d, determine the installed spacing: _____ ft

IIb. Input Data for *Analysis* of an Existing Heat Detection System (B-4)

> 1. Fill in the variables:
>   S = _____ ft (installed spacing of the existing heat detector)
> t_g = _____ sec, or α = _____ Btu/sec^3
> τ_o = _____ sec, or RTI = _____ $ft^{1/2}sec^{1/2}$
> ΔT = _____ °F and H = _____ ft
> 2. Using S and t_g (or α), select an analysis table (a) - (nn) from Table B-4: _____
> 3. Using τ_o (or RTI), ΔT, and H, determine the fire size at detector response Q_d = _____ Btu/sec
> 4. Calculate the time to detector response using: $t_d = \sqrt{\dfrac{Q_d}{\alpha}} = \sqrt{\dfrac{Btu/sec}{Btu/sec^3}}$ = _____ sec

***Figure B-1.4.4** Fire detection and analysis worksheet.*

B-2   Fire Development and Ceiling Height Considerations.

B-2.1   General. The purpose of this section is to discuss the effects of ceiling height and the selection of a threshold fire size that can be used as the basis for determination of type and spacing of automatic fire detectors in a specific situation.

B-2.1.1 A detector ordinarily operates sooner in detecting a fire if it is nearer to the fire.

B-2.1.2 Generally, height is the most important single dimension in cases where ceiling heights exceed 16 ft (4.9 m).

B-2.1.3 As smoke and heat rise from a fire, they tend to spread in the general form of an inverted cone. Therefore, the concentration within the cone varies inversely as a variable exponential function of the distance from the source. This effect is very significant in the early stages of a fire, because the angle of the cone is wide. As a fire intensifies, the angle of the cone narrows and the significance of the effect of height is lessened.

B-2.1.4   High Ceilings. As the ceiling height increases, a larger-size fire is necessary to actuate the same detector in the same length of time. In view of this, it is mandatory that the designer of a fire detection system calling for heat detectors consider the size of the fire and rate of heat release that may be permitted to develop before detection is ultimately obtained.

B-2.1.5 The most sensitive detectors suitable for the maximum ambient temperature at heights above 30 ft (9.1 m) should be employed.

B-2.1.6 The spacing recommended by testing laboratories for the location of detectors is an indication of their relative sensitivity. This applies with each detection principle; however, detectors operating on various physical principles have different inherent sensitivities to different types of fires and fuels.

B-2.1.7 Reduction of listed spacing may be required for any of the following purposes:

    (a) Faster response of the device to a fire;

    (b) Response of the device to a smaller fire;

    (c) Accommodation to room geometry;

    (d) Other special considerations, such as air movement or a ceiling or other obstruction.

B-2.2   Fire Development.

B-2.2.1 Fire development varies depending on the combustion characteristics of the fuels involved and the physical configuration of the fuels. After ignition, most fires grow in an accelerating pattern.

Fire growth is limited by factors such as quantity of fuel, arrangement of fuel, quantity of oxygen (to ensure the fire does not become ventilation controlled), and the effect of manual and automatic suppression systems.

B-2.2.2   Fire Size.

B-2.2.2.1 Fires can be characterized by their rate of heat release, measured in terms of the number of Btus per second (kW) generated. Typical maximum heat release rates, Q_m, for a number of different fuels and fuel configurations are provided in Tables B-2.2.2.1(a) and (b).

In Table B-2.2.2.1(a):

$$Q_m = qA$$

where:

    Q_m = the maximum or peak heat release rate in Btu/sec

    q = the heat release rate density per unit floor area in Btu/sec/ft^2

    A = the floor area of the fuel in ft^2.

B-2.2.2.2   Example. A particular hazard analysis is to be based on a fire scenario involving a 10-ft × 10-ft (3-m × 3-m) stack of wood pallets 5 ft (1.5 m) high. Approximately what peak heat release rate can be expected?

From Table B-2.2.2.1(a), the heat release rate density (q) for 5-ft (1.5-m) high wood pallets is about 330 Btu/sec/ft^2.

The area is 10 ft × 10 ft = 100 ft^2 (3 m × 3 m = 9 m^2)

Q_m = qA = 330 × 100 = 33,000 Btu/sec.

The fire would have a medium to fast fire growth rate reaching 1000 Btu/sec (1055 kW) in about 90 seconds to 190 seconds.

B-2.2.2.3 The National Institute of Standards and Technology (the former National Bureau of Standards) has developed a large-scale calorimeter for measuring the heat release rates of burning furniture. Two reports issued by NIST (*see references 3 and 14 in Appendix C*) describe the apparatus and data collected during two test series.

Table B-2.2.2.1(a)   Maximum Heat Release Rates

Warehouse Materials	Growth Time (t_g) (sec)	Heat Release Density (q) (Btu/sec/ft^2)	Classification (s = slow, m = medium, f = fast)
1. Wood pallets, stack, $1\frac{1}{2}$ ft high (6%–12% moisture)	325	125	f-m
2. Wood pallets, stack, 5 ft high (6%–12% moisture)	200	460	f-m
3. Wood pallets, stack, 10 ft high (6%–12% moisture)	125	940	f
4. Wood pallets, stack, 16 ft high (6%–12% moisture)	125	1500	f
5. Mail bags, filled, stored 5 ft high	190	35	m
6. Cartons, compartmented, stacked 15 ft high	60	150	f
7. Paper, vertical rolls, stacked 20 ft high	16–28	—	†
8. Cotton (also PE, PE/cot, acrylic/nylon/PE), garments in 12-ft high racks	21–42	—	†
9. Cartons on pallets, rack storage, 15 ft–30 ft high	40–275	—	f-m
10. Paper products, densely packed in cartons, rack storage, 20 ft high	480	—	s
11. PE letter trays, filled, stacked 5 ft high on cart	180	750	m
12. PE trash barrels in cartons, stacked 15 ft high	55	175	f
13. FRP shower stalls in cartons, stacked 15 ft high	85	125	f
14. PE bottles, packed in item 6	85	550	f
15. PE bottles in cartons, stacked 15 ft high	75	175	f
16. PE pallets, stacked 3 ft high	150	—	f
17. PE pallets, stacked 6 ft–8 ft high	30–60	—	f
18. PU mattress, single, horizontal	125	—	f
19. PE insulation board, rigid foam, stacked 15 ft high	8	170	†
20. PS jars, packed in item 6	55	1250	f
21. PS tubs nested in cartons, stacked 14 ft high	120	475	f
22. PS toy parts in cartons, stacked 15 ft high	125	180	f
23. PS insulation board, rigid, stacked 14 ft high	6	290	†
24. PVC bottles, packed in item 6	95	300	†
25. PP tubs, packed in item 6	100	390	†
26. PP and PE film in rolls, stacked 14 ft high	40	550	†
27. Distilled spirits in barrels, stacked 20 ft high	25–40	—	†
28. Methyl alcohol	—	65	—
29. Gasoline	—	290	—
30. Kerosene	—	290	—
31. Diesel oil	—	175	—

For SI units: 1 ft = 0.305 m.
NOTE: The heat release rates per unit floor area are for fully involved combustibles, assuming 100 percent combustion efficiency. The growth times shown are those required to exceed 1000 Btu/sec heat release rate for developing fires, assuming 100 percent combustion efficiency.
(PE = polyethylene; PS = polystyrene; PVC = polyvinyl chloride; PP = polypropylene; PU = polyurethane; FRP = fiberglass-reinforced polyester.)
†Fire growth rate exceeds design data.

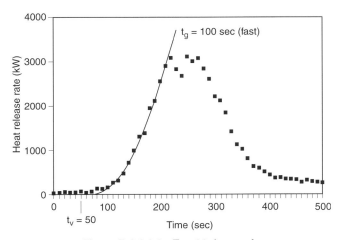

Figure B-2.2.2.3   *Test 38, foam sofa.*

Table B-2.2.2.1(b)   Maximum Heat Release Rates from Fire Detection Institute Analysis

Materials	Approximate Values (Btu/sec)
Medium wastebasket with milk cartons	100
Large barrel with milk cartons	140
Upholstered chair with polyurethane foam	350
Latex foam mattress (heat at room door)	1200
Furnished living room (heat at open door)	4000–8000

See Figure B.2.

Test data from 40 furniture calorimeter tests have been used to verify independently the power-law fire growth model,

a   *b*

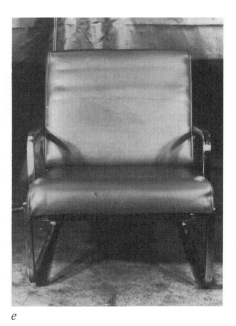

c   *d*   *e*

Figure B.2 *Samples of furniture styles used in fire tests included in Table B-2.2.2.3*
a) Test 49, polifoam filled seat and back
b) Test 45, easy chair with "California foam" polifilled cushions
c) Test 66, (cigarette ignition) wood frame with foam cushions
d) Test 56, wood frame chair with 90% latex foam and 10% cotton felt seat, 100% latex foam cushion and back
e) Test 53, metal frame with solid polifoam on plywood seat and back cushion
f) Test 64, molded flexible urethane frame, wood base, polyester batting covered with polifoam and imitation leather

f

$Q = \alpha t^2$ (*see reference 16 in Appendix C*). In this case, Q is the instantaneous heat release rate, α is the fire intensity coefficient, and t is time. The fire growth time, t_g, is arbitrarily defined as the time after established burning when the fire would reach a burning rate of 1000 Btu/sec (1055 kW). In terms of t_g:

$\alpha = 1000/t_g2$ Btu/sec^3

$\alpha = 1055/t_g2$ kW/sec^2

and

$Q = (1000/t_g2)t^2$ Btu/sec

$Q = (1055/t_g2)t^2$ kW.

Graphs of heat release data from the 40 furniture calorimeter tests can be found in reference 8 in Appendix C. Best fit power-law fire growth curves have been superimposed on the graphs. Data from these curves can be used with this guide to design or analyze fire detection systems that need to respond to similar items burning under a flat ceiling. Table B-2.2.2.3 is a summary of the data.

For reference, the table contains the test numbers used in the original NIST reports. The virtual time of origin, t_v, is the time at which the fires began to obey the power-law fire growth model. Prior to t_v, the fuels might have smoldered but did not burn vigorously with an open flame. The model curves are then predicted by the following:

$Q = \alpha(t - t_v)^2$ Btu/sec or kW

$Q = (1000/t_g2)(t - t_v)^2$ Btu/sec

$Q = (1055/t_g2)(t - t_v)^2$ kW.

Figure B-2.2.2.3 is an example of actual test data with a power-law curve superimposed. This shows how the model can be used to approximate the growth phase of the fire.

For tests 19, 21, 29, 42, and 67, different power-law curves were used to model the initial and the latter realms of burning. In examples such as these, engineers should choose the fire growth parameter that best describes the realm of burning to which the detection system is being designed to respond.

In addition to heat release rate data, the original NIST reports contain data on particulate conversion and radiation from the test specimens. These data can be used to determine the threshold fire size (heat release rate) at which tenability becomes endangered or the point at which additional fuel packages might become involved in the fire.

Table B-2.2.2.3 Furniture Heat Release Rates

Test No.	Item/Mass/Description	Growth Time (t_g) (sec)	Classification (s = slow, m = medium, f = fast)	Fuel Fire Intensity Coefficient (α) (kW/sec^2)	Virtual time (t_v) (sec)	Maximum Heat Release Rates (kW)
Test 15	Metal wardrobe, 41.4 kg (total)	50	f	0.4220	10	750
Test 18	Chair F33 (trial love seat), 39.2 kg	400	s	0.0066	140	950
Test 19	Chair F21, 28.15 kg (initial)	175	m	0.0344	110	350
Test 19	Chair F21, 28.15 kg (later)	50	f	0.4220	190	2000
Test 21	Metal wardrobe, 40.8 kg (total) (initial)	250	m	0.0169	10	250
Test 21	Metal wardrobe, 40.8 kg (total) (average)	120	f	0.0733	60	250
Test 21	Metal wardrobe, 40.8 kg (total) (later)	100	f	0.1055	30	140
Test 22	Chair F24, 28.3 kg	350	m	0.0086	400	700
Test 23	Chair F23, 31.2 kg	400	s	0.0066	100	700
Test 24	Chair F22, 31.9 kg	2000	s	0.0003	150	300
Test 25	Chair F26, 19.2 kg	200	m	0.0264	90	800
Test 26	Chair F27, 29.0 kg	200	m	0.0264	360	900
Test 27	Chair F29, 14.0 kg	100	f	0.1055	70	1850
Test 28	Chair F28, 29.2 kg	425	s	0.0058	90	700
Test 29	Chair F25, 27.8 kg (later)	60	f	0.2931	175	700
Test 29	Chair F25, 27.8 kg (initial)	100	f	0.1055	100	2000
Test 30	Chair F30, 25.2 kg	60	f	0.2931	70	950
Test 31	Chair F31 (love seat), 39.6 kg	60	F	0.2931	145	2600
Test 37	Chair F31 (love seat), 40.4 kg	80	f	0.1648	100	2750
Test 38	Chair F32 (sofa), 51.5 kg	100	f	0.1055	50	3000
Test 39	$1/2$-in. plywood wardrobe with fabrics, 68.5 kg	35	†	0.8612	20	3250
Test 40	$1/2$-in. plywood wardrobe with fabrics, 68.32 kg	35	†	0.8612	40	3500
Test 41	$1/8$-in. plywood wardrobe with fabrics, 36.0 kg	40	†	0.6594	40	6000
Test 42	$1/8$-in. plywood wardrobe with fire-retardant int. fin. (initial growth)	70	f	0.2153	50	2000

Table B-2.2.2.3   Furniture Heat Release Rates (cont.)

Test No.	Item/Mass/Description	Growth Time (t_g) (sec)	Classification (s = slow, m = medium, f = fast)	Fuel Fire Intensity Coefficient (α) (kW/sec^2)	Virtual time (t_v) (sec)	Maximum Heat Release Rates (kW)
Test 42	$^1/_8$-in. plywood wardrobe with fire-retardant int. fin. (later growth)	30	†	1.1722	100	5000
Test 43	Repeat of $^1/_2$-in. plywood wardrobe, 67.62 kg	30	†	1.1722	50	3000
Test 44	$^1/_8$-in. plywood wardrobe with fire-retardant latex paint, 37.26 kg	90	f	0.1302	30	2900
Test 45	Chair F21, 28.34 kg	100	f	0.1055	120	2100
Test 46	Chair F21, 28.34 kg	45	†	0.5210	130	2600
Test 47	Chair, adj. back metal frame, foam cushions, 20.82 kg	170	m	0.0365	30	250
Test 48	Easy chair C07, 11.52 kg	175	m	0.0344	90	950
Test 49	Easy chair F34, 15.68 kg	200	m	0.0264	50	200
Test 50	Chair, metal frame, minimum cushion, 16.52 kg	200	m	0.0264	120	3000
Test 51	Chair, molded fiberglass, no cushion, 5.28 kg	120	f	0.0733	20	35
Test 52	Molded plastic patient chair, 11.26 kg	275	m	0.0140	2090	700
Test 53	Chair, metal frame, padded seat and back, 15.54 kg	350	m	0.0086	50	280
Test 54	Love seat, metal frame, foam cushions, 27.26 kg	500	s	0.0042	210	300
Test 56	Chair, wood frame, latex foam cushions, 11.2 kg	500	s	0.0042	50	85
Test 57	Love seat, wood frame, foam cushions, 54.6 kg	350	m	0.0086	500	1000
Test 61	Wardrobe, $^3/_4$-in. particleboard, 120.33 kg	150	m	0.0469	0	1200
Test 62	Bookcase, plywood with aluminum frame, 30.39 kg	65	f	0.2497	40	25
Test 64	Easy chair, molded flexible urethane frame, 15.98 kg	1000	s	0.0011	750	450
Test 66	Easy chair, 23.02 kg	76	f	0.1827	3700	600
Test 67	Mattress and boxspring, 62.36 kg (later)	350	m	0.0086	400	500
Test 67	Mattress and boxspring, 62.36 kg (initial)	1100	s	0.0009	90	400

For SI units: 1 ft = 0.305 m; 1000 Btu/sec = 1055 kW; 1 lb = 0.453 kg.
†Fire growth exceeds design data.

B-2.2.2.4 A fire detection system can be designed to detect a fire of a certain size in terms of its heat release rate. This is called the threshold fire size (Q_d). The threshold size is the rate of heat release at which detection is desired.

B-2.2.2.5 Flame height can be used to assist in determining the fire size desired for detector response. As shown in Figure B-2.2.2.5, flame height and fire size are directly related (*see reference 2 in Appendix C*). In Figure B-2.2.2.5:

$$H_f = 0.584 \, (kQ)^{0.4}$$

where:

   H_f = flame height (ft)

   k = wall effect factor

use:

   k = 1 where there are no nearby walls

k = 2 where the fuel package is near a wall

k = 4 where the fuel package is in a corner

Q = heat release rate (Btu/sec).

B-2.2.2.6 **Example.** The design objective for a particular compartment is to detect a fire before it exceeds the height of the occupants [6 ft (1.83 m)]. What is the heat release rate that would be expected for flames 6 ft (1.83 m) high, assuming the fire is not near any walls?

From Figure B-2.2.2.5, the heat release rate is approximately 330 Btu/sec, or:

$$H_f = 0.584 \, (kQ)^{0.4}$$

$$6 = 0.584 \, (1{*}Q)^{0.4}$$

$$Q = 338 \text{ Btu/sec}$$

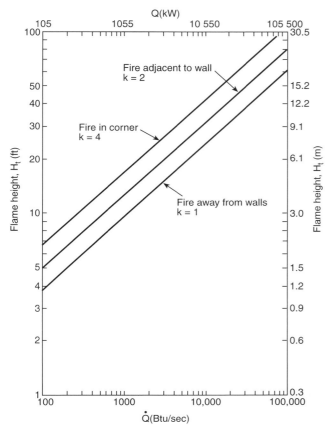

Figure B-2.2.2.5 *Heat release vs. flame height.*

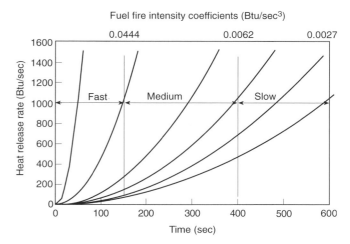

Figure B-2.2.3.2 *Power-law heat release rates.*

A 1,000 Btu/sec fire corresponds to a fire in the center of a room approximately 9 feet (3 m) high using the correlation from B-2.2.2.5.

B-2.2.3.2.2 A medium-developing fire takes 150 seconds (2 minutes, 30 seconds) or more and less than 400 seconds (6 minutes, 40 seconds) from the time that established burning occurs until the fire reaches a heat release rate of 1000 Btu/sec (1055 kW). Using the relationships discussed in B-2.2.2.3, this corresponds to $0.0444 \leq \alpha < 0.0062$ Btu/sec^3 ($0.0469 \leq \alpha < 0.0066$ kW/sec^2).

B-2.2.3.2.3 A fast-developing fire is a fire that takes less than 150 seconds (2 minutes, 30 seconds) from the time that established burning occurs until the fire reaches a heat release rate of 1000 Btu/sec (1055 kW). Using the relationships discussed in B-2.2.2.3, this corresponds to an α greater than 0.0444 Btu/sec^3 (0.0469 kW/sec^2).

B-2.2.3.3 The design fires used in this guide grow according to the following equation:

$$Q = (1000/t_g2)t^2$$

where:

Q is the heat release rate in Btu/sec

t_g is the fire growth time (149 sec = fast, 150 sec–399 sec = medium, 400 sec = slow)

t_g is the time (sec) after established burning occurs.

B-2.2.3 Fire Growth.

B-2.2.3.1 A second important consideration concerning fire development is the time (t_g) it takes for fire to reach a given heat release rate. Table B-2.2.2.1(a) and Table B-2.2.2.3 provide the times necessary to reach a heat release rate of 1000 Btu/sec (1055 kW) for a variety of materials in various configurations.

B-2.2.3.2 For purposes of this guide, fires are classified as being either slow-, medium-, or fast-developing. (*See Figure B-2.2.3.2.*)

B-2.2.3.2.1 A slow-developing fire is defined as a fire that takes 400 or more seconds (6 minutes, 40 seconds) from the time that established burning occurs until the fire reaches a heat release rate of 1000 Btu/sec (1055 kW). Using the relationships discussed in B-2.2.2.3, this corresponds to an α of 0.0062 Btu/sec^3 (0.0066 kW/sec^2) or less.

B-2.2.4 **Selection of Fire Size.**   The selection of threshold fire size, Q_d, should be based on an understanding of the characteristics of a specified space and the fire safety objectives for that space.

For example, in a particular installation it might be desirable to detect a typical wastebasket fire. Table B-2.2.2.1(b) includes data for a fire involving a comparable array of combustibles, specifically milk cartons in a wastebasket. Such a fire is indicated to produce a peak heat release rate of 100 Btu/sec.

See also discussion under B-1.2.1.2.

B-2.3   Ceiling Height.

B-2.3.1   The Fire Detection Institute data are based on the height of the ceiling above the fire. In this guide, it is recommended that the designer use the actual distance from floor to ceiling, since the ceiling height will thereby be more conservative and actual detector response will improve where the potential fuel in a room is above floor level.

B-2.3.2   Where the designer desires to consider the height of the potential fuel in the room, the distance between the fuel and the ceiling should be used in place of the ceiling height in the tables and graphs. This should be considered only where the minimum height of the potential fuel is always constant and where the concept is approved to the authority having jurisdiction.

B-2.3.3   The procedures presented in this guide are based on an analysis of test data for ceiling heights up to 30 ft (9.1 m). No data was analyzed for ceilings greater than 30 ft (9.1 m); therefore, in such installations, engineering judgment and the manufacturer's recommendations should be used.

B-2.4   Stratification.

B-2.4.1   When stratification occurs, the smoke/heat being transported from a fire might not be able to reach detectors mounted at a particular level above the fire. This is due to the fact that cooler air is continually entrained throughout the height of the fire plume, resulting in cooling of the smoke and fire plume gases and a reduction in buoyancy. To determine whether or not the rising smoke/heat from an axisymmetric fire plume will stratify below detectors, the following expression can be applied where the ambient temperature increases linearly with increasing elevation:

$$Z_m = 14.7 \, Q_c^{1/4} \, \Delta T_o^{-3/8} \text{ ft} = 5.54 \, Q_c^{1/4} \, \Delta T_o^{-3/8} \text{ m}$$

where:

Z_m = maximum height of smoke rise above the fire surface [ft (m)]

ΔT_o = difference between the ambient temperature at the location of detectors and the ambient temperature at the level of the fire surface [°F(°C)]

Q_c = convective portion of the heat release rate [Btu/sec (kW)].

The convective portion of the heat release rate, Q_c, can be estimated as 70 percent of the heat release rate.

As an alternative to using the noted expression to calculate directly the maximum height to which smoke/heat in the fire will rise, Figure B-2.4.1 can be used to determine Z_m for given fires.

Where Z_m, as calculated or determined graphically, is greater than the installed height of detectors, smoke/heat from a rising fire plume is predicted to reach the detectors. Where the compared values of Z_m and the installed height of detectors are comparable heights, the prediction that smoke/heat will reach the detectors might not be a reliable expectation.

B-2.4.2   The theoretical basis for the stratification calculation is based upon the works of Morton, Taylor, and Turner (*see reference 15 in Appendix C*); and Heskestad (*see reference 9 in Appendix C*). For further information regarding the derivation of the expression defining Z_m, the user is referred to the work of Klote and Milke (*see reference 13 in Appendix C*).

B-3   Heat Detector Spacing.

B-3.1   General.

B-3.1.1   This section discusses procedures for the determination of installed spacing of listed heat detectors used to detect flaming fires.

B-3.1.2   The determination of the installed spacing of heat detectors using these procedures adjusts the listed spacing to reflect the effects of ceiling height, threshold fire size, rate of fire development, and, for fixed temperature detectors, the ambient temperature and the temperature rating of the detector.

B-3.1.3   Other factors that affect detector response, such as beams and joists, are covered in Chapter 5.

B-3.1.4   The difference between the rated temperature of a fixed temperature detector (T_s) and the maximum ambient

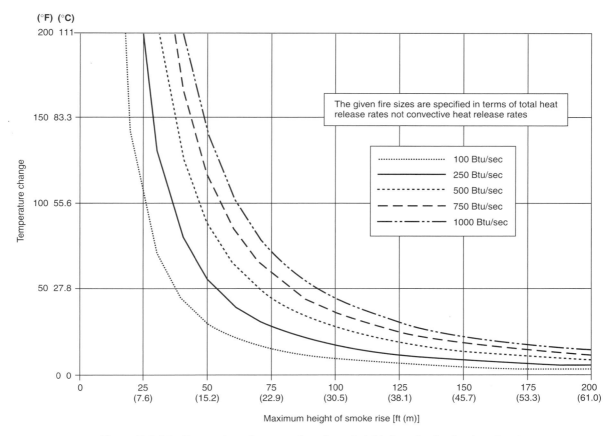

Figure B-2.4.1   *Temperature change and maximum height of smoke rise for given fire sizes.*

temperature (T_o) at the ceiling should be as small as possible. To reduce unwanted alarms, the difference between operating temperature and ambient temperature should be not less than 20°F (11°C).

B-3.1.5   Listed rate-of-rise heat detectors are designed to activate at a nominal rate of temperature rise of 15°F (8°C) per minute.

B-3.1.6   The listed spacing of a detector is an indicator of the detector's sensitivity. Given the same temperature rating, a detector listed for a 50-ft (15.2-m) spacing is more sensitive than a detector listed for a 20-ft (6.1-m) spacing.

B-3.1.7   Where using combination detectors incorporating both fixed temperature and rate-of-rise heat detection principles to detect a geometrically growing fire, the data herein for rate-of-rise detectors should be used in selecting an installed spacing, because the rate-of-rise principle controls the response.

B-3.1.8   Rate-compensated detectors are not specifically covered by this guide. However, a conservative approach to predicting their performance is to use the fixed-temperature heat detector guidance contained herein.

B-3.2   **Fixed-Temperature Heat Detector Spacing.**   Tables B-3.2.2 and B-3.2.4(a) through (y) are to be used to determine the installed spacing of fixed-temperature heat detectors. The analytical basis for the tables is presented in another portion of this guide. This subsection describes how the tables are to be used.

B-3.2.1   With the exception of ceiling height, the nearest value shown in the tables provides sufficient accuracy for these calculations. Interpolation is permitted but not necessary, except for ceiling height.

B-3.2.2   **Time Constant.**

(a) Given the detector's listed spacing and the detector's rated temperature (T_s), Table B-3.2.2 should be used to find

Table B-3.2.2   Time Constants (τ_o) for Any Listed Heat Detector†

Listed Spacing (ft)	ULI 128°	ULI 135°	ULI 145°	ULI 160°	ULI 170°	ULI 196°	FMRC All Temps.
10	400	330	262	195	160	97	196
15	250	190	156	110	89	45	110
20	165	135	105	70	52	17	70
25	124	100	78	48	32		48
30	95	80	61	36	22		36
40	71	57	41	18			
50	59	44	30				
70	36	24	9				

For SI units: 1 ft = 0.305 m.
NOTE 1:   These time constants are based on an analysis of the Underwriters Laboratories Inc. and Factory Mutual listing test procedures. Plunge test (*see reference 8 in Appendix C*) results performed on the detector to be used will give a more accurate time constant. See Section B-5 for a further discussion of detector time constants.
NOTE 2:   These time constants can be converted to response time index (RTI) values by multiplying by 5 ft/sec (1.5 m/sec). (*See B-3-3.*)
†At a reference velocity of 5 ft/sec (1.5 m/sec).

the detector time constant (τ_o). The time constant is a measure of the detector's sensitivity. (*See B-6.3.*)

(b) The response time index (RTI) can also be used to describe the sensitivity of a fixed-temperature heat detector. (*See B-6.3.*)

B-3.2.3   Minimum Ambient Temperature.

(a) The minimum ambient temperature (T_o) expected at the ceiling of the space to be protected is estimated. The temperature change (ΔT) of the detector required for detection ($\Delta T = T_s - T_o$) is calculated.

(b) Selection of the minimum ambient temperature necessitates engineering judgment. Use of the absolute minimum ambient temperature results in the most conservative designs. This is true because it is then assumed that the detector has to absorb enough energy to raise its temperature from the low ambient value up to its operating temperature. A review of historical data might show very low ambient temperatures that occur relatively infrequently, such as every 100 years.

Depending on actual design considerations, it might be more prudent to use an average minimum ambient temperature. In any case, a sensitivity analysis should be performed to determine the effect of changing the ambient temperature on the design results.

B-3.2.4   Having determined the detector's sensitivity (time constant or RTI) (*see B-3.2.2*), the temperature change of the detector required for detection (*see B-3.2.3*), the threshold fire size (*see B-2.2.4*), the fire growth rate (*see B-2.2.3*), and the ceiling height, Tables B-3.2.4(a) through (y) are used to deter-

mine the required installed spacing. Table B-3.2.4 provides an index to the tables.

Table B-3.2.4   Design Tables Index

	Threshold Fire Size (Btu/sec) Q_d	Fire Growth Period (sec) t_g	Fuel Fire Intensity Coefficient (Btu/sec³) α
Table B-3.2.4(a)	250	50	0.400
Table B-3.2.4(b)	250	150	0.044
Table B-3.2.4(c)	250	300	0.011
Table B-3.2.4(d)	250	500	0.004
Table B-3.2.4(e)	250	600	0.003
Table B-3.2.4(f)	500	50	0.400
Table B-3.2.4(g)	500	150	0.044
Table B-3.2.4(h)	500	300	0.011
Table B-3.2.4(i)	500	500	0.004
Table B-3.2.4(j)	500	600	0.003
Table B-3.2.4(k)	750	50	0.400
Table B-3.2.4(l)	750	150	0.044
Table B-3.2.4(m)	750	300	0.011
Table B-3.2.4(n)	750	500	0.004
Table B-3.2.4(o)	750	600	0.003
Table B-3.2.4(p)	1000	50	0.400
Table B-3.2.4(q)	1000	150	0.044
Table B-3.2.4(r)	1000	300	0.011
Table B-3.2.4(s)	1000	500	0.004
Table B-3.2.4(t)	1000	600	0.003
Table B-3.2.4(u)	2000	50	0.400
Table B-3.2.4(v)	2000	150	0.044
Table B-3.2.4(w)	2000	300	0.011
Table B-3.2.4(x)	2000	500	0.004
Table B-3.2.4(y)	2000	600	0.003

Table B-3.2.4(a) Q_d, Threshold Fire Size at Response: 250 Btu/sec t_g: 50 seconds to 1000 Btu/sec α: 0.400 Btu/sec³

τ	RTI	ΔT	Ceiling Height (ft)							τ	RTI	ΔT	Ceiling Height (ft)						
			4.0	8.0	12.0	16.0	20.0	24.0	28.0				4.0	8.0	12.0	16.0	20.0	24.0	28.0
			Installed Spacing of Detectors										Installed Spacing of Detectors						
25	56	40	7	5	2	0	0	0	0	225	503	40	2	0	0	0	0	0	0
25	56	60	6	3	1	0	0	0	0	225	503	60	1	0	0	0	0	0	0
25	56	80	5	2	0	0	0	0	0	225	503	80	0	0	0	0	0	0	0
25	56	100	4	2	0	0	0	0	0	225	503	100	0	0	0	0	0	0	0
25	56	120	4	1	0	0	0	0	0	225	503	120	0	0	0	0	0	0	0
25	56	140	3	1	0	0	0	0	0	225	503	140	0	0	0	0	0	0	0
50	112	40	5	3	1	0	0	0	0	250	559	40	2	0	0	0	0	0	0
50	112	60	4	2	0	0	0	0	0	250	559	60	0	0	0	0	0	0	0
50	112	80	3	1	0	0	0	0	0	250	559	80	0	0	0	0	0	0	0
50	112	100	3	0	0	0	0	0	0	250	559	100	0	0	0	0	0	0	0
50	112	120	2	0	0	0	0	0	0	250	559	120	0	0	0	0	0	0	0
50	112	140	2	0	0	0	0	0	0	250	559	140	0	0	0	0	0	0	0
75	168	40	4	2	0	0	0	0	0	275	615	40	1	0	0	0	0	0	0
75	168	60	3	1	0	0	0	0	0	275	615	60	0	0	0	0	0	0	0
75	168	80	2	0	0	0	0	0	0	275	615	80	0	0	0	0	0	0	0
75	168	100	2	0	0	0	0	0	0	275	615	100	0	0	0	0	0	0	0
75	168	120	2	0	0	0	0	0	0	275	615	120	0	0	0	0	0	0	0
75	168	140	1	0	0	0	0	0	0	275	615	140	0	0	0	0	0	0	0
100	224	40	3	1	0	0	0	0	0	300	671	40	1	0	0	0	0	0	0
100	224	60	2	0	0	0	0	0	0	300	671	60	0	0	0	0	0	0	0
100	224	80	2	0	0	0	0	0	0	300	671	80	0	0	0	0	0	0	0
100	224	100	1	0	0	0	0	0	0	300	671	100	0	0	0	0	0	0	0
100	224	120	1	0	0	0	0	0	0	300	671	120	0	0	0	0	0	0	0
100	224	140	1	0	0	0	0	0	0	300	671	140	0	0	0	0	0	0	0
125	280	40	3	0	0	0	0	0	0	325	727	40	1	0	0	0	0	0	0
125	280	60	2	0	0	0	0	0	0	325	727	60	0	0	0	0	0	0	0
125	280	80	1	0	0	0	0	0	0	325	727	80	0	0	0	0	0	0	0
125	280	100	1	0	0	0	0	0	0	325	727	100	0	0	0	0	0	0	0
125	280	120	0	0	0	0	0	0	0	325	727	120	0	0	0	0	0	0	0
125	280	140	0	0	0	0	0	0	0	325	727	140	0	0	0	0	0	0	0
150	335	40	2	0	0	0	0	0	0	350	783	40	1	0	0	0	0	0	0
150	335	60	2	0	0	0	0	0	0	350	783	60	0	0	0	0	0	0	0
150	335	80	1	0	0	0	0	0	0	350	783	80	0	0	0	0	0	0	0
150	335	100	0	0	0	0	0	0	0	350	783	100	0	0	0	0	0	0	0
150	335	120	0	0	0	0	0	0	0	350	783	120	0	0	0	0	0	0	0
150	335	140	0	0	0	0	0	0	0	350	783	140	0	0	0	0	0	0	0
175	391	40	2	0	0	0	0	0	0	375	839	40	0	0	0	0	0	0	0
175	391	60	1	0	0	0	0	0	0	375	839	60	0	0	0	0	0	0	0
175	391	80	1	0	0	0	0	0	0	375	839	80	0	0	0	0	0	0	0
175	391	100	0	0	0	0	0	0	0	375	839	100	0	0	0	0	0	0	0
175	391	120	0	0	0	0	0	0	0	375	839	120	0	0	0	0	0	0	0
175	391	140	0	0	0	0	0	0	0	375	839	140	0	0	0	0	0	0	0
200	447	40	2	0	0	0	0	0	0	400	894	40	0	0	0	0	0	0	0
200	447	60	1	0	0	0	0	0	0	400	894	60	0	0	0	0	0	0	0
200	447	80	0	0	0	0	0	0	0	400	894	80	0	0	0	0	0	0	0
200	447	100	0	0	0	0	0	0	0	400	894	100	0	0	0	0	0	0	0
200	447	120	0	0	0	0	0	0	0	400	894	120	0	0	0	0	0	0	0
200	447	140	0	0	0	0	0	0	0	400	894	140	0	0	0	0	0	0	0

NOTE: Detector time constant at a reference velocity of 5 ft/sec (1.5m/sec).
For SI units: 1 ft = 0.305 m; 1000 Btu/sec = 1055 kW.

Table B-3.2.4(b) Q_d, Threshold Fire Size at Response: 250 Btu/sec t_g: 150 seconds to 1000 Btu/sec α: 0.044 Btu/sec^3

τ	RTI	ΔT	Ceiling Height (ft)							τ	RTI	ΔT	Ceiling Height (ft)						
			4.0	8.0	12.0	16.0	20.0	24.0	28.0				4.0	8.0	12.0	16.0	20.0	24.0	28.0
			Installed Spacing of Detectors										Installed Spacing of Detectors						
25	56	40	15	12	9	6	3	0	0	225	503	40	5	3	1	0	0	0	0
25	56	60	12	9	6	3	0	0	0	225	503	60	4	2	0	0	0	0	0
25	56	80	10	7	4	1	0	0	0	225	503	80	3	1	0	0	0	0	0
25	56	100	9	6	2	0	0	0	0	225	503	100	2	0	0	0	0	0	0
25	56	120	8	4	1	0	0	0	0	225	503	120	2	0	0	0	0	0	0
25	56	140	7	4	1	0	0	0	0	225	503	140	2	0	0	0	0	0	0
50	112	40	11	9	6	3	1	0	0	250	559	40	5	2	0	0	0	0	0
50	112	60	9	6	3	1	0	0	0	250	559	60	3	1	0	0	0	0	0
50	112	80	7	5	2	0	0	0	0	250	559	80	3	0	0	0	0	0	0
50	112	100	6	4	1	0	0	0	0	250	559	100	2	0	0	0	0	0	0
50	112	120	6	3	1	0	0	0	0	250	559	120	2	0	0	0	0	0	0
50	112	140	5	2	0	0	0	0	0	250	559	140	1	0	0	0	0	0	0
75	168	40	9	7	4	2	0	0	0	275	615	40	4	2	0	0	0	0	0
75	168	60	7	5	2	0	0	0	0	275	615	60	3	1	0	0	0	0	0
75	168	80	6	3	1	0	0	0	0	275	615	80	2	0	0	0	0	0	0
75	168	100	5	3	0	0	0	0	0	275	615	100	2	0	0	0	0	0	0
75	168	120	4	2	0	0	0	0	0	275	615	120	2	0	0	0	0	0	0
75	168	140	4	1	0	0	0	0	0	275	615	140	1	0	0	0	0	0	0
100	224	40	8	6	3	1	0	0	0	300	671	40	4	2	0	0	0	0	0
100	224	60	6	4	2	0	0	0	0	300	671	60	3	1	0	0	0	0	0
100	224	80	5	3	1	0	0	0	0	300	671	80	2	0	0	0	0	0	0
100	224	100	4	2	0	0	0	0	0	300	671	100	2	0	0	0	0	0	0
100	224	120	4	1	0	0	0	0	0	300	671	120	1	0	0	0	0	0	0
100	224	140	3	1	0	0	0	0	0	300	671	140	1	0	0	0	0	0	0
125	280	40	7	5	2	1	0	0	0	325	727	40	4	2	0	0	0	0	0
125	280	60	5	3	1	0	0	0	0	325	727	60	3	1	0	0	0	0	0
125	280	80	4	2	0	0	0	0	0	325	727	80	2	0	0	0	0	0	0
125	280	100	4	1	0	0	0	0	0	325	727	100	2	0	0	0	0	0	0
125	280	120	3	1	0	0	0	0	0	325	727	120	1	0	0	0	0	0	0
125	280	140	3	0	0	0	0	0	0	325	727	140	1	0	0	0	0	0	0
150	335	40	6	4	2	0	0	0	0	350	783	40	4	2	0	0	0	0	0
150	335	60	5	2	1	0	0	0	0	350	783	60	3	0	0	0	0	0	0
150	335	80	4	2	0	0	0	0	0	350	783	80	2	0	0	0	0	0	0
150	335	100	3	1	0	0	0	0	0	350	783	100	2	0	0	0	0	0	0
150	335	120	3	0	0	0	0	0	0	350	783	120	1	0	0	0	0	0	0
150	335	140	2	0	0	0	0	0	0	350	783	140	1	0	0	0	0	0	0
175	391	40	6	3	1	0	0	0	0	375	839	40	3	1	0	0	0	0	0
175	391	60	4	2	0	0	0	0	0	375	839	60	2	0	0	0	0	0	0
175	391	80	3	1	0	0	0	0	0	375	839	80	2	0	0	0	0	0	0
175	391	100	3	1	0	0	0	0	0	375	839	100	1	0	0	0	0	0	0
175	391	120	2	0	0	0	0	0	0	375	839	120	1	0	0	0	0	0	0
175	391	140	2	0	0	0	0	0	0	375	839	140	0	0	0	0	0	0	0
200	447	40	5	3	1	0	0	0	0	400	894	40	3	1	0	0	0	0	0
200	447	60	4	2	0	0	0	0	0	400	894	60	2	0	0	0	0	0	0
200	447	80	3	1	0	0	0	0	0	400	894	80	2	0	0	0	0	0	0
200	447	100	3	0	0	0	0	0	0	400	894	100	1	0	0	0	0	0	0
200	447	120	2	0	0	0	0	0	0	400	894	120	1	0	0	0	0	0	0
200	447	140	2	0	0	0	0	0	0	400	894	140	0	0	0	0	0	0	0

NOTE: Detector time constant at a reference velocity of 5 ft/sec (1.5m/sec).
For SI units: 1 ft = 0.305 m; 1000 Btu/sec = 1055 kW.

Table B-3.2.4(c) Q_d, Threshold Fire Size at Response: 250 Btu/sec t_g: 300 seconds to 1000 Btu/sec α: 0.011 Btu/sec^3

τ	RTI	ΔT	*Ceiling Height (ft) 4.0	8.0	12.0	16.0	20.0	24.0	28.0
25	56	40	21	18	14	10	6	3	0
25	56	60	17	13	9	5	2	0	0
25	56	80	14	10	6	3	0	0	0
25	56	100	12	8	4	1	0	0	0
25	56	120	11	7	3	0	0	0	0
25	56	140	10	6	2	0	0	0	0
50	112	40	17	14	11	7	4	2	0
50	112	60	13	10	7	4	1	0	0
50	112	80	11	8	5	2	0	0	0
50	112	100	10	6	3	0	0	0	0
50	112	120	8	5	2	0	0	0	0
50	112	140	8	4	1	0	0	0	0
75	168	40	14	11	8	6	3	1	0
75	168	60	11	8	5	3	1	0	0
75	168	80	9	6	3	1	0	0	0
75	168	100	8	5	2	0	0	0	0
75	168	120	7	4	1	0	0	0	0
75	168	140	6	3	1	0	0	0	0
100	224	40	12	10	7	4	2	0	0
100	224	60	10	7	4	2	0	0	0
100	224	80	8	5	3	1	0	0	0
100	224	100	7	4	2	0	0	0	0
100	224	120	6	3	1	0	0	0	0
100	224	140	5	3	0	0	0	0	0
125	280	40	11	9	6	3	1	0	0
125	280	60	9	6	3	1	0	0	0
125	280	80	7	4	2	0	0	0	0
125	280	100	6	3	1	0	0	0	0
125	280	120	5	3	1	0	0	0	0
125	280	140	5	2	0	0	0	0	0
150	335	40	10	8	5	3	1	0	0
150	335	60	8	5	3	1	0	0	0
150	335	80	6	4	2	0	0	0	0
150	335	100	6	3	1	0	0	0	0
150	335	120	5	2	0	0	0	0	0
150	335	140	4	2	0	0	0	0	0
175	391	40	9	7	4	2	1	0	0
175	391	60	7	5	2	1	0	0	0
175	391	80	6	3	1	0	0	0	0
175	391	100	5	3	1	0	0	0	0
175	391	120	4	2	0	0	0	0	0
175	391	140	4	1	0	0	0	0	0
200	447	40	9	6	4	2	0	0	0
200	447	60	7	4	2	0	0	0	0
200	447	80	5	3	1	0	0	0	0
200	447	100	5	2	0	0	0	0	0
200	447	120	4	2	0	0	0	0	0
200	447	140	3	1	0	0	0	0	0

(*Installed Spacing of Detectors under Ceiling Height columns 4.0–28.0 ft)

τ	RTI	ΔT	*Ceiling Height (ft) 4.0	8.0	12.0	16.0	20.0	24.0	28.0
225	503	40	8	6	3	2	0	0	0
225	503	60	6	4	2	0	0	0	0
225	503	80	5	3	1	0	0	0	0
225	503	100	4	2	0	0	0	0	0
225	503	120	4	1	0	0	0	0	0
225	503	140	3	1	0	0	0	0	0
250	559	40	8	5	3	1	0	0	0
250	559	60	6	3	1	0	0	0	0
250	559	80	5	2	0	0	0	0	0
250	559	100	4	2	0	0	0	0	0
250	559	120	3	1	0	0	0	0	0
250	559	140	3	1	0	0	0	0	0
275	615	40	7	5	3	1	0	0	0
275	615	60	6	3	1	0	0	0	0
275	615	80	4	2	0	0	0	0	0
275	615	100	4	1	0	0	0	0	0
275	615	120	3	1	0	0	0	0	0
275	615	140	3	0	0	0	0	0	0
300	671	40	7	5	2	1	0	0	0
300	671	60	5	3	1	0	0	0	0
300	671	80	4	2	0	0	0	0	0
300	671	100	3	1	0	0	0	0	0
300	671	120	3	1	0	0	0	0	0
300	671	140	3	0	0	0	0	0	0
325	727	40	7	4	2	0	0	0	0
325	727	60	5	3	1	0	0	0	0
325	727	80	4	2	0	0	0	0	0
325	727	100	3	1	0	0	0	0	0
325	727	120	3	1	0	0	0	0	0
325	727	140	2	0	0	0	0	0	0
350	783	40	6	4	2	0	0	0	0
350	783	60	5	2	1	0	0	0	0
350	783	80	4	2	0	0	0	0	0
350	783	100	3	1	0	0	0	0	0
350	783	120	3	0	0	0	0	0	0
350	783	140	2	0	0	0	0	0	0
375	839	40	6	4	2	0	0	0	0
375	839	60	4	2	0	0	0	0	0
375	839	80	4	1	0	0	0	0	0
375	839	100	3	1	0	0	0	0	0
375	839	120	2	0	0	0	0	0	0
375	839	140	2	0	0	0	0	0	0
400	894	40	6	3	2	0	0	0	0
400	894	60	4	2	0	0	0	0	0
400	894	80	3	1	0	0	0	0	0
400	894	100	3	1	0	0	0	0	0
400	894	120	2	0	0	0	0	0	0
400	894	140	2	0	0	0	0	0	0

(*Installed Spacing of Detectors under Ceiling Height columns 4.0–28.0 ft)

NOTE: Detector time constant at a reference velocity of 5 ft/sec (1.5m/sec).
For SI units: 1 ft = 0.305 m; 1000 Btu/sec = 1055 kW.

Table B-3.2.4(d)   Q$_d$, Threshold Fire Size at Response: 250 Btu/sec   t$_g$: 500 seconds to 1000 Btu/sec   α: 0.004 Btu/sec3

τ	RTI	ΔT	Ceiling Height (ft)						
			4.0	8.0	12.0	16.0	20.0	24.0	28.0
			Installed Spacing of Detectors						
25	56	40	26	22	17	13	9	5	1
25	56	60	20	16	11	7	3	0	0
25	56	80	17	12	8	4	0	0	0
25	56	100	15	10	5	1	0	0	0
25	56	120	13	8	4	0	0	0	0
25	56	140	11	7	2	0	0	0	0
50	112	40	21	18	14	11	7	4	1
50	112	60	17	13	9	6	2	0	0
50	112	80	14	10	6	3	0	0	0
50	112	100	12	8	4	1	0	0	0
50	112	120	11	7	3	0	0	0	0
50	112	140	9	6	2	0	0	0	0
75	168	40	18	15	12	9	6	3	0
75	168	60	14	11	8	5	2	0	0
75	168	80	12	9	5	2	0	0	0
75	168	100	10	7	4	1	0	0	0
75	168	120	9	6	2	0	0	0	0
75	168	140	8	5	1	0	0	0	0
100	224	40	16	14	10	7	4	0	0
100	224	60	13	10	7	4	1	0	0
100	224	80	11	8	4	2	0	0	0
100	224	100	9	6	3	0	0	0	0
100	224	120	8	5	2	0	0	0	0
100	224	140	7	4	1	0	0	0	0
125	280	40	15	12	9	6	4	1	0
125	280	60	12	9	6	3	1	0	0
125	280	80	10	7	4	1	0	0	0
125	280	100	8	5	2	0	0	0	0
125	280	120	7	4	1	0	0	0	0
125	280	140	6	3	1	0	0	0	0
150	335	40	14	11	8	5	3	1	0
150	335	60	11	8	5	3	1	0	0
150	335	80	9	6	3	1	0	0	0
150	335	100	8	5	2	0	0	0	0
150	335	120	7	4	1	0	0	0	0
150	335	140	6	3	1	0	0	0	0
175	391	40	13	10	7	5	2	1	0
175	391	60	10	7	4	2	0	0	0
175	391	80	8	5	3	1	0	0	0
175	391	100	7	4	2	0	0	0	0
175	391	120	6	3	1	0	0	0	0
175	391	140	5	3	0	0	0	0	0
200	447	40	12	9	7	4	2	1	0
200	447	60	9	7	4	2	0	0	0
200	447	80	8	5	2	1	0	0	0
200	447	100	6	4	1	0	0	0	0
200	447	120	6	3	1	0	0	0	0
200	447	140	5	2	0	0	0	0	0

τ	RTI	ΔT	Ceiling Height (ft)						
			4.0	8.0	12.0	16.0	20.0	24.0	28.0
			Installed Spacing of Detectors						
225	503	40	11	9	6	4	2	0	0
225	503	60	9	6	4	1	0	0	0
225	503	80	7	5	2	0	0	0	0
225	503	100	6	3	1	0	0	0	0
225	503	120	5	3	1	0	0	0	0
225	503	140	5	2	0	0	0	0	0
250	559	40	11	8	6	3	1	0	0
250	559	60	8	6	3	1	0	0	0
250	559	80	7	4	2	0	0	0	0
250	559	100	6	3	1	0	0	0	0
250	559	120	5	2	0	0	0	0	0
250	559	140	4	2	0	0	0	0	0
275	615	40	10	8	5	3	1	0	0
275	615	60	8	5	3	1	0	0	0
275	615	80	6	4	2	0	0	0	0
275	615	100	5	3	1	0	0	0	0
275	615	120	5	2	0	0	0	0	0
275	615	140	4	2	0	0	0	0	0
300	671	40	10	1	0	0	7	5	3
300	671	60	7	0	0	0	5	3	1
300	671	80	6	0	0	0	4	1	0
300	671	100	5	0	0	0	3	1	0
300	671	120	4	0	0	0	2	0	0
300	671	140	4	0	0	0	2	0	0
325	727	40	9	1	0	0	7	4	2
325	727	60	7	0	0	0	5	2	1
325	727	80	6	0	0	0	3	1	0
325	727	100	5	0	0	0	2	0	0
325	727	120	4	0	0	0	2	0	0
325	727	140	4	0	0	0	1	0	0
350	783	40	9	1	0	0	6	4	2
350	783	60	7	0	0	0	4	2	0
350	783	80	6	0	0	0	3	1	0
350	783	100	5	0	0	0	2	0	0
350	783	120	4	0	0	0	2	0	0
350	783	140	3	0	0	0	1	0	0
375	839	40	9	0	0	0	6	4	2
375	839	60	6	0	0	0	4	2	0
375	839	80	5	0	0	0	3	1	0
375	839	100	4	0	0	0	2	0	0
375	839	120	4	0	0	0	2	0	0
375	839	140	3	0	0	0	1	0	0
400	894	40	8	0	0	0	6	4	2
400	894	60	6	0	0	0	4	2	0
400	894	80	5	0	0	0	3	1	0
400	894	100	4	0	0	0	2	0	0
400	894	120	4	0	0	0	1	0	0
400	894	140	3	0	0	0	1	0	0

NOTE: Detector time constant at a reference velocity of 5 ft/sec (1.5m/sec).
For SI units: 1 ft = 0.305 m; 1000 Btu/sec = 1055 kW.

Table B-3.2.4(e) Q_d, Threshold Fire Size at Response: 250 Btu/sec t_g: 600 seconds to 1000 Btu/sec α: 0.003 Btu/sec^3

τ	RTI	ΔT	4.0	8.0	12.0	16.0	20.0	24.0	28.0
			Installed Spacing of Detectors						
25	56	40	28	23	18	14	9	5	2
25	56	60	22	17	12	8	4	0	0
25	56	80	18	13	8	4	0	0	0
25	56	100	15	10	6	2	0	0	0
25	56	120	13	8	4	0	0	0	0
25	56	140	12	7	3	0	0	0	0
50	112	40	23	19	15	12	8	4	1
50	112	60	18	14	10	6	3	0	0
50	112	80	15	11	7	3	0	0	0
50	112	100	13	9	5	1	0	0	0
50	112	120	11	7	3	0	0	0	0
50	112	140	10	6	2	0	0	0	0
75	168	40	20	17	13	10	7	3	1
75	168	60	16	12	9	5	2	0	0
75	168	80	13	10	6	3	0	0	0
75	168	100	11	8	4	1	0	0	0
75	168	120	10	6	3	0	0	0	0
75	168	140	9	5	2	0	0	0	0
100	224	40	18	15	12	9	5	3	0
100	224	60	14	11	8	4	2	0	0
100	224	80	12	8	5	2	0	0	0
100	224	100	10	7	4	1	0	0	0
100	224	120	9	5	2	0	0	0	0
100	224	140	8	5	1	0	0	0	0
125	280	40	16	14	10	7	5	2	0
125	280	60	13	10	7	4	1	0	0
125	280	80	11	8	4	2	0	0	0
125	280	100	9	6	3	0	0	0	0
125	280	120	8	5	2	0	0	0	0
125	280	140	7	4	1	0	0	0	0
150	335	40	15	12	9	7	4	2	0
150	335	60	12	9	6	3	1	0	0
150	335	80	10	7	4	1	0	0	0
150	335	100	8	5	3	0	0	0	0
150	335	120	7	4	2	0	0	0	0
150	335	140	7	4	1	0	0	0	0
175	391	40	14	11	9	6	3	1	0
175	391	60	11	8	5	3	1	0	0
175	391	80	9	6	3	1	0	0	0
175	391	100	8	5	2	0	0	0	0
175	391	120	7	4	1	0	0	0	0
175	391	140	6	3	1	0	0	0	0
200	447	40	13	11	8	5	3	1	0
200	447	60	10	8	5	2	1	0	0
200	447	80	8	6	3	1	0	0	0
200	447	100	7	4	2	0	0	0	0
200	447	120	6	4	1	0	0	0	0
200	447	140	6	3	1	0	0	0	0

τ	RTI	ΔT	4.0	8.0	12.0	16.0	20.0	24.0	28.0
			Installed Spacing of Detectors						
225	503	40	12	10	7	5	2	1	0
225	503	60	10	7	4	2	0	0	0
225	503	80	8	5	3	1	0	0	0
225	503	100	7	4	2	0	0	0	0
225	503	120	6	3	1	0	0	0	0
225	503	140	5	3	0	0	0	0	0
250	559	40	12	9	7	4	2	1	0
250	559	60	9	7	4	2	0	0	0
250	559	80	8	5	2	1	0	0	0
250	559	100	6	4	1	0	0	0	0
250	559	120	6	3	1	0	0	0	0
250	559	140	5	2	0	0	0	0	0
275	615	40	11	9	6	4	2	0	0
275	615	60	9	6	4	1	0	0	0
275	615	80	7	5	2	0	0	0	0
275	615	100	6	3	1	0	0	0	0
275	615	120	5	3	1	0	0	0	0
275	615	140	5	2	0	0	0	0	0
300	671	40	11	8	6	3	1	0	0
300	671	60	8	6	3	1	0	0	0
300	671	80	7	4	2	0	0	0	0
300	671	100	6	3	1	0	0	0	0
300	671	120	5	2	0	0	0	0	0
300	671	140	4	2	0	0	0	0	0
325	727	40	10	8	5	3	1	0	0
325	727	60	8	5	3	1	0	0	0
325	727	80	6	4	2	0	0	0	0
325	727	100	6	3	1	0	0	0	0
325	727	120	5	2	0	0	0	0	0
325	727	140	4	2	0	0	0	0	0
350	783	40	10	7	5	3	1	0	0
350	783	60	8	5	3	1	0	0	0
350	783	80	6	4	2	0	0	0	0
350	783	100	5	3	1	0	0	0	0
350	783	120	5	2	0	0	0	0	0
350	783	140	4	2	0	0	0	0	0
375	839	40	10	7	5	3	1	0	0
375	839	60	7	5	3	1	0	0	0
375	839	80	6	3	1	0	0	0	0
375	839	100	5	3	1	0	0	0	0
375	839	120	4	2	0	0	0	0	0
375	839	140	4	1	0	0	0	0	0
400	894	40	9	7	4	2	1	0	0
400	894	60	7	5	2	1	0	0	0
400	894	80	6	3	1	0	0	0	0
400	894	100	5	2	0	0	0	0	0
400	894	120	4	2	0	0	0	0	0
400	894	140	4	1	0	0	0	0	0

NOTE: Detector time constant at a reference velocity of 5 ft/sec (1.5m/sec).
For SI units: 1 ft = 0.305 m; 1000 Btu/sec = 1055 kW.

Table B-3.2.4(f) Q_d, Threshold Fire Size at Response: 500 Btu/sec t_g: 50 seconds to 1000 Btu/sec α: 0.400 Btu/sec^3

τ	RTI	ΔT	Ceiling Height (ft)							τ	RTI	ΔT	Ceiling Height (ft)						
			4.0	8.0	12.0	16.0	20.0	24.0	28.0				4.0	8.0	12.0	16.0	20.0	24.0	28.0
			Installed Spacing of Detectors										Installed Spacing of Detectors						
25	56	40	13	11	8	5	2	1	0	225	503	40	4	2	0	0	0	0	0
25	56	60	11	8	5	3	1	0	0	225	503	60	3	1	0	0	0	0	0
25	56	80	9	6	4	1	0	0	0	225	503	80	2	0	0	0	0	0	0
25	56	100	8	5	3	1	0	0	0	225	503	100	2	0	0	0	0	0	0
25	56	120	7	4	2	0	0	0	0	225	503	120	2	0	0	0	0	0	0
25	56	140	7	4	1	0	0	0	0	225	503	140	1	0	0	0	0	0	0
50	112	40	10	7	5	2	1	0	0	250	559	40	4	2	0	0	0	0	0
50	112	60	8	5	3	1	0	0	0	250	559	60	3	1	0	0	0	0	0
50	112	80	7	4	2	0	0	0	0	250	559	80	2	0	0	0	0	0	0
50	112	100	6	3	1	0	0	0	0	250	559	100	2	0	0	0	0	0	0
50	112	120	5	3	0	0	0	0	0	250	559	120	1	0	0	0	0	0	0
50	112	140	5	2	0	0	0	0	0	250	559	140	1	0	0	0	0	0	0
75	168	40	8	6	3	1	0	0	0	275	615	40	4	2	0	0	0	0	0
75	168	60	6	4	2	0	0	0	0	275	615	60	3	0	0	0	0	0	0
75	168	80	5	3	1	0	0	0	0	275	615	80	2	0	0	0	0	0	0
75	168	100	4	2	0	0	0	0	0	275	615	100	2	0	0	0	0	0	0
75	168	120	4	2	0	0	0	0	0	275	615	120	1	0	0	0	0	0	0
75	168	140	3	1	0	0	0	0	0	275	615	140	1	0	0	0	0	0	0
100	224	40	7	4	2	0	0	0	0	300	671	40	3	1	0	0	0	0	0
100	224	60	5	3	1	0	0	0	0	300	671	60	2	0	0	0	0	0	0
100	224	80	4	2	0	0	0	0	0	300	671	80	2	0	0	0	0	0	0
100	224	100	4	1	0	0	0	0	0	300	671	100	1	0	0	0	0	0	0
100	224	120	3	1	0	0	0	0	0	300	671	120	1	0	0	0	0	0	0
100	224	140	3	0	0	0	0	0	0	300	671	140	0	0	0	0	0	0	0
125	280	40	6	4	2	0	0	0	0	325	727	40	3	1	0	0	0	0	0
125	280	60	5	2	0	0	0	0	0	325	727	60	2	0	0	0	0	0	0
125	280	80	4	2	0	0	0	0	0	325	727	80	2	0	0	0	0	0	0
125	280	100	3	1	0	0	0	0	0	325	727	100	1	0	0	0	0	0	0
125	280	120	3	0	0	0	0	0	0	325	727	120	1	0	0	0	0	0	0
125	280	140	2	0	0	0	0	0	0	325	727	140	0	0	0	0	0	0	0
150	335	40	5	3	1	0	0	0	0	350	783	40	3	1	0	0	0	0	0
150	335	60	4	2	0	0	0	0	0	350	783	60	2	0	0	0	0	0	0
150	335	80	3	1	0	0	0	0	0	350	783	80	2	0	0	0	0	0	0
150	335	100	3	0	0	0	0	0	0	350	783	100	1	0	0	0	0	0	0
150	335	120	2	0	0	0	0	0	0	350	783	120	0	0	0	0	0	0	0
150	335	140	2	0	0	0	0	0	0	350	783	140	0	0	0	0	0	0	0
175	391	40	5	3	1	0	0	0	0	375	839	40	3	1	0	0	0	0	0
175	391	60	4	2	0	0	0	0	0	375	839	60	2	0	0	0	0	0	0
175	391	80	3	1	0	0	0	0	0	375	839	80	1	0	0	0	0	0	0
175	391	100	2	0	0	0	0	0	0	375	839	100	1	0	0	0	0	0	0
175	391	120	2	0	0	0	0	0	0	375	839	120	0	0	0	0	0	0	0
175	391	140	2	0	0	0	0	0	0	375	839	140	0	0	0	0	0	0	0
200	447	40	5	2	0	0	0	0	0	400	894	40	3	0	0	0	0	0	0
200	447	60	3	1	0	0	0	0	0	400	894	60	2	0	0	0	0	0	0
200	447	80	3	0	0	0	0	0	0	400	894	80	1	0	0	0	0	0	0
200	447	100	2	0	0	0	0	0	0	400	894	100	1	0	0	0	0	0	0
200	447	120	2	0	0	0	0	0	0	400	894	120	0	0	0	0	0	0	0
200	447	140	1	0	0	0	0	0	0	400	894	140	0	0	0	0	0	0	0

NOTE: Detector time constant at a reference velocity of 5 ft/sec (1.5m/sec).
For SI units: 1 ft = 0.305 m; 1000 Btu/sec = 1055 kW.

Table B-3.2.4(g) Q_d, Threshold Fire Size at Response: 500 Btu/sec t_g: 150 seconds to 1000 Btu/sec α: 0.400 Btu/sec^3

τ	RTI	ΔT	Ceiling Height (ft) 4.0	8.0	12.0	16.0	20.0	24.0	28.0
			Installed Spacing of Detectors						
25	56	40	24	22	18	15	11	8	5
25	56	60	20	17	13	10	6	3	0
25	56	80	17	14	10	6	3	0	0
25	56	100	15	11	8	4	1	0	0
25	56	120	13	10	6	3	0	0	0
25	56	140	12	8	5	1	0	0	0
50	112	40	19	16	14	11	8	5	2
50	112	60	15	13	10	7	4	1	0
50	112	80	13	10	7	4	2	0	0
50	112	100	11	9	5	3	0	0	0
50	112	120	10	7	4	1	0	0	0
50	112	140	9	6	3	1	0	0	0
75	168	40	16	14	11	8	5	3	1
75	168	60	13	10	8	5	2	1	0
75	168	80	11	8	5	3	1	0	0
75	168	100	10	7	4	2	0	0	0
75	168	120	8	6	3	1	0	0	0
75	168	140	8	5	2	0	0	0	0
100	224	40	14	12	9	6	4	2	1
100	224	60	11	9	6	4	2	0	0
100	224	80	10	7	4	2	0	0	0
100	224	100	8	6	3	1	0	0	0
100	224	120	7	5	2	0	0	0	0
100	224	140	7	4	2	0	0	0	0
125	280	40	13	10	8	5	3	1	0
125	280	60	10	8	5	3	1	0	0
125	280	80	8	6	3	1	0	0	0
125	280	100	7	5	2	1	0	0	0
125	280	120	6	4	2	0	0	0	0
125	280	140	6	3	1	0	0	0	0
150	335	40	12	9	7	4	2	1	0
150	335	60	9	7	4	2	1	0	0
150	335	80	8	5	3	1	0	0	0
150	335	100	7	4	2	0	0	0	0
150	335	120	6	3	1	0	0	0	0
150	335	140	5	3	1	0	0	0	0
175	391	40	11	8	6	4	2	0	0
175	391	60	8	6	4	2	0	0	0
175	391	80	7	5	2	1	0	0	0
175	391	100	6	4	2	0	0	0	0
175	391	120	5	3	1	0	0	0	0
175	391	140	5	2	0	0	0	0	0
200	447	40	10	8	5	3	1	0	0
200	447	60	8	5	3	1	0	0	0
200	447	80	7	4	2	0	0	0	0
200	447	100	6	3	1	0	0	0	0
200	447	120	5	2	1	0	0	0	0
200	447	140	4	2	0	0	0	0	0
225	503	40	9	7	5	3	1	0	0
225	503	60	7	5	3	1	0	0	0
225	503	80	6	4	2	0	0	0	0
225	503	100	5	3	1	0	0	0	0
225	503	120	5	2	0	0	0	0	0
225	503	140	4	2	0	0	0	0	0
250	559	40	9	7	4	2	1	0	0
250	559	60	7	5	2	1	0	0	0
250	559	80	6	3	1	0	0	0	0
250	559	100	5	3	1	0	0	0	0
250	559	120	4	2	1	0	0	0	0
250	559	140	4	2	0	0	0	0	0
275	615	40	8	6	4	2	0	0	0
275	615	60	7	4	2	0	0	0	0
275	615	80	5	3	1	0	0	0	0
275	615	100	5	2	0	0	0	0	0
275	615	120	4	2	0	0	0	0	0
275	615	140	3	1	0	0	0	0	0
300	671	40	8	6	3	2	0	0	0
300	671	60	6	4	2	0	0	0	0
300	671	80	5	3	1	0	0	0	0
300	671	100	4	2	0	0	0	0	0
300	671	120	4	2	0	0	0	0	0
300	671	140	3	1	0	0	0	0	0
325	727	40	8	5	3	1	0	0	0
325	727	60	6	4	2	0	0	0	0
325	727	80	5	3	1	0	0	0	0
325	727	100	4	2	0	0	0	0	0
325	727	120	3	1	0	0	0	0	0
325	727	140	3	1	0	0	0	0	0
350	783	40	7	5	3	1	0	0	0
350	783	60	6	3	1	0	0	0	0
350	783	80	5	2	0	0	0	0	0
350	783	100	4	2	0	0	0	0	0
350	783	120	3	1	0	0	0	0	0
350	783	140	3	1	0	0	0	0	0
375	839	40	7	5	3	1	0	0	0
375	839	60	5	3	1	0	0	0	0
375	839	80	4	2	0	0	0	0	0
375	839	100	4	2	0	0	0	0	0
375	839	120	3	1	0	0	0	0	0
375	839	140	3	0	0	0	0	0	0
400	894	40	7	4	2	1	0	0	0
400	894	60	5	3	1	0	0	0	0
400	894	80	4	2	0	0	0	0	0
400	894	100	3	1	0	0	0	0	0
400	894	120	3	1	0	0	0	0	0
400	894	140	3	0	0	0	0	0	0

NOTE: Detector time constant at a reference velocity of 5 ft/sec (1.5m/sec).
For SI units: 1 ft = 0.305 m; 1000 Btu/sec = 1055 kW.

Table B-3.2.4(h) Q_d, Threshold Fire Size at Response: 500 Btu/sec t_g: 300 seconds to 1000 Btu/sec α: 0.011 Btu/sec³

τ	RTI	ΔT	Ceiling Height (ft) 4.0	8.0	12.0	16.0	20.0	24.0	28.0
			Installed Spacing of Detectors						
25	56	40	34	30	25	21	17	13	9
25	56	60	27	23	18	14	10	6	2
25	56	80	23	18	14	9	5	2	0
25	56	100	20	15	11	7	3	0	0
25	56	120	18	13	8	4	1	0	0
25	56	140	16	11	7	3	0	0	0
50	112	40	27	24	21	17	14	10	7
50	112	60	22	18	15	11	8	4	1
50	112	80	18	15	11	8	4	1	0
50	112	100	16	12	9	5	2	0	0
50	112	120	14	11	7	3	0	0	0
50	112	140	13	9	5	2	0	0	0
75	168	40	23	21	18	14	11	8	5
75	168	60	19	16	13	9	6	3	1
75	168	80	16	13	9	6	3	1	0
75	168	100	14	11	7	4	1	0	0
75	168	120	12	9	6	3	0	0	0
75	168	140	11	8	4	1	0	0	0
100	224	40	21	18	15	12	9	6	4
100	224	60	17	14	11	8	5	2	0
100	224	80	14	11	8	5	2	0	0
100	224	100	12	9	6	3	1	0	0
100	224	120	11	8	5	2	0	0	0
100	224	140	10	7	4	1	0	0	0
125	280	40	19	16	14	11	8	5	3
125	280	60	15	12	10	7	4	2	0
125	280	80	13	10	7	4	2	0	0
125	280	100	11	8	5	3	1	0	0
125	280	120	10	7	4	2	0	0	0
125	280	140	9	6	3	1	0	0	0
150	335	40	17	15	12	10	7	4	2
150	335	60	14	11	8	6	3	1	0
150	335	80	12	9	6	4	1	0	0
150	335	100	10	7	5	2	0	0	0
150	335	120	9	6	3	1	0	0	0
150	335	140	8	5	3	1	0	0	0
175	391	40	16	14	11	9	6	4	2
175	391	60	13	10	8	5	3	1	0
175	391	80	11	8	5	3	1	0	0
175	391	100	9	7	4	2	0	0	0
175	391	120	8	6	3	1	0	0	0
175	391	140	7	5	2	0	0	0	0
200	447	40	15	13	10	8	5	3	1
200	447	60	12	10	7	4	2	1	0
200	447	80	10	8	5	3	1	0	0
200	447	100	9	6	4	1	0	0	0
200	447	120	8	5	3	1	0	0	0
200	447	140	7	4	2	0	0	0	0
225	503	40	14	12	10	7	5	3	1
225	503	60	11	9	6	4	2	0	0
225	503	80	10	7	4	2	1	0	0
225	503	100	8	6	3	1	0	0	0
225	503	120	7	5	2	1	0	0	0
225	503	140	6	4	2	0	0	0	0
250	559	40	14	11	9	6	4	2	1
250	559	60	11	8	6	3	2	0	0
250	559	80	9	6	4	2	0	0	0
250	559	100	8	5	3	1	0	0	0
250	559	120	7	4	2	0	0	0	0
250	559	140	6	4	1	0	0	0	0
275	615	40	13	11	8	6	4	2	1
275	615	60	10	8	5	3	1	0	0
275	615	80	9	6	4	2	0	0	0
275	615	100	7	5	3	1	0	0	0
275	615	120	6	4	2	0	0	0	0
275	615	140	6	3	1	0	0	0	0
300	671	40	12	10	8	5	3	2	0
300	671	60	10	7	5	3	1	0	0
300	671	80	8	6	3	1	0	0	0
300	671	100	7	5	2	1	0	0	0
300	671	120	6	4	2	0	0	0	0
300	671	140	6	3	1	0	0	0	0
325	727	40	12	10	7	5	3	1	0
325	727	60	9	7	5	2	1	0	0
325	727	80	8	5	3	1	0	0	0
325	727	100	7	4	2	0	0	0	0
325	727	120	6	3	1	0	0	0	0
325	727	140	5	3	1	0	0	0	0
350	783	40	12	9	7	4	3	1	0
350	783	60	9	7	4	2	1	0	0
350	783	80	7	5	3	1	0	0	0
350	783	100	6	4	2	0	0	0	0
350	783	120	6	3	1	0	0	0	0
350	783	140	5	3	1	0	0	0	0
375	839	40	11	9	6	4	2	1	0
375	839	60	9	6	4	2	0	0	0
375	839	80	7	5	3	1	0	0	0
375	839	100	6	4	2	0	0	0	0
375	839	120	5	3	1	0	0	0	0
375	839	140	5	2	0	0	0	0	0
400	894	40	11	8	6	4	2	1	0
400	894	60	8	6	4	2	0	0	0
400	894	80	7	4	2	1	0	0	0
400	894	100	6	3	1	0	0	0	0
400	894	120	5	3	1	0	0	0	0
400	894	140	5	2	0	0	0	0	0

NOTE: Detector time constant at a reference velocity of 5 ft/sec (1.5m/sec).
For SI units: 1 ft = 0.305 m; 1000 Btu/sec = 1055 kW.

Table B-3.2.4(i) Q_d, Threshold Fire Size at Response: 500 Btu/sec t_g: 500 seconds to 1000 Btu/sec α: 0.004 Btu/sec³

τ	RTI	ΔT	Ceiling Height (ft) Installed Spacing of Detectors						
			4.0	8.0	12.0	16.0	20.0	24.0	28.0
25	56	40	41	35	30	25	20	16	11
25	56	60	32	26	21	16	12	7	3
25	56	80	27	21	16	11	7	3	0
25	56	100	23	17	12	8	4	0	0
25	56	120	20	15	10	5	1	0	0
25	56	140	18	13	8	3	0	0	0
50	112	40	34	30	26	22	18	14	10
50	112	60	27	23	18	14	10	6	3
50	112	80	23	18	14	10	6	2	0
50	112	100	20	15	11	7	3	0	0
50	112	120	17	13	9	5	1	0	0
50	112	140	16	11	7	3	0	0	0
75	168	40	30	26	23	19	15	12	8
75	168	60	24	20	16	13	9	5	2
75	168	80	20	16	12	9	5	2	0
75	168	100	17	14	10	6	2	0	0
75	168	120	15	11	8	4	1	0	0
75	168	140	14	10	6	2	0	0	0
100	224	40	27	24	20	17	14	10	7
100	224	60	21	18	15	11	8	4	2
100	224	80	18	15	11	8	4	1	0
100	224	100	16	12	9	5	2	0	0
100	224	120	14	10	7	5	3	0	0
100	224	140	13	9	5	2	0	0	0
125	280	40	25	22	19	15	12	9	6
125	280	60	20	17	13	10	7	4	1
125	280	80	16	13	10	7	4	1	0
125	280	100	14	11	8	5	2	0	0
125	280	120	13	9	6	3	0	0	0
125	280	140	11	8	5	2	0	0	0
150	335	40	23	20	17	14	11	8	5
150	335	60	18	15	12	9	6	3	1
150	335	80	15	12	9	6	3	1	0
150	335	100	13	10	7	4	1	0	0
150	335	120	12	9	5	3	0	0	0
150	335	140	11	7	4	1	0	0	0
175	391	40	21	19	16	13	10	7	4
175	391	60	17	14	11	8	5	3	1
175	391	80	14	11	8	5	3	1	0
175	391	100	12	9	6	3	1	0	0
175	391	120	11	8	5	2	0	0	0
175	391	140	10	7	4	1	0	0	0
200	447	40	20	18	15	12	9	6	4
200	447	60	16	13	10	7	5	2	1
200	447	80	13	11	8	5	2	0	0
200	447	100	12	9	6	3	1	0	0
200	447	120	10	7	4	2	0	0	0
200	447	140	9	6	3	1	0	0	0

τ	RTI	ΔT	Ceiling Height (ft) Installed Spacing of Detectors						
			4.0	8.0	12.0	16.0	20.0	24.0	28.0
225	503	40	19	17	14	11	8	6	3
225	503	60	15	13	10	7	4	2	0
225	503	80	13	10	7	4	2	0	0
225	503	100	11	8	5	3	1	0	0
225	503	120	10	7	4	2	0	0	0
225	503	140	9	6	3	1	0	0	0
250	559	40	18	16	13	10	8	5	3
250	559	60	14	12	9	6	4	2	0
250	559	80	12	9	7	4	2	0	0
250	559	100	10	8	5	2	1	0	0
250	559	120	9	6	4	1	0	0	0
250	559	140	8	6	3	1	0	0	0
275	615	40	17	15	12	10	7	5	2
275	615	60	14	11	8	6	3	1	0
275	615	80	12	9	6	3	1	0	0
275	615	100	10	7	5	2	0	0	0
275	615	120	9	6	3	1	0	0	0
275	615	140	8	5	3	1	0	0	0
300	671	40	17	14	12	9	6	4	2
300	671	60	13	11	8	5	3	1	0
300	671	80	11	8	6	3	1	0	0
300	671	100	10	7	4	2	0	0	0
300	671	120	8	6	3	1	0	0	0
300	671	140	8	5	2	0	0	0	0
325	727	40	16	14	11	9	6	4	2
325	727	60	13	10	8	5	3	1	0
325	727	80	11	8	5	3	1	0	0
325	727	100	9	7	4	2	0	0	0
325	727	120	8	5	3	1	0	0	0
325	727	140	7	5	2	0	0	0	0
350	783	40	16	13	11	8	6	3	2
350	783	60	12	10	7	5	2	1	0
350	783	80	10	8	5	3	1	0	0
350	783	100	9	6	4	2	0	0	0
350	783	120	8	5	3	1	0	0	0
350	783	140	7	4	2	0	0	0	0
375	839	40	15	13	10	8	5	3	1
375	839	60	12	9	7	4	2	1	0
375	839	80	10	7	5	2	1	0	0
375	839	100	8	6	3	1	0	0	0
375	839	120	7	5	2	1	0	0	0
375	839	140	7	4	2	0	0	0	0
400	894	40	14	12	10	7	5	3	1
400	894	60	11	9	6	4	2	1	0
400	894	80	9	7	4	2	1	0	0
400	894	100	8	6	3	1	0	0	0
400	894	120	7	5	2	1	0	0	0
400	894	140	6	4	2	0	0	0	0

NOTE: Detector time constant at a reference velocity of 5 ft/sec (1.5 m/sec).
For SI units: 1 ft = 0.305 m; 1000 Btu/sec = 1055 kW.

Table B-3.2.4(j) Q_d, Threshold Fire Size at Response: 500 Btu/sec t_g: 600 seconds to 1000 Btu/sec α: 0.003 Btu/sec³

τ	RTI	ΔT	\multicolumn{7}{c}{Ceiling Height (ft)}						
			4.0	8.0	12.0	16.0	20.0	24.0	28.0
			\multicolumn{7}{c}{Installed Spacing of Detectors}						
25	56	40	43	37	31	26	21	17	12
25	56	60	34	27	22	17	12	8	4
25	56	80	28	22	16	12	7	3	0
25	56	100	24	18	13	8	4	0	0
25	56	120	21	15	10	6	1	0	0
25	56	140	19	13	8	4	0	0	0
50	112	40	36	32	27	23	19	15	11
50	112	60	29	24	20	15	11	7	3
50	112	80	24	19	15	10	6	2	0
50	112	100	21	16	11	7	3	0	0
50	112	120	18	14	9	5	1	0	0
50	112	140	17	12	7	3	0	0	0
75	168	40	32	29	25	21	17	13	9
75	168	60	26	22	18	14	10	6	3
75	168	80	21	17	13	9	6	2	0
75	168	100	19	14	10	6	3	0	0
75	168	120	17	12	8	4	1	0	0
75	168	140	15	11	7	3	0	0	0
100	224	40	29	26	22	19	15	12	8
100	224	60	23	20	16	12	9	5	2
100	224	80	19	16	12	8	5	2	0
100	224	100	17	13	9	6	2	0	0
100	224	120	15	11	7	4	1	0	0
100	224	140	14	10	6	2	0	0	0
125	280	40	27	24	20	17	14	10	7
125	280	60	21	18	15	11	8	5	2
125	280	80	18	15	11	8	4	1	0
125	280	100	16	12	9	5	2	0	0
125	280	120	14	10	7	3	1	0	0
125	280	140	12	9	5	2	0	0	0
150	335	40	25	22	19	16	13	9	6
150	335	60	20	17	14	10	7	4	1
150	335	80	17	14	10	7	4	1	0
150	335	100	15	11	8	5	2	0	0
150	335	120	13	10	6	3	0	0	0
150	335	140	12	8	5	2	0	0	0
175	391	40	23	21	18	15	12	8	6
175	391	60	19	16	13	9	6	3	1
175	391	80	16	13	9	6	3	1	0
175	391	100	14	10	7	4	1	0	0
175	391	120	12	9	6	3	0	0	0
175	391	140	11	8	4	2	0	0	0
200	447	40	22	19	17	14	11	8	5
200	447	60	18	15	12	9	6	3	1
200	447	80	15	12	9	6	3	1	0
200	447	100	13	10	7	4	1	0	0
200	447	120	11	8	5	2	0	0	0
200	447	140	10	7	4	1	0	0	0
225	503	40	21	18	16	13	10	7	4
225	503	60	17	14	11	8	5	3	1
225	503	80	14	11	8	5	3	1	0
225	503	100	12	9	6	3	1	0	0
225	503	120	11	8	5	2	0	0	0
225	503	140	10	7	4	1	0	0	0
250	559	40	20	18	15	12	9	6	4
250	559	60	16	13	10	7	5	2	1
250	559	80	13	11	8	5	2	0	0
250	559	100	12	9	6	3	1	0	0
250	559	120	10	7	4	2	0	0	0
250	559	140	9	6	3	1	0	0	0
275	615	40	19	17	14	11	8	6	3
275	615	60	15	13	10	7	4	2	0
275	615	80	13	10	7	4	2	0	0
275	615	100	11	8	5	3	1	0	0
275	615	120	10	7	4	2	0	0	0
275	615	140	9	6	3	1	0	0	0
300	671	40	18	16	13	11	8	5	3
300	671	60	15	12	9	6	4	2	0
300	671	80	12	10	7	4	2	0	0
300	671	100	11	8	5	2	1	0	0
300	671	120	9	7	4	1	0	0	0
300	671	140	8	6	3	1	0	0	0
325	727	40	18	15	13	10	7	5	3
325	727	60	14	11	9	6	4	2	0
325	727	80	12	9	6	4	2	0	0
325	727	100	10	7	5	2	1	0	0
325	727	120	9	6	4	1	0	0	0
325	727	140	8	5	3	1	0	0	0
350	783	40	17	15	12	9	7	4	2
350	783	60	13	11	8	6	3	1	0
350	783	80	11	9	6	3	1	0	0
350	783	100	10	7	4	2	0	0	0
350	783	120	9	6	3	1	0	0	0
350	783	140	8	5	2	1	0	0	0
375	839	40	17	14	12	9	6	4	2
375	839	60	13	11	8	5	3	1	0
375	839	80	11	8	6	3	1	0	0
375	839	100	9	7	4	2	0	0	0
375	839	120	8	6	3	1	0	0	0
375	839	140	7	5	2	0	0	0	0
400	894	40	16	14	11	9	6	4	2
400	894	60	13	10	7	5	3	1	0
400	894	80	11	8	5	3	1	0	0
400	894	100	9	6	4	2	0	0	0
400	894	120	8	5	3	1	0	0	0
400	894	140	7	5	2	0	0	0	0

NOTE: Detector time constant at a reference velocity of 5 ft/sec (1.5m/sec).
For SI units: 1 ft = 0.305 m; 1000 Btu/sec = 1055 kW.

Table B-3.2.4(k) Q_d, Threshold Fire Size at Response: 750 Btu/sec t_g: 50 seconds to 1000 Btu/sec α: 0.400 Btu/sec^3

τ	RTI	ΔT	Ceiling Height (ft) 4.0	8.0	12.0	16.0	20.0	24.0	28.0
			Installed Spacing of Detectors						
25	56	40	18	15	13	10	7	4	2
25	56	60	15	12	9	6	4	1	0
25	56	80	13	10	7	4	2	0	0
25	56	100	11	9	6	3	1	0	0
25	56	120	10	7	4	2	0	0	0
25	56	140	9	6	4	1	0	0	0
50	112	40	14	11	9	6	3	2	0
50	112	60	11	9	6	3	1	0	0
50	112	80	9	7	4	2	0	0	0
50	112	100	8	6	3	1	0	0	0
50	112	120	7	5	2	0	0	0	0
50	112	140	7	4	2	0	0	0	0
75	168	40	11	9	6	4	2	0	0
75	168	60	9	7	4	2	0	0	0
75	168	80	8	5	3	1	0	0	0
75	168	100	7	4	2	0	0	0	0
75	168	120	6	3	1	0	0	0	0
75	168	140	5	3	1	0	0	0	0
100	224	40	10	7	5	3	1	0	0
100	224	60	8	5	3	1	0	0	0
100	224	80	7	4	2	0	0	0	0
100	224	100	6	3	1	0	0	0	0
100	224	120	5	3	1	0	0	0	0
100	224	140	4	2	0	0	0	0	0
125	280	40	9	6	4	2	0	0	0
125	280	60	7	5	2	1	0	0	0
125	280	80	6	3	1	0	0	0	0
125	280	100	5	3	1	0	0	0	0
125	280	120	4	2	0	0	0	0	0
125	280	140	4	2	0	0	0	0	0
150	335	40	8	6	3	1	0	0	0
150	335	60	6	4	2	0	0	0	0
150	335	80	5	3	1	0	0	0	0
150	335	100	4	2	0	0	0	0	0
150	335	120	4	2	0	0	0	0	0
150	335	140	3	1	0	0	0	0	0
175	391	40	7	5	3	1	0	0	0
175	391	60	6	3	1	0	0	0	0
175	391	80	5	2	0	0	0	0	0
175	391	100	4	2	0	0	0	0	0
175	391	120	3	1	0	0	0	0	0
175	391	140	3	1	0	0	0	0	0
200	447	40	7	4	2	1	0	0	0
200	447	60	5	3	1	0	0	0	0
200	447	80	4	2	0	0	0	0	0
200	447	100	4	1	0	0	0	0	0
200	447	120	3	1	0	0	0	0	0
200	447	140	3	0	0	0	0	0	0
225	503	40	6	4	2	0	0	0	0
225	503	60	5	3	1	0	0	0	0
225	503	80	4	2	0	0	0	0	0
225	503	100	3	1	0	0	0	0	0
225	503	120	3	1	0	0	0	0	0
225	503	140	2	0	0	0	0	0	0
250	559	40	6	4	2	0	0	0	0
250	559	60	4	2	0	0	0	0	0
250	559	80	4	2	0	0	0	0	0
250	559	100	3	1	0	0	0	0	0
250	559	120	3	0	0	0	0	0	0
250	559	140	2	0	0	0	0	0	0
275	615	40	6	3	1	0	0	0	0
275	615	60	4	2	0	0	0	0	0
275	615	80	3	1	0	0	0	0	0
275	615	100	3	1	0	0	0	0	0
275	615	120	2	0	0	0	0	0	0
275	615	140	2	0	0	0	0	0	0
300	671	40	5	3	1	0	0	0	0
300	671	60	4	2	0	0	0	0	0
300	671	80	3	1	0	0	0	0	0
300	671	100	3	0	0	0	0	0	0
300	671	120	2	0	0	0	0	0	0
300	671	140	2	0	0	0	0	0	0
325	727	40	5	3	1	0	0	0	0
325	727	60	4	2	0	0	0	0	0
325	727	80	3	1	0	0	0	0	0
325	727	100	2	0	0	0	0	0	0
325	727	120	2	0	0	0	0	0	0
325	727	140	2	0	0	0	0	0	0
350	783	40	5	3	1	0	0	0	0
350	783	60	4	1	0	0	0	0	0
350	783	80	3	1	0	0	0	0	0
350	783	100	2	0	0	0	0	0	0
350	783	120	2	0	0	0	0	0	0
350	783	140	2	0	0	0	0	0	0
375	839	40	5	2	0	0	0	0	0
375	839	60	3	1	0	0	0	0	0
375	839	80	3	0	0	0	0	0	0
375	839	100	2	0	0	0	0	0	0
375	839	120	2	0	0	0	0	0	0
375	839	140	1	0	0	0	0	0	0
400	894	40	4	2	0	0	0	0	0
400	894	60	3	1	0	0	0	0	0
400	894	80	2	0	0	0	0	0	0
400	894	100	2	0	0	0	0	0	0
400	894	120	2	0	0	0	0	0	0
400	894	140	1	0	0	0	0	0	0

NOTE: Detector time constant at a reference velocity of 5 ft/sec (1.5m/sec).
For SI units: 1 ft = 0.305 m; 1000 Btu/sec = 1055 kW.

Table B-3.2.4(I) Q_d, Threshold Fire Size at Response: 750 Btu/sec t_g: 150 seconds to 1000 Btu/sec α: 0.044 Btu/sec^3

τ	RTI	ΔT	4.0	8.0	12.0	16.0	20.0	24.0	28.0
			Installed Spacing of Detectors						
25	56	40	32	29	26	22	18	15	11
25	56	60	26	23	19	15	12	8	4
25	56	80	23	19	15	11	8	4	1
25	56	100	20	16	12	8	5	1	0
25	56	120	18	14	10	6	3	0	0
25	56	140	16	12	8	5	1	0	0
50	112	40	25	23	20	17	14	11	8
50	112	60	21	18	15	12	8	5	3
50	112	80	18	15	12	8	5	2	0
50	112	100	16	13	9	6	3	1	0
50	112	120	14	11	8	4	2	0	0
50	112	140	13	10	6	3	1	0	0
75	168	40	22	19	17	14	11	8	5
75	168	60	18	15	12	9	6	4	1
75	168	80	15	12	9	6	4	1	0
75	168	100	13	10	7	5	2	0	0
75	168	120	12	9	6	3	1	0	0
75	168	140	11	8	5	2	0	0	0
100	224	40	19	17	14	12	9	6	4
100	224	60	16	13	10	8	5	3	1
100	224	80	13	11	8	5	3	1	0
100	224	100	12	9	6	3	1	0	0
100	224	120	10	8	5	2	1	0	0
100	224	140	9	7	4	1	0	0	0
125	280	40	17	15	13	10	7	5	3
125	280	60	14	12	9	6	4	2	1
125	280	80	12	9	7	4	2	0	0
125	280	100	10	8	5	3	1	0	0
125	280	120	9	7	4	2	0	0	0
125	280	140	8	6	3	1	0	0	0
150	335	40	16	14	11	9	6	4	2
150	335	60	13	10	8	5	3	1	0
150	335	80	11	8	6	3	1	0	0
150	335	100	9	7	4	2	1	0	0
150	335	120	8	6	3	1	0	0	0
150	335	140	8	5	3	1	0	0	0
175	391	40	15	13	10	8	5	3	2
175	391	60	12	9	7	5	2	1	0
175	391	80	10	8	5	3	1	0	0
175	391	100	9	6	4	2	0	0	0
175	391	120	8	5	3	1	0	0	0
175	391	140	7	4	2	0	0	0	0
200	447	40	14	12	9	7	4	3	1
200	447	60	11	9	6	4	2	1	0
200	447	80	9	7	4	2	1	0	0
200	447	100	8	6	3	1	0	0	0
200	447	120	7	5	2	1	0	0	0
200	447	140	6	4	2	0	0	0	0
225	503	40	13	11	8	6	4	2	1
225	503	60	10	8	6	3	2	0	0
225	503	80	9	6	4	2	0	0	0
225	503	100	8	5	3	1	0	0	0
225	503	120	7	4	2	0	0	0	0
225	503	140	6	4	1	0	0	0	0
250	559	40	12	10	8	5	3	2	0
250	559	60	10	7	5	3	1	0	0
250	559	80	8	6	4	2	0	0	0
250	559	100	7	5	2	1	0	0	0
250	559	120	6	4	2	0	0	0	0
250	559	140	6	3	1	0	0	0	0
275	615	40	12	10	7	5	3	1	0
275	615	60	9	7	5	3	1	0	0
275	615	80	8	5	3	1	0	0	0
275	615	100	7	4	2	1	0	0	0
275	615	120	6	4	2	0	0	0	0
275	615	140	5	3	1	0	0	0	0
300	671	40	11	9	7	4	3	1	0
300	671	60	9	7	4	2	1	0	0
300	671	80	7	5	3	1	0	0	0
300	671	100	6	4	2	0	0	0	0
300	671	120	6	3	1	0	0	0	0
300	671	140	5	3	1	0	0	0	0
325	727	40	11	9	6	4	2	1	0
325	727	60	9	6	4	2	1	0	0
325	727	80	7	5	3	1	0	0	0
325	727	100	6	4	2	0	0	0	0
325	727	120	5	3	1	0	0	0	0
325	727	140	5	2	1	0	0	0	0
350	783	40	10	8	6	4	2	1	0
350	783	60	8	6	4	2	0	0	0
350	783	80	7	4	2	1	0	0	0
350	783	100	6	3	2	0	0	0	0
350	783	120	5	3	1	0	0	0	0
350	783	140	5	2	0	0	0	0	0
375	839	40	10	8	5	3	2	0	0
375	839	60	8	6	3	2	0	0	0
375	839	80	6	4	2	0	0	0	0
375	839	100	6	3	1	0	0	0	0
375	839	120	5	3	1	0	0	0	0
375	839	140	4	2	0	0	0	0	0
400	894	40	10	7	5	3	2	0	0
400	894	60	8	5	3	1	0	0	0
400	894	80	6	4	2	0	0	0	0
400	894	100	5	3	1	0	0	0	0
400	894	120	5	2	1	0	0	0	0
400	894	140	4	2	0	0	0	0	0

Column group header: Ceiling Height (ft) / Installed Spacing of Detectors

NOTE: Detector time constant at a reference velocity of 5 ft/sec (1.5m/sec).
For SI units: 1 ft = 0.305 m; 1000 Btu/sec = 1055 kW.

Table B-3.2.4(m) Q_d, Threshold Fire Size at Response: 750 Btu/sec t_g: 300 seconds to 1000 Btu/sec α: 0.011 Btu/sec^3

τ	RTI	ΔT	4.0	8.0	12.0	16.0	20.0	24.0	28.0
			\multicolumn{7}{Installed Spacing of Detectors}						
25	56	40	43	39	34	30	25	21	17
25	56	60	35	30	25	21	16	12	8
25	56	80	30	24	20	15	11	6	3
25	56	100	26	21	16	11	7	3	0
25	56	120	23	18	13	9	4	1	0
25	56	140	21	15	11	6	2	0	0
50	112	40	36	32	29	25	21	17	14
50	112	60	29	25	21	17	14	10	6
50	112	80	24	21	17	13	9	5	2
50	112	100	21	17	13	10	6	2	0
50	112	120	19	15	11	7	3	0	0
50	112	140	17	13	9	5	2	0	0
75	168	40	31	28	25	22	18	15	11
75	168	60	25	22	18	15	12	8	5
75	168	80	21	18	14	11	7	4	1
75	168	100	19	15	12	8	5	2	0
75	168	120	17	13	10	6	3	0	0
75	168	140	15	12	8	4	1	0	0
100	224	40	28	25	22	19	16	13	10
100	224	60	22	19	16	13	10	7	4
100	224	80	19	16	13	10	6	3	1
100	224	100	17	14	10	7	4	1	0
100	224	120	15	12	8	5	2	0	0
100	224	140	14	10	7	4	1	0	0
125	280	40	25	23	20	17	14	11	8
125	280	60	20	18	15	12	9	6	3
125	280	80	17	14	11	8	5	3	1
125	280	100	15	12	9	6	3	1	0
125	280	120	14	11	7	4	2	0	0
125	280	140	12	9	6	3	1	0	0
150	335	40	23	21	18	15	13	10	7
150	335	60	19	16	13	10	8	5	2
150	335	80	16	13	10	7	5	2	0
150	335	100	14	11	8	5	3	1	0
150	335	120	13	10	7	4	1	0	0
150	335	140	11	8	5	3	0	0	0
175	391	40	22	20	17	14	11	9	6
175	391	60	18	15	12	9	7	4	2
175	391	80	15	12	9	7	4	2	0
175	391	100	13	10	7	5	2	0	0
175	391	120	12	9	6	3	1	0	0
175	391	140	11	8	5	2	0	0	0
200	447	40	21	18	16	13	10	8	5
200	447	60	17	14	11	9	6	4	2
200	447	80	14	11	9	6	3	1	0
200	447	100	12	10	7	4	2	0	0
200	447	120	11	8	5	3	1	0	0
200	447	140	10	7	4	2	0	0	0
225	503	40	20	17	15	12	9	7	5
225	503	60	16	13	11	8	5	3	1
225	503	80	13	11	8	5	3	1	0
225	503	100	12	9	6	4	2	0	0
225	503	120	10	8	5	2	1	0	0
225	503	140	9	7	4	2	0	0	0
250	559	40	19	16	14	11	9	6	4
250	559	60	15	12	10	7	5	3	1
250	559	80	13	10	7	5	3	1	0
250	559	100	11	8	6	3	1	0	0
250	559	120	10	7	4	2	1	0	0
250	559	140	9	6	4	1	0	0	0
275	615	40	18	16	13	10	8	6	3
275	615	60	14	12	9	7	4	2	1
275	615	80	12	9	7	4	2	1	0
275	615	100	10	8	5	3	1	0	0
275	615	120	9	7	4	2	0	0	0
275	615	140	8	6	3	1	0	0	0
300	671	40	17	15	12	10	7	5	3
300	671	60	14	11	9	6	4	2	1
300	671	80	11	9	6	4	2	1	0
300	671	100	10	7	5	3	1	0	0
300	671	120	9	6	4	2	0	0	0
300	671	140	8	5	3	1	0	0	0
325	727	40	16	14	12	9	7	5	3
325	727	60	13	11	8	6	3	2	0
325	727	80	11	9	6	4	2	0	0
325	727	100	10	7	5	2	1	0	0
325	727	120	8	6	3	1	0	0	0
325	727	140	8	5	3	1	0	0	0
350	783	40	16	14	11	9	6	4	2
350	783	60	13	10	8	5	3	1	0
350	783	80	11	8	6	3	1	0	0
350	783	100	9	7	4	2	1	0	0
350	783	120	8	6	3	1	0	0	0
350	783	140	7	5	2	1	0	0	0
375	839	40	15	13	11	8	6	4	2
375	839	60	12	10	7	5	3	1	0
375	839	80	10	8	5	3	1	0	0
375	839	100	9	6	4	2	0	0	0
375	839	120	8	5	3	1	0	0	0
375	839	140	7	5	2	1	0	0	0
400	894	40	15	13	10	8	5	3	2
400	894	60	12	9	7	5	3	1	0
400	894	80	10	7	5	3	1	0	0
400	894	100	8	6	4	2	0	0	0
400	894	120	7	5	3	1	0	0	0
400	894	140	7	4	2	0	0	0	0

Ceiling Height (ft) — Installed Spacing of Detectors

NOTE: Detector time constant at a reference velocity of 5 ft/sec (1.5m/sec).
For SI units: 1 ft = 0.305 m; 1000 Btu/sec = 1055 kW.

Table B-3.2.4(n) Q_d, Threshold Fire Size at Response: 750 Btu/sec t_g: 500 seconds to 1000 Btu/sec α: 0.004 Btu/sec³

τ	RTI	ΔT	4.0	8.0	12.0	16.0	20.0	24.0	28.0
			Installed Spacing of Detectors						
25	56	40	52	45	39	34	29	24	20
25	56	60	41	34	28	23	18	14	9
25	56	80	34	28	22	17	12	8	4
25	56	100	29	23	18	13	8	4	0
25	56	120	26	20	14	10	5	1	0
25	56	140	23	17	12	7	3	0	0
50	112	40	44	40	35	30	26	22	18
50	112	60	35	30	26	21	17	12	8
50	112	80	30	25	20	15	11	7	3
50	112	100	26	21	16	12	7	3	0
50	112	120	23	18	13	9	5	1	0
50	112	140	21	16	11	7	3	0	0
75	168	40	39	35	31	27	24	20	16
75	168	60	31	27	23	19	15	11	7
75	168	80	26	22	18	14	10	6	3
75	168	100	23	19	15	10	7	3	0
75	168	120	20	16	12	8	4	1	0
75	168	140	18	14	10	6	2	0	0
100	224	40	35	32	29	25	21	18	14
100	224	60	28	25	21	17	14	10	6
100	224	80	24	20	16	13	9	5	2
100	224	100	21	17	13	10	6	2	0
100	224	120	19	15	11	7	4	0	0
100	224	140	17	13	9	5	2	0	0
125	280	40	32	30	26	23	20	16	13
125	280	60	26	23	19	16	12	9	6
125	280	80	22	19	15	12	8	5	2
125	280	100	19	16	12	9	5	2	0
125	280	120	17	14	10	7	3	0	0
125	280	140	16	12	8	5	2	0	0
150	335	40	30	28	25	21	18	15	12
150	335	60	24	21	18	15	11	8	5
150	335	80	21	17	14	11	7	4	1
150	335	100	18	15	11	8	5	2	0
150	335	120	16	13	9	6	3	0	0
150	335	140	15	11	8	4	1	0	0
175	391	40	28	26	23	20	17	14	10
175	391	60	23	20	17	14	10	7	4
175	391	80	19	16	13	10	7	4	1
175	391	100	17	14	11	7	4	1	0
175	391	120	15	12	9	5	2	0	0
175	391	140	14	10	7	4	1	0	0
200	447	40	27	24	22	18	16	12	10
200	447	60	22	19	16	13	10	7	4
200	447	80	18	15	12	9	6	3	1
200	447	100	16	13	10	7	4	1	0
200	447	120	14	11	8	5	2	0	0
200	447	140	13	10	7	4	1	0	0
225	503	40	26	23	20	17	14	12	9
225	503	60	20	18	15	12	9	6	3
225	503	80	17	14	11	8	6	3	1
225	503	100	15	12	9	6	3	1	0
225	503	120	13	11	7	4	2	0	0
225	503	140	12	9	6	3	1	0	0
250	559	40	24	22	19	16	14	11	8
250	559	60	19	17	14	11	8	5	3
250	559	80	16	14	11	8	5	3	1
250	559	100	14	12	9	6	3	1	0
250	559	120	13	10	7	4	2	0	0
250	559	140	12	9	6	3	1	0	0
275	615	40	23	21	18	16	13	10	7
275	615	60	19	16	13	10	8	5	3
275	615	80	16	13	10	7	5	2	1
275	615	100	14	11	8	5	3	1	0
275	615	120	12	9	7	4	1	0	0
275	615	140	11	8	5	3	1	0	0
300	671	40	22	20	18	15	12	9	7
300	671	60	18	15	13	10	7	5	2
300	671	80	15	13	10	7	4	2	0
300	671	100	13	11	8	5	2	1	0
300	671	120	12	9	6	3	1	0	0
300	671	140	11	8	5	2	0	0	0
325	727	40	22	19	17	14	11	9	6
325	727	60	17	15	12	9	7	4	2
325	727	80	15	12	9	6	4	2	0
325	727	100	13	10	7	5	2	1	0
325	727	120	11	9	6	3	1	0	0
325	727	140	10	7	5	2	0	0	0
350	783	40	21	19	16	13	11	8	6
350	783	60	17	14	12	9	6	4	2
350	783	80	14	12	9	6	4	2	0
350	783	100	12	10	7	4	2	0	0
350	783	120	11	8	5	3	1	0	0
350	783	140	10	7	4	2	0	0	0
375	839	40	20	18	16	13	10	8	5
375	839	60	16	14	11	8	6	3	2
375	839	80	14	11	8	6	3	1	0
375	839	100	12	9	7	4	2	0	0
375	839	120	11	8	5	3	1	0	0
375	839	140	10	7	4	2	0	0	0
400	894	40	20	17	15	12	10	7	5
400	894	60	16	13	11	8	5	3	1
400	894	80	13	11	8	5	3	1	0
400	894	100	11	9	6	4	2	0	0
400	894	120	10	8	5	3	1	0	0
400	894	140	9	7	4	2	0	0	0

(Columns above grouped under **Ceiling Height (ft)**; subheading: **Installed Spacing of Detectors**.)

NOTE: Detector time constant at a reference velocity of 5 ft/sec (1.5m/sec).
For SI units: 1 ft = 0.305 m; 1000 Btu/sec = 1055 kW.

Table B-3.2.4(o) Q_d, Threshold Fire Size at Response: 750 Btu/sec t_g: 600 seconds to 1000 Btu/sec α: 0.003 Btu/sec^3

τ	RTI	ΔT	Ceiling Height (ft)						
			4.0	8.0	12.0	16.0	20.0	24.0	28.0
			Installed Spacing of Detectors						
25	56	40	55	47	41	35	30	25	21
25	56	60	43	36	29	24	19	14	10
25	56	80	36	28	23	18	13	8	4
25	56	100	31	24	18	13	9	4	0
25	56	120	27	20	15	10	6	1	0
25	56	140	24	18	12	8	3	0	0
50	112	40	47	42	37	32	28	23	19
50	112	60	37	32	27	22	18	13	9
50	112	80	31	26	21	16	12	8	4
50	112	100	27	22	17	12	8	4	0
50	112	120	24	19	14	9	5	1	0
50	112	140	22	16	11	7	3	0	0
75	168	40	42	38	34	29	25	21	17
75	168	60	33	29	25	20	16	12	8
75	168	80	28	24	19	15	11	7	3
75	168	100	24	20	15	11	7	3	0
75	168	120	22	17	13	9	5	1	0
75	168	140	20	15	11	6	3	0	0
100	224	40	38	35	31	27	23	19	16
100	224	60	30	27	23	19	15	11	7
100	224	80	26	22	18	14	10	6	3
100	224	100	22	18	14	10	7	3	0
100	224	120	20	16	12	8	4	1	0
100	224	140	18	14	10	6	2	0	0
125	280	40	35	32	29	25	22	18	14
125	280	60	28	25	21	17	14	10	7
125	280	80	24	20	16	13	9	6	2
125	280	100	21	17	13	10	6	3	0
125	280	120	19	15	11	7	4	1	0
125	280	140	17	13	9	5	2	0	0
150	335	40	33	30	27	23	20	17	13
150	335	60	26	23	20	16	13	9	6
150	335	80	22	19	15	12	8	5	2
150	335	100	19	16	12	9	5	2	0
150	335	120	17	14	10	7	3	0	0
150	335	140	16	12	8	5	2	0	0
175	391	40	31	28	25	22	19	15	12
175	391	60	25	22	18	15	12	9	5
175	391	80	21	18	14	11	8	4	2
175	391	100	18	15	12	8	5	2	0
175	391	120	16	13	10	6	3	0	0
175	391	140	15	11	8	5	1	0	0
200	447	40	29	27	24	21	17	14	11
200	447	60	23	21	17	14	11	8	5
200	447	80	20	17	14	10	7	4	1
200	447	100	17	14	11	8	4	2	0
200	447	120	15	12	9	6	3	0	0
200	447	140	14	11	7	4	1	0	0
225	503	40	28	26	23	19	16	13	10
225	503	60	22	20	17	13	10	7	4
225	503	80	19	16	13	10	7	4	1
225	503	100	16	13	10	7	4	1	0
225	503	120	15	12	8	5	2	0	0
225	503	140	13	10	7	4	1	0	0
250	559	40	27	24	21	18	15	12	10
250	559	60	21	19	16	13	10	7	4
250	559	80	18	15	12	9	6	3	1
250	559	100	16	13	10	7	4	1	0
250	559	120	14	11	8	5	2	0	0
250	559	140	13	10	7	4	1	0	0
275	615	40	26	23	20	18	15	12	9
275	615	60	21	18	15	12	9	6	4
275	615	80	17	15	12	9	6	3	1
275	615	100	15	12	9	6	3	1	0
275	615	120	13	11	7	5	2	0	0
275	615	140	12	9	6	3	1	0	0
300	671	40	25	22	20	17	14	11	8
300	671	60	20	17	14	11	8	6	3
300	671	80	17	14	11	8	5	3	1
300	671	100	15	12	9	6	3	1	0
300	671	120	13	10	7	4	2	0	0
300	671	140	12	9	6	3	1	0	0
325	727	40	24	22	19	16	13	10	8
325	727	60	19	16	14	11	8	5	3
325	727	80	16	13	10	8	5	2	1
325	727	100	14	11	8	5	3	1	0
325	727	120	12	10	7	4	2	0	0
325	727	140	11	8	5	3	1	0	0
350	783	40	23	21	18	15	13	10	7
350	783	60	18	16	13	10	7	5	3
350	783	80	15	13	10	7	5	2	1
350	783	100	13	11	8	5	3	1	0
350	783	120	12	9	6	4	1	0	0
350	783	140	11	8	5	3	1	0	0
375	839	40	22	20	17	15	12	9	7
375	839	60	18	15	13	10	7	5	2
375	839	80	15	12	10	7	4	2	0
375	839	100	13	10	8	5	0	1	0
375	839	120	12	9	6	3	1	0	0
375	839	140	11	8	5	2	0	0	0
400	894	40	22	19	17	14	11	9	6
400	894	60	17	15	12	9	7	4	2
400	894	80	15	12	9	6	4	2	0
400	894	100	13	10	7	5	2	1	0
400	894	120	11	9	6	3	1	0	0
400	894	140	10	7	5	2	0	0	0

NOTE: Detector time constant at a reference velocity of 5 ft/sec (1.5m/sec).
For SI units: 1 ft = 0.305 m; 1000 Btu/sec = 1055 kW.

Table B-3.2.4(p) Q_d, Threshold Fire Size at Response: 1000 Btu/sec t_g: 50 seconds to 1000 Btu/sec α: 0.400 Btu/sec^3

| τ | RTI | ΔT | \multicolumn{7}{c}{Ceiling Height (ft)} |
| | | | 4.0 | 8.0 | 12.0 | 16.0 | 20.0 | 24.0 | 28.0 |
			\multicolumn{7}{c}{Installed Spacing of Detectors}
25	56	40	22	20	17	14	11	8	5
25	56	60	18	16	13	10	7	4	2
25	56	80	16	13	10	7	4	2	0
25	56	100	14	11	8	5	3	0	0
25	56	120	13	10	7	4	1	0	0
25	56	140	12	9	6	3	1	0	0
50	112	40	17	15	12	9	7	4	2
50	112	60	14	11	9	6	4	2	0
50	112	80	12	9	7	4	2	0	0
50	112	100	11	8	5	3	1	0	0
50	112	120	10	7	4	2	0	0	0
50	112	140	9	6	3	1	0	0	0
75	168	40	14	12	9	7	4	2	1
75	168	60	12	9	7	4	2	1	0
75	168	80	10	7	5	3	1	0	0
75	168	100	9	6	4	2	0	0	0
75	168	120	8	5	3	1	0	0	0
75	168	140	7	4	2	0	0	0	0
100	224	40	12	10	8	5	3	1	0
100	224	60	10	8	5	3	1	0	0
100	224	80	8	6	4	2	0	0	0
100	224	100	7	5	3	1	0	0	0
100	224	120	7	4	2	0	0	0	0
100	224	140	6	4	1	0	0	0	0
125	280	40	11	9	6	4	2	1	0
125	280	60	9	7	4	2	1	0	0
125	280	80	8	5	3	1	0	0	0
125	280	100	7	4	2	0	0	0	0
125	280	120	6	3	1	0	0	0	0
125	280	140	5	3	1	0	0	0	0
150	335	40	0	8	5	3	2	0	0
150	335	60	8	6	3	2	0	0	0
150	335	80	7	4	2	0	0	0	0
150	335	100	6	3	2	0	0	0	0
150	335	120	5	3	1	0	0	0	0
150	335	140	5	2	0	0	0	0	0
175	391	40	9	7	5	3	1	0	0
175	391	60	7	5	3	1	0	0	0
175	391	80	6	4	2	0	0	0	0
175	391	100	5	3	1	0	0	0	0
175	391	120	5	2	0	0	0	0	0
175	391	140	4	2	0	0	0	0	0
200	447	40	9	6	4	2	1	0	0
200	447	60	7	5	2	1	0	0	0
200	447	80	6	3	1	0	0	0	0
200	447	100	5	3	1	0	0	0	0
200	447	120	4	2	0	0	0	0	0
200	447	140	4	2	0	0	0	0	0
225	503	40	8	6	4	2	0	0	0
225	503	60	6	4	2	0	0	0	0
225	503	80	5	3	1	0	0	0	0
225	503	100	5	2	0	0	0	0	0
225	503	120	4	2	0	0	0	0	0
225	503	140	3	1	0	0	0	0	0
250	559	40	8	5	3	1	0	0	0
250	559	60	6	4	2	0	0	0	0
250	559	80	5	3	1	0	0	0	0
250	559	100	4	2	0	0	0	0	0
250	559	120	4	2	0	0	0	0	0
250	559	140	3	1	0	0	0	0	0
275	615	40	7	5	3	1	0	0	0
275	615	60	6	3	1	0	0	0	0
275	615	80	5	2	0	0	0	0	0
275	615	100	4	2	0	0	0	0	0
275	615	120	3	1	0	0	0	0	0
275	615	140	3	1	0	0	0	0	0
300	671	40	7	5	3	1	0	0	0
300	671	60	5	3	1	0	0	0	0
300	671	80	4	2	0	0	0	0	0
300	671	100	4	2	0	0	0	0	0
300	671	120	3	1	0	0	0	0	0
300	671	140	3	1	0	0	0	0	0
325	727	40	7	4	2	1	0	0	0
325	727	60	5	3	1	0	0	0	0
325	727	80	4	2	0	0	0	0	0
325	727	100	3	1	0	0	0	0	0
325	727	120	3	1	0	0	0	0	0
325	727	140	3	0	0	0	0	0	0
350	783	40	6	4	2	0	0	0	0
350	783	60	5	3	1	0	0	0	0
350	783	80	4	2	0	0	0	0	0
350	783	100	3	1	0	0	0	0	0
350	783	120	3	1	0	0	0	0	0
350	783	140	2	0	0	0	0	0	0
375	839	40	6	4	2	0	0	0	0
375	839	60	5	2	0	0	0	0	0
375	839	80	4	2	0	0	0	0	0
375	839	100	3	1	0	0	0	0	0
375	839	120	3	0	0	0	0	0	0
375	839	140	2	0	0	0	0	0	0
400	894	40	6	4	2	0	0	0	0
400	894	60	4	2	0	0	0	0	0
400	894	80	3	1	0	0	0	0	0
400	894	100	3	1	0	0	0	0	0
400	894	120	2	0	0	0	0	0	0
400	894	140	2	0	0	0	0	0	0

NOTE: Detector time constant at a reference velocity of 5 ft/sec (1.5m/sec).
For SI units: 1 ft = 0.305 m; 1000 Btu/sec = 1055 kW.

Table B-3.2.4(q) Q_d, Threshold Fire Size at Response: 1000 Btu/sec t_g: 150 seconds to 1000 Btu/sec α: 0.044 Btu/sec^3

τ	RTI	ΔT	Ceiling Height (ft)							τ	RTI	ΔT	Ceiling Height (ft)						
			4.0	8.0	12.0	16.0	20.0	24.0	28.0				4.0	8.0	12.0	16.0	20.0	24.0	28.0
			Installed Spacing of Detectors										Installed Spacing of Detectors						
25	56	40	39	36	32	28	25	21	17	225	503	40	16	14	12	9	7	5	3
25	56	60	32	28	24	21	17	13	9	225	503	60	13	11	8	6	4	2	1
25	56	80	27	24	20	16	12	8	4	225	503	80	11	9	6	4	2	1	0
25	56	100	24	20	16	12	8	4	1	225	503	100	10	7	5	3	1	0	0
25	56	120	22	18	14	10	6	2	0	225	503	120	9	6	4	2	0	0	0
25	56	140	20	16	12	8	4	0	0	225	503	140	8	5	3	1	0	0	0
50	112	40	31	29	26	22	19	16	13	250	559	40	16	13	11	9	6	4	2
50	112	60	25	23	19	16	13	9	6	250	559	60	12	10	8	5	3	2	0
50	112	80	22	19	15	12	9	6	3	250	559	80	11	8	6	3	2	0	0
50	112	100	19	16	13	9	6	3	0	250	559	100	9	7	4	2	1	0	0
50	112	120	17	14	11	7	4	1	0	250	559	120	8	6	3	1	0	0	0
50	112	140	16	13	9	6	2	0	0	250	559	140	7	5	3	1	0	0	0
75	168	40	27	24	22	19	16	13	10	275	615	40	15	13	10	8	6	4	2
75	168	60	22	19	16	13	10	7	4	275	615	60	12	10	7	5	3	1	0
75	168	80	19	16	13	10	7	4	2	275	615	80	10	8	5	3	1	0	0
75	168	100	16	14	11	7	4	2	0	275	615	100	9	6	4	2	0	0	0
75	168	120	15	12	9	6	3	1	0	275	615	120	8	5	3	1	0	0	0
75	168	140	13	10	7	4	2	0	0	275	615	140	7	5	2	1	0	0	0
100	224	40	24	22	19	16	13	10	8	300	671	40	14	12	10	7	5	3	2
100	224	60	19	17	14	11	8	6	3	300	671	60	11	9	7	4	2	1	0
100	224	80	16	14	11	8	5	3	1	300	671	80	10	7	5	3	1	0	0
100	224	100	14	12	9	6	3	1	0	300	671	100	8	6	4	2	0	0	0
100	224	120	13	10	7	5	2	0	0	300	671	120	7	5	3	1	0	0	0
100	224	140	12	9	6	3	1	0	0	300	671	140	7	4	2	0	0	0	0
125	280	40	21	19	17	14	11	9	6	325	727	40	14	12	9	7	5	3	1
125	280	60	17	15	12	10	7	4	2	325	727	60	11	9	6	4	2	1	0
125	280	80	15	12	10	7	4	2	1	325	727	80	9	7	4	2	1	0	0
125	280	100	13	10	8	5	3	1	0	325	727	100	8	6	3	1	0	0	0
125	280	120	12	9	6	4	1	0	0	325	727	120	7	5	2	1	0	0	0
125	280	140	11	8	5	3	1	0	0	325	727	140	6	4	2	0	0	0	0
150	335	40	20	18	15	13	10	7	5	350	783	40	13	11	9	6	4	2	1
150	335	60	16	14	11	8	6	4	2	350	783	60	10	8	6	4	2	1	0
150	335	80	14	11	9	6	3	2	0	350	783	80	9	6	4	2	1	0	0
150	335	100	12	9	7	4	2	1	0	350	783	100	8	5	3	1	0	0	0
150	335	120	11	8	5	3	1	0	0	350	783	120	7	4	2	1	0	0	0
150	335	140	10	7	4	2	1	0	0	350	783	140	6	4	2	0	0	0	0
175	391	40	18	16	14	11	9	6	4	375	839	40	13	11	8	6	4	2	1
175	391	60	15	13	10	7	5	3	1	375	839	60	10	8	5	3	2	0	0
175	391	80	13	10	8	5	3	1	0	375	839	80	8	6	4	2	0	0	0
175	391	100	11	9	6	4	2	0	0	375	839	100	7	5	3	1	0	0	0
175	391	120	10	7	5	2	1	0	0	375	839	120	6	4	2	0	0	0	0
175	391	140	9	6	4	2	0	0	0	375	839	140	6	3	1	0	0	0	0
200	447	40	17	15	13	10	8	5	3	400	894	40	12	10	8	5	3	2	1
200	447	60	14	12	9	7	4	2	1	400	894	60	10	7	5	3	1	0	0
200	447	80	12	9	7	4	2	1	0	400	894	80	8	6	4	2	0	0	0
200	447	100	10	8	5	3	1	0	0	400	894	100	7	5	3	1	0	0	0
200	447	120	9	7	4	2	1	0	0	400	894	120	6	4	2	0	0	0	0
200	447	140	8	6	3	1	0	0	0	400	894	140	6	3	1	0	0	0	0

NOTE: Detector time constant at a reference velocity of 5 ft/sec (1.5m/sec).
For SI units: 1 ft = 0.305 m; 1000 Btu/sec = 1055 kW.

Table B-3.2.4(r) Q_d, Threshold Fire Size at Response: 1000 Btu/sec t_g: 300 seconds to 1000 Btu/sec α: 0.011 Btu/sec^3

τ	RTI	ΔT	\multicolumn{7}{c}{Ceiling Height (ft)}	τ	RTI	ΔT	\multicolumn{7}{c}{Ceiling Height (ft)}
			4.0	8.0	12.0	16.0	20.0	24.0	28.0				4.0	8.0	12.0	16.0	20.0	24.0	28.0
			\multicolumn{7}{c}{Installed Spacing of Detectors}				\multicolumn{7}{c}{Installed Spacing of Detectors}
25	56	40	52	47	42	37	32	28	23	225	503	40	24	22	19	17	14	11	9
25	56	60	42	36	31	26	22	17	13	225	503	60	19	17	14	11	9	6	4
25	56	80	35	30	25	20	15	11	7	225	503	80	16	14	11	8	6	3	1
25	56	100	31	25	20	15	11	7	3	225	503	100	14	12	9	6	4	2	0
25	56	120	28	22	17	12	8	4	0	225	503	120	13	10	7	5	2	1	0
25	56	140	25	19	14	10	5	1	0	225	503	140	12	9	6	3	1	0	0
50	112	40	43	40	36	32	28	24	20	250	559	40	23	21	18	16	13	10	8
50	112	60	35	31	27	23	19	15	11	250	559	60	18	16	13	11	8	5	3
50	112	80	30	25	21	17	13	9	5	250	559	80	16	13	10	8	5	3	1
50	112	100	26	22	18	13	9	6	2	250	559	100	14	11	8	6	3	1	0
50	112	120	23	19	15	11	7	3	0	250	559	120	12	10	7	4	2	0	0
50	112	140	21	17	12	8	4	1	0	250	559	140	11	8	6	3	1	0	0
75	168	40	37	35	31	28	24	21	17	275	615	40	22	20	17	15	12	9	7
75	168	60	30	27	24	20	16	13	9	275	615	60	18	15	13	10	7	5	3
75	168	80	26	22	19	15	11	8	4	275	615	80	15	12	10	7	5	2	1
75	168	100	23	19	15	12	8	5	1	275	615	100	13	11	8	5	3	1	0
75	168	120	20	17	13	9	6	2	0	275	615	120	12	9	6	4	2	0	0
75	168	140	19	15	11	7	4	1	0	275	615	140	11	8	5	3	1	0	0
100	224	40	34	31	28	25	22	18	15	300	671	40	21	19	16	14	11	9	6
100	224	60	27	24	21	18	14	11	8	300	671	60	17	15	12	9	7	4	2
100	224	80	23	20	17	13	10	7	4	300	671	80	14	12	9	7	4	2	1
100	224	100	20	17	14	10	7	4	1	300	671	100	13	10	7	5	3	1	0
100	224	120	18	15	12	8	5	2	0	300	671	120	11	9	6	3	1	0	0
100	224	140	17	13	10	6	3	0	0	300	671	140	10	8	5	3	1	0	0
125	280	40	31	29	26	23	19	16	13	325	727	40	20	18	16	13	11	8	6
125	280	60	25	22	19	16	13	10	7	325	727	60	16	14	11	9	6	4	2
125	280	80	21	18	15	12	9	6	3	325	727	80	14	11	9	6	4	2	1
125	280	100	19	16	13	9	6	3	1	325	727	100	12	10	7	4	2	1	0
125	280	120	17	14	10	7	4	1	0	325	727	120	11	8	6	3	1	0	0
125	280	140	15	12	9	6	3	0	0	325	727	140	10	7	5	2	1	0	0
150	335	40	29	27	24	21	18	15	12	350	783	40	20	18	15	13	10	8	5
150	335	60	23	21	18	15	12	9	6	350	783	60	16	13	11	8	6	4	2
150	335	80	20	17	14	11	8	5	2	350	783	80	13	11	8	6	3	2	0
150	335	100	17	14	11	8	5	3	1	350	783	100	12	9	7	4	2	1	0
150	335	120	16	13	10	6	4	1	0	350	783	120	10	8	5	3	1	0	0
150	335	140	14	11	8	5	2	0	0	350	783	140	9	7	4	2	1	0	0
175	391	40	27	25	22	19	16	13	11	375	839	40	19	17	14	12	9	7	5
175	391	60	22	19	16	13	10	8	5	375	839	60	15	13	10	8	5	3	2
175	391	80	18	16	13	10	7	4	2	375	839	80	13	10	8	5	3	1	0
175	391	100	16	13	10	8	5	2	0	375	839	100	11	9	6	4	2	0	0
175	391	120	15	12	9	6	3	1	0	375	839	120	10	7	5	3	1	0	0
175	391	140	13	10	7	4	2	0	0	375	839	140	9	6	4	2	0	0	0
200	447	40	25	23	21	18	15	12	9	400	894	40	18	16	14	11	9	7	4
200	447	60	20	18	15	12	10	7	4	400	894	60	15	12	10	7	5	3	1
200	447	80	17	15	12	9	6	4	2	400	894	80	12	10	7	5	3	1	0
200	447	100	15	13	10	7	4	2	0	400	894	100	11	8	6	3	2	0	0
200	447	120	14	11	8	5	3	1	0	400	894	120	10	7	5	2	1	0	0
200	447	140	12	10	7	4	2	0	0	400	894	140	9	6	4	2	0	0	0

NOTE: Detector time constant at a reference velocity of 5 ft/sec (1.5m/sec).
For SI units: 1 ft = 0.305 m; 1000 Btu/sec = 1055 kW.

Table B-3.2.4(s) Q_d, Threshold Fire Size at Response: 1000 Btu/sec t_g: 500 seconds to 1000 Btu/sec α: 0.004 Btu/sec³

τ	RTI	ΔT	4.0	8.0	12.0	16.0	20.0	24.0	28.0	τ	RTI	ΔT	4.0	8.0	12.0	16.0	20.0	24.0	28.0
					Ceiling Height (ft)										Ceiling Height (ft)				
					Installed Spacing of Detectors										Installed Spacing of Detectors				
25	56	40	62	54	48	42	37	31	27	225	503	40	31	29	26	23	20	17	14
25	56	60	49	41	35	29	24	19	15	225	503	60	25	22	19	16	13	10	7
25	56	80	41	33	27	22	17	12	8	225	503	80	21	18	15	12	9	6	3
25	56	100	35	28	22	17	12	8	4	225	503	100	19	16	13	9	6	3	1
25	56	120	31	24	18	13	9	4	0	225	503	120	17	14	10	7	4	2	0
25	56	140	28	21	16	11	6	2	0	225	503	140	15	12	9	6	3	0	0
50	112	40	53	48	43	38	33	29	24	250	559	40	30	28	25	22	19	16	13
50	112	60	42	37	32	27	22	18	14	250	559	60	24	21	18	15	12	9	7
50	112	80	35	30	25	20	16	11	7	250	559	80	20	18	15	11	8	6	3
50	112	100	31	25	20	16	11	7	3	250	559	100	18	15	12	9	6	3	1
50	112	120	27	22	17	12	8	4	0	250	559	120	16	13	10	7	4	1	0
50	112	140	25	19	14	10	6	2	0	250	559	140	14	11	8	5	2	0	0
75	168	40	47	43	39	35	31	26	22	275	615	40	29	27	24	21	18	15	12
75	168	60	38	33	29	25	20	16	12	275	615	60	23	20	18	15	12	9	6
75	168	80	32	27	23	19	14	10	6	275	615	80	20	17	14	11	8	5	3
75	168	100	28	23	19	14	10	6	3	275	615	100	17	14	11	8	5	3	1
75	168	120	25	20	16	11	7	3	0	275	615	120	15	12	9	6	4	1	0
75	168	140	22	18	13	9	5	1	0	275	615	140	14	11	8	5	2	0	0
100	224	40	43	40	36	32	28	24	20	300	671	40	28	25	23	20	17	14	11
100	224	60	34	31	27	23	19	15	11	300	671	60	22	20	17	14	11	8	5
100	224	80	29	25	21	17	13	9	6	300	671	80	19	16	13	10	7	5	2
100	224	100	25	21	17	13	9	6	2	300	671	100	16	14	11	8	5	2	1
100	224	120	23	19	15	11	7	3	0	300	671	120	15	12	9	6	3	1	0
100	224	140	21	16	12	8	5	1	0	300	671	140	13	10	7	5	2	0	0
125	280	40	39	37	33	30	26	22	19	325	727	40	27	25	22	19	16	13	11
125	280	60	32	28	25	21	17	14	10	325	727	60	21	19	16	13	10	8	5
125	280	80	27	23	20	16	12	9	5	325	727	80	18	16	13	10	7	4	2
125	280	100	24	20	16	12	9	5	2	325	727	100	16	13	10	7	5	2	0
125	280	120	21	17	14	10	6	3	0	325	727	120	14	11	8	6	3	1	0
125	280	140	19	15	11	8	4	1	0	325	727	140	13	10	7	4	2	0	0
150	335	40	37	34	31	28	24	21	17	350	783	40	26	24	21	18	15	13	10
150	335	60	30	27	23	20	16	13	9	350	783	60	21	18	15	13	10	7	5
150	335	80	25	22	18	15	11	8	5	350	783	80	18	15	12	9	7	4	2
150	335	100	22	19	15	12	8	5	2	350	783	100	15	13	10	7	4	2	0
150	335	120	20	16	13	9	6	2	0	350	783	120	14	11	8	5	3	1	0
150	335	140	18	14	11	7	4	1	0	350	783	140	12	10	7	4	2	0	0
175	391	40	35	32	29	26	23	19	16	375	839	40	25	23	20	18	15	12	9
175	391	60	28	25	22	18	15	12	8	375	839	60	20	18	15	12	9	7	4
175	391	80	24	20	17	14	10	7	4	375	839	80	17	14	12	9	6	4	2
175	391	100	21	18	14	11	7	4	1	375	839	100	15	12	9	7	4	2	0
175	391	120	19	15	12	8	5	2	0	375	839	120	13	11	8	5	3	1	0
175	391	140	17	13	10	7	3	1	0	375	839	140	12	9	6	4	1	0	0
200	447	40	33	30	28	24	21	18	15	400	894	40	24	22	20	17	14	11	9
200	447	60	26	24	20	17	14	11	8	400	894	60	19	17	14	12	9	6	4
200	447	80	22	19	16	13	10	7	4	400	894	80	16	14	11	8	6	3	1
200	447	100	20	17	13	10	7	4	1	400	894	100	14	12	9	6	4	2	0
200	447	120	18	14	11	8	5	2	0	400	894	120	13	10	7	5	2	1	0
200	447	140	16	13	9	6	3	1	0	400	894	140	12	9	6	4	1	0	0

NOTE: Detector time constant at a reference velocity of 5 ft/sec (1.5m/sec).
For SI units: 1 ft = 0.305 m; 1000 Btu/sec = 1055 kW.

Table B-3.2.4(t)   Q$_d$, Threshold Fire Size at Response: 1000 Btu/sec   t$_g$: 600 seconds to 1000 Btu/sec   α: 0.003 Btu/sec^3

τ	RTI	ΔT	4.0	8.0	12.0	16.0	20.0	24.0	28.0	τ	RTI	ΔT	4.0	8.0	12.0	16.0	20.0	24.0	28.0
			Installed Spacing of Detectors										Installed Spacing of Detectors						
25	56	40	65	56	50	43	38	33	28	225	503	40	34	32	29	26	22	19	16
25	56	60	51	43	36	30	25	20	15	225	503	60	27	25	21	18	15	12	8
25	56	80	42	34	28	23	18	13	8	225	503	80	23	20	17	14	10	7	4
25	56	100	36	29	23	17	13	8	4	225	503	100	20	17	14	11	7	4	1
25	56	120	32	25	19	14	9	5	1	225	503	120	18	15	12	8	5	2	0
25	56	140	29	21	16	11	6	2	0	225	503	140	17	13	10	6	3	1	0
50	112	40	56	51	45	40	35	30	26	250	559	40	33	30	27	24	21	18	15
50	112	60	45	39	33	28	23	19	14	250	559	60	26	23	20	17	14	11	8
50	112	80	37	31	26	21	16	12	8	250	559	80	22	19	16	13	10	7	4
50	112	100	33	26	21	16	12	7	3	250	559	100	19	16	13	10	7	4	1
50	112	120	29	23	18	13	8	4	0	250	559	120	17	14	11	8	5	2	0
50	112	140	26	20	15	10	6	2	0	250	559	140	16	13	9	6	3	1	0
75	168	40	50	46	42	37	32	28	24	275	615	40	31	29	26	23	20	17	14
75	168	60	40	35	31	26	22	17	13	275	615	60	25	23	19	16	13	10	7
75	168	80	34	29	24	20	15	11	7	275	615	80	21	18	15	12	9	6	3
75	168	100	30	25	20	15	11	7	3	275	615	100	19	16	13	9	6	3	1
75	168	120	26	21	17	12	8	4	0	275	615	120	17	14	11	7	4	2	0
75	168	140	24	19	14	10	5	2	0	275	615	140	15	12	9	6	3	0	0
100	224	40	46	43	38	34	30	26	22	300	671	40	30	28	25	22	19	16	13
100	224	60	37	33	28	24	20	16	12	300	671	60	24	22	19	16	13	10	7
100	224	80	31	27	23	18	14	10	6	300	671	80	21	18	15	12	9	6	3
100	224	100	27	23	18	14	10	6	3	300	671	100	18	15	12	9	6	3	1
100	224	120	24	20	15	11	7	3	0	300	671	120	16	13	10	7	4	1	0
100	224	140	22	17	13	9	5	1	0	300	671	140	15	12	8	5	3	0	0
125	280	40	43	40	36	32	28	24	21	325	727	40	29	27	24	21	18	15	13
125	280	60	34	31	27	23	19	15	11	325	727	60	23	21	18	15	12	9	6
125	280	80	29	25	21	17	13	10	6	325	727	80	20	17	14	11	8	5	3
125	280	100	25	21	17	13	10	6	2	325	727	100	17	15	12	8	6	3	1
125	280	120	23	19	15	11	7	3	0	325	727	120	16	13	10	7	4	1	0
125	280	140	21	16	12	8	5	1	0	325	727	140	14	11	8	5	2	0	0
150	335	40	40	37	34	30	26	23	19	350	783	40	28	26	23	21	18	15	12
150	335	60	32	29	25	21	18	14	11	350	783	60	23	20	17	14	11	9	6
150	335	80	27	24	20	16	12	9	5	350	783	80	19	17	14	11	8	5	3
150	335	100	24	20	16	13	9	5	2	350	783	100	17	14	11	8	5	3	1
150	335	120	21	17	14	10	6	3	0	350	783	120	15	12	9	6	3	1	0
150	335	140	19	15	12	8	4	1	0	350	783	140	14	11	8	5	2	0	0
175	391	40	38	35	32	28	25	21	18	375	839	40	28	25	23	20	17	14	11
175	391	60	30	27	24	20	17	13	10	375	839	60	22	19	17	14	11	8	5
175	391	80	26	22	19	15	12	8	5	375	839	80	19	16	13	10	7	5	2
175	391	100	22	19	15	12	8	5	2	375	839	100	16	14	11	8	5	2	1
175	391	120	20	17	13	9	6	3	0	375	839	120	15	12	9	6	3	1	0
175	391	140	18	15	11	7	4	1	0	375	839	140	13	10	7	5	2	0	0
200	447	40	36	33	30	27	24	20	17	400	894	40	27	25	22	19	16	13	11
200	447	60	29	26	22	19	16	12	9	400	894	60	21	19	16	13	10	8	5
200	447	80	24	21	18	14	11	8	4	400	894	80	18	15	13	10	7	4	2
200	447	100	21	18	15	11	8	4	2	400	894	100	16	13	10	7	5	2	1
200	447	120	19	16	12	9	5	2	0	400	894	120	14	11	8	6	3	1	0
200	447	140	17	14	10	7	4	1	0	400	894	140	13	10	7	4	2	0	0

NOTE: Detector time constant at a reference velocity of 5 ft/sec (1.5m/sec).
For SI units: 1 ft = 0.305 m; 1000 Btu/sec = 1055 kW.

Table B-3.2.4(u) Q_d, Threshold Fire Size at Response: 1000 Btu/sec t_g: 50 seconds to 1000 Btu/sec α: 0.400 Btu/sec^3

τ	RTI	ΔT	Ceiling Height (ft)						
			4.0	8.0	12.0	16.0	20.0	24.0	28.0
			Installed Spacing of Detectors						
25	56	40	35	33	31	28	25	21	18
25	56	60	30	27	24	21	18	15	11
25	56	80	26	23	20	17	14	10	7
25	56	100	23	21	17	14	11	7	4
25	56	120	21	18	15	12	8	5	2
25	56	140	20	17	13	10	7	3	1
50	112	40	28	26	23	21	18	15	12
50	112	60	23	21	18	15	13	10	7
50	112	80	20	18	15	12	9	6	4
50	112	100	18	15	13	10	7	4	2
50	112	120	16	14	11	8	5	3	1
50	112	140	15	12	10	7	4	2	0
75	168	40	24	22	19	17	14	11	9
75	168	60	20	17	15	12	9	7	4
75	168	80	17	15	12	9	7	4	2
75	168	100	15	13	10	7	5	3	1
75	168	120	14	11	8	6	3	1	0
75	168	140	13	10	7	5	2	1	0
100	224	40	21	19	17	14	11	9	6
100	224	60	17	15	13	10	7	5	3
100	224	80	15	13	10	7	5	3	1
100	224	100	13	11	8	6	3	2	0
100	224	120	12	10	7	4	2	1	0
100	224	140	11	8	6	3	2	0	0
125	280	40	19	17	15	12	10	7	5
125	280	60	16	13	11	8	6	4	2
125	280	80	13	11	9	6	4	2	1
125	280	100	12	10	7	5	2	1	0
125	280	120	11	8	6	3	2	0	0
125	280	140	10	7	5	3	1	0	0
150	335	40	18	16	13	11	8	6	4
150	335	60	14	12	10	7	5	3	1
150	335	80	12	10	7	5	3	1	0
150	335	100	11	8	6	4	2	0	0
150	335	120	10	7	5	3	1	0	0
150	335	140	9	6	4	2	0	0	0
175	391	40	16	14	12	9	7	5	3
175	391	60	13	11	9	6	4	2	1
175	391	80	11	9	7	4	2	1	0
175	391	100	10	8	5	3	1	0	0
175	391	120	9	7	4	2	1	0	0
175	391	140	8	6	3	2	0	0	0
200	447	40	15	13	11	8	6	4	2
200	447	60	12	10	8	5	3	2	0
200	447	80	11	8	6	4	2	1	0
200	447	100	9	7	5	3	1	0	0
200	447	120	8	6	4	2	0	0	0
200	447	140	8	5	3	1	0	0	0
225	503	40	15	12	10	8	5	3	2
225	503	60	12	9	7	5	3	1	0
225	503	80	10	8	5	3	2	0	0
225	503	100	9	6	4	2	1	0	0
225	503	120	8	5	3	1	0	0	0
225	503	140	7	5	3	1	0	0	0
250	559	40	14	12	9	7	5	3	2
250	559	60	11	9	6	4	2	1	0
250	559	80	9	7	5	3	1	0	0
250	559	100	8	6	4	2	0	0	0
250	559	120	7	5	3	1	0	0	0
250	559	140	7	4	2	1	0	0	0
275	615	40	13	11	9	6	4	2	1
275	615	60	10	8	6	4	2	1	0
275	615	80	9	7	4	2	1	0	0
275	615	100	8	5	3	2	0	0	0
275	615	120	7	5	2	1	0	0	0
275	615	140	6	4	2	0	0	0	0
300	671	40	13	10	8	6	4	2	1
300	671	60	10	8	5	3	2	0	0
300	671	80	8	6	4	2	1	0	0
300	671	100	7	5	3	1	0	0	0
300	671	120	7	4	2	1	0	0	0
300	671	140	6	4	2	0	0	0	0
325	727	40	12	10	8	5	3	2	1
325	727	60	10	7	5	3	1	0	0
325	727	80	8	6	4	2	0	0	0
325	727	100	7	5	3	1	0	0	0
325	727	120	6	4	2	0	0	0	0
325	727	140	6	3	1	0	0	0	0
350	783	40	12	9	7	5	3	2	0
350	783	60	9	7	5	3	1	0	0
350	783	80	8	5	3	2	0	0	0
350	783	100	7	4	2	1	0	0	0
350	783	120	6	4	2	0	0	0	0
350	783	140	5	3	1	0	0	0	0
375	839	40	11	9	7	4	3	1	0
375	839	60	9	7	4	2	1	0	0
375	839	80	7	5	3	1	0	0	0
375	839	100	6	4	2	0	0	0	0
375	839	120	6	3	2	0	0	0	0
375	839	140	5	3	1	0	0	0	0
400	894	40	11	9	6	4	2	1	0
400	894	60	9	6	4	2	1	0	0
400	894	80	7	5	3	1	0	0	0
400	894	100	6	4	2	0	0	0	0
400	894	120	5	3	1	0	0	0	0
400	894	140	5	3	1	0	0	0	0

NOTE: Detector time constant at a reference velocity of 5 ft/sec (1.5m/sec).
For SI units: 1 ft = 0.305 m; 1000 Btu/sec = 1055 kW.

Table B-3.2.4(v)   Q$_d$, Threshold Fire Size at Response: 2000 Btu/sec   t$_g$: 150 seconds to 1000 Btu/sec α: 0.044 Btu/sec^3

τ	RTI	ΔT	Ceiling Height (ft)							τ	RTI	ΔT	Ceiling Height (ft)						
			4.0	8.0	12.0	16.0	20.0	24.0	28.0				4.0	8.0	12.0	16.0	20.0	24.0	28.0
			Installed Spacing of Detectors										Installed Spacing of Detectors						
25	56	40	60	57	53	49	44	40	36	225	503	40	27	26	23	21	18	15	13
25	56	60	50	46	41	37	32	28	23	225	503	60	22	20	18	15	12	10	7
25	56	80	43	38	34	29	25	20	16	225	503	80	19	17	14	11	9	6	4
25	56	100	38	33	28	24	19	15	11	225	503	100	17	14	12	9	7	4	2
25	56	120	34	29	25	20	15	11	7	225	503	120	15	13	10	7	5	3	1
25	56	140	31	26	21	17	13	8	4	225	503	140	14	11	9	6	4	2	0
50	112	40	49	47	44	40	37	33	30	250	559	40	26	24	22	19	17	14	12
50	112	60	40	38	34	31	27	23	19	250	559	60	21	19	16	14	11	9	6
50	112	80	35	32	28	24	21	17	13	250	559	80	18	16	13	11	8	6	4
50	112	100	31	28	24	20	16	12	9	250	559	100	16	14	11	8	6	4	2
50	112	120	28	24	21	17	13	9	6	250	559	120	14	12	9	7	4	2	1
50	112	140	26	22	18	14	10	7	3	250	559	140	13	11	8	5	3	1	0
75	168	40	43	41	38	35	32	28	25	275	615	40	25	23	21	18	16	13	11
75	168	60	35	33	30	26	23	20	16	275	615	60	20	18	16	13	11	8	6
75	168	80	30	28	24	21	18	14	11	275	615	80	17	15	13	10	7	5	3
75	168	100	27	24	21	17	14	10	7	275	615	100	15	13	10	8	5	3	2
75	168	120	24	21	18	14	11	8	4	275	615	120	14	11	9	6	4	2	1
75	168	140	22	19	16	12	9	5	2	275	615	140	12	10	7	5	3	1	0
100	224	40	38	36	34	31	28	25	22	300	671	40	24	22	20	17	15	12	10
100	224	60	31	29	26	23	20	17	14	300	671	60	19	17	15	12	10	7	5
100	224	80	27	25	22	18	15	12	9	300	671	80	17	14	12	9	7	5	3
100	224	100	24	21	18	15	12	9	6	300	671	100	15	12	10	7	5	3	1
100	224	120	22	19	16	13	9	6	3	300	671	120	13	11	8	6	3	2	0
100	224	140	20	17	14	11	7	4	2	300	671	140	12	10	7	5	2	1	0
125	280	40	35	33	31	28	25	22	19	325	727	40	23	21	19	16	14	11	9
125	280	60	29	27	24	21	18	15	12	325	727	60	19	17	14	12	9	7	5
125	280	80	25	22	19	16	13	11	8	325	727	80	16	14	11	9	6	4	2
125	280	100	22	19	16	13	10	7	5	325	727	100	14	12	9	7	4	2	1
125	280	120	20	17	14	11	8	5	3	325	727	120	13	10	8	5	3	1	0
125	280	140	18	15	12	9	6	4	1	325	727	140	11	9	7	4	2	1	0
150	335	40	32	31	28	26	23	20	17	350	783	40	22	20	18	16	13	11	8
150	335	60	27	24	22	19	16	13	10	350	783	60	18	16	13	11	9	6	4
150	335	80	23	20	18	15	12	9	6	350	783	80	15	13	11	8	6	4	2
150	335	100	20	18	15	12	9	6	4	350	783	100	13	11	9	6	4	2	1
150	335	120	18	16	13	10	7	4	2	350	783	120	12	10	7	5	3	1	0
150	335	140	17	14	11	8	5	3	1	350	783	140	11	9	6	4	2	1	0
175	391	40	30	29	26	24	21	18	15	375	839	40	22	20	17	15	12	10	8
175	391	60	25	23	20	17	15	12	9	375	839	60	17	15	13	10	8	6	4
175	391	80	21	19	16	14	11	8	6	375	839	80	15	13	10	8	5	3	2
175	391	100	19	16	14	11	8	6	3	375	839	100	13	11	8	6	4	2	1
175	391	120	17	15	12	9	6	4	2	375	839	120	12	9	7	5	3	1	0
175	391	140	16	13	10	7	5	2	1	375	839	140	11	8	6	3	2	0	0
200	447	40	29	27	25	22	19	17	14	400	894	40	21	19	17	14	12	9	7
200	447	60	23	21	19	16	13	11	8	400	894	60	17	15	12	10	8	5	3
200	447	80	20	18	15	12	10	7	5	400	894	80	14	12	10	7	5	3	2
200	447	100	18	15	13	10	7	5	3	400	894	100	13	10	8	5	3	2	1
200	447	120	16	14	11	8	5	3	1	400	894	120	11	9	6	4	2	1	0
200	447	140	15	12	9	7	4	2	1	400	894	140	10	8	5	3	2	0	0

NOTE: Detector time constant at a reference velocity of 5 ft/sec (1.5m/sec).
For SI units: 1 ft = 0.305 m; 1000 Btu/sec = 1055 kW.

Table B-3.2.4(w) Q_d, Threshold Fire Size at Response: 2000 Btu/sec t_g: 300 seconds to 1000 Btu/sec α: 0.011 Btu/sec^3

τ	RTI	ΔT	\multicolumn{7}{c}{Ceiling Height (ft)}							τ	RTI	ΔT	\multicolumn{7}{c}{Ceiling Height (ft)}
			4.0	8.0	12.0	16.0	20.0	24.0	28.0				4.0	8.0	12.0	16.0	20.0	24.0	28.0
			\multicolumn{7}{c}{Installed Spacing of Detectors}										\multicolumn{7}{c}{Installed Spacing of Detectors}
25	56	40	79	73	66	60	55	49	44	225	503	40	39	37	35	32	29	26	23
25	56	60	64	56	50	44	39	33	29	225	503	60	32	30	27	24	21	18	15
25	56	80	54	46	40	34	29	24	20	225	503	80	27	25	22	19	16	13	10
25	56	100	47	40	33	28	23	18	14	225	503	100	24	21	18	15	12	9	6
25	56	120	42	35	29	23	18	14	9	225	503	120	22	19	16	13	10	7	4
25	56	140	38	31	25	20	15	10	6	225	503	140	20	17	14	11	8	5	2
50	112	40	67	63	58	54	49	44	40	250	559	40	37	36	33	30	28	25	22
50	112	60	54	49	45	40	35	30	26	250	559	60	30	28	25	23	19	17	14
50	112	80	46	41	36	31	27	22	18	250	559	80	26	24	21	18	15	12	9
50	112	100	41	35	30	26	21	17	12	250	559	100	23	20	17	14	11	8	6
50	112	120	36	31	26	21	17	13	8	250	559	120	21	18	15	12	9	6	3
50	112	140	33	28	23	18	14	9	5	250	559	140	19	16	13	10	7	4	2
75	168	40	59	56	52	48	44	40	36	275	615	40	36	34	32	29	26	23	20
75	168	60	48	44	40	36	32	28	24	275	615	60	29	27	24	21	18	16	13
75	168	80	41	37	33	29	24	20	16	275	615	80	25	23	20	17	14	11	8
75	168	100	36	32	28	23	19	15	11	275	615	100	22	19	17	14	11	8	5
75	168	120	33	28	24	20	15	11	7	275	615	120	20	17	14	11	8	6	3
75	168	140	30	25	21	17	12	9	5	275	615	140	18	15	12	9	7	4	2
100	224	40	53	51	48	44	41	37	33	300	671	40	35	33	30	28	25	22	19
100	224	60	43	40	37	33	29	25	22	300	671	60	28	26	23	20	18	15	13
100	224	80	37	34	30	26	22	19	15	300	671	80	24	22	19	16	13	10	8
100	224	100	33	29	25	22	18	14	10	300	671	100	21	19	16	13	10	7	5
100	224	120	30	26	22	18	14	10	7	300	671	120	19	16	14	11	8	5	3
100	224	140	27	23	19	15	11	8	4	300	671	140	17	15	12	9	6	4	1
125	280	40	49	47	44	41	37	34	31	325	727	40	33	32	29	27	24	21	18
125	280	60	40	37	34	31	27	23	20	325	727	60	27	25	22	20	17	14	11
125	280	80	34	31	28	24	21	17	14	325	727	80	23	21	18	15	12	10	7
125	280	100	30	27	23	20	16	13	9	325	727	100	20	18	15	12	10	7	4
125	280	120	27	24	20	17	13	9	6	325	727	120	18	16	13	10	7	5	2
125	280	140	25	22	18	14	11	7	4	325	727	140	17	14	11	8	6	3	1
150	335	40	46	44	41	38	35	32	28	350	783	40	32	31	28	26	23	20	17
150	335	60	37	35	32	28	25	22	18	350	783	60	26	24	21	19	16	13	11
150	335	80	32	29	26	23	19	16	12	350	783	80	22	20	17	15	12	9	7
150	335	100	28	25	22	19	15	12	8	350	783	100	20	17	15	12	9	6	4
150	335	120	26	22	19	15	12	9	5	350	783	120	18	15	12	10	7	4	2
150	335	140	23	20	17	13	10	6	3	350	783	140	16	14	11	8	5	3	1
175	391	40	43	41	39	36	33	29	26	375	839	40	31	30	27	25	22	19	16
175	391	60	35	33	30	27	23	20	17	375	839	60	25	23	21	18	15	13	10
175	391	80	30	28	24	21	18	15	11	375	839	80	22	19	17	14	11	9	6
175	391	100	27	24	21	17	14	11	8	375	839	100	19	17	14	11	9	6	4
175	391	120	24	21	18	14	11	8	5	375	839	120	17	15	12	9	6	4	2
175	391	140	22	19	16	12	9	6	3	375	839	140	16	13	10	8	5	3	1
200	447	40	41	39	37	34	31	28	25	400	894	40	30	29	26	24	21	18	16
200	447	60	33	31	28	25	22	19	16	400	894	60	25	23	20	17	15	12	9
200	447	80	29	26	23	20	17	14	10	400	894	80	21	19	16	13	11	8	5
200	447	100	25	22	19	16	13	10	7	400	894	100	19	16	13	11	8	6	3
200	447	120	23	20	17	14	10	7	4	400	894	120	17	14	11	9	6	4	2
200	447	140	21	18	15	11	8	5	2	400	894	140	15	13	10	7	5	2	1

NOTE: Detector time constant at a reference velocity of 5 ft/sec (1.5m/sec).
For SI units: 1 ft = 0.305 m; 1000 Btu/sec = 1055 kW.

Table B-3.2.4(x) Q_d, Threshold Fire Size at Response: 2000 Btu/sec t_g: 500 seconds to 1000 Btu/sec α: 0.004 Btu/sec^3

τ	RTI	ΔT	Ceiling Height (ft)							τ	RTI	ΔT	Ceiling Height (ft)						
			4.0	8.0	12.0	16.0	20.0	24.0	28.0				4.0	8.0	12.0	16.0	20.0	24.0	28.0
			Installed Spacing of Detectors										Installed Spacing of Detectors						
25	56	40	92	82	74	67	60	54	49	225	503	40	50	48	45	42	38	35	31
25	56	60	72	62	55	48	42	36	31	225	503	60	40	38	34	31	27	24	20
25	56	80	61	51	43	37	32	26	22	225	503	80	35	31	28	24	21	17	14
25	56	100	52	43	36	30	25	20	15	225	503	100	30	27	24	20	16	13	9
25	56	120	46	37	31	25	20	15	10	225	503	120	27	24	20	17	13	10	6
25	56	140	42	33	27	21	16	11	7	225	503	140	25	21	18	14	11	7	4
50	112	40	81	74	68	62	56	51	46	250	559	40	48	46	43	40	37	33	30
50	112	60	64	57	51	45	40	35	30	250	559	60	39	36	33	30	26	23	19
50	112	80	54	47	41	35	30	25	20	250	559	80	33	30	27	23	20	17	13
50	112	100	47	40	34	29	23	19	14	250	559	100	29	26	23	19	16	12	9
50	112	120	42	35	29	24	19	14	10	250	559	120	26	23	20	16	13	9	6
50	112	140	38	31	25	20	15	11	6	250	559	140	24	21	17	14	10	7	4
75	168	40	73	68	63	58	53	48	43	275	615	40	46	44	41	38	35	32	29
75	168	60	58	53	47	42	37	33	28	275	615	60	37	35	32	28	25	22	19
75	168	80	49	44	38	33	28	24	19	275	615	80	32	29	26	22	19	16	13
75	168	100	43	37	32	27	22	18	13	275	615	100	28	25	22	18	15	12	8
75	168	120	39	33	27	22	18	13	9	275	615	120	25	22	19	15	12	9	5
75	168	140	35	29	24	19	14	10	6	275	615	140	23	20	16	13	10	6	3
100	224	40	67	63	58	54	50	45	41	300	671	40	45	43	40	37	34	31	27
100	224	60	54	49	45	40	35	31	27	300	671	60	36	34	31	27	24	21	18
100	224	80	46	41	36	31	27	23	18	300	671	80	31	28	25	22	18	15	12
100	224	100	40	35	30	26	21	17	13	300	671	100	27	24	21	18	14	11	8
100	224	120	36	31	26	21	17	13	9	300	671	120	24	21	18	15	11	8	5
100	224	140	33	27	23	18	14	10	6	300	671	140	22	19	16	13	9	6	3
125	280	40	62	59	55	51	47	43	38	325	727	40	43	41	39	36	33	30	26
125	280	60	50	46	42	38	33	29	25	325	727	60	35	32	30	26	23	20	17
125	280	80	43	38	34	30	25	21	17	325	727	80	30	27	24	21	18	14	11
125	280	100	37	33	29	24	20	16	12	325	727	100	26	23	20	17	14	11	8
125	280	120	34	29	25	20	16	12	8	325	727	120	24	21	17	14	11	8	5
125	280	140	31	26	22	17	13	9	5	325	727	140	22	19	15	12	9	6	3
150	335	40	58	55	52	48	44	40	36	350	783	40	42	40	37	35	31	28	25
150	335	60	47	44	40	36	32	28	24	350	783	60	34	31	29	26	22	19	16
150	335	80	40	36	32	28	24	20	16	350	783	80	29	26	23	20	17	14	11
150	335	100	35	31	27	23	19	15	11	350	783	100	25	23	19	16	13	10	7
150	335	120	32	27	23	19	15	11	8	350	783	120	23	20	17	14	11	7	5
150	335	140	29	25	20	16	12	9	5	350	783	140	21	18	15	12	8	5	3
175	391	40	55	53	49	46	42	38	35	375	839	40	41	39	36	33	30	27	24
175	391	60	44	41	38	34	30	26	23	375	839	60	33	31	28	25	22	19.	16
175	391	80	38	34	31	27	23	19	15	375	839	80	28	25	22	19	16	13	10
175	391	100	33	30	26	22	18	14	11	375	839	100	25	22	19	16	13	10	7
175	391	120	30	26	22	18	15	11	7	375	839	120	22	19	16	13	10	7	4
175	391	140	27	23	20	16	12	8	4	375	839	140	20	17	14	11	8	5	2
200	447	40	52	50	47	44	40	36	33	400	894	40	40	38	35	32	29	27	24
200	447	60	42	39	36	32	29	25	21	400	894	60	32	30	27	24	21	18	15
200	447	80	36	33	29	26	22	18	15	400	894	80	27	25	22	19	16	13	10
200	447	100	32	28	25	21	17	14	10	400	894	100	24	21	18	15	12	9	6
200	447	120	29	25	21	18	14	10	7	400	894	120	22	19	16	13	10	7	4
200	447	140	26	22	19	15	11	8	4	400	894	140	20	17	14	11	8	5	2

NOTE: Detector time constant at a reference velocity of 5 ft/sec (1.5m/sec).
For SI units: 1 ft = 0.305 m; 1000 Btu/sec = 1055 kW.

Table B-3.2.4(y) Q_d, Threshold Fire Size at Response: 2000 Btu/sec t_g: 600 seconds to 1000 Btu/sec α: 0.003 Btu/sec³

τ	RTI	ΔT	Ceiling Height (ft)						
			4.0	8.0	12.0	16.0	20.0	24.0	28.0
			Installed Spacing of Detectors						
25	56	40	96	85	78	68	62	56	50
25	56	60	75	64	56	49	43	37	32
25	56	80	63	52	44	38	32	27	22
25	56	100	54	44	37	31	25	20	15
25	56	120	48	38	31	25	20	15	11
25	56	140	43	34	27	21	16	12	7
50	112	40	86	78	71	64	58	53	48
50	112	60	68	60	53	47	41	36	31
50	112	80	57	49	42	36	31	26	21
50	112	100	49	41	35	29	24	19	15
50	112	120	44	36	30	24	19	15	10
50	112	140	40	32	26	21	16	11	7
75	168	40	78	72	66	61	55	50	45
75	168	60	62	56	50	44	39	34	29
75	168	80	52	46	40	34	30	25	20
75	168	100	46	39	33	28	23	19	14
75	168	120	41	34	28	23	19	14	10
75	168	140	37	30	25	20	15	11	6
100	224	40	72	67	62	57	52	48	43
100	224	60	57	52	47	42	37	32	28
100	224	80	49	43	38	33	28	24	19
100	224	100	43	37	32	27	22	18	13
100	224	120	38	32	27	22	18	13	9
100	224	140	35	29	24	19	14	10	6
125	280	40	67	63	59	54	50	45	41
125	280	60	54	49	45	40	35	31	27
125	280	80	46	41	36	31	27	23	18
125	280	100	40	35	30	26	21	17	13
125	280	120	36	31	26	21	17	13	9
125	280	140	33	27	23	18	14	10	6
150	335	40	63	60	56	52	47	43	39
150	335	60	51	47	42	38	34	30	26
150	335	80	43	39	34	30	26	22	18
150	335	100	38	33	29	25	20	16	12
150	335	120	34	29	25	21	16	12	8
150	335	140	31	26	22	17	13	9	5
175	391	40	60	57	53	49	45	41	37
175	391	60	48	44	41	36	32	28	24
175	391	80	41	37	33	29	25	21	17
175	391	100	36	32	28	24	19	15	12
175	391	120	32	28	24	20	16	12	8
175	391	140	29	25	21	17	13	9	5
200	447	40	57	54	51	47	43	40	36
200	447	60	46	43	39	35	31	27	23
200	447	80	39	35	32	28	24	20	16
200	447	100	34	30	26	23	19	15	11
200	447	120	31	27	23	19	15	11	7
200	447	140	28	24	20	16	12	8	5

τ	RTI	ΔT	Ceiling Height (ft)						
			4.0	8.0	12.0	16.0	20.0	24.0	28.0
			Installed Spacing of Detectors						
225	503	40	54	52	49	45	42	38	34
225	503	60	44	41	37	34	30	26	22
225	503	80	37	34	30	26	23	19	15
225	503	100	33	29	25	22	18	14	11
225	503	120	30	26	22	18	14	11	7
225	503	140	27	23	19	15	12	8	4
250	559	40	52	50	47	43	40	36	33
250	559	60	42	39	36	32	29	25	21
250	559	80	36	33	29	25	22	18	15
250	559	100	32	28	24	21	17	14	10
250	559	120	28	25	21	17	14	10	7
250	559	140	26	22	19	15	11	8	4
275	615	40	50	48	45	42	39	35	32
275	615	60	41	38	35	31	28	24	21
275	615	80	35	32	28	25	21	18	14
275	615	100	30	27	24	20	17	13	10
275	615	120	27	24	20	17	13	10	6
275	615	140	25	22	18	14	11	7	4
300	671	40	49	47	44	40	37	34	30
300	671	60	39	37	33	30	27	23	20
300	671	80	33	31	27	24	20	17	13
300	671	100	29	26	23	19	16	12	9
300	671	120	26	23	20	16	13	9	6
300	671	140	24	21	17	14	10	7	4
325	727	40	47	45	42	39	36	33	29
325	727	60	38	35	32	29	26	22	19
325	727	80	32	30	26	23	20	16	13
325	727	100	29	25	22	19	15	12	9
325	727	120	26	22	19	16	12	9	6
325	727	140	23	20	17	13	10	7	3
350	783	40	46	44	41	38	35	32	28
350	783	60	37	34	31	28	25	22	18
350	783	80	31	29	25	22	19	16	12
350	783	100	28	25	21	18	15	12	8
350	783	120	25	22	18	15	12	9	5
350	783	140	23	20	16	13	9	6	3
375	839	40	44	42	40	37	34	31	27
375	839	60	36	33	30	27	24	21	18
375	839	80	31	28	25	21	18	15	12
375	839	100	27	24	21	18	14	11	8
375	839	120	24	21	18	15	11	8	5
375	839	140	22	19	16	12	9	6	3
400	894	40	43	41	39	36	33	30	26
400	894	60	35	32	30	26	23	20	17
400	894	80	30	27	24	21	18	14	11
400	894	100	26	23	20	17	14	11	8
400	894	120	23	21	17	14	11	8	5
400	894	140	21	18	15	12	9	6	3

NOTE: Detector time constant at a reference velocity of 5 ft/sec (1.5m/sec).
For SI units: 1 ft = 0.305 m; 1000 Btu/sec = 1055 kW.

B-3.2.5 Installed spacings listed as zero (0) in the tables indicate that the detector chosen will not respond within the design objectives.

B-3.2.6 Example.

Input:

Ceiling height: 8 ft (2.4 m)

Detector type: Fixed temperature 135°F (57°C)

Listed spacing: 30 ft (9.1 m)

Fire:

Q_d: 500 Btu/sec (527 kW)

Fire growth rate: slow

t_g: 600 sec

α: 0.003 Btu/sec^3

Environmental conditions:

T_o: 55°F (12.8°C)

Required installed spacing:

From Table B-3.2.2, the detector time constant (τ_o) is 80 seconds.

$(RTI = 80 \sqrt{5} = 180 \text{ ft}^{1/2} \text{ sec}^{1/2})$

$\Delta T = T_s - T_o = 135 - 55 = 80°F$

From Table B-3.2.4(j):

For τ_o = 75 sec − spacing = 17 ft (5.18 m)

For τ_o = 100 sec − spacing = 16 ft (4.9 m)

By interpolation:

Spacing = 17 − [(17 − 16/80 − 75/100 − 75)] = 16.8, rounded to 17.0 ft.

For SI units: 1 ft = 0.305 m.

NOTE: Interpolation for τ_o = 80 seconds was not required, but was included for demonstration. Where the ceiling height is 16 ft (4.9 m), the required spacing would be 8.8 ft (2.68 m). Using the detector in the above example, at a ceiling height of 28 ft (8.53 m), no practical spacing would ensure detection of the fire at the threshold fire size of 500 Btu/sec (527 kW). A more sensitive detector would need to be used. Alternatively, the design objectives could be changed to accept a larger fire. These results clearly illustrate the need to consider ceiling height in the design of a detection system.

B-3.3 Rate-of-Rise Heat Detector Spacing.

B-3.3.1 Tables B-3.3.2(a) and B-3.3.2(b) are to be used to determine the installed spacing of rate-of-rise heat detectors. The analytical basis for the tables is presented in Section B-6. This section shows how the tables are to be used.

B-3.3.2 Installed Spacings.

(a) Table B-3.3.2(a) provides installed spacings for rate-of-rise heat detectors required to achieve detection for a specific threshold fire size, fire growth rate, and ceiling height. This table may be permitted to be used directly to determine installed spacings for 50-ft (15.2-m) listed spacing detectors.

(b) Tables B-3.3.2(a) and B-3.3.2(b) use the following values for t_g:

Fast fire growth rate, t_g = 150 sec

Medium fire growth rate, t_g = 300 sec

Slow fire growth rate, t_g = 600 sec.

B-3.3.3 For rate-of-rise heat detectors with a listed spacing of other than 50 ft (15.2 m), installed spacing obtained from Table B-3.3.2(a) is to be multiplied by the modifier shown in Table B-3.3.2(b) for the appropriate listed spacing and fire growth rate. This takes into account the difference in sensitivity between the detector and a 50-ft (15.2-m) listed detector.

B-3.3.4 Having determined the threshold fire size (*see B-2.2.2*), the fire growth rate (*see B-2.2.3*), the detector's listed spacing, and the ceiling height, Table B-3.3.2(a) is used to determine the correct spacing for 50-ft (15.2-m) listed detectors. Table B-3.3.2(b) is used to determine the spacing modifier. The required installed spacing is determined by multiplying the correct spacing by the spacing modifier.

B-3.3.5 Example.

Input:

Ceiling height: 12 ft (3.7 m)

Detector type: Combination rate-of-rise, fixed temperature, 30-ft (9.1-m) listed spacing

Q_d: 500 Btu/sec

Fire growth rate: medium

Spacing:

From Table B-3.3.2(a), installed spacing = 18 ft (5.5 m)

From Table B-3.3.2(b), spacing modifier = 0.86

I. Input Data for Design and Analysis (*B-2*)

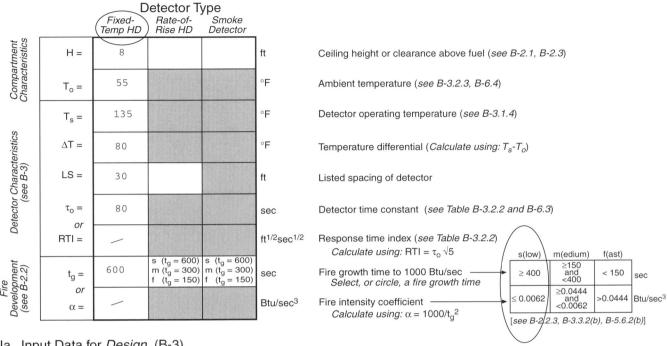

IIa. Input Data for *Design* (B-3)

IIa.1 Establish Design Objectives

> Determine the size of the fire at which detector response is desired using either 1 or 2:
> 1. *Select* Q_d = ___500___ Btu/sec   [see B-2.2.4, Table B-2.2.2.1(a), (b), or B-2.2.2.3], or
>
> 2. *Calculate* Q_d using the time after ignition at which detector response is desired, t_d, using: $Q_d = \alpha t_d^2$ = _____ Btu/sec

IIa.2 Calculate Detector Response

Fixed-Temperature HD (B-3.2)	*Rate-of-Rise HD (B-3.3)*	*Smoke Detector (B-5)*
1. Fill in the variables: Q_d = ___500___ Btu/sec t_g = ___600___ sec, or α = _____ Btu/sec³ τ_o = ___80___ sec, or RTI = _____ ft^{1/2} sec^{1/2} ΔT = ___80___ °F and H = ___8___ ft 2. Using Q_d and t_g (or α), select the appropriate design table (a) through (y) from Table B-3.2.4: __B-3.2.4__ (j) 3. Using τ_o (or RTI), ΔT, and H, determine the installed spacing: (___17___ ft)	1. Fill in the variables: Fire growth (s,m,f) = _____ H = _____ ft Q_d = _____ Btu/sec 2. Select installed spacing from Table B-3.3.2(a): _____ ft 3. Select spacing modifier from Table B-3.3.2(b): x _____ 4. Calculate installed spacing: _____ ft	1. Fill in the variables: Fire growth (s,m,f) = _____ H = _____ ft Q_d = _____ Btu/sec 2. Using fire growth (s, m, or f), select Figure B-5.5.1(a), (b), or (c): _____ 3. Using H and Q_d, determine the installed spacing: _____ ft

IIb. Input Data for *Analysis* of an Existing Heat Detection System (B-4)

> 1. Fill in the variables:
>   S = _____ ft (installed spacing of the existing heat detector)
> t_g = _____ sec, or α = _____ Btu/sec³
> τ_o = _____ sec, or RTI = _____ ft^{1/2}sec^{1/2}
> ΔT = _____ °F and H = _____ ft
> 2. Using S and t_g (or α), select an analysis table (a) - (nn) from Table B-4: _____
> 3. Using τ_o (or RTI), ΔT, and H, determine the fire size at detector response Q_d = _____ Btu/sec
> 4. Calculate the time to detector response using: $t_d = \sqrt{\dfrac{Q_d}{\alpha}} = \sqrt{\dfrac{\text{Btu/sec}}{\text{Btu/sec}^3}}$ = _____ sec

Figure B-3.2.6 *Completed worksheet for B-3.2.6, Example.*

Table B-3.3.2(a)   Installed Spacings for Rate-of-Rise Heat Detectors (Threshold Fire Size and Growth Rate)

Ceiling Height (ft)	Q_d = 1000 Btu/sec			Q_d = 750 Btu/sec			Q_d = 500 Btu/sec			Q_d = 250 Btu/sec			Q_d = 100 Btu/sec		
	s	m	f	s	m	f	s	m	f	s	m	f	s	m	f
4	28	32	32	26	28	27	22	24	23	16	17	16	11	11	10
5	27	31	31	25	27	27	21	23	22	15	16	15	10	10	9
6	26	30	31	24	26	27	20	22	22	15	15	15	9	9	9
7	25	29	30	23	26	26	19	21	21	14	14	14	9	9	8
8	24	29	30	22	25	26	18	21	21	13	13	14	8	8	8
9	23	28	29	21	24	25	17	20	20	12	13	13	7	7	7
10	22	27	29	20	23	25	16	19	20	12	12	13	7	7	7
11	21	27	28	18	23	24	15	19	19	11	12	12	6	6	6
12	20	26	26	17	22	24	15	18	19	10	11	12	5	5	5
13	19	25	27	16	22	23	14	18	18	9	11	11	5	5	5
14	18	24	27	15	21	22	13	17	18	9	10	11		4	
15	16	24	26	14	20	21	12	17	17	8	10	10			
16	15	23	25	13	19	21	11	16	16	7	9	10			
17	14	22	25	12	19	20	10	15	16	6	9	9			
18	13	22	24	11	18	20	9	14	15		8	8			
19	12	21	23	10	17	19	8	14	14		8	8			
20	11	20		9	16	19	7	13	14		7	7			
21	10	19		8	15	18		12	13		7				
22	9	19		7	15	17		12	13		6				
23	8	18			14	17		11	12		5				
24		17			13	16		11	11		5				
25		16			12	15		10	10		4				
26		15			12	15		9	10						
27		14			11	14		9							
28		13			11	13		8							
29		13			10			8							
30		12			10			7							

s = slow fire, m = medium fire, f = fast fire.
For SI units: 1 ft = 0.305 m.

Table B-3.3.2(b)   Spacing Modifiers for Rate-of-Rise Heat Detectors

Listed Spacing (ft)	Fire Growth Rate		
	Slow	Medium	Fast
15	0.57	0.55	0.45
20	0.72	0.63	0.62
25	0.84	0.78	0.76
30	0.92	0.86	0.85
40	0.98	0.96	0.95
50	1.00	1.00	1.00
70	1.00	1.01	1.02

For SI units: 1 ft = 0.305 m.

Installed spacing = 18 × 0.86 = 15.5 ft (4.7 m)

NOTE: This answer may be permitted to be rounded to either 15 ft (4.6 m) or 16 ft (4.9 m). Use of 15 ft (4.6 m) would be slightly conservative. However, depending on field conditions, use of 16 ft (4.9 m) might fit the space better.

B-3.4 Design Curves.

B-3.4.1 The design curves [Figures B-3.4.1 (a) through (i)] also may be permitted to be used to determine the installed spacings of heat detectors. However, they are not as comprehensive as the tables, because the tables include additional fire growth rates, fire sizes, and detector sensitivities.

There are slight variations between the results depicted in the graphs versus those depicted in the heat detector tables. The graphs are based on the original data correlations of Heskestad and Delichatsios[10], while slightly different correlations developed by Beyler[4] were used to generate the tables.

I.   Input Data for Design and Analysis (*B-2*)

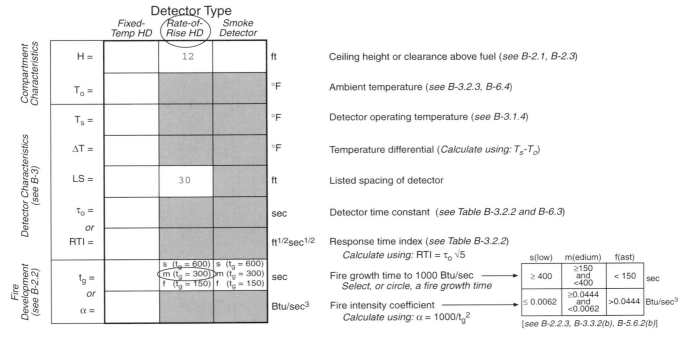

IIa. Input Data for *Design* (B-3)

IIa.1 Establish Design Objectives

Determine the size of the fire at which detector response is desired using either 1 or 2:

1. *Select* Q_d = ___500___ Btu/sec   [see B-2.2.4, Table B-2.2.2.1(a), (b), or B-2.2.2.3], or

2. *Calculate* Q_d using the time after ignition at which detector response is desired, t_d, using: $Q_d = \alpha t_d{}^2$ =_____ Btu/sec

IIa.2 Calculate Detector Response

Fixed-Temperature HD (B-3.2)	*Rate-of-Rise HD (B-3.3)*	*Smoke Detector (B-5)*
1. Fill in the variables: Q_d =_____ Btu/sec t_g =_____ sec, or α = _____ Btu/sec³ τ_o =_____ sec, or RTI = _____ ft¹/² sec¹/² ΔT = _____ °F and H = _____ ft 2. Using Q_d and t_g (or α), select the appropriate design table (a) through (y) from Table B-3.2.4:_____ 3. Using τ_o (or RTI), ΔT, and H, determine the installed spacing: _____ ft	1. Fill in the variables: Fire growth (s,ⓜ,f) = ___m___ H = ___12___ ft Q_d = ___500___ Btu/sec 2. Select installed spacing from Table B-3.3.2(a): ___18___ ft 3. Select spacing modifier from Table B-3.3.2(b): x ___0.86___ 4. Calculate installed spacing: ⟨ 15.5 ft ⟩	1. Fill in the variables: Fire growth (s,m,f) = _____ H = _____ ft Q_d = _____ Btu/sec 2. Using fire growth (s, m, or f), select Figure B-5.5.1(a), (b), or (c): _____ 3. Using H and Q_d, determine the installed spacing: _____ ft

IIb. Input Data for *Analysis* of an Existing Heat Detection System (B-4)

1. Fill in the variables:
   S = _____ ft (installed spacing of the existing heat detector)
   t_g = _____ sec, or α = _____ Btu/sec³
   τ_o = _____ sec, or RTI = _____ ft¹/²sec¹/²
   ΔT = _____ °F and H = _____ ft

2. Using S and t_g (or α), select an analysis table (a) - (nn) from Table B-4: _____

3. Using τ_o (or RTI), ΔT, and H, determine the fire size at detector response Q_d = _____ Btu/sec

4. Calculate the time to detector response using: $t_d = \sqrt{\dfrac{Q_d}{\alpha}} = \sqrt{\dfrac{\text{Btu/sec}}{\text{Btu/sec}^3}} = $ _____ sec

Figure B-3.3.5 *Completed worksheet for B-3.3.5, Example.*

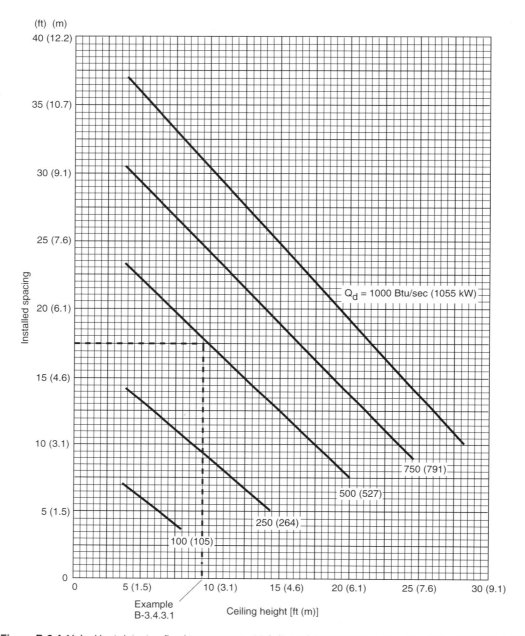

Figure B-3.4.1(a)   *Heat detector, fixed temperature, 30-ft (9.1-m) listed spacing, slow fire [DT = 65°F (36.1°C)].*

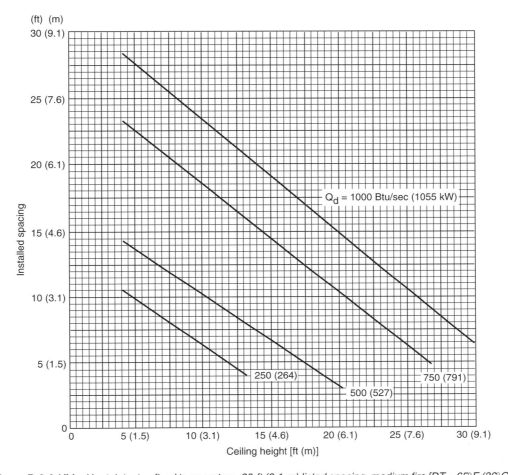

Figure B-3.4.1(b) *Heat detector, fixed temperature, 30-ft (9.1-m) listed spacing, medium fire [DT = 65° F (36° C)].*

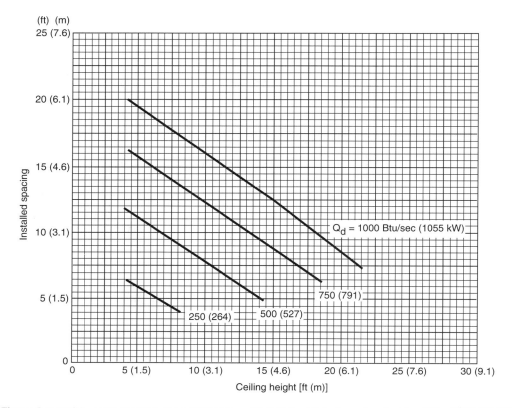

Figure B-3.4.1(c)   *Heat detector, fixed temperature, 30-ft (9.1-m) listed spacing, fast fire [DT = 65°F (36°C)].*

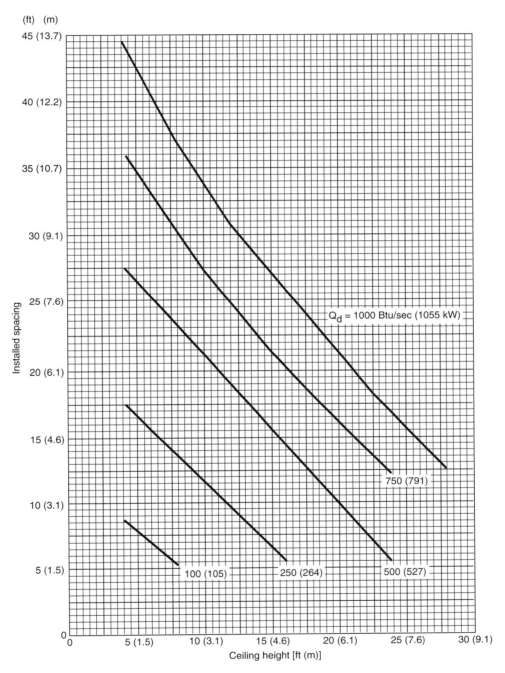

Figure B-3.4.1(d)   *Heat detector, fixed temperature, 50-ft (15.2-m) listed spacing, slow fire [DT = 65°F (36°C)].*

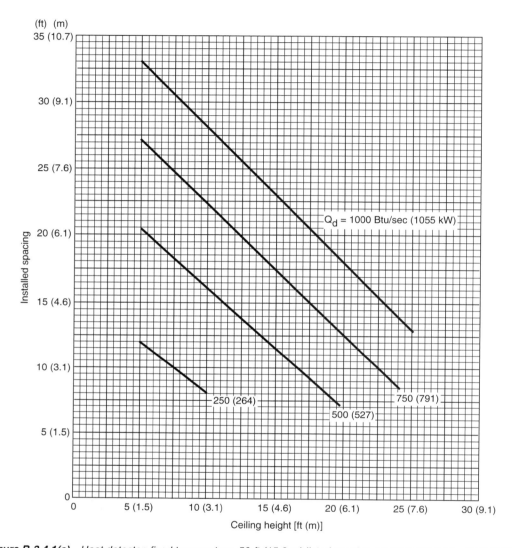

Figure B-3.4.1(e)   *Heat detector, fixed temperature, 50-ft (15.2-m) listed spacing, medium fire [DT = 65°F (36°C)].*

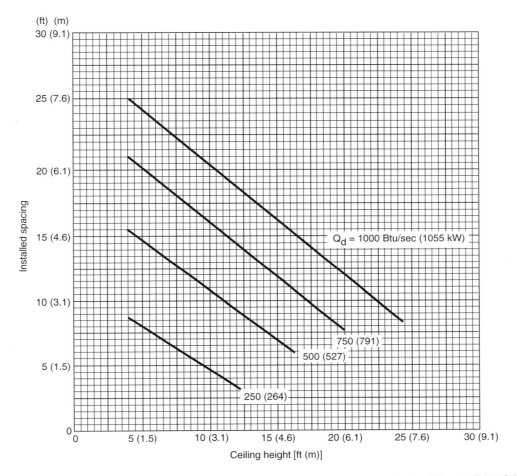

Figure B-3.4.1(f) *Heat detector, fixed temperature, 50-ft (15.2-m) listed spacing, fast fire [DT = 65°F (36°C)].*

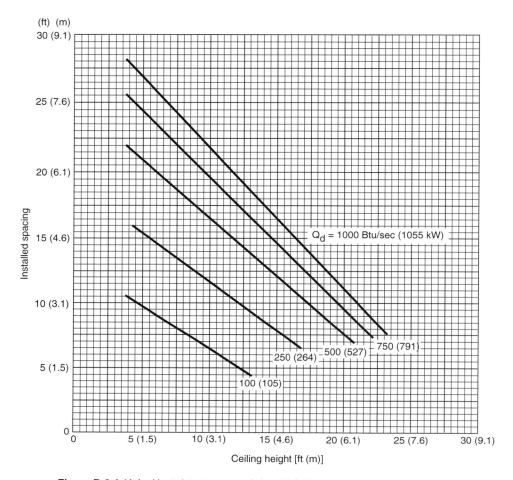

Figure B-3.4.1(g)   *Heat detector, rate-of-rise, 50-ft (15.2-m) listed spacing, slow fire.*

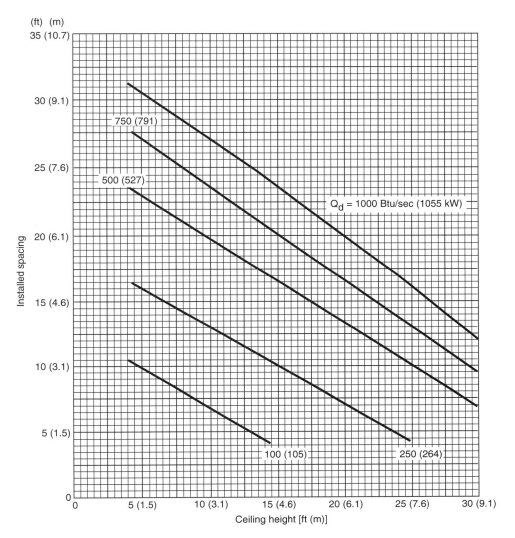

Figure B-3.4.1(h)   *Heat detector, rate-of-rise, 50-ft (15.2-m) listed spacing, medium fire.*

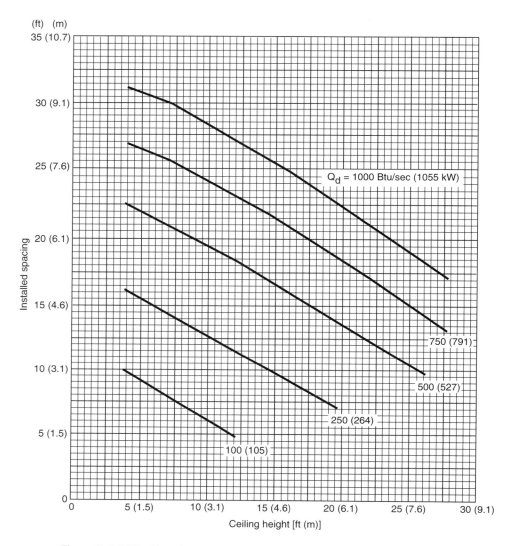

Figure B-3.4.1(i)   *Heat detector, rate-of-rise, 50-ft (15.2-m) listed spacing, fast fire.*

B-3.4.1.1 **Fixed-Temperature Heat Detectors.** Figures B-3.4.1(a) through (f) can be used directly to determine the installed spacing for fixed-temperature heat detectors having listed spacings of 30 ft and 50 ft (9.1 m and 15.2 m), respectively, where the difference between the detectors' rated temperature (T_s) and the ambient temperature (T_o) is 65°F (36°C). Where ΔT is not 65°F (36°C), the tables previously discussed in B-3.2 should be used.

B-3.4.1.2 **Rate-of-Rise Heat Detectors.** Figures B-3.4.1(g), (h), and (i) can be used directly to determine the installed spacing for rate-of-rise heat detectors having a listed spacing of 50 ft (15.2 m).

B-3.4.2 To use the curves, the same format is to be followed as is used with the tables. The designer first determines how large a fire can be tolerated before detection can occur. This is the threshold fire size, Q_d. Curves are presented, in most cases, for values of Q_d = 1000, 750, 500, 250, and 100 Btu/sec (1055, 791, 527, 264, and 105 kW). Interpolation between values of Q_d on a given graph is permitted. Table B-2.2.2.1(a) and Table B-2.2.2.3 also contain examples of various fuels and their fire growth rates under specified conditions.

B-3.4.3 Once a threshold size and expected fire growth rate have been selected, an installed detector spacing can be obtained from Figures B-3.4.1(a) through (i) for a specific detector's listed spacing, ambient temperature, and ceiling height. As in B-3.2.6, to determine the installed spacing of 135°F (57°C) fixed-temperature heat detectors with a listed spacing of 30 ft (9.1 m) and to detect a slowly developing fire at a threshold fire size of 500 Btu/sec (527 kW) in a room 10 ft (3 m) high with an ambient temperature of 70°F (21°C), the examples outlined in B-3.4.3.1 and B-3.4.3.2 are used.

B-3.4.3.1 **Example 1.**

Input:

Ceiling height: 10 ft (3 m)

Detector type: Fixed temperature 135°F (57°C)

Listed spacing: 30 ft (9.1 m)

Fire:

Q_d: 500 Btu/sec (527 kW)

Fire growth rate: slow

t_g: 600 sec

Environmental conditions:

T_o: 70°F (21°C)

$\Delta T = 135 - 70 = 65°F$ (36°C)

Required installed spacing:

From Figure B-3.4.1(a), an installed spacing of 18 ft (5.2 m) (17.5 ft rounded to 18 ft) is used.

It should be noted that, where the ceiling height is 15 ft (4.6 m), the same graph gives an installed spacing of 12 ft (3.5 m). A ceiling height of 20 ft (6.1 m) would require a spacing of 8 ft (2.4 m). This change in spacing clearly illustrates the need to consider ceiling height in the design of a detection system.

B-3.4.3.2 **Example 2.**

Input:

Ceiling height: 10 ft (3 m)

Detector type: Combination rate-of-rise and fixed temperature

Listed spacing: 50 ft (15.2 m)

Fire:

Q_d: 500 Btu/sec (527 kW)

Fire growth rate: fast

t_g: 150 sec

Environmental conditions:

T_o: 70°F (21°C)

ΔT: 65°F (36°C)

Spacing:

From Figure B-3.4.1(i), an installed spacing of 20 ft (6.1 m) (19.5 ft rounded to 20 ft) is used.

A 30-ft (9.1-m) fixed temperature detector would require a 7.5-ft (2.5-m) spacing.

If the fire growth rate is slow, as in Example 1, the rate-of-rise detector would require an installed spacing of 16 ft (4.9 m).

B-4   Analysis of Existing Heat Detection Systems.

(a) Use of Tables B-4(a) through (nn). Tables B-4(a) through (nn) can be used to determine the size fire (heat release rate) to which existing fixed-temperature heat detection systems will respond. Table B-4 provides an index to Tables B-4 (a) through (nn).

The use of the analysis tables is similar to that described for new designs. The difference is that the spacing of the existing detectors needs to be known. An estimate of the fire intensity coefficient (() or the fire growth time (t_g) also has to be made for the fuel that is expected to burn.

(b) **Example.**

Input:

Ceiling height: 8 ft (2.4 m)

Detector type: Fixed temperature 135°F (57°C)

Listed spacing: 30 ft (9.1 m)

Installed spacing: 15 ft (4.6 m)

Fire:

Fire growth rate: slow

t_g: 600 sec

α: 0.003 Btu/sec^3

Environmental conditions:

T_o: 55°F (12.8°C)

Threshold fire size (Q_d) at detector response:

From Table B-3.2.2, the detector time constant (τ_o) is 80 seconds.

$\Delta T = T_s - T_o = 135 - 55 = 80°F$

From Table B-4(t):

For $\tau_o = 75$ sec: $Q_d = 418$ Btu/sec

For $\tau_o = 100$ sec: $Q_d = 472$ Btu/sec

By interpolation:

$Q_d = 418 - [(75 - 80)\ (418 - 472)/(75 - 100)]$

$Q_d = 429$ Btu/sec

Table B-4   Analysis Tables Index

	Installed Spacing (ft)	Fire Growth Rate (sec) τ_g	Fuel Fire Intensity Coefficient (Btu/sec^3) a
Table B-4(a)	8	50	0.400
Table B-4(b)	8	150	0.044
Table B-4(c)	8	300	0.011
Table B-4(d)	8	500	0.004
Table B-4(e)	8	600	0.003
Table B-4(f)	10	50	0.400
Table B-4(g)	10	150	0.044
Table B-4(h)	10	300	0.011
Table B-4(i)	10	500	0.004
Table B-4(j)	10	600	0.003
Table B-4(k)	12	50	0.400
Table B-4(l)	12	150	0.044
Table B-4(m)	12	300	0.011
Table B-4(n)	12	500	0.004
Table B-4(o)	12	600	0.003
Table B-4(p)	15	50	0.400
Table B-4(q)	15	150	0.044
Table B-4(r)	15	300	0.011
Table B-4(s)	15	500	0.004
Table B-4(t)	15	600	0.003
Table B-4(u)	20	50	0.400
Table B-4(v)	20	150	0.044
Table B-4(w)	20	300	0.011
Table B-4(x)	20	500	0.004
Table B-4(y)	20	600	0.003
Table B-4(z)	25	50	0.400
Table B-4(aa)	25	150	0.044
Table B-4(bb)	25	300	0.011
Table B-4(cc)	25	500	0.004
Table B-4(dd)	25	600	0.003
Table B-4(ee)	30	50	0.400
Table B-4(ff)	30	150	0.044
Table B-4(gg)	30	300	0.011
Table B-4(hh)	30	500	0.004
Table B-4(ii)	30	600	0.003
Table B-4(jj)	50	50	0.400
Table B-4(kk)	50	150	0.044
Table B-4(ll)	50	300	0.011
Table B-4(mm)	50	500	0.004
Table B-4(nn)	50	600	0.003

I.   Input Data for Design and Analysis (*B-2*)

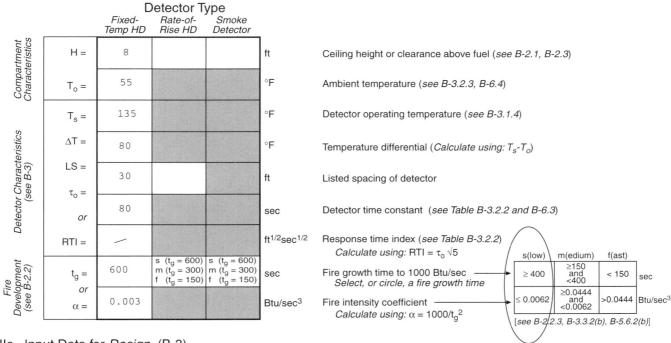

IIa. Input Data for *Design* (B-3)

IIa.1 Establish Design Objectives

> Determine the size of the fire at which detector response is desired using either 1 or 2:
> 1. *Select* Q_d = _____ Btu/sec   [see B-2.2.4, Table B-2.2.2.1(a), (b), or B-2.2.2.3], or
>
> 2. *Calculate* Q_d using the time after ignition at which detector response is desired, t_d, using: $Q_d = \alpha t_d^2$ = _____ Btu/sec

IIa.2 Calculate Detector Response

Fixed-Temperature HD (B-3.2)	*Rate-of-Rise HD (B-3.3)*	*Smoke Detector (B-5)*
1. Fill in the variables: Q_d = _____ Btu/sec t_g = _____ sec, or α = _____ Btu/sec^3 τ_o = _____ sec, or RTI = _____ ft$^{1/2}$ sec$^{1/2}$ ΔT = _____ °F and H = _____ ft 2. Using Q_d and t_g (or α), select the appropriate design table (a) through (y) from Table B-3.2.4: _____ 3. Using τ_o (or RTI), ΔT, and H, determine the installed spacing: _____ ft	1. Fill in the variables: Fire growth (s,m,f) = _____ H = _____ ft Q_d = _____ Btu/sec 2. Select installed spacing from Table B-3.3.2(a): _____ ft 3. Select spacing modifier from Table B-3.3.2(b): x _____ 4. Calculate installed spacing: _____ft	1. Fill in the variables: Fire growth (s,m,f) = _____ H = _____ ft Q_d = _____ Btu/sec 2. Using fire growth (s, m, or f), select Figure B-5.5.1(a), (b), or (c): _____ 3. Using H and Q_d, determine the installed spacing: _____ ft

IIb. Input Data for *Analysis* of an Existing Heat Detection System (B-4)

> 1. Fill in the variables:
>   S = ___15___ ft (installed spacing of the existing heat detector)
> t_g = ___600___ sec, or α = __0.003__ Btu/sec^3
> τ_o = ___80___ sec, or RTI = ___╱___ ft$^{1/2}$sec$^{1/2}$
> ΔT = ___80___ °F and H = ___8___ ft
> 2. Using S and t_g (or α), select an analysis table (a) - (nn) from Table B-4: __B-4(t)__
> 3. Using τ_o (or RTI), ΔT, and H, determine the fire size at detector response Q_d = ___429___ Btu/sec
> 4. Calculate the time to detector response using: $t_d = \sqrt{\dfrac{Q_d}{\alpha}} = \sqrt{\dfrac{429 \ \text{Btu/sec}}{0.003 \ \text{Btu/sec}^3}} = $ ___379___ sec

Figure B-4 *Completed worksheet for B-4(b), Example.*

Table B-4(a)   Installed Spacing of Heat Detector: 8ft t_g: 50 seconds to 1000 Btu/sec α: 0.400 Btu/sec³

τ	RTI	ΔT	4.0	8.0	12.0	16.0	20.0	24.0	28.0	τ	RTI	ΔT	4.0	8.0	12.0	16.0	20.0	24.0	28.0
			Ceiling Height (ft) / Fire Size at Detector Response (Btu/sec)										Ceiling Height (ft) / Fire Size at Detector Response (Btu/sec)							
25	56	40	300	402	535	668	832	1016	1219	225	503	40	968	1337	1754	2111	2537	2991	3468
25	56	60	368	508	687	877	1106	1365	1657	225	503	60	1254	1747	2294	2774	3342	3949	4590
25	56	80	450	618	838	1102	1381	1722	2110	225	503	80	1527	2129	2794	3392	4096	4851	5653
25	56	100	512	716	985	1308	1661	2090	2585	225	503	100	1794	2494	3268	3980	4819	5720	6681
25	56	120	573	815	1132	1517	1949	2473	3082	225	503	120	2057	2845	3724	4549	5520	6567	7689
25	56	140	654	919	1282	1730	2265	2870	3601	225	503	140	2317	3185	4168	5104	6206	7400	8683
50	112	40	422	571	755	926	1136	1366	1614	250	559	40	1011	1417	1865	2247	2698	3177	3681
50	112	60	546	738	976	1211	1496	1811	2157	250	559	60	1339	1866	2447	2955	3556	4197	4873
50	112	80	642	883	1181	1484	1846	2251	2699	250	559	80	1637	2278	2982	3614	4358	5155	5999
50	112	100	754	1033	1383	1752	2194	2692	3248	250	559	100	1928	2669	3489	4241	5126	6076	7087
50	112	120	865	1179	1582	2018	2542	3138	3810	250	559	120	2215	3046	3890	4842	5870	6972	8150
50	112	140	928	1305	1773	2318	2895	3592	4386	250	559	140	2499	3412	4356	5431	6597	7852	9197
75	168	40	542	722	908	1137	1389	1659	1948	275	615	40	1093	1513	1981	2380	2854	3358	3887
75	168	60	702	932	1219	1492	1826	2193	2589	275	615	60	1424	1982	2596	3131	3763	4437	5147
75	168	80	813	1111	1472	1824	2245	2710	3217	275	615	80	1746	2422	3165	3829	4612	5449	6334
75	168	100	931	1289	1718	2146	2656	3221	3844	275	615	100	2061	2840	3618	4488	5424	6421	7479
75	168	120	1016	1451	1955	2464	3063	3733	4475	275	615	120	2371	3242	4128	5129	6209	7365	8597
75	168	140	1149	1629	2193	2778	3470	4247	5115	275	615	140	2679	3633	4622	5753	6977	8291	9697
100	224	40	625	841	1101	1332	1614	1920	2246	300	671	40	1151	1595	2089	2508	3005	3533	4087
100	224	60	802	1087	1427	1742	2122	2535	2978	300	671	60	1507	2096	2740	3301	3964	4670	5413
100	224	80	944	1305	1728	2128	2604	3125	3687	300	671	80	1853	2563	3259	4032	4859	5735	6661
100	224	100	1050	1503	2012	2501	3074	3703	4388	300	671	100	2192	3007	3820	4734	5714	6756	7862
100	224	120	1222	1723	2298	2867	3537	4276	5088	300	671	120	2526	3434	4359	5409	6540	7748	9033
100	224	140	1360	1925	2573	3226	3995	4849	5791	300	671	140	2859	3849	4881	6066	7346	8718	10183
125	280	40	729	967	1208	1501	1820	2160	2519	325	727	40	1208	1677	2194	2633	3152	3704	4282
125	280	60	912	1238	1622	1972	2394	2850	3337	325	727	60	1589	2207	2804	3461	4160	4898	5672
125	280	80	1036	1472	1959	2409	2936	3508	4123	325	727	80	1959	2701	3428	4236	5100	6014	6978
125	280	100	1233	1730	2294	2830	3461	4150	4895	325	727	100	2322	3171	4018	4973	5996	7084	8234
125	280	120	1398	1968	2614	3240	3976	4782	5661	325	727	120	2680	3623	4585	5682	6862	8121	9457
125	280	140	1561	2201	2926	3642	4484	5411	6246	325	727	140	3038	4061	5133	6371	7706	9135	10657
150	335	40	793	1066	1340	1664	2013	2384	2775	350	783	40	1265	1756	2297	2754	3296	3871	4472
150	335	60	979	1362	1797	2187	2649	3145	3674	350	783	60	1671	2315	2937	3623	4352	5119	5925
150	335	80	1185	1656	2186	2673	3247	3868	4533	350	783	80	2064	2836	3592	4435	5335	6287	7289
150	335	100	1378	1933	2554	3138	3825	4570	5373	350	783	100	2451	3331	4211	5207	6272	7403	8599
150	335	120	1568	2201	2911	3590	4389	5259	6202	350	783	120	2834	3808	4805	5949	7177	8485	9872
150	335	140	1757	2462	3257	4033	4944	5942	7027	350	783	140	3218	4270	5380	6669	8058	9542	11121
175	391	40	882	1175	1468	1818	2195	2595	3016	375	839	40	1321	1835	2398	2874	3437	4034	4658
175	391	60	1046	1483	1965	2391	2890	3425	3993	375	839	60	1751	2422	3069	3782	4539	5336	6172
175	391	80	1301	1819	2397	2923	3542	4210	4923	375	839	80	2169	2969	3753	4630	5565	6553	7592
175	391	100	1520	2127	2802	3431	4170	4970	5827	375	839	100	2579	3489	4401	5436	6543	7716	8955
175	391	120	1734	2423	3193	3923	4782	5713	6718	375	839	120	2987	3990	5021	6210	7486	8842	10279
175	391	140	1947	2712	3573	4405	5382	6447	7601	375	839	140	3303	4445	5620	6961	8403	9941	11575
200	447	40	925	1257	1586	1964	2369	2797	3247	400	894	40	1377	1912	2423	2982	3574	4193	4840
200	447	60	1168	1625	2136	2587	3121	3692	4298	400	894	60	1831	2527	3197	3937	4723	5549	6415
200	447	80	1415	1977	2599	3162	3825	4537	5295	400	894	80	2272	3100	3911	4821	5791	6814	7890
200	447	100	1658	2313	3040	3711	4501	5352	6262	400	894	100	2707	3645	4586	5660	6807	8023	9304
200	447	120	1897	2637	3464	4242	5158	6148	7212	400	894	120	3141	4169	5233	6466	7788	9192	10677
200	447	140	2133	2952	3875	4761	5802	6932	8152	400	894	140	3456	4640	5857	7247	8741	10332	12020

NOTE: Detector time constant at a reference velocity of 5 ft/sec (1.5m/sec).
For SI units: 1 ft = 0.305 m; 1000 Btu/sec = 1055 kW.

Table B-4(b)   Installed Spacing of Heat Detector: 8ft t_g: 50 seconds to 1000 Btu/sec   α: 0.044 Btu/sec³

τ	RTI	ΔT	Ceiling Height (ft) Fire Size at Detector Response (Btu/sec)							τ	RTI	ΔT	Ceiling Height (ft) Fire Size at Detector Response (Btu/sec)						
			4.0	8.0	12.0	16.0	20.0	24.0	28.0				4.0	8.0	12.0	16.0	20.0	24.0	28.0
25	56	40	118	167	232	311	400	507	631	225	503	40	425	570	726	906	1102	1312	1538
25	56	60	154	226	322	440	584	752	952	225	503	60	558	759	992	1227	1500	1797	2119
25	56	80	194	286	415	579	781	1026	1309	225	503	80	693	939	1231	1533	1885	2271	2694
25	56	100	228	346	512	726	993	1319	1699	225	503	100	809	1108	1461	1833	2265	2744	3272
25	56	120	263	409	614	883	1221	1633	2118	225	503	120	926	1274	1686	2130	2645	3220	3859
25	56	140	299	473	721	1049	1462	1969	2573	225	503	140	1026	1434	1909	2425	3025	3700	4456
50	112	40	171	237	320	410	517	638	775	250	559	40	453	608	774	964	1170	1392	1628
50	112	60	224	317	435	574	728	913	1126	250	559	60	596	810	1041	1304	1591	1902	2239
50	112	80	281	397	550	735	949	1205	1504	250	559	80	738	1002	1311	1629	1997	2401	2841
50	112	100	329	474	666	901	1185	1516	1912	250	559	100	876	1185	1556	1946	2397	2896	3444
50	112	120	377	552	785	1074	1428	1846	2347	250	559	120	982	1358	1794	2258	2795	3392	4053
50	112	140	424	630	906	1254	1683	2202	2808	250	559	140	1107	1531	2029	2567	3192	3891	4671
75	168	40	216	296	395	498	620	756	906	275	615	40	480	645	820	1021	1237	1469	1716
75	168	60	283	395	533	683	861	1063	1291	275	615	60	646	863	1103	1379	1680	2005	2355
75	168	80	352	492	668	876	1107	1381	1696	275	615	80	783	1063	1389	1722	2107	2527	2984
75	168	100	413	585	803	1063	1360	1714	2125	275	615	100	921	1256	1647	2054	2525	3043	3611
75	168	120	472	678	939	1255	1622	2063	2578	275	615	120	1039	1440	1898	2382	2941	3560	4242
75	168	140	531	770	1076	1451	1901	2427	3055	275	615	140	1177	1624	2146	2706	3355	4078	4881
100	224	40	255	349	462	577	713	863	1027	300	671	40	507	681	865	1075	1301	1543	1801
100	224	60	343	467	622	788	983	1202	1446	300	671	60	680	911	1163	1452	1766	2105	2469
100	224	80	416	578	776	996	1254	1548	1880	300	671	80	827	1123	1445	1811	2213	2650	3123
100	224	100	488	685	929	1214	1530	1904	2333	300	671	100	967	1325	1736	2161	2650	3187	3774
100	224	120	559	792	1081	1424	1811	2273	2806	300	671	120	1109	1523	2000	2503	3083	3723	4427
100	224	140	636	898	1234	1637	2101	2656	3301	300	671	140	1246	1715	2260	2842	3514	4261	5087
125	280	40	291	397	523	650	799	962	1140	325	727	40	533	717	909	1129	1365	1616	1884
125	280	60	391	532	704	885	1097	1333	1593	325	727	60	714	959	1222	1524	1851	2203	2580
125	280	80	476	657	877	1114	1392	1705	2056	325	727	80	881	1184	1517	1899	2317	2770	3259
125	280	100	558	779	1046	1342	1690	2086	2534	325	727	100	1014	1393	1823	2264	2772	3328	3933
125	280	120	647	899	1214	1571	1992	2476	3029	325	727	120	1169	1601	2100	2622	3222	3884	4608
125	280	140	723	1017	1382	1813	2300	2878	3543	325	727	140	1314	1803	2371	2974	3670	4440	5288
150	335	40	325	443	581	719	880	1056	1246	350	783	40	559	751	952	1181	1426	1688	1965
150	335	60	435	593	781	976	1204	1456	1733	350	783	60	747	1005	1280	1594	1933	2298	2688
150	335	80	531	732	971	1226	1523	1855	2224	350	783	80	917	1239	1589	1986	2418	2887	3392
150	335	100	634	869	1157	1473	1842	2259	2728	350	783	100	1072	1462	1885	2365	2892	3466	4089
150	335	120	720	999	1340	1719	2164	2671	3245	350	783	120	1228	1679	2197	2737	3359	4041	4786
150	335	140	805	1128	1522	1967	2491	3093	3780	350	783	140	1380	1890	2480	3104	3822	4615	5486
175	391	40	357	486	637	784	957	1145	1347	375	839	40	584	785	994	1232	1486	1757	2045
175	391	60	478	650	854	1063	1307	1574	1866	375	839	60	780	1050	1336	1662	2014	2391	2795
175	391	80	584	803	1061	1332	1649	1999	2386	375	839	80	953	1294	1658	2070	2518	3002	3523
175	391	100	694	952	1262	1598	1989	2427	2915	375	839	100	1122	1528	1967	2464	3009	3601	4242
175	391	120	790	1094	1460	1861	2330	2860	3456	375	839	120	1286	1754	2292	2851	3492	4195	4960
175	391	140	892	1236	1656	2125	2675	3301	4011	375	839	140	1446	1975	2587	3231	3971	4787	5681
200	447	40	396	530	676	846	1031	1230	1444	400	894	40	609	818	1036	1282	1545	1826	2122
200	447	60	519	705	924	1146	1405	1687	1995	400	894	60	813	1094	1392	1729	2093	2483	2899
200	447	80	646	873	1148	1435	1769	2138	2543	400	894	80	989	1348	1726	2153	2615	3114	3651
200	447	100	752	1031	1363	1718	2129	2588	3096	400	894	100	1171	1593	2048	2562	3123	3733	4393
200	447	120	870	1188	1576	1998	2490	3042	3660	400	894	120	1343	1829	2359	2962	3623	4346	5132
200	447	140	959	1337	1785	2277	2852	3503	4236	400	894	140	1511	2058	2692	3356	4118	4956	5872

NOTE: Detector time constant at a reference velocity of 5 ft/sec (1.5m/sec).
For SI units: 1 ft = 0.305 m; 1000 Btu/sec = 1055 kW.

Table B-4(c)   Installed Spacing of Heat Detector: 8ft t_g: 300 seconds to 1000 Btu/sec α: 0.011 Btu/sec^3

τ	RTI	ΔT	Ceiling Height (ft) Fire Size at Detector Response (Btu/sec) 4.0	8.0	12.0	16.0	20.0	24.0	28.0
25	56	40	70	104	152	211	285	374	477
25	56	60	95	149	223	321	443	592	767
25	56	80	122	196	302	442	620	838	1099
25	56	100	148	246	387	575	815	1110	1463
25	56	120	174	299	479	719	1027	1405	1858
25	56	140	201	354	576	873	1253	1721	2283
50	112	40	101	144	200	267	345	439	547
50	112	60	136	200	284	389	517	668	849
50	112	80	172	257	372	520	703	926	1187
50	112	100	205	315	465	661	906	1206	1560
50	112	120	239	374	563	811	1124	1508	1963
50	112	140	273	437	666	970	1356	1830	2399
75	168	40	127	178	242	318	402	502	616
75	168	60	170	245	339	453	586	746	931
75	168	80	215	311	438	595	786	1012	1280
75	168	100	255	378	540	745	998	1303	1661
75	168	120	296	445	646	903	1223	1612	2072
75	168	140	336	514	756	1069	1461	1942	2510
100	224	40	150	209	281	361	455	561	682
100	224	60	201	286	390	514	654	821	1013
100	224	80	253	361	500	667	864	1099	1374
100	224	100	300	436	611	827	1088	1397	1764
100	224	120	347	511	725	993	1322	1714	2183
100	224	140	393	587	843	1167	1568	2055	2628
125	280	40	171	237	317	403	504	618	745
125	280	60	230	323	437	567	719	893	1093
125	280	80	289	408	557	736	941	1184	1466
125	280	100	342	490	678	906	1173	1493	1867
125	280	120	395	573	801	1081	1420	1819	2294
125	280	140	447	656	926	1262	1674	2163	2748
150	335	40	192	264	350	443	551	671	805
150	335	60	261	360	482	619	780	963	1170
150	335	80	322	451	612	801	1015	1267	1557
150	335	100	381	542	742	981	1258	1587	1969
150	335	120	440	631	873	1166	1511	1923	2406
150	335	140	497	721	1006	1356	1778	2276	2867
175	391	40	211	289	383	481	596	723	863
175	391	60	287	394	525	670	839	1030	1245
175	391	80	353	493	664	859	1087	1348	1646
175	391	100	419	591	803	1055	1341	1679	2070
175	391	120	482	687	943	1248	1603	2025	2517
175	391	140	545	784	1083	1447	1875	2387	2987
200	447	40	229	314	413	518	639	772	919
200	447	60	311	426	566	719	896	1095	1318
200	447	80	384	533	715	918	1156	1427	1733
200	447	100	454	638	862	1119	1421	1770	2169
200	447	120	523	741	1010	1328	1694	2126	2626
200	447	140	596	844	1158	1535	1974	2497	3106
225	503	40	247	337	443	553	680	820	973
225	503	60	335	458	606	766	951	1158	1389
225	503	80	413	572	763	976	1223	1503	1819
225	503	100	489	684	919	1186	1499	1858	2267
225	503	120	569	794	1075	1406	1782	2225	2735
225	503	140	639	902	1230	1621	2072	2605	3224
250	559	40	264	360	471	587	720	866	1025
250	559	60	357	488	644	811	1005	1220	1458
250	559	80	441	610	811	1032	1289	1578	1902
250	559	100	522	728	975	1252	1576	1945	2363
250	559	120	607	844	1138	1474	1868	2322	2842
250	559	140	682	959	1301	1705	2167	2712	3341
275	615	40	280	382	499	620	759	911	1076
275	615	60	380	518	681	856	1057	1280	1525
275	615	80	469	646	856	1086	1352	1651	1984
275	615	100	555	771	1028	1316	1650	2029	2457
275	615	120	643	893	1199	1546	1953	2418	2948
275	615	140	723	1014	1369	1779	2261	2817	3456
300	671	40	297	403	520	652	797	955	1126
300	671	60	401	546	717	899	1108	1338	1591
300	671	80	496	682	901	1140	1415	1722	2064
300	671	100	593	813	1081	1378	1723	2113	2550
300	671	120	679	941	1259	1617	2035	2512	3052
300	671	140	763	1067	1436	1858	2352	2921	3571
325	727	40	317	425	546	684	834	998	1174
325	727	60	422	574	753	941	1157	1395	1656
325	727	80	522	716	945	1192	1476	1792	2143
325	727	100	623	854	1132	1439	1795	2194	2641
325	727	120	713	988	1317	1687	2116	2604	3155
325	727	140	802	1119	1502	1936	2442	3023	3684
350	783	40	332	445	571	714	871	1040	1222
350	783	60	443	602	787	982	1206	1451	1719
350	783	80	548	750	987	1242	1535	1861	2220
350	783	100	653	894	1182	1499	1865	2274	2731
350	783	120	747	1033	1375	1755	2196	2694	3256
350	783	140	841	1171	1566	2012	2531	3123	3795
375	839	40	347	465	596	744	906	1081	1269
375	839	60	463	629	821	1023	1253	1506	1781
375	839	80	573	783	1029	1292	1594	1928	2296
375	839	100	682	933	1232	1558	1933	2353	2820
375	839	120	781	1078	1431	1822	2274	2784	3356
375	839	140	885	1221	1629	2086	2618	3222	3906
400	894	40	362	485	620	774	941	1121	1314
400	894	60	483	655	846	1062	1300	1560	1842
400	894	80	604	817	1070	1341	1651	1994	2371
400	894	100	710	971	1280	1615	2001	2430	2907
400	894	120	814	1122	1486	1888	2351	2872	3454
400	894	140	919	1270	1690	2160	2704	3320	4015

NOTE: Detector time constant at a reference velocity of 5 ft/sec (1.5m/sec).
For SI units: 1 ft = 0.305 m; 1000 Btu/sec = 1055 kW.

Table B-4(d)   Installed Spacing of Heat Detector: 8ft t_g: 500 seconds to 1000 Btu/sec α: 0.004 Btu/sec^3

τ	RTI	ΔT	Ceiling Height (ft) 4.0	8.0	12.0	16.0	20.0	24.0	28.0
			Fire Size at Detector Response (Btu/sec)						
25	56	40	49	78	118	171	237	318	413
25	56	60	70	116	183	272	385	523	689
25	56	80	91	158	256	386	553	759	1005
25	56	100	113	203	336	513	740	1020	1357
25	56	120	136	252	422	651	944	1306	1741
25	56	140	160	304	515	799	1163	1613	2154
50	112	40	70	104	148	204	273	355	453
50	112	60	96	149	220	312	427	568	735
50	112	80	124	196	297	431	601	809	1058
50	112	100	150	246	381	562	792	1075	1415
50	112	120	178	298	471	703	1000	1365	1803
50	112	140	206	353	567	855	1222	1676	2221
75	168	40	87	126	176	236	309	393	493
75	168	60	120	178	255	351	470	612	782
75	168	80	152	231	338	476	649	860	1109
75	168	100	184	286	427	612	845	1131	1471
75	168	120	215	343	521	757	1057	1425	1867
75	168	140	247	402	621	912	1283	1740	2289
100	224	40	103	147	201	266	341	430	533
100	224	60	141	205	288	389	512	658	830
100	224	80	179	264	378	521	698	912	1163
100	224	100	214	324	472	661	898	1188	1530
100	224	120	250	386	571	811	1114	1486	1928
100	224	140	286	449	674	970	1345	1805	2354
125	280	40	118	166	225	295	374	466	573
125	280	60	160	231	319	426	552	702	878
125	280	80	203	295	415	564	746	961	1218
125	280	100	243	360	515	711	952	1245	1590
125	280	120	282	427	619	865	1173	1548	1993
125	280	140	322	494	727	1028	1407	1871	2423
150	335	40	131	184	248	320	404	501	611
150	335	60	179	255	349	462	592	747	926
150	335	80	226	325	452	607	793	1013	1273
150	335	100	270	395	557	759	1005	1300	1650
150	335	120	313	466	666	918	1231	1610	2058
150	335	140	356	537	778	1085	1469	1937	2492
175	391	40	144	201	269	345	434	535	649
175	391	60	196	278	378	496	631	790	973
175	391	80	247	353	487	648	837	1063	1327
175	391	100	295	428	597	806	1058	1356	1711
175	391	120	342	503	711	971	1289	1669	2123
175	391	140	389	579	828	1142	1532	2004	2562
200	447	40	157	217	290	370	463	568	685
200	447	60	213	300	405	526	668	832	1020
200	447	80	268	380	520	688	882	1113	1381
200	447	100	320	460	637	852	1107	1413	1771
200	447	120	370	539	755	1022	1346	1731	2189
200	447	140	421	620	877	1199	1594	2067	2633
225	503	40	169	233	310	394	491	599	721
225	503	60	232	321	432	558	705	873	1065
225	503	80	288	406	553	727	926	1162	1435
225	503	100	343	490	675	897	1158	1469	1832
225	503	120	397	574	799	1073	1403	1792	2254
225	503	140	451	659	925	1254	1656	2134	2703
250	559	40	181	249	329	417	518	630	756
250	559	60	247	342	458	588	740	914	1110
250	559	80	307	432	584	765	970	1211	1488
250	559	100	366	520	712	941	1208	1524	1892
250	559	120	424	609	841	1122	1455	1853	2320
250	559	140	480	697	971	1309	1717	2200	2774
275	615	40	192	264	348	439	544	660	789
275	615	60	263	362	483	618	775	953	1154
275	615	80	326	456	615	798	1012	1258	1541
275	615	100	388	549	748	984	1257	1578	1951
275	615	120	449	642	882	1171	1510	1914	2386
275	615	140	509	734	1017	1362	1773	2266	2844
300	671	40	203	278	366	461	569	690	823
300	671	60	277	381	507	647	809	992	1198
300	671	80	345	481	645	834	1053	1305	1593
300	671	100	410	578	783	1026	1305	1632	2010
300	671	120	474	674	922	1219	1564	1974	2451
300	671	140	541	770	1062	1415	1833	2331	2915
325	727	40	214	292	384	482	594	719	855
325	727	60	292	400	531	676	843	1030	1240
325	727	80	362	504	675	869	1094	1351	1644
325	727	100	431	606	818	1063	1352	1685	2069
325	727	120	498	706	961	1266	1617	2033	2516
325	727	140	568	806	1106	1467	1892	2396	2985
350	783	40	224	306	401	503	619	747	887
350	783	60	306	419	555	704	875	1068	1282
350	783	80	380	527	704	903	1134	1397	1694
350	783	100	452	633	852	1104	1398	1738	2126
350	783	120	522	737	1000	1312	1670	2092	2580
350	783	140	594	841	1149	1518	1949	2460	3055
375	839	40	235	320	419	523	643	774	918
375	839	60	320	437	577	731	907	1104	1324
375	839	80	397	550	732	937	1173	1442	1744
375	839	100	472	659	885	1143	1444	1790	2184
375	839	120	545	767	1038	1357	1722	2150	2644
375	839	140	619	875	1191	1569	2007	2524	3125
400	894	40	245	333	435	543	666	801	949
400	894	60	333	455	600	758	939	1141	1364
400	894	80	414	572	760	970	1212	1486	1794
400	894	100	492	685	918	1182	1489	1841	2241
400	894	120	572	798	1075	1396	1773	2208	2707
400	894	140	645	908	1233	1619	2063	2588	3194

NOTE: Detector time constant at a reference velocity of 5 ft/sec (1.5m/sec).
For SI units: 1 ft = 0.305 m; 1000 Btu/sec = 1055 kW.

Table B-4(e)   Installed Spacing of Heat Detector: 8ft t_g: 600 seconds to 1000 Btu/sec α: 0.003 Btu/sec^3

τ	RTI	ΔT	Ceiling Height (ft) 4.0	8.0	12.0	16.0	20.0	24.0	28.0	τ	RTI	ΔT	Ceiling Height (ft) 4.0	8.0	12.0	16.0	20.0	24.0	28.0
			Fire Size at Detector Response (Btu/sec)										Fire Size at Detector Response (Btu/sec)						
25	56	40	44	71	110	160	225	303	397	225	503	40	148	205	274	351	440	541	654
25	56	60	63	108	173	259	370	505	668	225	503	60	202	284	385	504	639	798	981
25	56	80	83	148	244	372	536	738	980	225	503	80	254	362	496	658	848	1073	1336
25	56	100	104	192	323	497	721	997	1328	225	503	100	303	438	609	818	1070	1368	1720
25	56	120	126	240	408	633	922	1279	1709	225	503	120	352	515	725	985	1302	1681	2132
25	56	140	149	291	499	780	1139	1584	2119	225	503	140	401	593	843	1158	1546	2013	2572
50	112	40	62	93	135	188	254	334	429	250	559	40	158	219	291	371	463	567	684
50	112	60	86	135	203	292	404	542	705	250	559	60	215	302	407	528	670	832	1019
50	112	80	111	180	278	409	575	779	1024	250	559	80	271	384	524	691	885	1114	1380
50	112	100	136	228	360	537	763	1042	1376	250	559	100	323	464	641	856	1110	1414	1770
50	112	120	161	279	448	676	968	1328	1761	250	559	120	375	545	761	1027	1350	1732	2187
50	112	140	188	332	542	826	1188	1636	2175	250	559	140	426	626	884	1204	1598	2069	2630
75	168	40	77	112	158	215	284	365	462	275	615	40	168	232	307	390	486	593	713
75	168	60	106	160	233	325	440	579	744	275	615	60	231	320	429	554	700	866	1056
75	168	80	136	210	312	446	614	821	1066	275	615	80	287	405	550	723	921	1155	1425
75	168	100	165	262	398	578	807	1088	1425	275	615	100	343	490	673	893	1152	1460	1820
75	168	120	194	316	489	720	1015	1378	1813	275	615	120	397	574	797	1069	1397	1783	2241
75	168	140	224	373	586	872	1238	1689	2231	275	615	140	451	658	923	1250	1649	2123	2689
100	224	40	91	130	180	241	312	396	495	300	671	40	178	244	323	409	508	619	742
100	224	60	124	184	261	357	475	616	784	300	671	60	244	337	451	579	729	900	1093
100	224	80	158	238	346	483	654	863	1110	300	671	80	303	426	577	754	957	1195	1469
100	224	100	191	295	436	619	851	1134	1471	300	671	100	362	514	703	930	1193	1506	1869
100	224	120	224	353	531	765	1062	1428	1866	300	671	120	419	602	831	1110	1440	1834	2296
100	224	140	257	413	631	920	1288	1742	2287	300	671	140	475	690	962	1295	1699	2178	2747
125	280	40	103	147	201	265	339	427	528	325	727	40	187	257	339	428	530	644	770
125	280	60	141	206	288	388	510	653	824	325	727	60	256	353	471	604	758	933	1130
125	280	80	180	265	378	520	695	905	1155	325	727	80	319	447	602	782	992	1234	1512
125	280	100	216	326	472	660	895	1181	1520	325	727	100	380	538	733	965	1234	1551	1919
125	280	120	252	388	571	810	1110	1478	1916	325	727	120	440	630	865	1150	1485	1884	2350
125	280	140	289	451	675	968	1340	1796	2341	325	727	140	499	721	999	1340	1746	2233	2806
150	335	40	115	162	220	288	366	456	561	350	783	40	196	269	354	446	551	668	797
150	335	60	158	227	313	419	542	690	863	350	783	60	269	370	492	628	786	965	1166
150	335	80	200	291	409	556	734	947	1200	350	783	80	334	467	627	811	1026	1273	1555
150	335	100	239	355	508	701	939	1229	1570	350	783	100	398	562	763	1000	1274	1596	1968
150	335	120	279	421	611	854	1158	1530	1970	350	783	120	461	657	899	1190	1530	1934	2404
150	335	140	318	488	718	1016	1391	1851	2398	350	783	140	523	751	1037	1384	1796	2287	2864
175	391	40	127	177	239	309	391	485	593	375	839	40	205	281	369	464	572	692	824
175	391	60	173	247	338	448	575	727	903	375	839	60	281	386	512	652	814	997	1201
175	391	80	218	315	439	591	774	989	1246	375	839	80	349	486	652	841	1060	1312	1598
175	391	100	261	384	543	741	983	1274	1620	375	839	100	416	585	792	1035	1313	1640	2016
175	391	120	304	454	650	898	1206	1581	2024	375	839	120	481	683	932	1229	1574	1983	2458
175	391	140	347	524	761	1064	1443	1906	2456	375	839	140	549	781	1073	1427	1844	2341	2922
200	447	40	138	192	257	330	416	513	624	400	894	40	214	292	383	481	592	715	851
200	447	60	187	266	362	476	607	763	942	400	894	60	292	401	531	675	841	1028	1236
200	447	80	237	339	468	625	810	1031	1291	400	894	80	364	506	676	869	1094	1350	1640
200	447	100	283	412	576	780	1027	1321	1670	400	894	100	433	608	820	1065	1352	1684	2065
200	447	120	329	485	688	942	1255	1630	2078	400	894	120	501	709	964	1268	1618	2032	2511
200	447	140	374	559	802	1111	1495	1961	2513	400	894	140	571	810	1110	1470	1893	2395	2981

NOTE: Detector time constant at a reference velocity of 5 ft/sec (1.5m/sec).
For SI units: 1 ft = 0.305 m; 1000 Btu/sec = 1055 kW.

Table B-4(f) Installed Spacing of Heat Detector: 10ft t_g: 50 seconds to 1000 Btu/sec α: 0.400 Btu/sec^3

τ	RTI	ΔT	Ceiling Height (ft) Fire Size at Detector Response (Btu/sec)							τ	RTI	ΔT	Ceiling Height (ft) Fire Size at Detector Response (Btu/sec)						
			4.0	8.0	12.0	16.0	20.0	24.0	28.0				4.0	8.0	12.0	16.0	20.0	24.0	28.0
25	56	40	376	499	623	779	956	1152	1366	225	503	40	1234	1667	2041	2468	2925	3406	3911
25	56	60	486	641	811	1027	1273	1550	1860	225	503	60	1622	2186	2681	3249	3858	4503	5183
25	56	80	570	769	1013	1271	1591	1957	2371	225	503	80	1998	2671	3274	3976	4731	5535	6386
25	56	100	675	902	1193	1517	1916	2377	2906	225	503	100	2367	3132	3838	4670	5569	6529	7551
25	56	120	751	1024	1371	1766	2249	2814	3465	225	503	120	2731	3577	4380	5340	6382	7499	8692
25	56	140	827	1146	1551	2040	2593	3268	4050	225	503	140	3096	3911	4902	5994	7178	8452	9818
50	112	40	552	692	904	1084	1306	1550	1813	250	559	40	1317	1778	2175	2628	3111	3620	4152
50	112	60	712	895	1171	1418	1723	2059	2426	250	559	60	1740	2337	2860	3461	4105	4786	5503
50	112	80	828	1106	1391	1738	2129	2561	3038	250	559	80	2150	2859	3494	4237	5035	5882	6778
50	112	100	945	1282	1633	2055	2531	3064	3658	250	559	100	2553	3355	4096	4976	5925	6936	8011
50	112	120	1035	1446	1905	2371	2936	3575	4292	250	559	120	2953	3739	4669	5689	6787	7963	9215
50	112	140	1177	1626	2143	2685	3343	4094	4942	250	559	140	3267	4270	5234	6384	7631	8970	10402
75	168	40	673	890	1088	1331	1598	1885	2191	275	615	40	1398	1886	2305	2783	3291	3826	4386
75	168	60	885	1157	1418	1744	2104	2495	2916	275	615	60	1855	2484	3033	3667	4345	5061	5813
75	168	80	1000	1370	1725	2136	2590	3086	3625	275	615	80	2300	3042	3707	4489	5329	6219	7159
75	168	100	1174	1602	2024	2518	3066	3670	4333	275	615	100	2737	3485	4340	5272	6270	7331	8456
75	168	120	1330	1822	2315	2893	3538	4255	5047	275	615	120	3175	3975	4954	6027	7181	8413	9723
75	168	140	1484	2037	2602	3264	4010	4843	5769	275	615	140	3488	4440	5548	6761	8070	9472	10968
100	224	40	801	1051	1276	1554	1858	2183	2528	300	671	40	1478	1992	2432	2933	3467	4027	4613
100	224	60	985	1337	1662	2037	2446	2885	3356	300	671	60	1970	2628	3202	3867	4578	5328	6115
100	224	80	1195	1623	2026	2494	3005	3560	4158	300	671	80	2448	3221	3915	4735	5615	6547	7529
100	224	100	1389	1893	2375	2934	3550	4221	4951	300	671	100	2922	3687	4584	5560	6605	7716	8890
100	224	120	1581	2155	2714	3365	4086	4877	5742	300	671	120	3301	4199	5231	6356	7563	8850	10217
100	224	140	1771	2411	3046	3790	4618	5532	6537	300	671	140	3705	4696	5858	7129	8498	9962	11520
125	280	40	908	1193	1447	1759	2097	2457	2837	325	727	40	1557	2096	2555	3079	3637	4222	4834
125	280	60	1130	1528	1889	2307	2761	3246	3763	325	727	60	2084	2769	3367	4063	4805	5588	6409
125	280	80	1364	1853	2303	2823	3389	3998	4652	325	727	80	2596	3397	4117	4975	5894	6866	7889
125	280	100	1595	2164	2699	3319	3997	4732	5525	325	727	100	3106	3883	4821	5842	6933	8090	9313
125	280	120	1822	2466	3083	3803	4594	5456	6392	325	727	120	3489	4421	5501	6677	7937	9278	10699
125	280	140	2046	2759	3458	4277	5183	6175	7257	325	727	140	3919	4945	6160	7488	8915	10439	12059
150	335	40	971	1308	1605	1949	2320	2713	3127	350	783	40	1635	2198	2676	3222	3804	4413	5049
150	335	60	1257	1702	2101	2560	3055	3583	4144	350	783	60	2196	2908	3528	4253	5027	5842	6696
150	335	80	1528	2070	2564	3132	3749	4410	5117	350	783	80	2744	3476	4308	5209	6166	7178	8241
150	335	100	1794	2421	3004	3681	4418	5213	6067	350	783	100	3291	4076	5052	6117	7253	8456	9726
150	335	120	2055	2760	3430	4214	5072	6002	7006	350	783	120	3675	4639	5765	6991	8302	9695	11170
150	335	140	2314	3089	3845	4737	5716	6782	7939	350	783	140	4130	5189	6456	7839	9323	10905	12585
175	391	40	1035	1419	1755	2130	2531	2954	3400	375	839	40	1713	2298	2794	3362	3966	4600	5260
175	391	60	1381	1869	2303	2800	3335	3903	4505	375	839	60	2308	3044	3685	4440	5244	6090	6976
175	391	80	1688	2277	2811	3426	4091	4801	5558	375	839	80	2891	3636	4501	5438	6433	7483	8586
175	391	100	1988	2666	3294	4025	4818	5670	6582	375	839	100	3366	4257	5279	6386	7566	8814	10130
175	391	120	2284	3042	3760	4606	5527	6521	7591	375	839	120	3858	4853	6024	7298	8659	10104	11631
175	391	140	2577	3406	4214	5173	6223	7361	8590	375	839	140	4339	5428	6745	8183	9723	11362	13100
200	447	40	1151	1552	1902	2302	2732	3185	3661	400	894	40	1790	2396	2909	3499	4126	4782	5466
200	447	60	1503	2030	2495	3029	3601	4209	4851	400	894	60	2419	3178	3840	4622	5457	6334	7251
200	447	80	1844	2477	3047	3706	4417	5175	5980	400	894	80	3038	3793	4691	5662	6694	7782	8923
200	447	100	2179	2903	3571	4353	5201	6108	7076	400	894	100	3522	4440	5501	6649	7872	9165	10527
200	447	120	2509	3313	4076	4980	5962	7019	8151	400	894	120	4040	5062	6278	7599	9009	10504	12083
200	447	140	2837	3712	4567	5591	6709	7917	9215	400	894	140	4546	5662	7029	8519	10115	11810	13606

NOTE: Detector time constant at a reference velocity of 5 ft/sec (1.5m/sec).
For SI units: 1 ft = 0.305 m; 1000 Btu/sec = 1055 kW.

Table B-4(g)   Installed Spacing of Heat Detector: 10ft t_g: 150 seconds to 1000 Btu/sec α: 0.044 Btu/sec^3

τ	RTI	ΔT	Ceiling Height (ft)							τ	RTI	ΔT	Ceiling Height (ft)						
			4.0	8.0	12.0	16.0	20.0	24.0	28.0				4.0	8.0	12.0	16.0	20.0	24.0	28.0
			Fire Size at Detector Response (Btu/sec)										Fire Size at Detector Response (Btu/sec)						
25	56	40	152	209	280	362	461	576	709	225	503	40	544	694	871	1064	1273	1497	1737
25	56	60	205	284	389	519	671	856	1071	225	503	60	727	931	1173	1441	1734	2052	2395
25	56	80	252	358	502	683	904	1164	1473	225	503	80	895	1170	1460	1803	2181	2595	3046
25	56	100	298	435	621	858	1151	1504	1912	225	503	100	1034	1378	1738	2157	2623	3136	3701
25	56	120	345	514	746	1044	1416	1864	2385	225	503	120	1195	1587	2011	2507	3063	3680	4634
25	56	140	392	595	877	1241	1696	2247	2890	225	503	140	1345	1790	2280	2856	3504	4230	5040
50	112	40	221	296	380	481	596	727	872	250	559	40	580	742	928	1132	1352	1588	1840
50	112	60	297	398	525	669	841	1040	1268	250	559	60	775	994	1250	1532	1840	2173	2531
50	112	80	362	496	664	860	1097	1374	1695	250	559	80	948	1246	1555	1915	2311	2743	3212
50	112	100	426	593	805	1064	1366	1730	2154	250	559	100	1116	1473	1849	2289	2775	3310	3895
50	112	120	490	691	950	1269	1649	2107	2645	250	559	120	1279	1694	2138	2658	3237	3878	4585
50	112	140	561	791	1098	1482	1952	2506	3166	250	559	140	1440	1909	2422	3024	3698	4450	5284
75	168	40	277	369	467	584	715	861	1021	275	615	40	616	787	984	1198	1429	1676	1939
75	168	60	372	495	635	803	995	1212	1456	275	615	60	822	1055	1324	1621	1943	2290	2663
75	168	80	454	614	807	1022	1280	1576	1913	275	615	80	1001	1322	1647	2024	2438	2888	3375
75	168	100	535	731	971	1245	1574	1957	2397	275	615	100	1187	1564	1958	2417	2924	3479	4085
75	168	120	622	848	1135	1472	1879	2355	2908	275	615	120	1362	1797	2261	2804	3406	4070	4800
75	168	140	697	963	1302	1713	2194	2772	3446	275	615	140	1534	2025	2560	3187	3887	4664	5523
100	224	40	328	435	545	677	823	983	1158	300	671	40	663	832	1038	1263	1504	1762	2036
100	224	60	439	582	738	926	1136	1371	1632	300	671	60	880	1115	1397	1707	2043	2405	2792
100	224	80	537	720	926	1171	1450	1766	2122	300	671	80	1065	1398	1736	2130	2561	3028	3533
100	224	100	641	857	1122	1418	1770	2174	2633	300	671	100	1256	1652	2063	2542	3069	3644	4270
100	224	120	730	990	1307	1667	2098	2596	3168	300	671	120	1443	1898	2382	2946	3571	4258	5010
100	224	140	819	1121	1492	1920	2434	3034	3726	300	671	140	1625	2138	2695	3346	4071	4873	5757
125	280	40	383	487	617	763	923	1097	1286	325	727	40	695	875	1091	1325	1577	1845	2130
125	280	60	501	662	835	1040	1268	1520	1799	325	727	60	918	1172	1467	1791	2141	2516	2918
125	280	80	625	822	1044	1310	1611	1947	2322	325	727	80	1123	1471	1823	2234	2681	3166	3687
125	280	100	729	973	1249	1580	1956	2382	2862	325	727	100	1325	1738	2166	2664	3210	3805	4451
125	280	120	831	1121	1467	1850	2307	2828	3421	325	727	120	1522	1996	2499	3085	3732	4441	5216
125	280	140	935	1268	1670	2123	2665	3288	4002	325	727	140	1716	2249	2827	3502	4252	5078	5986
150	335	40	426	543	685	844	1017	1204	1406	350	783	40	727	918	1142	1387	1648	1927	2222
150	335	60	559	738	925	1147	1392	1662	1957	350	783	60	957	1229	1537	1873	2236	2626	3041
150	335	80	694	914	1155	1442	1762	2119	2513	350	783	80	1179	1542	1909	2335	2799	3300	3839
150	335	100	812	1081	1380	1733	2133	2581	3082	350	783	100	1392	1822	2266	2783	3348	3963	4628
150	335	120	930	1246	1602	2025	2507	3052	3667	350	783	120	1600	2093	2614	3222	3890	4621	5418
150	335	140	1034	1406	1839	2317	2886	3535	4271	350	783	140	1805	2330	2955	3655	4428	5279	6211
175	391	40	467	595	750	920	1106	1306	1521	375	839	40	758	959	1193	1447	1718	2007	2312
175	391	60	628	798	1011	1249	1511	1797	2108	375	839	60	996	1304	1605	1954	2330	2733	3162
175	391	80	760	1002	1261	1567	1907	2283	2697	375	839	80	1234	1589	1992	2434	2914	3432	3987
175	391	100	899	1186	1504	1880	2302	2772	3295	375	839	100	1459	1880	2364	2900	3484	4118	4802
175	391	120	1013	1362	1744	2192	2699	3268	3906	375	839	120	1678	2161	2726	3355	4045	4798	5616
175	391	140	1146	1540	1981	2503	3099	3773	4534	375	839	140	1894	2434	3081	3804	4601	5476	6431
200	447	40	506	646	811	994	1191	1403	1631	400	894	40	789	1000	1242	1505	1787	2085	2400
200	447	60	678	866	1093	1347	1625	1927	2254	400	894	60	1035	1359	1671	2032	2422	2837	3280
200	447	80	825	1086	1362	1687	2047	2442	2874	400	894	80	1289	1657	2074	2532	3027	3561	4132
200	447	100	966	1284	1623	2021	2465	2957	3501	400	894	100	1524	1959	2461	3014	3617	4269	4973
200	447	120	1109	1477	1880	2352	2884	3478	4138	400	894	120	1754	2251	2837	3486	4197	4971	5810
200	447	140	1247	1667	2133	2682	3305	4005	4790	400	894	140	1982	2536	3205	3951	4771	5669	6648

NOTE: Detector time constant at a reference velocity of 5 ft/sec (1.5m/sec).
For SI units: 1 ft = 0.305 m; 1000 Btu/sec = 1055 kW.

Table B-4(h)   Installed Spacing of Heat Detector: 10ft t_g: 300 seconds to 1000 Btu/sec α: 0.011 Btu/sec^3

τ	RTI	ΔT	Ceiling Height (ft) Fire Size at Detector Response (Btu/sec) 4.0	8.0	12.0	16.0	20.0	24.0	28.0
25	56	40	91	131	183	249	330	424	537
25	56	60	126	187	271	379	513	675	863
25	56	80	160	246	367	524	720	957	1234
25	56	100	195	309	471	682	947	1268	1645
25	56	120	230	377	582	853	1193	1605	2091
25	56	140	267	448	701	1036	1456	1967	2574
50	112	40	131	180	241	313	399	500	616
50	112	60	179	250	344	459	597	762	956
50	112	80	224	322	451	615	816	1054	1338
50	112	100	269	395	565	782	1051	1377	1758
50	112	120	314	470	684	960	1305	1723	2213
50	112	140	359	549	810	1150	1575	2092	2700
75	168	40	164	222	293	371	465	572	694
75	168	60	224	306	410	532	679	851	1050
75	168	80	279	389	530	703	909	1155	1444
75	168	100	333	473	654	881	1158	1485	1874
75	168	120	386	558	784	1068	1420	1839	2338
75	168	140	440	645	918	1266	1697	2220	2833
100	224	40	194	260	335	424	526	640	769
100	224	60	265	357	471	602	757	937	1143
100	224	80	328	451	604	784	1001	1255	1550
100	224	100	391	545	739	977	1258	1596	1991
100	224	120	453	639	879	1175	1534	1959	2464
100	224	140	519	735	1022	1380	1821	2344	2967
125	280	40	222	296	377	474	583	705	841
125	280	60	302	404	528	667	832	1020	1234
125	280	80	374	508	673	863	1090	1353	1656
125	280	100	445	612	820	1065	1360	1706	2108
125	280	120	514	716	969	1278	1642	2079	2591
125	280	140	587	820	1122	1493	1938	2473	3103
150	335	40	248	329	417	521	637	766	909
150	335	60	366	448	577	729	903	1100	1321
150	335	80	417	563	739	939	1176	1448	1759
150	335	100	495	676	897	1153	1458	1814	2224
150	335	120	577	789	1056	1372	1752	2199	2718
150	335	140	651	901	1218	1603	2058	2602	3239
175	391	40	273	361	455	566	689	825	975
175	391	60	370	490	627	789	972	1177	1407
175	391	80	458	615	796	1012	1259	1541	1860
175	391	100	544	737	971	1237	1554	1920	2339
175	391	120	631	858	1140	1468	1859	2316	2844
175	391	140	712	979	1311	1703	2175	2730	3375
200	447	40	301	386	491	609	739	882	1038
200	447	60	401	531	676	846	1038	1251	1489
200	447	80	497	665	856	1082	1339	1631	1959
200	447	100	595	796	1042	1319	1647	2023	2452
200	447	120	684	925	1220	1560	1964	2432	2969
200	447	140	771	1053	1400	1806	2290	2856	3510

τ	RTI	ΔT	Ceiling Height (ft) Fire Size at Detector Response (Btu/sec) 4.0	8.0	12.0	16.0	20.0	24.0	28.0
225	503	40	323	415	526	650	787	936	1099
225	503	60	432	570	723	901	1101	1324	1570
225	503	80	535	713	914	1150	1417	1718	2056
225	503	100	639	853	1103	1399	1738	2125	2563
225	503	120	734	990	1299	1650	2066	2545	3092
225	503	140	828	1125	1487	1906	2403	2980	3645
250	559	40	345	443	560	690	833	989	1159
250	559	60	461	608	768	955	1163	1394	1649
250	559	80	572	759	969	1215	1493	1804	2151
250	559	100	682	907	1168	1476	1826	2224	2672
250	559	120	783	1052	1374	1738	2166	2656	3214
250	559	140	888	1196	1572	2003	2514	3103	3778
275	615	40	366	470	593	729	878	1041	1217
275	615	60	490	644	812	1007	1224	1463	1725
275	615	80	613	806	1024	1280	1567	1888	2244
275	615	100	723	961	1232	1551	1913	2321	2780
275	615	120	831	1113	1439	1823	2264	2766	3334
275	615	140	938	1264	1655	2098	2622	3223	3910
300	671	40	387	496	624	767	922	1091	1273
300	671	60	518	680	855	1058	1283	1530	1800
300	671	80	647	850	1077	1342	1639	1969	2335
300	671	100	763	1012	1294	1624	1997	2416	2885
300	671	120	883	1173	1510	1907	2360	2873	3453
300	671	140	988	1330	1725	2191	2728	3342	4040
325	727	40	407	521	656	804	965	1140	1328
325	727	60	546	715	897	1107	1340	1595	1873
325	727	80	680	892	1128	1403	1710	2050	2425
325	727	100	803	1063	1355	1696	2080	2510	2989
325	727	120	925	1230	1579	1989	2454	2979	3570
325	727	140	1038	1395	1803	2283	2833	3459	4168
350	783	40	427	546	686	840	1007	1188	1382
350	783	60	572	749	937	1156	1396	1659	1945
350	783	80	713	934	1179	1463	1779	2128	2513
350	783	100	842	1113	1415	1766	2161	2602	3091
350	783	120	967	1287	1647	2069	2546	3083	3685
350	783	140	1092	1459	1879	2372	2935	3574	4295
375	839	40	446	571	716	875	1048	1235	1435
375	839	60	606	774	977	1203	1451	1722	2016
375	839	80	745	976	1228	1521	1847	2205	2599
375	839	100	886	1162	1473	1835	2241	2692	3192
375	839	120	1009	1342	1714	2148	2637	3185	3798
375	839	140	1141	1521	1953	2460	3036	3688	4421
400	894	40	465	595	745	910	1089	1281	1487
400	894	60	631	807	1017	1250	1505	1784	2085
400	894	80	776	1016	1277	1579	1913	2281	2684
400	894	100	920	1209	1530	1903	2319	2781	3291
400	894	120	1055	1397	1780	2225	2726	3286	3910
400	894	140	1190	1582	2027	2547	3136	3800	4544

NOTE: Detector time constant at a reference velocity of 5 ft/sec (1.5m/sec).
For SI units: 1 ft = 0.305 m; 1000 Btu/sec = 1055 kW.

Table B-4(i)   Installed Spacing of Heat Detector: 10ft t_g: 500 seconds to 1000 Btu/sec α: 0.004 Btu/sec^3

τ	RTI	ΔT	Ceiling Height (ft) Fire Size at Detector Response (Btu/sec)						
			4.0	8.0	12.0	16.0	20.0	24.0	28.0
25	56	40	65	98	143	202	274	362	465
25	56	60	92	146	222	322	447	598	775
25	56	80	121	200	312	458	643	867	1133
25	56	100	150	258	410	609	861	1166	1530
25	56	120	181	320	516	773	1098	1493	1963
25	56	140	214	386	630	950	1352	1844	2430
50	112	40	91	130	179	241	316	405	510
50	112	60	127	186	266	369	495	649	828
50	112	80	162	246	361	511	698	924	1191
50	112	100	198	309	464	666	920	1229	1593
50	112	120	235	376	574	835	1162	1561	2031
50	112	140	272	447	692	1015	1422	1917	2506
75	168	40	114	158	212	279	356	449	556
75	168	60	158	223	308	415	545	699	882
75	168	80	199	290	410	564	754	981	1251
75	168	100	241	359	519	725	982	1293	1660
75	168	120	283	431	634	898	1228	1630	2103
75	168	140	326	506	756	1082	1492	1991	2580
100	224	40	134	183	243	312	395	491	602
100	224	60	185	257	348	460	592	751	937
100	224	805	233	330	458	616	810	1040	1313
100	224	100	280	406	572	783	1044	1356	1727
100	224	120	328	484	693	961	1295	1700	2176
100	224	140	376	564	819	1150	1563	2065	2657
125	280	40	153	207	272	345	432	533	647
125	280	60	210	288	386	501	640	803	992
125	280	80	264	369	503	667	863	1099	1375
125	280	100	317	451	624	841	1106	1421	1795
125	280	120	370	534	751	1024	1362	1768	2250
125	280	140	422	619	882	1218	1635	2141	2736
150	335	40	171	229	297	377	468	572	690
150	335	60	234	318	422	542	686	853	1046
150	335	80	293	405	546	717	918	1158	1437
150	335	100	351	493	674	897	1165	1486	1864
150	335	120	409	582	807	1087	1430	1838	2324
150	335	140	466	673	944	1285	1707	2213	2815
175	391	40	187	250	322	407	503	611	733
175	391	60	257	346	456	582	731	903	1100
175	391	80	321	440	588	762	971	1216	1499
175	391	100	384	534	723	953	1225	1551	1933
175	391	120	447	629	861	1148	1494	1909	2398
175	391	140	512	725	1004	1352	1780	2289	2895
200	447	40	204	271	347	435	536	648	774
200	447	60	279	374	489	620	774	951	1152
200	447	80	348	474	629	809	1023	1273	1561
200	447	100	416	574	770	1007	1285	1616	2002
200	447	120	483	674	915	1209	1560	1980	2473
200	447	140	552	775	1062	1419	1848	2365	2975

τ	RTI	ΔT	Ceiling Height (ft) Fire Size at Detector Response (Btu/sec)						
			4.0	8.0	12.0	16.0	20.0	24.0	28.0
225	503	40	219	291	370	463	568	685	815
225	503	60	300	400	518	657	817	998	1204
225	503	80	374	507	668	854	1074	1329	1622
225	503	100	446	612	816	1056	1343	1680	2070
225	503	120	518	717	966	1268	1625	2051	2548
225	503	140	591	823	1120	1484	1919	2441	3055
250	559	40	234	310	393	490	599	720	854
250	559	60	320	426	549	693	858	1045	1255
250	559	80	399	538	706	898	1124	1385	1683
250	559	100	476	649	861	1107	1401	1743	2139
250	559	120	555	760	1017	1323	1689	2121	2623
250	559	140	628	870	1176	1548	1990	2517	3135
275	615	40	249	328	415	517	630	755	893
275	615	60	339	451	578	728	898	1090	1305
275	615	80	423	569	743	941	1173	1440	1742
275	615	100	505	685	904	1157	1458	1806	2206
275	615	120	588	801	1066	1379	1753	2190	2697
275	615	140	665	916	1231	1611	2059	2593	3216
300	671	40	266	343	437	542	659	788	930
300	671	60	358	475	607	762	938	1135	1355
300	671	80	447	599	755	983	1221	1493	1801
300	671	100	533	720	947	1207	1513	1868	2273
300	671	120	619	841	1115	1435	1815	2259	2772
300	671	140	701	961	1285	1668	2128	2668	3296
325	727	40	280	361	458	567	688	821	967
325	727	60	377	499	636	796	976	1178	1403
325	727	80	470	628	810	1024	1269	1546	1860
325	727	100	564	755	988	1255	1568	1929	2340
325	727	120	650	880	1162	1489	1877	2327	2845
325	727	140	735	1005	1338	1729	2196	2742	3376
350	783	40	293	378	479	592	717	854	1003
350	783	60	395	522	664	829	1014	1221	1451
350	783	80	493	657	844	1064	1315	1598	1917
350	783	100	591	789	1024	1302	1622	1989	2406
350	783	120	681	919	1209	1543	1938	2394	2919
350	783	140	770	1048	1389	1788	2263	2816	3456
375	839	40	306	394	499	616	744	885	1039
375	839	60	413	544	691	861	1051	1263	1498
375	839	80	515	685	878	1104	1360	1649	1974
375	839	100	616	822	1064	1348	1675	2048	2471
375	839	120	710	957	1254	1596	1998	2461	2991
375	839	140	803	1091	1440	1847	2329	2889	3535
400	894	40	319	410	519	639	772	916	1074
400	894	60	431	567	718	893	1088	1305	1544
400	894	80	537	712	911	1143	1405	1700	2030
400	894	100	642	854	1102	1394	1728	2107	2536
400	894	120	740	994	1299	1648	2057	2527	3063
400	894	140	836	1132	1490	1905	2395	2692	3614

NOTE: Detector time constant at a reference velocity of 5 ft/sec (1.5m/sec).
For SI units: 1 ft = 0.305 m; 1000 Btu/sec = 1055 kW.

Table B-4(j)   Installed Spacing of Heat Detector: 10ft t_g: 600 seconds to 1000 Btu/sec α: 0.003 Btu/sec^3

τ	RTI	ΔT	4.0	8.0	12.0	16.0	20.0	24.0	28.0	τ	RTI	ΔT	4.0	8.0	12.0	16.0	20.0	24.0	28.0
			Ceiling Height (ft)										Ceiling Height (ft)						

τ	RTI	ΔT	Fire Size at Detector Response (Btu/sec)							τ	RTI	ΔT	Fire Size at Detector Response (Btu/sec)						
25	56	40	58	89	133	190	260	346	446	225	503	40	192	256	328	413	509	618	739
25	56	60	84	136	210	307	429	577	752	225	503	60	264	354	465	591	741	912	1108
25	56	80	110	188	298	442	623	843	1105	225	503	80	330	451	600	775	984	1227	1510
25	56	100	139	244	394	591	838	1139	1498	225	503	100	395	547	737	968	1240	1564	1944
25	56	120	168	305	499	753	1073	1463	1928	225	503	120	460	644	878	1165	1510	1924	2410
25	56	140	200	371	611	927	1325	1811	2391	225	503	140	526	742	1022	1371	1797	2304	2906
50	112	40	81	116	163	222	295	381	483	250	559	40	205	272	348	437	536	648	773
50	112	60	114	170	246	346	469	619	795	250	559	60	281	377	492	623	776	952	1152
50	112	80	146	226	338	485	668	890	1153	250	559	80	352	478	633	813	1026	1274	1561
50	112	100	179	287	439	637	887	1191	1550	250	559	100	421	579	776	1009	1289	1618	2001
50	112	120	214	352	547	803	1126	1519	1987	250	559	120	489	681	921	1215	1564	1982	2472
50	112	140	249	420	662	981	1382	1871	2454	250	559	140	559	783	1070	1425	1853	2367	2972
75	168	40	100	140	191	254	328	417	521	275	615	40	218	289	367	459	563	678	806
75	168	60	140	201	282	384	510	661	839	275	615	60	298	398	515	653	811	991	1194
75	168	80	178	264	379	528	714	938	1203	275	615	80	373	505	665	850	1068	1321	1611
75	168	100	216	329	484	685	937	1244	1605	275	615	100	446	611	813	1052	1337	1671	2058
75	168	120	256	398	596	855	1180	1576	2043	275	615	120	518	716	964	1264	1618	2040	2534
75	168	140	295	470	715	1036	1440	1932	2515	275	615	140	591	823	1118	1479	1912	2430	3039
100	224	40	118	162	218	283	361	452	559	300	671	40	230	304	386	482	588	707	839
100	224	60	164	230	316	422	550	704	884	300	671	60	315	419	541	683	845	1029	1237
100	224	80	207	298	419	572	760	985	1253	300	671	80	394	531	697	886	1110	1367	1661
100	224	100	250	369	529	734	988	1297	1660	300	671	100	471	641	850	1094	1385	1723	2114
100	224	120	294	443	645	907	1235	1633	2103	300	671	120	549	752	1006	1309	1671	2098	2596
100	224	140	338	519	767	1091	1498	1994	2579	300	671	140	622	862	1164	1532	1970	2493	3106
125	280	40	134	183	242	311	393	487	596	325	727	40	242	320	405	503	614	736	871
125	280	60	186	257	348	459	590	746	930	325	727	60	331	440	565	712	879	1067	1278
125	280	80	234	332	458	615	806	1034	1304	325	727	80	414	557	728	922	1150	1412	1711
125	280	100	282	408	573	782	1040	1349	1716	325	727	100	495	671	886	1136	1432	1775	2170
125	280	120	330	486	694	959	1290	1692	2164	325	727	120	576	786	1047	1356	1724	2156	2658
125	280	140	379	566	820	1148	1557	2056	2643	325	727	140	652	900	1210	1585	2028	2556	3173
150	335	40	150	202	266	338	423	521	633	350	783	40	257	332	423	525	638	764	902
150	335	60	207	283	379	493	629	789	975	350	783	60	347	460	589	740	911	1104	1319
150	335	80	260	363	495	657	851	1083	1356	350	783	80	434	581	754	957	1190	1457	1760
150	335	100	312	445	616	829	1091	1403	1773	350	783	100	518	701	922	1177	1478	1826	2226
150	335	120	365	527	741	1012	1346	1747	2225	350	783	120	602	819	1087	1402	1776	2213	2719
150	335	140	417	612	872	1204	1617	2118	2708	350	783	140	682	937	1254	1632	2085	2618	3239
175	391	40	164	221	286	364	453	554	669	375	839	40	268	346	440	546	662	791	933
175	391	60	226	308	409	527	667	831	1020	375	839	60	363	480	613	768	944	1140	1359
175	391	80	284	394	531	698	896	1131	1407	375	839	80	453	606	783	991	1230	1501	1808
175	391	100	341	480	657	876	1140	1457	1830	375	839	100	545	730	957	1217	1524	1877	2281
175	391	120	398	568	788	1064	1402	1806	2286	375	839	120	628	852	1127	1447	1828	2270	2781
175	391	140	454	657	923	1260	1677	2178	2774	375	839	140	711	974	1299	1683	2142	2680	3306
200	447	40	178	239	308	389	482	587	705	400	894	40	280	361	458	566	686	818	963
200	447	60	245	331	437	559	704	872	1064	400	894	60	378	499	636	796	975	1176	1399
200	447	80	308	423	566	736	940	1180	1459	400	894	80	472	630	812	1025	1268	1544	1856
200	447	100	369	514	698	922	1190	1511	1887	400	894	100	567	758	991	1257	1569	1927	2336
200	447	120	429	606	833	1115	1458	1865	2348	400	894	120	654	884	1166	1492	1878	2326	2842
200	447	140	489	700	973	1316	1737	2241	2840	400	894	140	740	1010	1342	1733	2198	2742	3372

NOTE: Detector time constant at a reference velocity of 5 ft/sec (1.5m/sec).
For SI units: 1 ft = 0.305 m; 1000 Btu/sec = 1055 kW.

Table B-4(k)   Installed Spacing of Heat Detector: 12ft t_g: 50 seconds to 1000 Btu/sec α: 0.400 Btu/sec^3

τ	RTI	ΔT	____ Ceiling Height (ft) ____							τ	RTI	ΔT	____ Ceiling Height (ft) ____						
			4.0	8.0	12.0	16.0	20.0	24.0	28.0				4.0	8.0	12.0	16.0	20.0	24.0	28.0
			Fire Size at Detector Response (Btu/sec)										Fire Size at Detector Response (Btu/sec)						
25	56	40	482	585	730	897	1085	1291	1518	225	503	40	1535	2023	2398	2845	3325	3831	4360
25	56	60	593	751	952	1184	1446	1740	2069	225	503	60	2046	2586	3145	3747	4389	5068	5783
25	56	80	722	913	1168	1467	1810	2200	2640	225	503	80	2543	3159	3843	4588	5386	6234	7130
25	56	100	821	1090	1384	1753	2182	2675	3238	225	503	100	3036	3702	4507	5391	6342	7357	8435
25	56	120	927	1243	1598	2043	2565	3169	3864	225	503	120	3416	4215	5145	6168	7271	8452	9712
25	56	140	1007	1389	1836	2339	2959	3683	4518	225	503	140	3834	4714	5765	6925	8180	9529	10973
50	112	40	668	831	1026	1244	1482	1740	2017	250	559	40	1643	2161	2556	3029	3537	4071	4630
50	112	60	878	1078	1338	1633	1958	2314	2702	250	559	60	2021	2763	3355	3992	4671	5388	6142
50	112	80	993	1328	1635	2006	2422	2880	3385	250	559	80	2746	3378	4101	4889	5732	6626	7569
50	112	100	1162	1552	1923	2374	2882	3450	4079	250	559	100	3289	3959	4809	5744	6748	7816	8949
50	112	120	1316	1766	2206	2739	3345	4026	4788	250	559	120	3671	4505	5490	6570	7733	8976	10298
50	112	140	1469	1977	2486	3105	3812	4613	5516	250	559	140	4123	5039	6151	7375	8697	10114	11627
75	168	40	867	1043	1273	1533	1815	2117	2438	275	615	40	1750	2294	2709	3208	3742	4304	4892
75	168	60	1021	1370	1663	2011	2392	2804	3248	275	615	60	2354	2936	3558	4230	4944	5698	6489
75	168	80	1255	1668	2030	2466	2947	3472	4042	275	615	80	2947	3590	4351	5181	6067	7006	7995
75	168	100	1461	1946	2382	2907	3491	4132	4834	275	615	100	3423	4200	5102	6087	7141	8262	9448
75	168	120	1665	2216	2726	3342	4031	4793	5632	275	615	120	3921	4788	5824	6961	8182	9484	10866
75	168	140	1867	2480	3065	3774	4570	5458	6441	275	615	140	4408	5355	6525	7811	9198	10681	12261
100	224	40	957	1257	1497	1790	2111	2452	2814	300	671	40	1856	2425	2858	3381	3942	4531	5145
100	224	60	1226	1625	1954	2349	2781	3245	3741	300	671	60	2507	3104	3756	4461	5210	5999	6826
100	224	80	1486	1971	2383	2878	3420	4006	4638	300	671	80	3149	3797	4594	5465	6393	7375	8409
100	224	100	1742	2303	2795	3388	4042	4753	5525	300	671	100	3635	4441	5387	6420	7524	8696	9933
100	224	120	1993	2624	3195	3888	4654	5494	6410	300	671	120	4168	5063	6150	7341	8618	9978	11419
100	224	140	2242	2937	3587	4380	5263	6234	7300	300	671	140	4690	5663	6889	8236	9686	11233	12879
125	280	40	1047	1409	1695	2026	2382	2761	3160	325	727	40	1960	2475	2995	3550	4136	4751	5392
125	280	60	1399	1852	2220	2661	3139	3651	4195	325	727	60	2658	3268	3949	4686	5469	6292	7155
125	280	80	1708	2253	2709	3258	3857	4501	5190	325	727	80	3259	3992	4831	5741	6711	7736	8813
125	280	100	2011	2636	3176	3833	4552	5330	6168	325	727	100	3844	4676	5666	6745	7897	9118	10407
125	280	120	2309	2933	3625	4393	5233	6147	7137	325	727	120	4412	5331	6467	7711	9044	10460	11959
125	280	140	2605	3283	4067	4943	5907	6960	8106	325	727	140	4840	5956	7244	8651	10162	11772	13482
150	335	40	1199	1584	1886	2246	2636	3049	3483	350	783	40	2064	2594	3136	3715	4326	4966	5634
150	335	60	1566	2067	2470	2953	3475	4032	4622	350	783	60	2809	3428	4138	4906	5721	6579	7476
150	335	80	1923	2520	3015	3615	4267	4965	5710	350	783	80	3430	4188	5062	6012	7022	8087	9207
150	335	100	2272	2874	3530	4250	5031	5872	6774	350	783	100	4050	4906	5938	7062	8262	9531	10869
150	335	120	2618	3277	4032	4868	5778	6763	7824	350	783	120	4653	5594	6778	8074	9460	10931	12486
150	335	140	2961	3668	4522	5473	6514	7646	8870	350	783	140	5089	6249	7591	9056	10627	12299	14071
175	391	40	1313	1736	2064	2455	2876	3321	3789	375	839	40	2166	2712	3275	3876	4511	5176	5869
175	391	60	1729	2273	2707	3229	3793	4392	5026	375	839	60	3599	4379	5289	6276	7326	8432	9593
175	391	80	2133	2699	3300	3953	4656	5406	6204	375	839	80	3599	4379	5289	6276	7326	8432	9593
175	391	100	2529	3161	3869	4647	5487	6388	7350	375	839	100	4254	5132	6204	7373	8619	9936	11322
175	391	120	2922	3605	4419	5319	6297	7350	8479	375	839	120	4765	5844	7081	8428	9867	11392	13003
175	391	140	3317	4035	4954	5977	7092	8299	9599	375	839	140	5334	6535	7931	9453	11082	12815	14649
200	447	40	1425	1882	2234	2654	3105	3581	4081	400	894	40	2268	2827	3410	4034	4693	5382	6100
200	447	60	1889	2471	2934	3493	4097	4737	5412	400	894	60	3112	3740	4503	5328	6211	7134	8098
200	447	80	2339	2933	3577	4277	5028	5828	6676	400	894	80	3766	4567	5511	6535	7623	8769	9970
200	447	100	2783	3436	4194	5026	5923	6882	7903	400	894	100	4456	5353	6465	7677	8968	10332	11766
200	447	120	3226	3919	4789	5752	6793	7911	9107	400	894	120	4979	6095	7379	8776	10266	11845	13509
200	447	140	3538	4378	5367	6460	7646	8925	10298	400	894	140	5575	6817	8264	9842	11529	13321	15215

NOTE: Detector time constant at a reference velocity of 5 ft/sec (1.5m/sec).
For SI units: 1 ft = 0.305 m; 1000 Btu/sec = 1055 kW.

Table B-4(I) Installed Spacing of Heat Detector: 12ft t_g: 150 seconds to 1000 Btu/sec α: 0.044 Btu/sec^3

τ	RTI	ΔT	Ceiling Height (ft)							τ	RTI	ΔT	Ceiling Height (ft)						
			4.0	8.0	12.0	16.0	20.0	24.0	28.0				4.0	8.0	12.0	16.0	20.0	24.0	28.0
			Fire Size at Detector Response (Btu/sec)										Fire Size at Detector Response (Btu/sec)						
25	56	40	190	254	327	418	525	648	790	225	503	40	685	837	1023	1228	1450	1687	1940
25	56	60	256	345	461	602	766	964	1194	225	503	60	905	1122	1380	1665	1976	2313	2676
25	56	80	316	437	596	794	1030	1313	1644	225	503	80	1104	1390	1718	2085	2488	2927	3405
25	56	100	375	531	738	999	1318	1693	2136	225	503	100	1304	1647	2047	2496	2993	3540	4139
25	56	120	435	628	888	1217	1623	2100	2665	225	503	120	1499	1898	2369	2902	3497	4155	4882
25	56	140	501	729	1045	1448	1946	2542	3230	225	503	140	1691	2143	2688	3307	4002	4778	5639
50	112	40	275	359	447	556	679	818	972	250	559	40	729	894	1091	1307	1540	1789	2054
50	112	60	368	483	613	774	959	1173	1415	250	559	60	958	1197	1470	1771	2097	2450	2829
50	112	80	452	602	779	997	1253	1550	1893	250	559	80	1182	1483	1829	2214	2636	3095	3592
50	112	100	534	722	954	1228	1561	1953	2407	250	559	100	1396	1757	2178	2648	3167	3736	4357
50	112	120	621	842	1127	1475	1886	2380	2957	250	559	120	1606	2024	2519	3076	3695	4378	5129
50	112	140	700	963	1303	1725	2227	2832	3540	250	559	140	1813	2284	2855	3501	4223	5025	5913
75	168	40	353	439	549	675	814	969	1139	275	615	40	772	949	1156	1383	1628	1889	2166
75	168	60	462	590	747	929	1134	1366	1625	275	615	60	1012	1271	1557	1873	2215	2583	2977
75	168	80	567	745	943	1184	1461	1778	2137	275	615	80	1257	1574	1937	2340	2780	3258	3774
75	168	100	675	889	1138	1443	1798	2209	2679	275	615	100	1487	1864	2305	2796	3336	3927	4569
75	168	120	772	1030	1334	1708	2148	2660	3251	275	615	120	1712	2146	2664	3245	3888	4595	5370
75	168	140	878	1173	1543	1980	2510	3132	3854	275	615	140	1934	2421	3017	3690	4439	5267	6180
100	224	40	415	516	641	782	937	1107	1292	300	671	40	814	1002	1220	1458	1713	1985	2274
100	224	60	545	692	869	1070	1295	1545	1822	300	671	60	1079	1343	1643	1972	2329	2712	3121
100	224	80	678	862	1091	1355	1655	1993	2371	300	671	80	1332	1662	2042	2463	2921	3417	3951
100	224	100	794	1039	1311	1642	2021	2454	2943	300	671	100	1576	1968	2428	2940	3501	4113	4776
100	224	120	914	1202	1531	1932	2397	2932	3542	300	671	120	1816	2265	2805	3409	4076	4807	5605
100	224	140	1017	1361	1752	2227	2783	3427	4168	300	671	140	2027	2554	3175	3874	4649	5503	6442
125	280	40	472	587	726	881	1050	1235	1435	325	727	40	870	1055	1282	1530	1796	2080	2379
125	280	60	634	788	982	1202	1445	1714	2009	325	727	60	1137	1413	1726	2069	2440	2838	3262
125	280	80	771	978	1230	1516	1837	2196	2594	325	727	80	1404	1749	2145	2582	3058	3572	4124
125	280	100	910	1164	1473	1829	2233	2688	3199	325	727	100	1664	2070	2549	3081	3663	4295	4979
125	280	120	1029	1359	1715	2144	2635	3194	3826	325	727	120	1919	2381	2943	3570	4260	5014	5836
125	280	140	1168	1540	1957	2462	3045	3714	4476	325	727	140	2136	2684	3330	4054	4855	5735	6699
150	335	40	526	654	806	974	1157	1356	1570	350	783	40	904	1105	1342	1601	1878	2172	2483
150	335	60	704	877	1088	1326	1587	1873	2186	350	783	60	1193	1481	1807	2164	2549	2961	3400
150	335	80	871	1089	1360	1668	2010	2390	2809	350	783	80	1476	1833	2245	2699	3192	3723	4293
150	335	100	1004	1293	1626	2006	2434	2913	3446	350	783	100	1750	2169	2667	3218	3820	4473	5178
150	335	120	1159	1493	1889	2345	2863	3446	4101	350	783	120	2021	2495	3078	3727	4440	5218	6062
150	335	140	1306	1708	2152	2686	3297	3992	4778	350	783	140	2243	2812	3481	4230	5056	5961	6951
175	391	40	578	717	881	1062	1259	1471	1698	375	839	40	938	1155	1401	1670	1957	2262	2583
175	391	60	771	962	1190	1443	1722	2026	2356	375	839	60	1249	1548	1887	2257	2656	3082	3535
175	391	80	944	1193	1484	1812	2176	2576	3014	375	839	80	1547	1916	2343	2814	3324	3872	4459
175	391	100	1111	1416	1772	2176	2627	3129	3684	375	839	100	1836	2267	2783	3353	3975	4648	5373
175	391	120	1276	1633	2056	2538	3081	3690	4369	375	839	120	2088	2606	3210	3881	4617	5417	6284
175	391	140	1438	1847	2337	2900	3450	4262	5072	375	839	140	2348	2937	3629	4402	5253	6184	7198
200	447	40	640	779	954	1147	1356	1581	1821	400	894	40	973	1203	1460	1738	2035	2350	2682
200	447	60	836	1043	1286	1556	1852	2172	2519	400	894	60	1304	1614	1965	2348	2760	3200	3668
200	447	80	1017	1293	1604	1951	2334	2754	3213	400	894	80	1617	1997	2440	2926	3453	4018	4622
200	447	100	1209	1534	1912	2339	2813	3338	3915	400	894	100	1921	2362	2896	3486	4127	4819	5564
200	447	120	1389	1768	2215	2723	3292	3926	4629	400	894	120	2180	2715	3340	4033	4790	5612	6502
200	447	140	1566	1998	2515	3107	3775	4523	5359	400	894	140	2451	3059	3775	4572	5447	6403	7441

NOTE: Detector time constant at a reference velocity of 5 ft/sec (1.5m/sec).
For SI units: 1 ft = 0.305 m; 1000 Btu/sec = 1055 kW.

Table B-4(m)   Installed Spacing of Heat Detector: 12ft t_g: 300 seconds to 1000 Btu/sec α: 0.011 Btu/sec³

τ	RTI	ΔT	4.0	8.0	12.0	16.0	20.0	24.0	28.0	τ	RTI	ΔT	4.0	8.0	12.0	16.0	20.0	24.0	28.0
			Fire Size at Detector Response (Btu/sec)										Fire Size at Detector Response (Btu/sec)						
25	56	40	115	160	218	290	375	479	599	225	503	40	401	500	619	751	897	1056	1229
25	56	60	159	228	322	442	588	761	964	225	503	60	538	682	851	1043	1257	1494	1756
25	56	80	203	302	437	611	825	1082	1380	225	503	80	672	855	1077	1332	1619	1941	2301
25	56	100	247	380	562	797	1087	1435	1840	225	503	100	796	1025	1301	1621	1987	2401	2869
25	56	120	294	463	697	998	1370	1818	2339	225	503	120	919	1201	1525	1914	2363	2877	3462
25	56	140	341	551	840	1212	1673	2228	2875	225	503	140	1034	1366	1750	2212	2750	3370	4081
50	112	40	164	219	283	363	456	564	688	250	559	40	428	534	659	798	950	1116	1295
50	112	60	224	305	408	534	682	861	1069	250	559	60	575	726	904	1105	1328	1574	1844
50	112	80	282	392	536	716	932	1192	1496	250	559	80	717	911	1143	1408	1705	2037	2407
50	112	100	339	482	672	911	1206	1554	1967	250	559	100	849	1090	1378	1710	2087	2513	2991
50	112	120	396	575	815	1121	1498	1946	2476	250	559	120	977	1266	1613	2015	2477	3003	3598
50	112	140	458	672	966	1343	1809	2369	3022	250	559	140	1106	1451	1848	2324	2875	3508	4230
75	168	40	205	270	342	429	530	645	775	275	615	40	455	566	697	843	1002	1174	1360
75	168	60	280	372	482	617	776	961	1174	275	615	60	617	770	956	1165	1397	1651	1930
75	168	80	349	473	629	814	1040	1305	1615	275	615	80	761	964	1207	1482	1789	2132	2511
75	168	100	418	576	777	1025	1322	1679	2096	275	615	100	905	1153	1453	1797	2186	2623	3111
75	168	120	487	680	931	1245	1628	2080	2616	275	615	120	1035	1339	1699	2114	2588	3126	3733
75	168	140	558	787	1092	1476	1948	2508	3170	275	615	140	1174	1522	1944	2434	2999	3644	4377
100	224	40	243	312	395	491	600	722	859	300	671	40	481	598	735	886	1052	1231	1423
100	224	60	330	433	552	697	865	1058	1278	300	671	60	651	812	1006	1224	1464	1726	2013
100	224	80	411	548	711	910	1145	1418	1734	300	671	80	804	1017	1269	1554	1872	2224	2613
100	224	100	490	662	877	1132	1440	1804	2227	300	671	100	952	1215	1526	1882	2282	2730	3229
100	224	120	573	778	1042	1367	1753	2216	2757	300	671	120	1097	1409	1782	2211	2698	3248	3865
100	224	140	649	895	1213	1608	2083	2653	3321	300	671	140	1240	1601	2037	2542	3120	3778	4523
125	280	40	281	354	445	548	665	795	939	325	727	40	506	628	771	929	1101	1286	1485
125	280	60	376	485	618	773	950	1151	1379	325	727	60	685	854	1056	1281	1529	1800	2096
125	280	80	468	617	791	1001	1246	1528	1852	325	727	80	846	1068	1329	1624	1952	2315	2713
125	280	100	562	744	966	1237	1556	1928	2359	325	727	100	999	1275	1598	1965	2376	2836	3346
125	280	120	648	871	1149	1480	1880	2352	2899	325	727	120	1155	1478	1864	2305	2805	3367	3996
125	280	140	734	998	1331	1732	2220	2798	3473	325	727	140	1305	1679	2128	2647	3239	3910	4667
150	335	40	313	393	491	602	727	864	1015	350	783	40	531	658	807	971	1148	1340	1546
150	335	60	419	537	680	845	1031	1241	1477	350	783	60	717	894	1104	1337	1593	1873	2176
150	335	80	521	683	867	1089	1344	1636	1967	350	783	80	892	1118	1389	1693	2031	2403	2812
150	335	100	624	821	1055	1338	1668	2050	2489	350	783	100	1047	1334	1668	2046	2469	2939	3460
150	335	120	720	958	1245	1594	2005	2486	3042	350	783	120	1211	1546	1943	2398	2910	3484	4125
150	335	140	814	1095	1444	1857	2356	2943	3626	350	783	140	1368	1754	2218	2751	3356	4040	4809
175	391	40	344	430	536	654	786	930	1089	375	839	40	555	688	842	1011	1195	1393	1605
175	391	60	461	587	739	913	1109	1328	1573	375	839	60	749	934	1151	1392	1656	1944	2255
175	391	80	579	739	940	1172	1438	1740	2081	375	839	80	929	1167	1447	1761	2109	2490	2908
175	391	100	684	895	1140	1435	1777	2170	2617	375	839	100	1099	1392	1736	2125	2560	3041	3573
175	391	120	788	1042	1342	1704	2127	2618	3183	375	839	120	1267	1612	2022	2488	3013	3600	4252
175	391	140	895	1189	1546	1979	2490	3087	3778	375	839	140	1431	1828	2305	2852	3472	4169	4950
200	447	40	373	466	578	704	842	994	1160	400	894	40	579	717	876	1052	1241	1445	1663
200	447	60	500	635	796	979	1184	1413	1666	400	894	60	781	972	1197	1446	1718	2013	2333
200	447	80	626	798	1010	1253	1530	1842	2192	400	894	80	966	1215	1504	1827	2185	2576	3003
200	447	100	741	958	1222	1530	1883	2287	2744	400	894	100	1146	1448	1803	2204	2649	3141	3684
200	447	120	861	1115	1435	1811	2247	2749	3323	400	894	120	1321	1677	2099	2578	3115	3714	4378
200	447	140	964	1279	1649	2097	2621	3230	3930	400	894	140	1493	1901	2391	2952	3585	4295	5089

NOTE: Detector time constant at a reference velocity of 5 ft/sec (1.5m/sec).
For SI units: 1 ft = 0.305 m; 1000 Btu/sec = 1055 kW.

Table B-4(n)   Installed Spacing of Heat Detector: 12ft t_g: 500 seconds to 1000 Btu/sec α: 0.004 Btu/sec^3

τ	RTI	ΔT	Ceiling Height (ft)							τ	RTI	ΔT	Ceiling Height (ft)						
			4.0	8.0	12.0	16.0	20.0	24.0	28.0				4.0	8.0	12.0	16.0	20.0	24.0	28.0
			Fire Size at Detector Response (Btu/sec)										Fire Size at Detector Response (Btu/sec)						
25	56	40	82	120	170	235	314	408	519	225	503	40	276	348	436	536	648	773	911
25	56	60	117	180	266	376	512	676	866	225	503	60	374	481	611	762	933	1128	1348
25	56	80	154	246	373	536	738	982	1265	225	503	80	468	615	786	991	1229	1503	1816
25	56	100	192	318	491	713	989	1321	1710	225	503	100	563	743	962	1227	1538	1900	2318
25	56	120	233	395	618	906	1262	1692	2195	225	503	120	652	872	1146	1471	1861	2320	2854
25	56	140	275	478	755	1113	1556	2090	2717	225	503	140	740	1001	1328	1723	2200	2764	3422
50	112	40	115	158	213	280	361	457	570	250	559	40	295	371	463	567	684	813	955
50	112	60	160	227	317	430	568	732	926	250	559	60	399	512	647	803	980	1180	1405
50	112	80	205	301	431	596	801	1046	1332	250	559	80	499	653	830	1042	1286	1565	1884
50	112	100	251	379	554	779	1057	1392	1783	250	559	100	599	788	1014	1286	1604	1972	2394
50	112	120	298	462	687	976	1336	1768	2274										
50	112	140	347	549	828	1188	1634	2173	2803	275	615	40	313	393	489	598	719	852	998
										275	615	60	424	541	682	844	1026	1231	1461
75	168	40	143	191	252	322	407	506	621	275	615	80	530	685	873	1091	1342	1627	1951
75	168	60	198	271	366	483	623	790	987	275	615	100	635	832	1064	1344	1668	2042	2470
75	168	80	251	353	488	657	864	1110	1400	275	615	120	734	972	1258	1601	2007	2478	3021
75	168	100	304	439	618	846	1127	1462	1858	275	615	140	833	1113	1459	1868	2359	2934	3602
75	168	120	358	528	756	1048	1411	1846	2355	300	671	40	330	414	515	627	752	890	1041
75	168	140	413	620	902	1265	1714	2256	2889	300	671	60	447	570	716	883	1071	1282	1516
100	224	40	168	222	286	362	451	555	673	300	671	80	563	721	915	1140	1396	1688	2017
100	224	60	232	312	413	533	678	849	1048	300	671	100	669	874	1114	1400	1732	2112	2546
100	224	80	292	402	543	717	926	1176	1469	300	671	120	774	1021	1314	1666	2078	2555	3104
100	224	100	353	495	680	913	1197	1535	1933	300	671	140	881	1168	1517	1939	2437	3019	3692
100	224	120	414	591	825	1121	1487	1922	2437										
100	224	140	477	690	976	1342	1795	2336	2976	325	727	40	348	435	539	656	785	927	1082
										325	727	60	471	598	750	922	1115	1331	1570
125	280	40	191	251	319	400	494	601	723	325	727	80	592	755	956	1187	1450	1747	2082
125	280	60	263	350	455	582	732	907	1109	325	727	100	703	910	1162	1456	1794	2181	2620
125	280	80	331	449	596	774	989	1243	1539	325	727	120	813	1068	1369	1729	2148	2632	3187
125	280	100	398	549	741	979	1265	1608	2010	325	727	140	922	1221	1578	2009	2514	3103	3782
125	280	120	466	651	892	1194	1563	2001	2519	350	783	40	364	455	564	685	818	964	1122
125	280	140	535	756	1050	1421	1877	2421	3064	350	783	60	493	626	782	960	1158	1379	1624
150	335	40	213	275	350	436	535	646	772	350	783	80	619	789	996	1234	1503	1806	2146
150	335	60	293	386	496	629	784	964	1170	350	783	100	736	950	1209	1510	1855	2249	2694
150	335	80	368	493	647	830	1051	1309	1609	350	783	120	851	1115	1423	1791	2218	2708	3269
150	335	100	441	600	800	1042	1335	1682	2087	350	783	140	963	1272	1638	2077	2591	3186	3871
150	335	120	518	709	958	1266	1637	2081	2602										
150	335	140	589	820	1122	1498	1959	2506	3153	375	839	40	381	475	587	712	849	999	1162
										375	839	60	516	653	814	997	1201	1427	1676
175	391	40	237	301	380	471	574	690	819	375	839	80	646	823	1036	1280	1555	1864	2210
175	391	60	321	420	536	675	835	1020	1230	375	839	100	768	989	1256	1564	1916	2316	2767
175	391	80	402	535	693	885	1111	1374	1678	375	839	120	891	1154	1476	1852	2286	2784	3350
175	391	100	482	650	857	1105	1403	1755	2164	375	839	140	1005	1323	1697	2145	2667	3269	3960
175	391	120	564	765	1022	1337	1712	2161	2686	400	894	40	397	495	611	739	881	1034	1201
175	391	140	641	883	1192	1576	2038	2592	3242	400	894	60	538	679	846	1034	1243	1474	1728
200	447	40	257	325	409	504	612	732	866	400	894	80	673	856	1075	1325	1606	1921	2273
200	447	60	348	450	574	719	885	1074	1289	400	894	100	800	1028	1301	1617	1976	2382	2840
200	447	80	436	575	740	939	1171	1439	1747	400	894	120	926	1198	1528	1912	2354	2858	3431
200	447	100	522	697	913	1167	1471	1828	2241	400	894	140	1047	1373	1756	2212	2741	3351	4048
200	447	120	609	819	1085	1403	1787	2241	2770										
200	447	140	691	943	1261	1652	2119	2678	3332										

NOTE: Detector time constant at a reference velocity of 5 ft/sec (1.5m/sec).
For SI units: 1 ft = 0.305 m; 1000 Btu/sec = 1055 kW.

Table B-4(o)   Installed Spacing of Heat Detector: 12ft t_g: 600 seconds to 1000 Btu/sec α: 0.003 Btu/sec^3

τ	RTI	ΔT	Ceiling Height (ft) 4.0	8.0	12.0	16.0	20.0	24.0	28.0	τ	RTI	ΔT	Ceiling Height (ft) 4.0	8.0	12.0	16.0	20.0	24.0	28.0
			Fire Size at Detector Response (Btu/sec)										Fire Size at Detector Response (Btu/sec)						
25	56	40	74	110	159	221	299	391	498	225	503	40	242	307	387	478	582	698	827
25	56	60	106	167	251	359	493	654	841	225	503	60	329	427	547	686	847	1031	1240
25	56	80	141	232	356	517	716	955	1235	225	503	80	414	547	707	900	1126	1388	1690
25	56	100	178	302	472	692	964	1291	1675	225	503	100	496	665	874	1122	1420	1770	2177
25	56	120	217	378	598	882	1234	1658	2156	225	503	120	580	783	1041	1353	1731	2177	2699
25	56	140	258	459	734	1087	1525	2054	2675	225	503	140	659	903	1214	1597	2057	2609	3255
50	112	40	102	142	194	259	337	430	540	250	559	40	259	327	410	506	613	732	865
50	112	60	143	207	293	403	538	699	889	250	559	60	351	454	578	722	887	1076	1289
50	112	80	185	278	404	566	767	1008	1290	250	559	80	441	581	746	944	1174	1441	1747
50	112	100	228	353	524	745	1020	1350	1736	250	559	100	532	704	916	1173	1476	1830	2241
50	112	120	273	433	654	940	1294	1721	2222	250	559	120	616	828	1093	1410	1792	2243	2769
50	112	140	319	518	793	1149	1590	2121	2745	250	559	140	700	952	1270	1660	2125	2680	3330
75	168	40	126	171	227	294	375	471	582	275	615	40	274	346	433	532	643	766	902
75	168	60	176	245	335	447	585	747	939	275	615	60	373	479	608	757	927	1120	1337
75	168	80	224	322	451	616	819	1061	1346	275	615	80	467	613	783	986	1222	1494	1804
75	168	100	274	403	577	800	1077	1409	1797	275	615	100	562	742	960	1222	1531	1890	2304
75	168	120	324	488	711	999	1356	1786	2288	275	615	120	651	871	1143	1466	1854	2309	2838
75	168	140	375	577	854	1212	1655	2190	2816	275	615	140	740	1000	1326	1718	2191	2751	3404
100	224	40	148	197	258	328	413	511	625	300	671	40	290	365	455	558	672	798	938
100	224	60	205	280	374	491	630	796	990	300	671	60	393	504	638	791	966	1163	1384
100	224	80	261	364	498	666	872	1115	1403	300	671	80	493	644	819	1028	1269	1545	1860
100	224	100	316	451	630	856	1134	1467	1859	300	671	100	592	779	1002	1271	1585	1949	2368
100	224	120	372	542	769	1059	1418	1848	2355	300	671	120	686	913	1188	1521	1914	2375	2908
100	224	140	429	636	916	1276	1722	2259	2888	300	671	140	779	1047	1380	1779	2258	2822	3479
125	280	40	168	222	285	361	449	550	667	325	727	40	305	383	477	583	701	831	974
125	280	60	233	312	412	531	675	844	1040	325	727	60	414	529	667	825	1004	1205	1431
125	280	80	294	404	543	716	922	1170	1460	325	727	80	518	671	855	1069	1316	1597	1915
125	280	100	355	497	681	911	1193	1527	1922	325	727	100	622	815	1044	1319	1639	2008	2430
125	280	120	417	593	825	1119	1481	1913	2423	325	727	120	720	954	1235	1575	1974	2440	2977
125	280	140	481	692	977	1340	1789	2326	2961	325	727	140	817	1093	1434	1838	2323	2893	3554
150	335	40	187	246	312	392	484	589	708	350	783	40	319	401	498	607	729	862	1009
150	335	60	259	344	447	572	719	892	1091	350	783	60	434	553	695	858	1041	1247	1476
150	335	80	326	442	587	762	974	1225	1517	350	783	80	543	700	890	1110	1361	1647	1970
150	335	100	393	541	731	966	1249	1588	1985	350	783	100	650	850	1085	1366	1692	2066	2493
150	335	120	460	643	881	1180	1545	1979	2491	350	783	120	753	994	1282	1628	2034	2504	3046
150	335	140	528	747	1038	1405	1857	2396	3033	350	783	140	854	1138	1482	1897	2389	2963	3629
175	391	40	206	266	338	422	517	626	749	375	839	40	334	418	519	631	756	893	1043
175	391	60	283	373	481	611	73	939	1141	375	839	60	453	576	723	890	1078	1288	1522
175	391	80	356	478	629	809	1026	1280	1575	375	839	80	571	729	924	1150	1406	1697	2024
175	391	100	429	584	780	1020	1306	1649	2049	375	839	100	679	880	1126	1413	1744	2123	2555
175	391	120	503	691	936	1239	1605	2045	2560	375	839	120	785	1034	1328	1681	2092	2568	3115
175	391	140	573	801	1097	1469	1925	2467	3107	375	839	140	894	1183	1533	1956	2453	3033	3703
200	447	40	225	287	363	451	550	662	788	400	894	40	348	435	539	655	783	924	1077
200	447	60	307	402	514	649	805	985	1191	400	894	60	472	599	750	922	1114	1329	1566
200	447	80	385	513	668	855	1076	1334	1633	400	894	80	594	758	958	1189	1450	1746	2078
200	447	100	463	625	827	1070	1364	1710	2113	400	894	100	706	914	1165	1458	1795	2180	2617
200	447	120	542	738	989	1298	1668	2111	2629	400	894	120	817	1073	1373	1733	2150	2632	3183
200	447	140	617	853	1156	1534	1990	2538	3181	400	894	140	928	1226	1584	2013	2517	3103	3777

NOTE: Detector time constant at a reference velocity of 5 ft/sec (1.5m/sec).
For SI units: 1 ft = 0.305 m; 1000 Btu/sec = 1055 kW.

Table B-4(p)   Installed Spacing of Heat Detector: 15ft t_g: 50 seconds to 1000 Btu/sec α: 0.400 Btu/sec^3

τ	RTI	ΔT	Ceiling Height (ft) Fire Size at Detector Response (Btu/sec) 4.0	8.0	12.0	16.0	20.0	24.0	28.0
25	56	40	618	745	903	1085	1287	1509	1753
25	56	60	790	962	1181	1434	1720	2040	2395
25	56	80	935	1169	1452	1781	2157	2583	3061
25	56	100	1046	1393	1722	2131	2605	3146	3759
25	56	120	1215	1600	1992	2488	3066	3731	4490
25	56	140	1357	1799	2266	2852	3541	4340	5255
50	112	40	894	1063	1269	1503	1758	2034	2331
50	112	60	1098	1370	1656	1976	2327	2711	3128
50	112	80	1321	1664	2025	2431	2883	3380	3925
50	112	100	1541	1949	2385	2880	3435	4052	4735
50	112	120	1758	2226	2739	3327	3991	4734	5563
50	112	140	1974	2499	3091	3775	4552	5429	6412
75	168	40	1035	1356	1577	1852	2153	2476	2819
75	168	60	1370	1712	2055	2432	2843	3286	3763
75	168	80	1668	2083	2511	2986	3507	4074	4687
75	168	100	1960	2439	2951	3525	4159	4853	5611
75	168	120	2248	2783	3380	4055	4806	5634	6543
75	168	140	2533	3119	3804	4582	5453	6420	7487
100	224	40	1245	1547	1843	2162	2505	2870	3256
100	224	60	1625	2019	2412	2841	3305	3803	4334
100	224	80	1994	2460	2947	3484	4069	4701	5379
100	224	100	2356	2880	3459	4105	4813	5582	6413
100	224	120	2714	3285	3958	4714	5547	6457	7446
100	224	140	3071	3680	4447	5315	6276	7332	8485
125	280	40	1417	1760	2091	2447	2827	3231	3657
125	280	60	1870	2304	2741	3218	3731	4280	4862
125	280	80	2309	2809	3348	3944	4589	5281	6021
125	280	100	2741	3289	3930	4643	5420	6259	7160
125	280	120	3172	3752	4493	5325	6236	7224	8291
125	280	140	3485	4195	5044	5996	7043	8184	9422
150	335	40	1584	1960	2323	2713	3129	3569	4032
150	335	60	2107	2571	3049	3570	4130	4726	5357
150	335	80	2616	3137	3726	4375	5076	5827	6625
150	335	100	3121	3674	4371	5148	5990	6896	7865
150	335	120	3501	4183	4995	5900	6884	7947	9090
150	335	140	3927	4681	5605	6637	7766	8989	10309
175	391	40	1746	2151	2543	2965	3414	3889	4387
175	391	60	2339	2825	3341	3905	4509	5150	5827
175	391	80	2920	3449	4083	4784	5539	6345	7199
175	391	100	3390	4034	4790	5627	6532	7502	8535
175	391	120	3879	4598	5473	6446	7501	8636	9851
175	391	140	4356	5145	6138	7247	8454	9757	11157
200	447	40	1906	2333	2753	3205	3687	4194	4726
200	447	60	2569	3068	3620	4224	4870	5554	6276
200	447	80	3224	3749	4426	5176	5982	6840	7748
200	447	100	3707	4382	5192	6086	7051	8082	9177
200	447	120	4247	4995	5930	6969	8092	9296	10581
200	447	140	4670	5583	6649	7831	9113	10493	11970
225	503	40	2062	2508	2955	3436	3948	4487	5051
225	503	60	2796	3303	3889	4531	5218	5943	6707
225	503	80	3407	4029	4755	5553	6408	7317	8276
225	503	100	4017	4717	5578	6528	7551	8640	9796
225	503	120	4609	5378	6371	7472	8661	9932	11285
225	503	140	5045	6009	7141	8393	9749	11203	12755
250	559	40	2217	2678	3150	3659	4200	4769	5364
250	559	60	3023	3530	4148	4828	5553	6319	7123
250	559	80	3660	4305	5073	5917	6820	7777	8787
250	559	100	4321	5042	5951	6955	8033	9180	10394
250	559	120	4835	5740	6796	7959	9211	10547	11966
250	559	140	5409	6421	7617	8938	10363	11889	13515
275	615	40	2370	2843	3339	3875	4445	5043	5668
275	615	60	3251	3751	4440	5115	5878	6683	7527
275	615	80	3909	4573	5381	6269	7219	8224	9282
275	615	100	4622	5357	6313	7369	8502	9704	10974
275	615	120	5151	6098	7209	8432	9745	11144	12627
275	615	140	5765	6821	8079	9466	10960	12556	14254
300	671	40	2522	3003	3522	4085	4682	5309	5963
300	671	60	3362	3956	4644	5395	6194	7037	7919
300	671	80	4153	4834	5682	6613	7607	8658	9763
300	671	100	4785	5655	6666	7772	8957	10214	11539
300	671	120	5461	6446	7612	8891	10265	11725	13271
300	671	140	6114	7211	8529	9980	11541	13206	14973
325	727	40	2673	3160	3701	4289	4913	5567	6250
325	727	60	3552	4163	4882	5667	6502	7381	8302
325	727	80	4396	5089	5974	6947	7985	9081	10232
325	727	100	5049	5655	6666	7772	8957	10214	11539
325	727	120	5764	6787	8004	9340	10772	12292	13898
325	727	140	6456	7591	8968	10482	12108	13839	15674
350	783	40	2824	3313	3876	4489	5139	5820	6530
350	783	60	3739	4366	5115	5933	6803	7718	8675
350	783	80	4522	5330	6260	7274	8355	9495	10691
350	783	100	5308	6244	7345	8550	9836	11196	12628
350	783	120	6062	7119	8387	9778	11267	12846	14512
350	783	140	6793	7963	9397	10972	12662	14459	16360
375	839	40	2975	3463	4047	4684	5359	6067	6804
375	839	60	3925	4564	5342	6193	7097	8047	9040
375	839	80	4736	5573	6540	7594	8717	9899	11139
375	839	100	5563	6530	7674	8926	10261	11671	13154
375	839	120	6356	7445	8763	10207	11752	13388	15113
375	839	140	7126	8328	9817	11452	13204	15065	17032
400	894	40	3126	3611	4214	4875	5575	6309	7072
400	894	60	4108	4759	5566	6449	7386	8370	9398
400	894	80	4948	5811	6814	7908	9071	10296	11579
400	894	100	5814	6809	7997	9294	10677	12137	13670
400	894	120	6647	7764	9131	10628	12227	13920	15702
400	894	140	7455	8685	10229	11923	13736	15660	17690

NOTE: Detector time constant at a reference velocity of 5 ft/sec (1.5m/sec).
For SI units: 1 ft = 0.305 m; 1000 Btu/sec = 1055 kW.

Table B-4(q)   Installed Spacing of Heat Detector: 15ft t_g: 150 seconds to 1000 Btu/sec α: 0.044 Btu/sec^3

τ	RTI	ΔT	Ceiling Height (ft) Fire Size at Detector Response (Btu/sec) 4.0	8.0	12.0	16.0	20.0	24.0	28.0
25	56	40	259	322	407	509	627	762	917
25	56	60	341	447	572	731	918	1137	1390
25	56	80	423	567	749	968	1237	1552	1916
25	56	100	511	691	930	1228	1581	2003	2492
25	56	120	592	819	1120	1498	1950	2488	3112
25	56	140	674	951	1321	1785	2348	3004	3774
50	112	40	371	451	555	675	810	961	1129
50	112	60	488	610	763	942	1147	1381	1645
50	112	80	610	767	972	1216	1500	1829	2204
50	112	100	718	923	1186	1501	1873	2307	2806
50	112	120	826	1089	1404	1798	2266	2815	3449
50	112	140	936	1247	1637	2107	2679	3352	4132
75	168	40	463	560	681	818	970	1138	1322
75	168	60	622	755	929	1129	1354	1607	1889
75	168	80	758	943	1173	1441	1747	2095	2487
75	168	100	900	1129	1419	1759	2154	2606	3121
75	168	120	1021	1314	1667	2086	2575	3142	3791
75	168	140	1161	1499	1918	2421	3014	3703	4497
100	224	40	546	658	794	947	1116	1300	1500
100	224	60	729	886	1079	1300	1545	1818	2118
100	224	80	900	1103	1357	1648	1978	2347	2759
100	224	100	1044	1316	1633	2000	2419	2894	3428
100	224	120	1212	1527	1909	2356	2872	3461	4129
100	224	140	1368	1736	2187	2719	3337	4049	4861
125	280	40	635	750	899	1067	1251	1450	1666
125	280	60	829	1006	1219	1458	1724	2016	2335
125	280	80	1011	1251	1528	1842	2195	2586	3019
125	280	100	1203	1490	1832	2226	2670	3169	3726
125	280	120	1384	1725	2136	2612	3154	3768	4458
125	280	140	1563	1957	2440	3002	3649	4385	5219
150	335	40	703	834	998	1180	1378	1592	1822
150	335	60	927	1120	1350	1608	1893	2203	2541
150	335	80	1138	1391	1689	2026	2400	2814	3268
150	335	100	1345	1654	2021	2440	2910	3433	4012
150	335	120	1549	1912	2351	2855	3425	4064	4779
150	335	140	1750	2166	2679	3272	3948	4712	5570
175	391	40	768	915	1091	1287	1499	1727	1972
175	391	60	1008	1228	1475	1751	2053	2382	2738
175	391	80	1252	1524	1842	2201	2597	3032	3507
175	391	100	1482	1810	2201	2645	3139	3687	4289
175	391	120	1708	2090	2556	3088	3685	4351	5090
175	391	140	1931	2366	2908	3532	4237	5028	5912
200	447	40	832	992	1181	1389	1615	1857	2115
200	447	60	1103	1331	1594	1888	2208	2554	2928
200	447	80	1363	1651	1990	2369	2786	3242	3738
200	447	100	1615	1960	2374	2842	3360	3932	4558
200	447	120	1863	2261	2753	3312	3936	4628	5392
200	447	140	2081	2557	3129	3782	4516	5335	6246
225	503	40	902	1067	1267	1488	1726	1981	2253
225	503	60	1189	1431	1710	2019	2356	2720	3111
225	503	80	1471	1774	2131	2531	2968	3445	3962
225	503	100	1745	2105	2541	3032	3574	4169	4818
225	503	120	2015	2426	2944	3529	4179	4897	5687
225	503	140	2242	2742	3342	4023	4786	5634	6571
250	559	40	953	1138	1350	1583	1834	2101	2386
250	559	60	1273	1527	1821	2147	2500	2881	3288
250	559	80	1576	1893	2269	2687	3145	3642	4179
250	559	100	1873	2244	2703	3216	3782	4399	5072
250	559	120	2131	2586	3128	3739	4415	5159	5974
250	559	140	2398	2920	3548	4258	5050	5925	6890
275	615	40	1006	1208	1431	1675	1938	2218	2515
275	615	60	1354	1621	1929	2270	2640	3037	3461
275	615	80	1680	2008	2402	2840	3317	3834	4390
275	615	100	1999	2380	2859	3395	3983	4624	5319
275	615	120	2266	2741	3307	3943	4645	5414	6254
275	615	140	2550	3094	3748	4487	5306	6209	7201
300	671	40	1071	1276	1509	1765	2040	2332	2641
300	671	60	1434	1712	2034	2391	2776	3188	3629
300	671	80	1782	2121	2532	2988	3484	4020	4597
300	671	100	2089	2512	3012	3570	4180	4843	5560
300	671	120	2398	2892	3482	4142	4869	5664	6528
300	671	140	2699	3264	3944	4709	5556	6487	7506
325	727	40	1128	1342	1586	1853	2139	2443	2764
325	727	60	1513	1801	2137	2508	2908	3337	3793
325	727	80	1883	2230	2658	3133	3648	4203	4798
325	727	100	2202	2641	3161	3740	4372	5057	5796
325	727	120	2528	3040	3652	4337	5088	5907	6797
325	727	140	2846	3429	4135	4927	5801	6759	7804
350	783	40	1184	1407	1661	1938	2235	2551	2884
350	783	60	1591	1888	2238	2623	3038	3482	3953
350	783	80	1983	2338	2782	3274	3808	4381	4995
350	783	100	2312	2767	3307	3906	4560	5266	6027
350	783	120	2655	3184	3819	4527	5303	6146	7060
350	783	140	2989	3591	4322	5140	6041	7026	8098
375	839	40	1239	1470	1734	2022	2330	2657	3001
375	839	60	1667	1973	2336	2735	3165	3623	4111
375	839	80	2049	2442	2903	3413	3964	4556	5188
375	839	100	2421	2891	3450	4070	4744	5472	6254
375	839	120	2780	3325	3982	4713	5513	6380	7318
375	839	140	3131	3749	4505	5349	6276	7287	8385
400	894	40	1293	1532	1806	2104	2423	2761	3116
400	894	60	1743	2056	2432	2845	3289	3763	4265
400	894	80	2139	2545	3022	3549	4118	4727	5378
400	894	100	2528	3012	3590	4230	4925	5673	6477
400	894	120	2904	3464	4143	4896	5719	6610	7572
400	894	140	3271	3905	4684	5554	6507	7544	8669

NOTE: Detector time constant at a reference velocity of 5 ft/sec (1.5m/sec).
For SI units: 1 ft = 0.305 m; 1000 Btu/sec = 1055 kW.

Table B-4(r)   Installed Spacing of Heat Detector: 15ft t_g: 300 seconds to 1000 Btu/sec α: 0.011 Btu/sec³

τ	RTI	ΔT	4.0	8.0	12.0	16.0	20.0	24.0	28.0	τ	RTI	ΔT	4.0	8.0	12.0	16.0	20.0	24.0	28.0
			Ceiling Height (ft)										Ceiling Height (ft)						
			Fire Size at Detector Response (Btu/sec)										Fire Size at Detector Response (Btu/sec)						
25	56	40	156	207	273	353	451	565	697	225	503	40	530	639	768	912	1070	1242	1429
25	56	60	215	297	407	543	706	900	1125	225	503	60	718	871	1057	1268	1502	1760	2044
25	56	80	275	394	553	754	997	1280	1612	225	503	80	894	1095	1340	1621	1937	2289	2681
25	56	100	337	498	713	984	1314	1700	2151	225	503	100	1058	1313	1621	1976	2379	2835	3346
25	56	120	403	608	885	1234	1657	2155	2736	225	503	120	1222	1529	1902	2335	2833	3399	4040
25	56	140	469	725	1069	1501	2026	2648	3365	225	503	140	1385	1744	2184	2701	3299	3984	4764
50	112	40	221	279	353	442	545	664	801	250	559	40	567	681	817	968	1133	1313	1506
50	112	60	300	394	508	650	819	1017	1246	250	559	60	766	928	1123	1343	1586	1854	2147
50	112	80	378	508	674	876	1121	1411	1747	250	559	80	950	1165	1421	1713	2039	2403	2805
50	112	100	456	626	847	1121	1450	1843	2298	250	559	100	1129	1396	1716	2083	2499	2966	3488
50	112	120	538	749	1029	1381	1808	2309	2895	250	559	120	1305	1623	2010	2458	2968	3547	4198
50	112	140	617	877	1222	1657	2186	2808	3535	250	559	140	1478	1849	2305	2837	3449	4147	4938
75	168	40	276	342	426	523	634	760	902	275	615	40	609	722	864	1022	1194	1381	1582
75	168	60	373	475	602	753	930	1134	1368	275	615	60	813	983	1187	1416	1668	1945	2247
75	168	80	467	611	784	996	1249	1544	1885	275	615	80	1006	1233	1500	1802	2140	2514	2926
75	168	100	564	745	976	1254	1591	1988	2449	275	615	100	1199	1476	1808	2188	2616	3095	3628
75	168	120	656	882	1172	1526	1957	2466	3057	275	615	120	1385	1715	2116	2577	3101	3692	4355
75	168	140	748	1022	1376	1816	2345	2975	3707	275	615	140	1569	1952	2423	2970	3596	4306	5109
100	224	40	326	399	491	597	716	850	999	300	671	40	642	762	910	1075	1254	1448	1655
100	224	60	439	551	688	850	1036	1248	1489	300	671	60	867	1037	1249	1486	1748	2033	2344
100	224	80	554	700	888	1112	1373	1676	2023	300	671	80	1066	1299	1576	1890	2238	2622	3045
100	224	100	658	850	1093	1385	1731	2135	2601	300	671	100	1267	1554	1899	2291	2731	3221	3765
100	224	120	763	1006	1303	1671	2109	2624	3221	300	671	120	1464	1805	2219	2694	3231	3834	4509
100	224	140	874	1160	1525	1969	2508	3144	3882	300	671	140	1658	2053	2538	3100	3740	4464	5279
125	280	40	371	452	552	666	794	935	1092	325	727	40	674	801	956	1127	1312	1513	1727
125	280	60	500	621	769	941	1137	1358	1606	325	727	60	907	1090	1310	1556	1826	2120	2440
125	280	80	627	787	987	1222	1493	1805	2159	325	727	80	1122	1364	1651	1975	2334	2728	3161
125	280	100	746	951	1207	1512	1868	2280	2753	325	727	100	1334	1631	1987	2391	2843	3345	3901
125	280	120	870	1117	1432	1812	2260	2783	3386	325	727	120	1541	1892	2319	2808	3358	3975	4661
125	280	140	980	1289	1663	2123	2671	3314	4059	325	727	140	1746	2151	2651	3227	3881	4619	5446
150	335	40	414	502	610	732	867	1017	1181	350	783	40	706	839	1000	1177	1369	1576	1798
150	335	60	557	688	846	1028	1233	1464	1721	350	783	60	947	1141	1369	1624	1902	2205	2534
150	335	80	697	869	1081	1327	1610	1931	2294	350	783	80	1177	1428	1725	2058	2427	2833	3276
150	335	100	829	1047	1317	1633	2001	2423	2904	350	783	100	1399	1706	2073	2489	2953	3467	4034
150	335	120	958	1226	1556	1949	2408	2941	3552	350	783	120	1617	1978	2418	2920	3483	4113	4812
150	335	140	1088	1405	1800	2273	2832	3484	4236	350	783	140	1819	2247	2761	3352	4021	4772	5612
175	391	40	454	549	665	794	938	1095	1267	375	839	40	738	877	1043	1226	1425	1638	1866
175	391	60	617	752	919	1111	1326	1566	1832	375	839	60	988	1191	1427	1690	1977	2289	2626
175	391	80	764	947	1170	1428	1722	2054	2426	375	839	80	1231	1490	1796	2140	2520	2935	3388
175	391	100	911	1139	1422	1751	2130	2563	3054	375	839	100	1464	1779	2158	2586	3061	3587	4165
175	391	120	1047	1331	1675	2081	2553	3096	3716	375	839	120	1693	2062	2515	3030	3606	4248	4960
175	391	140	1191	1522	1932	2420	2991	3653	4413	375	839	140	1900	2341	2870	3475	4158	4923	5775
200	447	40	493	595	717	854	1005	1170	1349	400	894	40	769	913	1085	1275	1480	1700	1934
200	447	60	668	812	990	1191	1415	1665	1940	400	894	60	1030	1241	1484	1755	2051	2371	2716
200	447	80	828	1022	1257	1526	1831	2173	2555	400	894	80	1284	1551	1867	2221	2610	3036	3499
200	447	100	982	1228	1523	1865	2256	2700	3201	400	894	100	1528	1851	2241	2680	3168	3705	4294
200	447	120	1137	1431	1790	2210	2695	3249	3879	400	894	120	1767	2144	2610	3138	3727	4382	5106
200	447	140	1289	1635	2060	2562	3146	3820	4589	400	894	140	1980	2433	2976	3596	4293	5071	5937

NOTE: Detector time constant at a reference velocity of 5 ft/sec (1.5m/sec).
For SI units: 1 ft = 0.305 m; 1000 Btu/sec = 1055 kW.

Table B-4(s)   Installed Spacing of Heat Detector: 15ft t_g: 500 seconds to 1000 Btu/sec α: 0.004 Btu/sec^3

τ	RTI	ΔT	Ceiling Height (ft)							τ	RTI	ΔT	Ceiling Height (ft)						
			4.0	8.0	12.0	16.0	20.0	24.0	28.0				4.0	8.0	12.0	16.0	20.0	24.0	28.0
			Fire Size at Detector Response (Btu/sec)										Fire Size at Detector Response (Btu/sec)						
25	56	40	112	156	215	289	379	484	606	225	503	40	365	445	542	652	775	911	1061
25	56	60	160	235	337	464	619	801	1013	225	503	60	497	617	761	928	1117	1331	1571
25	56	80	210	323	474	664	894	1164	1480	225	503	80	626	784	980	1210	1474	1776	2119
25	56	100	264	419	625	884	1198	1571	2001	225	503	100	748	951	1202	1500	1847	2248	2707
25	56	120	321	523	789	1124	1530	2013	2570	225	503	120	873	1119	1429	1800	2238	2748	3335
25	56	140	381	633	965	1381	1887	2488	3183	225	503	140	988	1292	1661	2112	2647	3275	4001
50	112	40	155	204	267	342	433	541	665	250	559	40	390	474	575	690	817	958	1112
50	112	60	216	295	399	528	683	868	1082	250	559	60	530	655	806	978	1173	1393	1638
50	112	80	278	392	544	735	967	1239	1558	250	559	80	667	832	1035	1271	1542	1850	2198
50	112	100	341	496	702	962	1279	1651	2087	250	559	100	796	1007	1266	1571	1925	2332	2796
50	112	120	408	606	872	1208	1617	2099	2663	250	559	120	925	1182	1501	1880	2325	2840	3432
50	112	140	474	722	1053	1471	1980	2580	3282	250	559	140	1052	1359	1741	2200	2742	3375	4106
75	168	40	192	245	313	393	488	598	724	275	615	40	414	502	607	726	858	1004	1162
75	168	60	265	351	460	590	749	936	1152	275	615	60	567	693	849	1027	1228	1453	1703
75	168	80	337	458	614	808	1040	1316	1637	275	615	80	706	878	1088	1331	1608	1922	2276
75	168	100	411	571	780	1042	1358	1735	2173	275	615	100	843	1062	1328	1641	2001	2415	2885
75	168	120	486	688	956	1294	1705	2190	2756	275	615	120	978	1245	1572	1959	2411	2932	3529
75	168	140	561	811	1143	1563	2074	2677	3382	275	615	140	1112	1428	1820	2287	2836	3475	4210
100	224	40	226	283	356	442	541	654	784	300	671	40	437	529	639	762	899	1048	1212
100	224	60	309	400	514	652	814	1004	1223	300	671	60	598	729	891	1075	1282	1512	1767
100	224	80	392	521	682	878	1115	1393	1717	300	671	80	745	923	1140	1389	1673	1993	2353
100	224	100	477	642	857	1123	1441	1820	2261	300	671	100	893	1115	1389	1709	2077	2496	2972
100	224	120	559	768	1041	1381	1792	2281	2851	300	671	120	1030	1305	1641	2036	2495	3023	3626
100	224	140	642	898	1234	1656	2170	2775	3484	300	671	140	1172	1496	1897	2372	2929	3574	4315
125	280	40	257	319	397	487	591	709	842	325	727	40	460	555	669	797	938	1092	1260
125	280	60	351	448	568	711	878	1072	1294	325	727	60	628	765	932	1122	1334	1570	1830
125	280	80	443	579	745	948	1189	1471	1798	325	727	80	783	968	1191	1447	1737	2063	2429
125	280	100	537	710	931	1199	1524	1906	2350	325	727	100	935	1167	1449	1776	2151	2577	3059
125	280	120	627	844	1123	1469	1882	2374	2947	325	727	120	1083	1365	1709	2112	2579	3113	3722
125	280	140	718	982	1324	1750	2262	2874	3586	325	727	140	1230	1563	1972	2456	3021	3673	4419
150	335	40	286	353	436	531	640	762	899	350	783	40	482	581	699	831	976	1135	1307
150	335	60	390	492	619	768	941	1139	1365	350	783	60	658	800	972	1168	1386	1627	1893
150	335	80	491	632	807	1016	1262	1549	1879	350	783	80	820	1011	1240	1503	1800	2132	2504
150	335	100	593	776	1000	1277	1606	1992	2439	350	783	100	977	1218	1507	1842	2224	2657	3145
150	335	120	691	918	1204	1551	1972	2467	3043	350	783	120	1134	1423	1776	2187	2661	3203	3818
150	335	140	790	1063	1412	1840	2359	2973	3689	350	783	140	1287	1628	2046	2539	3112	3770	4523
175	391	40	313	385	472	573	686	813	954	375	839	40	504	607	729	865	1014	1177	1353
175	391	60	427	535	668	823	1001	1204	1435	375	839	60	687	834	1012	1213	1436	1683	1954
175	391	80	541	684	866	1082	1334	1626	1960	375	839	80	863	1053	1289	1558	1862	2201	2578
175	391	100	647	834	1070	1353	1687	2078	2529	375	839	100	1020	1268	1565	1907	2296	2736	3230
175	391	120	753	989	1279	1636	2061	2561	3140	375	839	120	1183	1480	1841	2261	2742	3291	3912
175	391	140	863	1142	1499	1931	2455	3074	3792	375	839	140	1343	1692	2119	2621	3201	3867	4626
200	447	40	340	416	508	613	731	863	1008	400	894	40	525	632	758	898	1051	1218	1399
200	447	60	463	577	715	876	1060	1269	1504	400	894	60	715	868	1051	1257	1486	1738	2014
200	447	80	584	735	924	1147	1405	1701	2040	400	894	80	895	1095	1337	1613	1922	2268	2651
200	447	100	698	894	1137	1427	1768	2163	2618	400	894	100	1064	1317	1621	1971	2367	2814	3314
200	447	120	812	1053	1355	1719	2150	2654	3238	400	894	120	1232	1536	1905	2333	2823	3379	4006
200	447	140	926	1218	1579	2022	2552	3174	3896	400	894	140	1398	1754	2191	2702	3290	3964	4729

NOTE: Detector time constant at a reference velocity of 5 ft/sec (1.5m/sec).
For SI units: 1 ft = 0.305 m; 1000 Btu/sec = 1055 kW.

Table B-4(t)   Installed Spacing of Heat Detector: 15ft t_g: 600 seconds to 1000 Btu/sec α: 0.003 Btu/sec³

τ	RTI	ΔT	\multicolumn{7}{c}{Ceiling Height (ft) — Fire Size at Detector Response (Btu/sec)}						
			4.0	8.0	12.0	16.0	20.0	24.0	28.0
25	56	40	100	143	201	273	360	462	582
25	56	60	145	220	319	444	596	775	983
25	56	80	193	305	454	640	867	1135	1445
25	56	100	245	399	603	858	1168	1536	1961
25	56	120	300	501	765	1095	1496	1973	2525
25	56	140	358	610	938	1350	1850	2445	3134
50	112	40	137	184	244	316	405	509	630
50	112	60	193	270	371	497	650	830	1040
50	112	80	251	363	512	699	927	1195	1509
50	112	100	310	462	666	922	1234	1602	2032
50	112	120	372	569	832	1164	1568	2045	2602
50	112	140	437	683	1011	1424	1927	2525	3216
75	168	40	170	219	283	360	450	557	679
75	168	60	236	317	421	548	703	885	1097
75	168	80	303	418	570	759	987	1258	1574
75	168	100	370	525	730	988	1301	1671	2103
75	168	120	441	638	902	1235	1641	2119	2679
75	168	140	511	756	1085	1499	2004	2601	3299
100	224	40	199	252	320	401	495	604	728
100	224	60	275	361	468	600	757	942	1156
100	224	80	350	472	627	819	1049	1322	1640
100	224	100	426	586	794	1054	1367	1741	2175
100	224	120	504	705	972	1307	1714	2195	2757
100	224	140	581	829	1160	1576	2083	2682	3382
125	280	40	226	283	355	440	538	650	777
125	280	60	311	401	514	650	811	998	1215
125	280	80	394	522	683	876	1110	1386	1706
125	280	100	480	644	857	1121	1436	1811	2248
125	280	120	563	770	1041	1379	1786	2271	2836
125	280	140	646	901	1234	1653	2163	2763	3466
150	335	40	251	313	389	478	579	695	825
150	335	60	345	440	558	699	864	1055	1273
150	335	80	436	570	734	934	1172	1451	1773
150	335	100	529	701	919	1184	1504	1882	2322
150	335	120	618	834	1110	1451	1860	2348	2915
150	335	140	709	970	1309	1731	2239	2845	3551
175	391	40	275	340	421	514	619	738	872
175	391	60	377	477	601	746	915	1110	1332
175	391	80	476	614	785	990	1233	1515	1840
175	391	100	576	755	976	1248	1573	1954	2395
175	391	120	672	895	1177	1520	1935	2425	2995
175	391	140	769	1038	1382	1808	2319	2928	3636
200	447	40	298	367	452	549	658	781	918
200	447	60	408	513	642	792	966	1164	1390
200	447	80	515	658	835	1046	1293	1579	1907
200	447	100	621	803	1034	1311	1640	2025	2470
200	447	120	723	954	1239	1590	2009	2503	3075
200	447	140	826	1104	1454	1881	2399	3011	3722

τ	RTI	ΔT	\multicolumn{7}{c}{Ceiling Height (ft) — Fire Size at Detector Response (Btu/sec)}						
			4.0	8.0	12.0	16.0	20.0	24.0	28.0
225	503	40	321	393	481	582	696	822	963
225	503	60	438	548	681	837	1015	1218	1447
225	503	80	555	700	883	1100	1351	1642	1974
225	503	100	664	853	1090	1373	1707	2096	2544
225	503	120	773	1012	1303	1659	2083	2580	3156
225	503	140	885	1168	1521	1957	2479	3094	3808
250	559	40	342	418	510	615	732	863	1007
250	559	60	467	581	720	880	1063	1270	1504
250	559	80	590	742	931	1152	1409	1704	2040
250	559	100	706	902	1145	1434	1773	2166	2618
250	559	120	822	1063	1365	1727	2156	2658	3237
250	559	140	937	1230	1590	2032	2559	3177	3895
275	615	40	363	442	538	647	768	902	1051
275	615	60	495	614	757	923	1110	1322	1559
275	615	80	625	782	977	1204	1466	1766	2106
275	615	100	747	949	1199	1494	1839	2236	2692
275	615	120	873	1117	1426	1795	2229	2735	3318
275	615	140	989	1286	1658	2106	2638	3261	3982
300	671	40	383	466	565	678	803	941	1093
300	671	60	523	646	794	964	1157	1373	1614
300	671	80	658	821	1022	1255	1522	1827	2171
300	671	100	787	996	1252	1553	1903	2306	2766
300	671	120	916	1170	1486	1861	2301	2812	3398
300	671	140	1041	1345	1724	2179	2716	3344	4068
325	727	40	403	489	592	708	837	979	1134
325	727	60	554	677	830	1005	1202	1423	1668
325	727	80	691	860	1066	1305	1577	1887	2235
325	727	100	826	1041	1303	1611	1967	2375	2839
325	727	120	959	1222	1544	1926	2372	2888	3479
325	727	140	1092	1403	1789	2251	2794	3427	4155
350	783	40	423	512	618	738	871	1016	1175
350	783	60	580	707	865	1045	1246	1472	1722
350	783	80	723	898	1109	1353	1631	1946	2299
350	783	100	869	1085	1354	1668	2030	2443	2911
350	783	120	1003	1272	1602	1991	2443	2964	3559
350	783	140	1141	1460	1854	2322	2871	3509	4242
375	839	40	442	534	644	767	903	1053	1215
375	839	60	605	737	899	1084	1290	1520	1774
375	839	80	755	934	1151	1401	1685	2004	2362
375	839	100	904	1129	1404	1724	2092	2510	2984
375	839	120	1047	1322	1659	2054	2512	3039	3639
375	839	140	1190	1515	1917	2392	2948	3591	4329
400	894	40	460	555	669	796	936	1088	1255
400	894	60	630	767	933	1122	1333	1567	1826
400	894	80	786	971	1193	1448	1737	2062	2425
400	894	100	939	1171	1453	1779	2153	2577	3055
400	894	120	1089	1371	1715	2117	2581	3113	3718
400	894	140	1237	1570	1979	2462	3024	3673	4415

NOTE: Detector time constant at a reference velocity of 5 ft/sec (1.5m/sec).
For SI units: 1 ft = 0.305 m; 1000 Btu/sec = 1055 kW.

Table B-4(u)   Installed Spacing of Heat Detector: 20ft t_g: 50 seconds to 1000 Btu/sec α: 0.400 Btu/sec^3

τ	RTI	ΔT	Ceiling Height (ft) 4.0	8.0	12.0	16.0	20.0	24.0	28.0
			Fire Size at Detector Response (Btu/sec)						
25	56	40	906	1047	1221	1424	1650	1898	2169
25	56	60	1122	1352	1602	1890	2214	2575	2974
25	56	80	1351	1649	1975	2354	2786	3271	3813
25	56	100	1578	1940	2347	2825	3373	3994	4693
25	56	120	1804	2229	2722	3305	3979	4747	5616
25	56	140	2029	2519	3103	3797	4605	5532	6583
50	112	40	1236	1478	1711	1969	2251	2556	2883
50	112	60	1603	1922	2239	2596	2989	3417	3881
50	112	80	1960	2340	2744	3202	3710	4270	4881
50	112	100	2311	2744	3236	3800	4431	5129	5898
50	112	120	2657	3139	3723	4397	5156	6002	6940
50	112	140	3002	3527	4207	4996	5891	6893	8010
75	168	40	1543	1841	2119	2425	2756	3110	3487
75	168	60	2035	2402	2775	3192	3648	4139	4666
75	168	80	2514	2926	3396	3925	4508	5141	5824
75	168	100	2986	3429	3995	4641	5355	6135	6983
75	168	120	3362	3912	4582	5347	6197	7131	8153
75	168	140	3768	4387	5162	6050	7041	8136	9340
100	224	40	1831	2165	2483	2830	3205	3605	4028
100	224	60	2445	2833	3255	3726	4239	4789	5374
100	224	80	3048	3454	3980	4576	5227	5929	6682
100	224	100	3518	4039	4678	5400	6192	7051	7977
100	224	120	4022	4608	5358	6208	7145	8166	9273
100	224	140	4514	5164	6026	7007	8094	9282	10577
125	280	40	2109	2465	2816	3202	3618	4059	4525
125	280	60	2845	3231	3696	4219	4784	5388	6029
125	280	80	3451	3934	4520	5177	5892	6660	7478
125	280	100	4060	4606	5309	6103	6989	7903	8904
125	280	120	4651	5254	6076	7007	8027	9131	10321
125	280	140	5095	5877	6826	7897	9074	10354	11739
150	335	40	2379	2746	3129	3550	4004	4484	4990
150	335	60	3243	3607	4110	4680	5295	5950	6643
150	335	80	3886	4389	5026	5742	6517	7346	8228
150	335	100	4583	5139	5902	6763	7700	8705	9778
150	335	120	5111	5853	6751	7758	8857	10042	11312
150	335	140	5716	6553	7580	8736	10000	11368	12841
175	391	40	2646	3014	3424	3880	4369	4887	5430
175	391	60	3469	3952	4502	5117	5780	6483	7226
175	391	80	4309	4821	5507	6277	7110	7999	8940
175	391	100	4948	5637	6465	7391	8394	9468	10610
175	391	120	5642	6428	7392	8473	9648	10910	12257
175	391	140	6313	7194	8296	9533	10882	12335	13893
200	447	40	2910	3270	3706	4194	4718	5271	5850
200	447	60	3821	4289	4877	5535	6243	6993	7783
200	447	80	4603	5227	5966	6790	7678	8622	9621
200	447	100	5395	6120	7004	7991	9059	10198	11408
200	447	120	6155	6978	8006	9157	10405	11741	13164
200	447	140	6891	7809	8983	10298	11727	13263	14903
225	503	40	3174	3517	3978	4496	5052	5639	6254
225	503	60	4140	4613	5237	5937	6688	7483	8318
225	503	80	4969	5623	6408	7282	8223	9223	10277
225	503	100	5828	6585	7523	8569	9699	10902	12176
225	503	120	6654	7508	8598	9816	11134	12542	14038
225	503	140	7454	8402	9644	11034	12541	14157	15879
250	559	40	3323	3744	4239	4788	5375	5995	6643
250	559	60	4454	4927	5586	6325	7118	7956	8836
250	559	80	5325	6006	6835	7759	8751	9803	10911
250	559	100	6251	7034	8024	9128	10317	11582	12919
250	559	120	7142	8021	9170	10453	11839	13317	14885
250	559	140	8006	8976	10283	11746	13329	15023	16823
275	615	40	3543	3973	4492	5070	5688	6339	7019
275	615	60	4628	5221	5923	6701	7534	8414	9337
275	615	80	5672	6377	7250	8221	9262	10366	11526
275	615	100	6665	7471	8511	9670	10918	12242	13640
275	615	120	7621	8519	9725	11072	12524	14070	15706
275	615	140	8373	9525	10904	12438	14095	15863	17740
300	671	40	3759	4195	4738	5344	5992	6674	7385
300	671	60	4901	5516	6251	7066	7939	8860	9824
300	671	80	6013	6739	7653	8670	9760	10913	12124
300	671	100	7071	7896	8984	10198	11502	12885	14342
300	671	120	7931	8996	10265	11674	13190	14802	16506
300	671	140	8867	10066	11508	13111	14839	16682	18634
325	727	40	3972	4411	4978	5612	6288	7000	7742
325	727	60	5169	5803	6571	7423	8334	9294	10299
325	727	80	6348	7092	8046	9108	10245	11446	12707
325	727	100	7472	8311	9446	10712	12071	13511	15026
325	727	120	8362	9468	10792	12261	13840	15517	17286
325	727	140	9350	10593	12097	13767	15566	17481	19506
350	783	40	4182	4623	5213	5873	6577	7318	8089
350	783	60	5431	6083	6883	7771	8719	9718	10763
350	783	80	6678	7437	8431	9536	10719	11967	13277
350	783	100	7706	8707	9897	11215	12628	14123	15695
350	783	120	8784	9929	11307	12835	14475	16215	18049
350	783	140	9824	11109	12674	14409	16277	18261	20359
375	839	40	4392	4830	5442	6128	6860	7629	8430
375	839	60	5690	6358	7189	8112	9097	10133	11216
375	839	80	7003	7775	8807	9955	11182	12477	13834
375	839	100	8067	9103	10339	11707	13173	14722	16349
375	839	120	9199	10381	11811	13396	15097	16899	18796
375	839	140	10289	11615	13238	15038	16972	19026	21194
400	894	40	4470	5021	5666	6378	7137	7933	8763
400	894	60	5945	6627	7489	8445	9466	10540	11661
400	894	80	7326	8107	9175	10365	11637	12977	14380
400	894	100	8423	9491	10772	12190	13707	15309	16991
400	894	120	9606	10824	12306	13947	15707	17569	19528
400	894	140	10747	12111	13791	15654	17654	19776	22013

NOTE: Detector time constant at a reference velocity of 5 ft/sec (1.5m/sec).
For SI units: 1 ft = 0.305 m; 1000 Btu/sec = 1055 kW.

Table B-4(v) Installed Spacing of Heat Detector: 20ft t_g: 150 seconds to 1000 Btu/sec α: 0.044 Btu/sec^3

τ	RTI	ΔT	Ceiling Height (ft) Fire Size at Detector Response (Btu/sec)						
			4.0	8.0	12.0	16.0	20.0	24.0	28.0
25	56	40	379	456	556	675	812	969	1145
25	56	60	506	629	786	975	1197	1452	1743
25	56	80	635	805	1028	1299	1618	1988	2410
25	56	100	759	994	1285	1646	2074	2571	3139
25	56	120	889	1182	1561	2017	2564	3199	3926
25	56	140	1009	1378	1845	2417	3085	3869	4766
50	112	40	540	635	754	891	1045	1217	1406
50	112	60	723	863	1041	1249	1487	1756	2058
50	112	80	898	1087	1331	1619	1952	2333	2764
50	112	100	1059	1312	1628	2004	2444	2950	3526
50	112	120	1226	1540	1934	2407	2963	3606	4341
50	112	140	1393	1771	2250	2828	3510	4301	5207
75	168	40	684	787	923	1078	1250	1439	1645
75	168	60	906	1063	1263	1493	1751	2040	2359
75	168	80	1111	1331	1600	1911	2266	2666	3114
75	168	100	1319	1597	1939	2339	2800	3324	3914
75	168	120	1525	1862	2283	2780	3355	4014	4762
75	168	140	1729	2128	2634	3234	3933	4739	5656
100	224	40	801	923	1075	1247	1436	1642	1866
100	224	60	1061	1244	1465	1716	1996	2304	2643
100	224	80	1312	1553	1846	2182	2560	2982	3450
100	224	100	1558	1856	2225	2653	3138	3684	4294
100	224	120	1801	2157	2607	3132	3732	4413	5179
100	224	140	2043	2457	2992	3621	4345	5171	6105
125	280	40	915	1049	1216	1404	1609	1832	2072
125	280	60	1211	1412	1652	1923	2224	2553	2911
125	280	80	1501	1759	2075	2435	2837	3282	3772
125	280	100	1784	2098	2493	2947	3459	4029	4662
125	280	120	2064	2432	2911	3465	4093	4798	5587
125	280	140	2304	2763	3330	3989	4742	5592	6548
150	335	40	1007	1167	1348	1551	1772	2011	2266
150	335	60	1355	1570	1828	2119	2440	2789	3167
150	335	80	1681	1953	2291	2675	3101	3569	4080
150	335	100	2002	2326	2747	3227	3765	4361	5017
150	335	120	2278	2691	3199	3782	4438	5171	5984
150	335	140	2570	3053	3651	4342	5124	6002	6982
175	391	40	1114	1280	1474	1691	1927	2181	2451
175	391	60	1492	1720	1996	2306	2646	3014	3412
175	391	80	1856	2138	2497	2904	3352	3843	4377
175	391	100	2178	2542	2988	3495	4059	4680	5361
175	391	120	2504	2939	3475	4086	4771	5531	6369
175	391	140	2824	3330	3959	4680	5493	6399	7406
200	447	40	1211	1387	1594	1825	2075	2343	2629
200	447	60	1626	1864	2156	2485	2843	3231	3647
200	447	80	2028	2315	2695	3123	3594	4108	4664
200	447	100	2367	2750	3220	3753	4342	4988	5694
200	447	120	2722	3176	3739	4379	5092	5880	6744
200	447	140	3070	3596	4254	5007	5850	6785	7819
225	503	40	1305	1490	1710	1954	2218	2500	2800
225	503	60	1757	2002	2311	2657	3034	3440	3875
225	503	80	2157	2485	2885	3335	3828	4364	4942
225	503	100	2551	2951	3444	4001	4616	5287	6017
225	503	120	2934	3405	3995	4663	5404	6218	7110
225	503	140	3309	3852	4540	5323	6196	7161	8224
250	559	40	1396	1590	1821	2078	2356	2652	2966
250	559	60	1886	2136	2461	2824	3218	3642	4095
250	559	80	2308	2650	3070	3540	4055	4612	5212
250	559	100	2730	3145	3661	4242	4881	5578	6332
250	559	120	3140	3627	4242	4937	5706	6548	7466
250	559	140	3542	4100	4817	5630	6533	7528	8619
275	615	40	1486	1687	1930	2199	2490	2799	3126
275	615	60	2013	2267	2606	2985	3397	3839	4310
275	615	80	2454	2810	3249	3740	4275	4853	5474
275	615	100	2905	3334	3872	4477	5140	5860	6639
275	615	120	3342	3843	4483	5205	6000	6869	7814
275	615	140	3729	4341	5086	5929	6861	7886	9005
300	671	40	1574	1782	2035	2317	2620	2942	3283
300	671	60	2100	2392	2747	3143	3571	4030	4518
300	671	80	2598	2966	3423	3934	4489	5088	5730
300	671	100	3076	3517	4077	4705	5392	6136	6939
300	671	120	3541	4053	4717	5465	6288	7184	8155
300	671	140	3942	4577	5349	6220	7182	8236	9383
325	727	40	1661	1874	2138	2431	2747	3082	3435
325	727	60	2213	2516	2885	3297	3741	4217	4722
325	727	80	2739	3119	3593	4123	4699	5318	5981
325	727	100	3244	3697	4277	4928	5638	6406	7232
325	727	120	3690	4257	4946	5720	6568	7491	8488
325	727	140	4150	4806	5605	6505	7496	8578	9755
350	783	40	1746	1964	2239	2543	2871	3218	3584
350	783	60	2324	2637	3020	3447	3908	4400	4922
350	783	80	2877	3268	3760	4308	4904	5543	6226
350	783	100	3411	3872	4473	5146	5879	6670	7520
350	783	120	3871	4458	5170	5969	6843	7792	8815
350	783	140	4354	5031	5856	6784	7803	8914	10119
375	839	40	1831	2052	2337	2653	2992	3352	3730
375	839	60	2432	2755	3152	3594	4071	4578	5117
375	839	80	3014	3413	3923	4490	5105	5763	6466
375	839	100	3530	4043	4665	5360	6115	6929	7802
375	839	120	4050	4654	5390	6213	7113	8087	9136
375	839	140	4554	5251	6102	7057	8105	9244	10477
400	894	40	1916	2139	2433	2761	3111	3483	3873
400	894	60	2539	2871	3282	3738	4230	4754	5308
400	894	80	3149	3557	4082	4668	5301	5980	6702
400	894	100	3683	4212	4854	5569	6347	7183	8079
400	894	120	4225	4847	5605	6453	7378	8377	9452
400	894	140	4751	5467	6343	7325	8401	9568	10829

NOTE: Detector time constant at a reference velocity of 5 ft/sec (1.5m/sec).
For SI units: 1 ft = 0.305 m; 1000 Btu/sec = 1055 kW.

Table B-4(w)   Installed Spacing of Heat Detector: 20ft t_g: 300 seconds to 1000 Btu/sec α: 0.011 Btu/sec^3

τ	RTI	ΔT	Ceiling Height (ft) Fire Size at Detector Response (Btu/sec) 4.0	8.0	12.0	16.0	20.0	24.0	28.0	τ	RTI	ΔT	Ceiling Height (ft) Fire Size at Detector Response (Btu/sec) 4.0	8.0	12.0	16.0	20.0	24.0	28.0
25	56	40	232	294	375	473	589	723	876	225	503	40	777	895	1039	1202	1380	1573	1782
25	56	60	322	429	566	731	928	1157	1419	225	503	60	1045	1223	1435	1675	1942	2235	2555
25	56	80	415	572	774	1022	1311	1649	2037	225	503	80	1308	1539	1822	2146	2509	2912	3357
25	56	100	513	726	1002	1338	1733	2193	2720	225	503	100	1561	1850	2208	2621	3088	3612	4195
25	56	120	613	890	1247	1681	2190	2784	3464	225	503	120	1812	2157	2596	3104	3684	4338	5072
25	56	140	717	1065	1508	2047	2685	3418	4264	225	503	140	2040	2464	2987	3597	4296	5092	5989
50	112	40	326	395	484	588	709	847	1003	250	559	40	830	954	1105	1275	1460	1661	1878
50	112	60	445	557	700	870	1071	1303	1567	250	559	60	1120	1302	1523	1773	2049	2352	2682
50	112	80	567	725	929	1177	1471	1812	2203	250	559	80	1397	1637	1931	2266	2640	3055	3511
50	112	100	685	902	1174	1509	1908	2371	2902	250	559	100	1668	1964	2336	2762	3242	3777	4372
50	112	120	805	1082	1436	1865	2378	2976	3660	250	559	120	1919	2288	2741	3264	3857	4523	5269
50	112	140	928	1270	1709	2248	2881	3623	4473	250	559	140	2173	2611	3149	3774	4487	5295	6204
75	168	40	405	483	580	693	821	966	1128	275	615	40	887	1011	1169	1346	1539	1748	1972
75	168	60	556	673	824	1004	1211	1448	1718	275	615	60	1189	1379	1609	1868	2154	2467	2806
75	168	80	695	864	1077	1333	1632	1977	2372	275	615	80	1484	1731	2037	2383	2769	3194	3661
75	168	100	835	1059	1342	1683	2085	2552	3087	275	615	100	1773	2076	2460	2899	3391	3939	4545
75	168	120	975	1259	1618	2054	2570	3171	3860	275	615	120	2034	2416	2883	3419	4026	4706	5463
75	168	140	1118	1469	1907	2445	3085	3831	4685	275	615	140	2303	2753	3307	3947	4675	5496	6416
100	224	40	477	562	668	790	927	1079	1249	300	671	40	932	1066	1231	1415	1615	1831	2063
100	224	60	650	778	940	1129	1346	1591	1866	300	671	60	1257	1453	1692	1961	2256	2578	2927
100	224	80	812	992	1217	1482	1790	2142	2542	300	671	80	1570	1824	2140	2497	2894	3331	3808
100	224	100	971	1208	1502	1852	2262	2735	3275	300	671	100	1859	2185	2581	3033	3538	4098	4716
100	224	120	1133	1427	1797	2240	2763	3369	4062	300	671	120	2147	2540	3021	3572	4192	4885	5654
100	224	140	1294	1650	2102	2647	3292	4041	4901	300	671	140	2430	2892	3461	4117	4859	5694	6626
125	280	40	543	636	750	880	1026	1187	1364	325	727	40	977	1120	1292	1483	1690	1913	2152
125	280	60	737	876	1049	1248	1474	1728	2012	325	727	60	1324	1526	1774	2052	2357	2688	3046
125	280	80	922	1112	1349	1625	1943	2304	2711	325	727	80	1654	1914	2240	2609	3017	3465	3953
125	280	100	1101	1348	1655	2017	2436	2916	3462	325	727	100	1956	2291	2700	3164	3682	4254	4884
125	280	120	1280	1586	1968	2423	2954	3567	4265	325	727	120	2258	2662	3156	3721	4355	5061	5843
125	280	140	1459	1827	2291	2846	3498	4254	5118	325	727	140	2554	3029	3612	4283	5040	5889	6834
150	335	40	611	705	827	966	1120	1289	1474	350	783	40	1023	1173	1351	1549	1763	1994	2240
150	335	60	820	968	1151	1361	1598	1861	2154	350	783	60	1389	1598	1854	2140	2454	2795	3163
150	335	80	1020	1226	1474	1763	2091	2462	2877	350	783	80	1737	2002	2339	2718	3138	3596	4095
150	335	100	1223	1481	1801	2175	2605	3095	3649	350	783	100	2050	2395	2815	3292	3823	4407	5049
150	335	120	1420	1738	2133	2601	3142	3763	4469	350	783	120	2366	2780	3288	3867	4515	5235	6029
150	335	140	1616	1995	2473	3040	3702	4466	5337	350	783	140	2676	3162	3760	4446	5218	6081	7039
175	391	40	668	771	901	1048	1210	1387	1580	375	839	40	1073	1225	1409	1613	1835	2072	2325
175	391	60	903	1057	1250	1470	1716	1989	2291	375	839	60	1453	1668	1932	2227	2550	2900	3277
175	391	80	1122	1334	1595	1895	2235	2616	3041	375	839	80	1803	2088	2435	2826	3256	3725	4235
175	391	100	1339	1609	1942	2328	2770	3271	3834	375	839	100	2142	2497	2929	3419	3961	4558	5212
175	391	120	1554	1883	2293	2773	3326	3958	4672	375	839	120	2472	2897	3418	4011	4673	5406	6213
175	391	140	1768	2157	2650	3230	3904	4677	5555	375	839	140	2795	3292	3906	4607	5394	6270	7242
200	447	40	724	834	971	1126	1296	1482	1682	400	894	40	1119	1276	1466	1677	1905	2149	2409
200	447	60	973	1141	1344	1574	1831	2114	2425	400	894	60	1517	1736	2008	2313	2645	3003	3389
200	447	80	1216	1439	1711	2022	2374	2766	3201	400	894	80	1880	2173	2530	2931	3372	3852	4373
200	447	100	1452	1731	2077	2477	2931	3443	4016	400	894	100	2233	2597	3041	3542	4097	4707	5372
200	447	120	1685	2022	2447	2941	3507	4149	4873	400	894	120	2576	3011	3546	4152	4828	5574	6394
200	447	140	1902	2313	2821	3416	4102	4886	5773	400	894	140	2913	3421	4048	4764	5566	6457	7442

NOTE: Detector time constant at a reference velocity of 5 ft/sec (1.5m/sec).
For SI units: 1 ft = 0.305 m; 1000 Btu/sec = 1055 kW.

Table B-4(x)   Installed Spacing of Heat Detector: 20ft t_g: 500 seconds to 1000 Btu/sec α: 0.004 Btu/sec^3

τ	RTI	ΔT	4.0	8.0	12.0	16.0	20.0	24.0	28.0	τ	RTI	ΔT	4.0	8.0	12.0	16.0	20.0	24.0	28.0
			Ceiling Height (ft) Fire Size at Detector Response (Btu/sec)																
25	56	40	168	266	300	390	497	622	764	225	503	40	535	625	736	862	1002	1157	1326
25	56	60	242	343	473	631	816	1033	1280	225	503	60	732	869	1037	1231	1450	1696	1970
25	56	80	322	474	668	904	1184	1505	1875	225	503	80	921	1108	1340	1609	1918	2268	2663
25	56	100	407	617	884	1207	1589	2030	2537	225	503	100	1103	1347	1647	2001	2409	2877	3407
25	56	120	497	772	1118	1537	2032	2603	3260	225	503	120	1286	1587	1963	2407	2925	3522	4203
25	56	140	592	938	1370	1891	2508	3220	4040	225	503	140	1468	1831	2286	2830	3467	4204	5048
50	112	40	230	290	366	458	567	693	836	250	559	40	575	665	781	911	1056	1216	1390
50	112	60	323	425	553	711	898	1116	1365	250	559	60	780	922	1097	1297	1522	1773	2052
50	112	80	418	568	760	996	1273	1598	1970	250	559	80	978	1174	1413	1689	2004	2361	2760
50	112	100	517	721	985	1308	1688	2132	2642	250	559	100	1174	1424	1733	2093	2509	2982	3517
50	112	120	618	884	1227	1645	2138	2714	3374	250	559	120	1367	1676	2059	2511	3036	3638	4323
50	112	140	723	1057	1485	2006	2625	3339	4162	250	559	140	1559	1929	2393	2944	3587	4330	5177
75	168	40	284	348	429	525	636	764	910	275	615	40	609	704	824	959	1109	1273	1452
75	168	60	394	499	632	792	981	1201	1452	275	615	60	827	974	1155	1361	1592	1849	2133
75	168	80	506	656	850	1087	1367	1693	2067	275	615	80	1036	1238	1484	1767	2089	2451	2857
75	168	100	617	823	1087	1407	1790	2237	2749	275	615	100	1243	1500	1816	2184	2606	3086	3626
75	168	120	730	996	1337	1755	2249	2827	3489	275	615	120	1446	1762	2154	2614	3145	3753	4443
75	168	140	846	1177	1603	2124	2739	3459	4285	275	615	140	1649	2026	2498	3057	3707	4454	5306
100	224	40	332	401	487	587	703	835	983	300	671	40	643	742	866	1006	1161	1330	1513
100	224	60	458	569	707	872	1064	1286	1540	300	671	60	877	1025	1211	1423	1660	1923	2213
100	224	805	585	741	939	1179	1462	1790	2166	300	671	80	1094	1301	1553	1844	2172	2541	2952
100	224	100	709	918	1185	1510	1894	2343	2857	300	671	100	1310	1574	1898	2273	2703	3188	3735
100	224	120	834	1105	1448	1864	2361	2941	3606	300	671	120	1524	1846	2246	2714	3253	3867	4562
100	224	140	962	1296	1722	2240	2859	3581	4410	300	671	140	1737	2120	2601	3168	3825	4579	5435
125	280	40	377	450	541	647	767	903	1055	325	727	40	675	779	907	1052	1211	1385	1573
125	280	60	522	634	778	948	1145	1371	1627	325	727	60	919	1075	1266	1484	1727	1996	2291
125	280	80	658	820	1025	1269	1556	1887	2266	325	727	80	1150	1362	1621	1919	2254	2629	3046
125	280	100	795	1011	1283	1611	1999	2450	2966	325	727	100	1376	1646	1977	2361	2797	3290	3842
125	280	120	933	1206	1553	1975	2474	3056	3724	325	727	120	1601	1928	2337	2813	3359	3980	4681
125	280	140	1072	1412	1837	2359	2980	3704	4536	325	727	140	1814	2212	2702	3277	3942	4702	5564
150	335	40	419	497	593	704	829	969	1126	350	783	40	707	815	947	1097	1260	1439	1632
150	335	60	578	697	847	1022	1225	1455	1715	350	783	60	960	1123	1320	1544	1793	2068	2368
150	335	80	728	896	1107	1357	1649	1984	2365	350	783	80	1204	1422	1688	1993	2335	2716	3139
150	335	100	880	1099	1378	1711	2103	2557	3076	350	783	100	1441	1716	2056	2447	2891	3390	3948
150	335	120	1025	1307	1659	2085	2588	3173	3843	350	783	120	1676	2009	2426	2911	3465	4092	4799
150	335	140	1176	1519	1952	2478	3102	3829	4663	350	783	140	1896	2302	2801	3385	4058	4824	5692
175	391	40	459	541	643	759	889	1034	1194	375	839	40	738	850	987	1140	1309	1492	1689
175	391	60	631	756	912	1094	1302	1537	1801	375	839	60	1002	1171	1374	1603	1858	2138	2444
175	391	80	794	969	1187	1444	1740	2080	2465	375	839	80	1257	1481	1754	2065	2414	2802	3230
175	391	100	955	1185	1470	1810	2206	2664	3186	375	839	100	1505	1786	2133	2532	2983	3489	4054
175	391	120	1116	1403	1763	2194	2701	3289	3962	375	839	120	1750	2088	2514	3007	3568	4203	4917
175	391	140	1277	1626	2066	2597	3224	3954	4790	375	839	140	1977	2390	2899	3492	4172	4946	5820
200	447	40	498	584	690	811	946	1096	1261	400	894	40	769	884	1025	1183	1356	1544	1746
200	447	60	683	813	976	1164	1377	1617	1886	400	894	60	1045	1217	1426	1661	1921	2207	2519
200	447	80	863	1040	1265	1527	1830	2175	2564	400	894	80	1310	1538	1818	2136	2492	2886	3321
200	447	100	1030	1267	1560	1906	2309	2771	3297	400	894	100	1568	1854	2208	2615	3074	3587	4158
200	447	120	1202	1497	1864	2302	2814	3406	4083	400	894	120	1811	2166	2600	3101	3671	4313	5033
200	447	140	1374	1730	2177	2714	3346	4079	4919	400	894	140	2056	2478	2995	3597	4286	5067	5947

NOTE: Detector time constant at a reference velocity of 5 ft/sec (1.5m/sec).
For SI units: 1 ft = 0.305 m; 1000 Btu/sec = 1055 kW.

Table B-4(y) Installed Spacing of Heat Detector: 20ft t_g: 600 seconds to 1000 Btu/sec α: 0.003 Btu/sec^3

τ	RTI	ΔT	Ceiling Height (ft) — Fire Size at Detector Response (Btu/sec)						
			4.0	8.0	12.0	16.0	20.0	24.0	28.0
25	56	40	151	208	280	369	474	595	735
25	56	60	222	321	449	605	787	1000	1244
25	56	80	298	449	641	874	1149	1466	1831
25	56	100	380	590	854	1173	1551	1986	2487
25	56	120	467	743	1085	1499	1989	2554	3205
25	56	140	560	906	1334	1850	2461	3167	3979
50	112	40	205	262	336	425	531	654	794
50	112	60	290	390	517	671	855	1069	1314
50	112	80	379	527	717	949	1223	1543	1910
50	112	100	472	675	936	1255	1632	2071	2574
50	112	120	569	834	1174	1588	2079	2646	3299
50	112	140	669	1003	1428	1945	2558	3265	4080
75	168	40	251	312	389	481	588	713	854
75	168	60	352	453	582	738	923	1138	1385
75	168	80	455	603	793	1024	1300	1622	1990
75	168	100	558	760	1021	1340	1716	2157	2662
75	168	120	664	927	1265	1679	2167	2739	3394
75	168	140	774	1103	1525	2042	2652	3365	4182
100	224	40	293	358	439	534	645	771	915
100	224	60	408	513	645	804	992	1209	1457
100	224	80	524	674	866	1101	1378	1701	2071
100	224	100	638	844	1105	1422	1802	2244	2751
100	224	120	754	1019	1357	1768	2260	2833	3491
100	224	140	876	1202	1623	2140	2750	3465	4285
125	280	40	332	400	486	585	699	829	975
125	280	60	460	569	706	869	1060	1279	1529
125	280	80	587	742	939	1176	1456	1781	2153
125	280	100	713	921	1185	1507	1888	2332	2841
125	280	120	839	1108	1448	1860	2353	2928	3587
125	280	140	968	1300	1722	2235	2850	3566	4389
150	335	40	369	441	530	634	752	885	1035
150	335	60	509	623	765	933	1127	1349	1602
150	335	80	648	808	1010	1251	1534	1861	2235
150	335	100	784	997	1266	1591	1974	2420	2931
150	335	120	921	1192	1535	1952	2447	3024	3685
150	335	140	1059	1395	1817	2334	2950	3668	4493
175	391	40	404	480	573	681	803	940	1093
175	391	60	559	675	822	994	1192	1418	1674
175	391	80	705	871	1078	1324	1611	1941	2318
175	391	100	852	1071	1345	1674	2061	2509	3022
175	391	120	998	1275	1623	2044	2541	3120	3784
175	391	140	1145	1484	1913	2433	3051	3771	4598
200	447	40	437	517	615	727	853	994	1150
200	447	60	604	725	877	1054	1257	1486	1746
200	447	80	761	932	1145	1396	1687	2021	2401
200	447	100	918	1142	1422	1756	2146	2598	3114
200	447	120	1073	1355	1709	2134	2635	3217	3883
200	447	140	1229	1573	2007	2531	3152	3875	4704
225	503	40	470	553	654	771	901	1046	1206
225	503	60	647	773	930	1112	1319	1554	1816
225	503	80	815	991	1210	1466	1762	2100	2483
225	503	100	980	1211	1497	1836	2232	2687	3206
225	503	120	1146	1434	1794	2224	2729	3313	3982
225	503	140	1311	1661	2100	2629	3253	3979	4810
250	559	40	501	587	693	814	948	1097	1260
250	559	60	688	819	982	1169	1381	1620	1886
250	559	80	871	1049	1273	1535	1836	2178	2565
250	559	100	1042	1278	1571	1916	2316	2775	3297
250	559	120	1216	1510	1877	2312	2822	3410	4082
250	559	140	1390	1746	2192	2726	3354	4083	4917
275	615	40	531	621	731	855	994	1146	1314
275	615	60	729	865	1032	1224	1441	1684	1955
275	615	80	918	1105	1335	1602	1908	2255	2646
275	615	100	1102	1344	1643	1994	2399	2863	3389
275	615	120	1285	1585	1958	2400	2914	3506	4182
275	615	140	1467	1829	2282	2822	3455	4187	5024
300	671	40	565	654	767	896	1038	1195	1366
300	671	60	769	909	1081	1278	1500	1748	2023
300	671	80	966	1159	1395	1668	1980	2332	2727
300	671	100	1160	1408	1713	2070	2481	2949	3479
300	671	120	1352	1658	2038	2486	3006	3602	4281
300	671	140	1543	1911	2370	2916	3554	4291	5131
325	727	40	594	686	803	936	1082	1243	1418
325	727	60	808	952	1129	1331	1558	1811	2091
325	727	80	1014	1212	1454	1733	2050	2407	2807
325	727	100	1217	1471	1782	2146	2562	3036	3570
325	727	120	1418	1730	2116	2571	3096	3698	4381
325	727	140	1617	1990	2457	3010	3653	4394	5239
350	783	40	621	717	838	975	1125	1289	1469
350	783	60	846	994	1176	1383	1615	1873	2157
350	783	80	1062	1265	1512	1797	2119	2482	2886
350	783	100	1274	1532	1850	2220	2642	3121	3660
350	783	120	1482	1800	2194	2655	3186	3793	4480
350	783	140	1691	2069	2543	3103	3752	4497	5346
375	839	40	648	748	873	1013	1167	1335	1518
375	839	60	886	1036	1222	1434	1671	1933	2222
375	839	80	1108	1316	1569	1860	2188	2555	2964
375	839	100	1329	1593	1917	2293	2721	3205	3749
375	839	120	1546	1869	2270	2737	3275	3887	4579
375	839	140	1763	2146	2628	3194	3849	4600	5453
400	894	40	675	778	906	1050	1208	1381	1567
400	894	60	921	1077	1268	1485	1726	1993	2287
400	894	80	1154	1366	1625	1921	2255	2628	3042
400	894	100	1383	1652	1983	2365	2799	3289	3838
400	894	120	1609	1937	2344	2819	3363	3980	4677
400	894	140	1824	2222	2711	3284	3946	4702	5559

NOTE: Detector time constant at a reference velocity of 5 ft/sec (1.5m/sec).
For SI units: 1 ft = 0.305 m; 1000 Btu/sec = 1055 kW.

Table B-4(z)   Installed Spacing of Heat Detector: 25ft t_g: 50 seconds to 1000 Btu/sec α: 0.400 Btu/sec^3

τ	RTI	ΔT	4.0	8.0	12.0	16.0	20.0	24.0	28.0	τ	RTI	ΔT	4.0	8.0	12.0	16.0	20.0	24.0	28.0
			Ceiling Height (ft)										**Ceiling Height (ft)**						
			Fire Size at Detector Response (Btu/sec)										Fire Size at Detector Response (Btu/sec)						
25	56	40	1187	1381	1575	1797	2046	2319	2617	225	503	40	4308	4621	5097	5646	6240	6868	7527
25	56	60	1529	1795	2072	2393	2756	3158	3601	225	503	60	5557	6060	6716	7463	8270	9127	10027
25	56	80	1864	2194	2560	2990	3477	4023	4630	225	503	80	6816	7399	8224	9163	10179	11261	12402
25	56	100	2194	2586	3050	3596	4220	4924	5711	225	503	100	7852	8660	9660	10790	12016	13323	14706
25	56	120	2520	2977	3545	4217	4989	5865	6847	225	503	120	8944	9877	11047	12368	13804	15339	16968
25	56	140	2846	3368	4048	4855	5786	6846	8038	225	503	140	9998	11057	12397	13911	15559	17325	19205
50	112	40	1687	1953	2201	2479	2786	3117	3472	250	559	40	4498	4918	5430	6011	6638	7301	7995
50	112	60	2226	2545	2885	3276	3708	4179	4688	250	559	60	5955	6469	7161	7949	8801	9703	10650
50	112	80	2752	3103	3541	4048	4613	5233	5908	250	559	80	7316	7902	8770	9760	10830	11968	13165
50	112	100	3272	3642	4183	4813	5518	6298	7152	250	559	100	8404	9247	10301	11491	12779	14151	15601
50	112	120	3655	4164	4818	5578	6432	7382	8429	250	559	120	9576	10547	11778	13168	14675	16283	17988
50	112	140	4098	4682	5452	6348	7360	8490	9742	250	559	140	10706	11807	13215	14805	16532	18380	20343
75	168	40	2138	2432	2722	3049	3407	3791	4199	275	615	40	4783	5216	5754	6365	7025	7721	8449
75	168	60	2869	3178	3570	4023	4520	5056	5631	275	615	60	6345	6866	7593	8421	9315	10261	11254
75	168	80	3467	3868	4373	4954	5595	6291	7041	275	615	80	7635	8379	9300	10339	11462	12653	13906
75	168	100	4071	4532	5151	5866	6656	7519	8455	275	615	100	8942	9818	10923	12171	13521	14956	16470
75	168	120	4657	5177	5915	6767	7713	8752	9884	275	615	120	10192	11198	12487	13943	15520	17200	18978
75	168	140	5114	5806	6670	7665	8774	9997	11336	275	615	140	11397	12534	14008	15672	17477	19404	21449
100	224	40	2567	2861	3187	3557	3961	4392	4849	300	671	40	5061	5505	6069	6709	7400	8128	8890
100	224	60	3380	3738	4183	4692	5249	5846	6483	300	671	60	6728	7252	8012	8880	9815	10804	11841
100	224	80	4147	4555	5120	5770	6482	7250	8073	300	671	80	8080	8852	9815	10903	12076	13320	14627
100	224	100	4770	5330	6024	6817	7688	8633	9650	300	671	100	9468	10373	11528	12833	14242	15738	17316
100	224	120	5433	6084	6906	7845	8882	10010	11230	300	671	120	10794	11831	13178	14698	16343	18093	19942
100	224	140	6076	6820	7773	8864	10071	11390	12823	300	671	140	12074	13242	14780	16516	18397	20402	22526
125	280	40	2988	3258	3614	4024	4470	4945	5447	325	727	40	5335	5787	6375	7044	7765	8525	9319
125	280	60	3894	4257	4747	5309	5921	6576	7270	325	727	60	7107	7628	8421	9327	10302	11333	12413
125	280	80	4675	5183	5810	6523	7302	8139	9031	325	727	80	8516	9312	10317	11452	12675	13970	15330
125	280	100	5471	6070	6830	7697	8646	9670	10765	325	727	100	9982	10914	12119	13478	14945	16502	18140
125	280	120	6236	6926	7823	8846	9969	11185	12491	325	727	120	11385	12449	13852	15435	17145	18963	20882
125	280	140	6978	7760	8796	9979	11280	12694	14220	325	727	140	12737	13933	15534	17340	19294	21376	23577
150	335	40	3296	3620	4012	4460	4946	5462	6006	350	783	40	5604	6063	6675	7372	8122	8913	9738
150	335	60	4390	4745	5276	5887	6551	7260	8010	350	783	60	7308	7985	8820	9763	10778	11850	12972
150	335	80	5239	5777	6457	7231	8073	8975	9933	350	783	80	8943	9763	10809	11989	13260	14605	16016
150	335	100	6139	6765	7588	8525	9548	10646	11818	350	783	100	10488	11443	12696	14109	15633	17248	18946
150	335	120	7003	7718	8686	9788	10994	12293	13685	350	783	120	11965	13053	14510	16155	17930	19815	21801
150	335	140	7841	8645	9759	11029	12422	13928	15546	350	783	140	13391	14609	16271	18146	20171	22327	24605
175	391	40	3641	3968	4390	4873	5397	5952	6536	375	839	40	5869	6333	6968	7692	8471	9292	10148
175	391	60	4728	5197	5776	6435	7149	7909	8711	375	839	60	7647	8344	9211	10190	11244	12356	13518
175	391	80	5781	6340	7071	7902	8805	9769	10791	375	839	80	9363	10204	11289	12514	13833	15227	16688
175	391	100	6781	7426	8308	9312	10405	11575	12820	375	839	100	10985	11961	13260	14726	16305	17978	19735
175	391	120	7744	8471	9506	10684	11969	13350	14823	375	839	120	12537	13645	15155	16859	18698	20648	22702
175	391	140	8515	9479	10675	12030	13510	15106	16813	375	839	140	14034	15271	16992	18934	21031	23260	25611
200	447	40	3977	4301	4750	5267	5827	6420	7042	400	894	40	6130	6597	7255	8005	8813	9663	10549
200	447	60	5149	5637	6256	6959	7721	8530	9382	400	894	60	7979	8695	9594	10609	11700	12851	14054
200	447	80	6305	6880	7659	8545	9506	10529	11612	400	894	80	9776	10636	11760	13029	14394	15836	17346
200	447	100	7405	8059	8997	10065	11225	12465	13780	400	894	100	11475	12470	13814	15331	16965	18694	20509
200	447	120	8293	9185	10291	11542	12903	14363	15915	400	894	120	13101	14225	15787	17551	19451	21465	23584
200	447	140	9269	10283	11552	12988	14553	16235	18031	400	894	140	14670	15921	17700	19707	21873	24174	26598

NOTE: Detector time constant at a reference velocity of 5 ft/sec (1.5m/sec).
For SI units: 1 ft = 0.305 m; 1000 Btu/sec = 1055 kW.

Table B-4(aa)   Installed Spacing of Heat Detector: 25ft t_g: 150 seconds to 1000 Btu/sec α: 0.044 Btu/sec³

τ	RTI	ΔT	Ceiling Height (ft) 4.0	8.0	12.0	16.0	20.0	24.0	28.0
			Fire Size at Detector Response (Btu/sec)						
25	56	40	520	609	724	861	1018	1195	1394
25	56	60	701	843	1029	1250	1507	1801	2131
25	56	80	881	1084	1352	1672	2045	2472	2954
25	56	100	1045	1335	1695	2126	2629	3205	3855
25	56	120	1231	1596	2058	2611	3256	3994	4828
25	56	140	1414	1875	2449	3126	3924	4836	5867
50	112	40	742	843	976	1130	1303	1495	1707
50	112	60	981	1148	1352	1590	1861	2166	2507
50	112	80	1223	1450	1734	2068	2452	2887	3376
50	112	100	1460	1755	2127	2568	3078	3659	4316
50	112	120	1696	2065	2534	3092	3740	4483	5322
50	112	140	1933	2380	2955	3640	4439	5355	6392
75	168	40	925	1042	1192	1364	1555	1765	1993
75	168	60	1228	1410	1635	1895	2186	2510	2867
75	168	80	1526	1770	2076	2433	2837	3290	3794
75	168	100	1820	2127	2522	2985	3513	4110	4779
75	168	120	2092	2484	2976	3554	4219	4974	5824
75	168	140	2371	2845	3440	4143	4955	5881	6927
100	224	40	1084	1220	1386	1576	1785	2012	2258
100	224	60	1451	1647	1893	2174	2487	2831	3208
100	224	80	1806	2060	2390	2771	3198	3673	4197
100	224	100	2129	2466	2887	3375	3928	4546	5234
100	224	120	2456	2869	3388	3993	4681	5456	6322
100	224	140	2778	3273	3896	4624	5458	6403	7463
125	280	40	1237	1385	1566	1772	1998	2243	2506
125	280	60	1661	1867	2132	2434	2768	3134	3531
125	280	80	2047	2329	2683	3088	3539	4037	4583
125	280	100	2426	2781	3229	3744	4322	4965	5675
125	280	120	2796	3228	3776	4409	5123	5923	6811
125	280	140	3160	3673	4326	5084	5945	6913	7993
150	335	40	1382	1540	1735	1957	2199	2460	2740
150	335	60	1863	2074	2357	2679	3034	3421	3838
150	335	80	2285	2583	2959	3388	3864	4385	4954
150	335	100	2707	3079	3552	4094	4699	5368	6101
150	335	120	3119	3568	4143	4805	5549	6374	7287
150	335	140	3524	4052	4735	5524	6414	7409	8513
175	391	40	1522	1687	1896	2132	2390	2667	2963
175	391	60	2031	2270	2571	2913	3288	3695	4133
175	391	80	2513	2825	3222	3674	4174	4719	5311
175	391	100	2978	3363	3861	4430	5062	5756	6514
175	391	120	3431	3891	4495	5186	5958	6812	7750
175	391	140	3839	4413	5128	5948	6868	7891	9021
200	447	40	1658	1828	2049	2300	2573	2866	3177
200	447	60	2207	2458	2776	3137	3532	3959	4416
200	447	80	2733	3056	3475	3950	4473	5041	5656
200	447	100	3239	3635	4158	4752	5411	6131	6915
200	447	120	3695	4202	4833	5553	6354	7236	8201
200	447	140	4164	4761	5505	6356	7308	8360	9518

τ	RTI	ΔT	Ceiling Height (ft) 4.0	8.0	12.0	16.0	20.0	24.0	28.0
			Fire Size at Detector Response (Btu/sec)						
225	503	40	1791	1963	2197	2462	2749	3057	3383
225	503	60	2378	2639	2974	3353	3767	4213	4690
225	503	80	2946	3279	3718	4215	4761	5353	5991
225	503	100	3494	3898	4444	5064	5748	6495	7304
225	503	120	3975	4501	5159	5908	6738	7648	8640
225	503	140	4477	5096	5870	6752	7734	8817	10004
250	559	40	1922	2095	2340	2618	2919	3241	3583
250	559	60	2544	2815	3166	3562	3995	4459	4956
250	559	80	3154	3495	3954	4473	5041	5655	6316
250	559	100	3697	4151	4721	5366	6076	6848	7683
250	559	120	4246	4792	5476	6252	7111	8049	9069
250	559	140	4781	5421	6223	7136	8150	9263	10479
275	615	40	2018	2221	2478	2770	3085	3421	3776
275	615	60	2706	2985	3351	3765	4216	4699	5214
275	615	80	3359	3705	4183	4722	5312	5949	6632
275	615	100	3927	4398	4990	5660	6395	7193	8053
275	615	120	4510	5074	5783	6587	7474	8440	9488
275	615	140	5077	5737	6567	7511	8555	9698	10944
300	671	40	2134	2344	2613	2917	3245	3595	3965
300	671	60	2865	3151	3532	3963	4431	4932	5466
300	671	80	3514	3909	4405	4966	5577	6236	6941
300	671	100	4152	4638	5252	5946	6706	7529	8414
300	671	120	4767	5348	6083	6914	7828	8823	9898
300	671	140	5366	6044	6903	7876	8951	10124	11399
325	727	40	2248	2465	2745	3061	3402	3765	4148
325	727	60	3022	3312	3709	4156	4641	5160	5711
325	727	80	3700	4108	4623	5204	5836	6516	7242
325	727	100	4371	4873	5508	6226	7010	7857	8767
325	727	120	5019	5617	6375	7233	8175	9197	10299
325	727	140	5650	6345	7230	8233	9338	10541	11846
350	783	40	2359	2583	2873	3201	3555	3932	4328
350	783	60	3176	3470	3881	4344	4846	5383	5952
350	783	80	3882	4303	4836	5436	6089	6790	7538
350	783	100	4586	5102	5759	6499	7308	8179	9114
350	783	120	5266	5879	6662	7546	8514	9563	10693
350	783	140	5928	6639	7551	8583	9717	10951	12285
375	839	40	2469	2698	2999	3339	3705	4094	4504
375	839	60	3288	3624	4050	4529	5048	5601	6187
375	839	80	4061	4494	5044	5664	6337	7059	7827
375	839	100	4798	5327	6004	6767	7599	8495	9453
375	839	120	5509	6136	6942	7852	8847	9923	11080
375	839	140	6201	6927	7865	8925	10089	11352	12716
400	894	40	2577	2811	3123	3474	3852	4254	4676
400	894	60	3429	3776	4215	4710	5245	5815	6418
400	894	80	4236	4681	5248	5887	6580	7322	8111
400	894	100	5006	5548	6245	7030	7885	8805	9787
400	894	120	5748	6388	7218	8153	9174	10277	11460
400	894	140	6471	7209	8174	9262	10455	11747	13140

NOTE: Detector time constant at a reference velocity of 5 ft/sec (1.5m/sec).
For SI units: 1 ft = 0.305 m; 1000 Btu/sec = 1055 kW.

Table B-4(bb)   Installed Spacing of Heat Detector: 25ft t_g: 300 seconds to 1000 Btu/sec α: 0.011 Btu/sec^3

τ	RTI	ΔT	Ceiling Height (ft) Fire Size at Detector Response (Btu/sec)						
			4.0	8.0	12.0	16.0	20.0	24.0	28.0
25	56	40	321	396	493	609	744	899	1074
25	56	60	451	580	746	946	1178	1444	1744
25	56	80	582	781	1030	1324	1669	2063	2508
25	56	100	719	994	1337	1740	2210	2747	3354
25	56	120	866	1223	1667	2194	2797	3491	4275
25	56	140	1013	1467	2021	2675	3426	4289	5265
50	112	40	447	527	630	751	890	1048	1225
50	112	60	616	747	916	1118	1351	1619	1921
50	112	80	781	976	1223	1518	1863	2258	2706
50	112	100	948	1214	1550	1952	2422	2961	3571
50	112	120	1119	1468	1898	2418	3025	3721	4509
50	112	140	1294	1729	2271	2914	3670	4535	5514
75	168	40	560	642	753	882	1028	1192	1374
75	168	60	757	897	1075	1283	1522	1794	2100
75	168	80	954	1156	1410	1710	2059	2457	2907
75	168	100	1151	1422	1761	2166	2638	3178	3791
75	168	120	1350	1695	2130	2650	3258	3956	4746
75	168	140	1550	1978	2517	3163	3918	4785	5767
100	224	40	654	746	865	1003	1157	1329	1518
100	224	60	890	1035	1222	1439	1687	1966	2278
100	224	80	1113	1323	1587	1896	2251	2655	3110
100	224	100	1338	1616	1964	2376	2853	3398	4014
100	224	120	1563	1914	2356	2881	3492	4193	4987
100	224	140	1789	2218	2762	3411	4169	5038	6024
125	280	40	742	842	969	1116	1279	1459	1656
125	280	60	1002	1163	1360	1588	1845	2133	2452
125	280	80	1260	1480	1755	2074	2439	2851	3312
125	280	100	1513	1799	2158	2580	3065	3617	4238
125	280	120	1765	2120	2573	3107	3725	4431	5230
125	280	140	2004	2447	3001	3657	4420	5294	6283
150	335	40	825	932	1068	1223	1395	1584	1789
150	335	60	1118	1284	1491	1729	1997	2294	2622
150	335	80	1401	1629	1914	2245	2621	3042	3511
150	335	100	1680	1972	2344	2777	3273	3833	4462
150	335	120	1944	2318	2783	3328	3955	4669	5473
150	335	140	2213	2667	3232	3898	4669	5549	6544
175	391	40	908	1019	1162	1325	1506	1703	1917
175	391	60	1226	1399	1616	1865	2143	2450	2787
175	391	80	1536	1771	2068	2411	2797	3228	3707
175	391	100	1842	2139	2523	2969	3475	4045	4683
175	391	120	2122	2507	2985	3543	4181	4904	5715
175	391	140	2412	2877	3456	4135	4915	5803	6805
200	447	40	979	1101	1252	1424	1613	1818	2040
200	447	60	1330	1510	1737	1996	2284	2601	2948
200	447	80	1667	1907	2216	2570	2968	3410	3899
200	447	100	1979	2299	2696	3154	3673	4254	4901
200	447	120	2294	2689	3181	3752	4402	5135	5956
200	447	140	2605	3081	3674	4365	5158	6056	7065

τ	RTI	ΔT	Ceiling Height (ft) Fire Size at Detector Response (Btu/sec)						
			4.0	8.0	12.0	16.0	20.0	24.0	28.0
225	503	40	1054	1181	1339	1519	1715	1929	2159
225	503	60	1430	1617	1853	2122	2420	2748	3104
225	503	80	1794	2039	2359	2725	3134	3588	4087
225	503	100	2124	2453	2863	3335	3866	4459	5116
225	503	120	2460	2866	3371	3956	4619	5364	6195
225	503	140	2791	3278	3885	4591	5396	6305	7324
250	559	40	1126	1258	1423	1610	1815	2037	2275
250	559	60	1529	1720	1966	2245	2553	2890	3257
250	559	80	1900	2166	2497	2875	3297	3761	4271
250	559	100	2265	2603	3026	3511	4055	4659	5328
250	559	120	2621	3037	3557	4155	4832	5589	6430
250	559	140	2972	3469	4092	4812	5631	6552	7581
275	615	40	1196	1333	1505	1699	1912	2142	2388
275	615	60	1625	1821	2075	2364	2683	3030	3406
275	615	80	2016	2290	2632	3022	3455	3931	4452
275	615	100	2401	2749	3184	3682	4239	4856	5536
275	615	120	2778	3204	3737	4350	5040	5810	6663
275	615	140	3149	3656	4293	5028	5861	6795	7835
300	671	40	1264	1405	1584	1786	2007	2244	2498
300	671	60	1720	1919	2182	2481	2809	3166	3552
300	671	80	2129	2411	2764	3165	3610	4097	4629
300	671	100	2535	2891	3339	3850	4420	5049	5741
300	671	120	2932	3366	3913	4541	5245	6027	6893
300	671	140	3323	3837	4490	5241	6088	7035	8087
325	727	40	1331	1476	1662	1871	2099	2344	2606
325	727	60	1797	2014	2286	2594	2932	3299	3695
325	727	80	2239	2528	2892	3305	3761	4260	4803
325	727	100	2666	3030	3490	4014	4597	5239	5943
325	727	120	3083	3525	4086	4727	5446	6242	7120
325	727	140	3493	4015	4682	5449	6310	7271	8336
350	783	40	1397	1546	1738	1954	2189	2442	2711
350	783	60	1884	2107	2388	2706	3053	3430	3835
350	783	80	2347	2644	3018	3442	3910	4420	4975
350	783	100	2794	3166	3638	4175	4771	5426	6142
350	783	120	3231	3680	4255	4911	5643	6453	7343
350	783	140	3638	4189	4871	5653	6530	7505	8583
375	839	40	1461	1614	1812	2035	2277	2537	2814
375	839	60	1970	2199	2488	2815	3172	3558	3973
375	839	80	2453	2757	3141	3576	4055	4577	5143
375	839	100	2920	3299	3783	4333	4942	5609	6337
375	839	120	3377	3833	4421	5091	5837	6660	7564
375	839	140	3797	4360	5057	5854	6746	7735	8827
400	894	40	1525	1680	1885	2115	2364	2631	2915
400	894	60	2053	2288	2586	2922	3288	3684	4108
400	894	80	2558	2868	3262	3708	4199	4732	5308
400	894	100	3045	3430	3926	4488	5110	5790	6531
400	894	120	3497	3982	4584	5268	6028	6865	7782
400	894	140	3953	4528	5239	6051	6958	7962	9067

NOTE: Detector time constant at a reference velocity of 5 ft/sec (1.5m/sec).
For SI units: 1 ft = 0.305 m; 1000 Btu/sec = 1055 kW.

Table B-4(cc)   Installed Spacing of Heat Detector: 25ft t_g: 300 seconds to 1000 Btu/sec α: 0.011 Btu/sec^3

τ	RTI	ΔT	Ceiling Height (ft) Fire Size at Detector Response (Btu/sec) 4.0	8.0	12.0	16.0	20.0	24.0	28.0
25	56	40	234	307	397	505	632	776	939
25	56	60	341	470	631	820	1041	1293	1578
25	56	80	456	652	895	1181	1510	1887	2314
25	56	100	578	853	1186	1579	2030	2548	3134
25	56	120	708	1069	1503	2013	2599	3270	4030
25	56	140	847	1301	1843	2479	3210	4048	4996
50	112	40	317	389	481	589	716	861	1026
50	112	60	450	573	730	919	1140	1393	1680
50	112	80	583	772	1009	1291	1621	2000	2428
50	112	100	722	984	1312	1699	2153	2672	3259
50	112	120	869	1211	1638	2146	2731	3404	4166
50	112	140	1017	1452	1987	2621	3352	4191	5142
75	168	40	389	465	559	671	801	948	1113
75	168	60	545	670	829	1020	1241	1495	1783
75	168	80	698	885	1121	1404	1735	2114	2544
75	168	100	855	1113	1436	1824	2277	2797	3386
75	168	120	1016	1354	1776	2277	2865	3540	4304
75	168	140	1182	1606	2134	2761	3495	4336	5289
100	224	40	455	534	633	750	882	1032	1201
100	224	60	631	761	924	1118	1342	1598	1887
100	224	80	804	994	1233	1518	1849	2230	2661
100	224	100	978	1237	1562	1950	2403	2925	3515
100	224	120	1156	1491	1910	2413	3001	3677	4443
100	224	140	1337	1758	2278	2906	3640	4483	5438
125	280	40	515	598	703	824	961	1115	1287
125	280	60	712	846	1015	1213	1441	1700	1992
125	280	80	905	1098	1341	1630	1964	2347	2780
125	280	100	1095	1357	1685	2075	2530	3053	3645
125	280	120	1289	1625	2046	2550	3139	3816	4583
125	280	140	1484	1902	2425	3053	3787	4631	5588
150	335	40	576	659	769	895	1037	1196	1371
150	335	60	788	927	1102	1305	1538	1801	2096
150	335	80	996	1197	1446	1739	2077	2463	2899
150	335	100	1206	1472	1805	2199	2657	3182	3777
150	335	120	1415	1755	2180	2686	3277	3955	4725
150	335	140	1626	2044	2571	3200	3935	4780	5739
175	391	40	629	718	832	963	1110	1274	1454
175	391	60	865	1005	1186	1395	1633	1900	2199
175	391	80	1087	1292	1548	1846	2189	2579	3018
175	391	100	1312	1583	1922	2321	2783	3311	3908
175	391	120	1536	1880	2310	2821	3415	4096	4867
175	391	140	1763	2183	2714	3346	4083	4930	5891
200	447	40	680	773	893	1029	1181	1349	1534
200	447	60	931	1080	1266	1482	1725	1998	2301
200	447	80	1173	1384	1646	1951	2299	2694	3136
200	447	100	1414	1691	2036	2441	2908	3440	4040
200	447	120	1654	2002	2438	2954	3552	4236	5011
200	447	140	1887	2318	2855	3491	4231	5081	6044
225	503	40	729	827	951	1093	1250	1423	1612
225	503	60	996	1152	1344	1566	1815	2093	2402
225	503	80	1257	1473	1741	2053	2407	2807	3254
225	503	100	1513	1795	2147	2559	3031	3568	4172
225	503	120	1769	2120	2564	3086	3689	4377	5154
225	503	140	2012	2450	2993	3634	4379	5232	6198
250	559	40	778	879	1008	1155	1317	1495	1689
250	559	60	1063	1222	1420	1648	1903	2187	2501
250	559	80	1338	1559	1835	2153	2514	2919	3371
250	559	100	1610	1896	2256	2674	3153	3695	4303
250	559	120	1871	2235	2686	3215	3824	4516	5298
250	559	140	2134	2578	3128	3776	4526	5383	6352
275	615	40	824	930	1063	1215	1382	1565	1764
275	615	60	1126	1290	1494	1728	1989	2279	2598
275	615	80	1418	1643	1925	2250	2618	3029	3486
275	615	100	1705	1995	2362	2788	3273	3821	4434
275	615	120	1976	2347	2807	3343	3957	4655	5441
275	615	140	2253	2703	3261	3916	4672	5534	6507
300	671	40	875	980	1117	1273	1446	1633	1837
300	671	60	1188	1357	1567	1806	2074	2369	2694
300	671	80	1495	1725	2014	2346	2720	3138	3601
300	671	100	1788	2091	2466	2899	3391	3945	4564
300	671	120	2080	2457	2924	3468	4090	4793	5584
300	671	140	2369	2826	3392	4055	4817	5684	6661
325	727	40	916	1028	1170	1331	1508	1701	1909
325	727	60	1249	1422	1637	1883	2156	2458	2788
325	727	80	1571	1805	2101	2440	2821	3245	3714
325	727	100	1876	2185	2567	3008	3508	4068	4692
325	727	120	2181	2565	3040	3591	4220	4930	5726
325	727	140	2482	2946	3521	4191	4960	5833	6815
350	783	40	958	1075	1221	1387	1569	1766	1980
350	783	60	1308	1485	1706	1958	2238	2545	2881
350	783	80	1647	1884	2186	2532	2920	3350	3825
350	783	100	1962	2277	2667	3116	3623	4189	4820
350	783	120	2279	2670	3153	3713	4349	5066	5868
350	783	140	2593	3064	3647	4326	5102	5981	6968
375	839	40	1000	1121	1272	1442	1628	1831	2049
375	839	60	1367	1547	1774	2032	2317	2630	2972
375	839	80	1721	1960	2269	2623	3017	3454	3935
375	839	100	2047	2368	2765	3222	3736	4310	4946
375	839	120	2376	2774	3265	3833	4477	5200	6008
375	839	140	2702	3179	3771	4459	5243	6128	7121
400	894	40	1042	1166	1321	1496	1687	1894	2117
400	894	60	1424	1608	1841	2104	2396	2715	3062
400	894	80	1781	2036	2352	2712	3113	3557	4044
400	894	100	2130	2457	2862	3326	3848	4428	5072
400	894	120	2472	2875	3375	3951	4603	5334	6148
400	894	140	2809	3293	3894	4590	5382	6274	7273

NOTE: Detector time constant at a reference velocity of 5 ft/sec (1.5m/sec).
For SI units: 1 ft = 0.305 m; 1000 Btu/sec = 1055 kW.

Table B-4(dd) Installed Spacing of Heat Detector: 25ft t_g: 600 seconds to 1000 Btu/sec α: 0.003 Btu/sec^3

τ	RTI	ΔT	Ceiling Height (ft) Fire Size at Detector Response (Btu/sec)						
			4.0	8.0	12.0	16.0	20.0	24.0	28.0
25	56	40	211	283	373	479	603	744	904
25	56	60	313	441	601	789	1005	1254	1534
25	56	80	423	620	860	1143	1468	1841	2262
25	56	100	542	817	1148	1536	1983	2495	3074
25	56	120	669	1030	1460	1965	2546	3211	3964
25	56	140	804	1259	1797	2427	3153	3983	4923
50	112	40	283	353	442	548	672	814	975
50	112	60	404	529	683	869	1087	1336	1618
50	112	80	531	719	954	1233	1559	1933	2356
50	112	100	663	925	1250	1638	2084	2597	3178
50	112	120	801	1146	1571	2075	2655	3322	4076
50	112	140	946	1383	1915	2544	3269	4102	5044
75	168	40	345	417	509	617	742	885	1047
75	168	60	488	610	766	952	1170	1419	1702
75	168	80	630	814	1047	1326	1653	2027	2451
75	168	100	776	1034	1356	1738	2186	2701	3283
75	168	120	928	1265	1685	2183	2766	3434	4190
75	168	140	1085	1509	2036	2663	3387	4221	5166
100	224	40	402	477	572	683	811	956	1119
100	224	60	563	688	846	1034	1253	1504	1788
100	224	80	721	907	1141	1421	1747	2122	2548
100	224	100	885	1138	1458	1841	2290	2805	3389
100	224	120	1048	1383	1796	2295	2878	3546	4304
100	224	140	1218	1636	2158	2779	3507	4342	5288
125	280	40	454	533	632	747	878	1026	1191
125	280	60	633	761	923	1115	1336	1589	1874
125	280	80	807	996	1233	1514	1842	2219	2645
125	280	100	983	1240	1561	1946	2395	2911	3496
125	280	120	1162	1494	1910	2408	2991	3661	4420
125	280	140	1344	1762	2278	2900	3628	4464	5412
150	335	40	504	586	689	808	942	1094	1262
150	335	60	699	832	998	1193	1418	1673	1961
150	335	80	891	1082	1322	1607	1937	2315	2743
150	335	100	1079	1339	1663	2049	2500	3017	3604
150	335	120	1271	1605	2022	2521	3105	3776	4537
150	335	140	1466	1880	2399	3021	3750	4587	5536
175	391	40	555	637	743	866	1005	1160	1332
175	391	60	762	899	1070	1270	1498	1757	2047
175	391	80	966	1164	1409	1698	2031	2412	2842
175	391	100	1171	1435	1763	2152	2605	3124	3712
175	391	120	1376	1713	2133	2634	3219	3891	4654
175	391	140	1583	1999	2520	3143	3872	4710	5662
200	447	40	599	685	796	923	1066	1225	1400
200	447	60	822	964	1140	1344	1577	1839	2133
200	447	80	1042	1243	1494	1787	2124	2508	2941
200	447	100	1260	1527	1860	2254	2710	3231	3822
200	447	120	1478	1817	2242	2746	3334	4008	4772
200	447	140	1697	2114	2639	3265	3995	4835	5788
225	503	40	642	732	847	978	1126	1288	1467
225	503	60	884	1027	1208	1417	1654	1921	2218
225	503	80	1114	1320	1576	1874	2216	2603	3039
225	503	100	1345	1617	1956	2354	2814	3338	3931
225	503	120	1576	1919	2349	2857	3448	4124	4891
225	503	140	1809	2228	2757	3385	4118	4960	5915
250	559	40	684	777	896	1032	1183	1350	1533
250	559	60	938	1088	1273	1488	1730	2001	2302
250	559	80	1185	1395	1656	1960	2306	2698	3137
250	559	100	1429	1705	2049	2452	2916	3445	4040
250	559	120	1673	2019	2454	2967	3561	4241	5010
250	559	140	1908	2338	2872	3505	4241	5085	6042
275	615	40	724	822	944	1084	1240	1411	1597
275	615	60	992	1147	1338	1557	1804	2079	2384
275	615	80	1253	1468	1735	2044	2395	2791	3234
275	615	100	1510	1791	2141	2549	3018	3550	4150
275	615	120	1767	2116	2557	3076	3674	4357	5129
275	615	140	2011	2447	2986	3624	4363	5210	6170
300	671	40	764	865	991	1135	1295	1470	1661
300	671	60	1047	1205	1400	1625	1877	2157	2466
300	671	80	1320	1540	1812	2126	2483	2884	3331
300	671	100	1590	1874	2230	2645	3119	3656	4258
300	671	120	1850	2212	2659	3183	3786	4473	5248
300	671	140	2112	2553	3099	3742	4485	5336	6298
325	727	40	803	907	1037	1185	1349	1528	1723
325	727	60	1100	1261	1462	1691	1948	2233	2547
325	727	80	1386	1609	1887	2207	2570	2975	3427
325	727	100	1669	1956	2318	2739	3219	3760	4367
325	727	120	1938	2305	2759	3289	3897	4588	5367
325	727	140	2210	2657	3209	3858	4606	5461	6426
350	783	40	841	948	1082	1234	1402	1585	1783
350	783	60	1151	1316	1522	1756	2018	2308	2626
350	783	80	1451	1677	1961	2287	2655	3065	3522
350	783	100	1746	2036	2405	2832	3317	3863	4474
350	783	120	2024	2396	2857	3393	4007	4703	5485
350	783	140	2307	2759	3318	3973	4727	5585	6554
375	839	40	882	988	1126	1282	1453	1641	1843
375	839	60	1202	1371	1581	1820	2087	2382	2705
375	839	80	1514	1744	2034	2365	2739	3155	3616
375	839	100	1812	2115	2490	2923	3414	3966	4581
375	839	120	2108	2486	2953	3496	4116	4817	5603
375	839	140	2402	2859	3425	4087	4846	5709	6682
400	894	40	916	1027	1169	1329	1504	1696	1902
400	894	60	1251	1424	1639	1883	2155	2455	2782
400	894	80	1577	1810	2105	2443	2822	3243	3709
400	894	100	1884	2193	2574	3013	3510	4067	4688
400	894	120	2191	2575	3049	3598	4224	4930	5721
400	894	140	2495	2958	3531	4199	4965	5833	6809

NOTE: Detector time constant at a reference velocity of 5 ft/sec (1.5m/sec).
For SI units: 1 ft = 0.305 m; 1000 Btu/sec = 1055 kW.

Table B-4(ee)   Installed Spacing of Heat Detector: 30ft t_g: 50 seconds to 1000 Btu/sec α: 0.400 Btu/sec³

τ	RTI	ΔT	4.0	8.0	12.0	16.0	20.0	24.0	28.0	τ	RTI	ΔT	4.0	8.0	12.0	16.0	20.0	24.0	28.0
			Ceiling Height (ft) — Fire Size at Detector Response (Btu/sec)										Ceiling Height (ft) — Fire Size at Detector Response (Btu/sec)						
25	56	40	1541	1757	1963	2204	2475	2773	3096	225	503	40	5480	5821	6309	6882	7509	8174	8873
25	56	60	2013	2288	2589	2944	3344	3788	4276	225	503	60	7117	7642	8320	9106	9963	10875	11835
25	56	80	2472	2801	3207	3687	4231	4839	5512	225	503	80	8688	9332	10193	11189	12273	13430	14652
25	56	100	2925	3307	3828	4445	5148	5936	6813	225	503	100	10171	10936	11980	13185	14498	15902	17388
25	56	120	3311	3811	4458	5223	6098	7084	8183	225	503	120	11590	12477	13707	15122	16667	18321	20077
25	56	140	3718	4318	5100	6024	7084	8283	9621	225	503	140	12958	13973	15390	17018	18797	20707	22739
50	112	40	2231	2480	2735	3032	3361	3718	4101	250	559	40	5868	6207	6721	7327	7988	8689	9425
50	112	60	2986	3235	3592	4014	4484	4997	5550	250	559	60	7610	8154	8869	9698	10600	11560	12569
50	112	80	3587	3941	4414	4970	5590	6270	7009	250	559	80	9298	9962	10868	11915	13056	14271	15552
50	112	100	4208	4627	5222	5918	6698	7558	8499	250	559	100	10891	11674	12772	14038	15416	16888	18443
50	112	120	4810	5300	6023	6868	7819	8873	10031	250	559	120	12415	13320	14610	16095	17713	19444	21279
50	112	140	5293	5963	6823	7827	8959	10219	11610	250	559	140	13884	14914	16399	18105	19967	21961	24080
75	168	40	2867	3087	3379	3725	4106	4517	4955	275	615	40	6248	6582	7121	7758	8452	9188	9959
75	168	60	3727	4024	4437	4922	5458	6038	6659	275	615	60	8091	8651	9402	10272	11218	12224	13280
75	168	80	4572	4905	5441	6070	6768	7525	8340	275	615	80	9892	10572	11522	12620	13815	15086	16425
75	168	100	5231	5745	6416	7196	8061	9006	10029	275	615	100	11594	12391	13540	14865	16307	17844	19467
75	168	120	5958	6567	7375	8311	9354	10496	11739	275	615	120	13221	14137	15486	17039	18729	20535	22446
75	168	140	6664	7375	8324	9424	10652	12003	13479	275	615	140	14789	15828	17379	19161	21102	23180	25383
100	224	40	3365	3621	3952	4342	4771	5232	5719	300	671	40	6623	6947	7510	8176	8902	9672	10478
100	224	60	4462	4734	5193	5736	6333	6976	7661	300	671	60	8560	9135	9919	10830	11819	12870	13972
100	224	80	5310	5761	6363	7062	7832	8664	9554	300	671	80	10474	11166	12158	13306	14553	15879	17274
100	224	100	6213	6751	7492	8352	9300	10329	11434	300	671	100	12283	13089	14287	15670	17174	18775	20464
100	224	120	7081	7711	8596	9621	10755	11989	13321	300	671	120	14013	14933	16339	17957	19718	21596	23582
100	224	140	7923	8650	9684	10880	12207	13655	15225	300	671	140	15680	16719	18333	20188	22208	24366	26652
125	280	40	3870	4115	4479	4909	5383	5888	6423	325	727	40	6827	7290	7888	8584	9342	10144	10984
125	280	60	4990	5377	5889	6486	7141	7843	8588	325	727	60	9019	9606	10424	11373	12405	13499	14646
125	280	80	6091	6557	7214	7978	8817	9720	10682	325	727	80	11045	11746	12779	13974	15273	16652	18102
125	280	100	7137	7682	8488	9423	10450	11560	12747	325	727	100	12961	13769	15016	16455	18019	19683	21436
125	280	120	8143	8770	9728	10838	12060	13383	14805	325	727	120	14793	15710	17170	18853	20683	22632	24690
125	280	140	8954	9826	10946	12235	13658	15203	16869	325	727	140	16294	17578	19263	21190	23287	25524	27891
150	335	40	4248	4569	4970	5440	5954	6503	7082	350	783	40	7161	7635	8257	8982	9770	10605	11478
150	335	60	5589	5987	6541	7189	7898	8856	9459	350	783	60	9470	10067	10917	11905	12977	14114	15305
150	335	80	6837	7304	8013	8838	9743	10713	11744	350	783	80	11607	12312	13385	14627	15976	17408	18912
150	335	100	7873	8550	9423	10430	11533	12720	13986	350	783	100	13629	14435	15728	17223	18845	20570	22386
150	335	120	8963	9758	10793	11984	13291	14701	16210	350	783	120	15303	16459	17983	19729	21626	23645	25774
150	335	140	10016	10935	12134	13513	15029	16669	18430	350	783	140	17103	18426	20172	22170	24341	26656	29103
175	391	40	4673	5004	5436	5942	6496	7085	7706	375	839	40	7488	7973	8619	9371	10190	11055	11961
175	391	60	6166	6565	7160	7856	8616	9428	10285	375	839	60	9914	10518	11400	12424	13537	14715	15949
175	391	80	7411	8005	8771	9655	10622	11657	12754	375	839	80	12161	12867	13978	15266	16664	18147	19703
175	391	100	8667	9379	10311	11387	12563	13825	15166	375	839	100	14054	15076	16425	17974	19654	21439	23316
175	391	120	9870	10702	11805	13073	14462	15956	17550	375	839	120	16012	17201	18779	20587	22549	24636	26835
175	391	140	11031	11989	13264	14729	16336	18068	19921	375	839	140	17897	19256	21063	23129	25374	27765	30289
200	447	40	5082	5421	5881	6422	7012	7641	8301	400	894	40	7810	8304	8973	9753	10600	11497	12433
200	447	60	6726	7119	7752	8493	9303	10165	11075	400	894	60	10351	10960	11873	12934	14085	15304	16580
200	447	80	8060	8681	9496	10437	11464	12561	13721	400	894	80	12709	13411	14560	15893	17339	18872	20480
200	447	100	9432	10172	11162	12303	13549	14883	16298	400	894	100	14664	15714	17108	18710	20447	22291	24228
200	447	120	10744	11607	12774	14117	15585	17160	18837	400	894	120	16709	17929	19559	21427	23454	25608	27876
200	447	140	12010	12999	14347	15895	17589	19411	21355	400	894	140	18677	20070	21936	24070	26387	28853	31453

NOTE: Detector time constant at a reference velocity of 5 ft/sec (1.5m/sec).
For SI units: 1 ft = 0.305 m; 1000 Btu/sec = 1055 kW.

Table B-4(ff) Installed Spacing of Heat Detector: 30ft t_g: 150 seconds to 1000 Btu/sec α: 0.044 Btu/sec^3

τ	RTI	ΔT	Ceiling Height (ft) Fire Size at Detector Response (Btu/sec)							τ	RTI	ΔT	Ceiling Height (ft) Fire Size at Detector Response (Btu/sec)						
			4.0	8.0	12.0	16.0	20.0	24.0	28.0				4.0	8.0	12.0	16.0	20.0	24.0	28.0
25	56	40	684	780	911	1066	1243	1443	1665	225	503	40	2301	2481	2727	3010	3319	3650	4002
25	56	60	919	1084	1300	1556	1850	2183	2555	225	503	60	3092	3339	3697	4107	4556	5040	5558
25	56	80	1150	1399	1715	2088	2518	3005	3550	225	503	80	3788	4151	4626	5169	5766	6413	7110
25	56	100	1387	1728	2157	2663	3245	3904	4641	225	503	100	4480	4938	5535	6217	6971	7791	8680
25	56	120	1629	2072	2626	3278	4026	4872	5819	225	503	120	5150	5709	6433	7262	8180	9185	10279
25	56	140	1877	2432	3122	3931	4858	5906	7077	225	503	140	5805	6468	7326	8308	9400	10601	11914
50	112	40	958	1073	1220	1391	1583	1796	2030	250	559	40	2461	2645	2903	3200	3524	3870	4237
50	112	60	1284	1465	1696	1965	2270	2613	2992	250	559	60	3312	3559	3933	4361	4829	5333	5871
50	112	80	1603	1856	2182	2563	3000	3492	4042	250	559	80	4047	4423	4917	5482	6102	6772	7492
50	112	100	1919	2251	2683	3192	3775	4436	5177	250	559	100	4786	5257	5877	6585	7364	8211	9126
50	112	120	2217	2654	3203	3852	4598	5445	6394	250	559	120	5501	6073	6823	7680	8627	9662	10783
50	112	140	2526	3066	3743	4544	5467	6514	7688	250	559	140	6198	6876	7762	8775	9898	11130	12472
75	168	40	1194	1323	1486	1675	1885	2116	2366	275	615	40	2616	2805	3075	3385	3722	4083	4465
75	168	60	1604	1795	2044	2334	2659	3019	3414	275	615	60	3479	3772	4162	4608	5095	5618	6175
75	168	80	2004	2257	2602	3004	3459	3967	4529	275	615	80	4299	4686	5200	5786	6428	7122	7865
75	168	100	2361	2716	3168	3694	4294	4967	5717	275	615	100	5083	5567	6209	6942	7747	8620	956
75	168	120	2732	3178	3745	4408	5166	6022	6978	275	615	120	5842	6427	7203	8087	9063	10126	11276
75	168	140	3100	3645	4337	5148	6078	7131	8311	275	615	140	6582	7272	8186	9230	10385	11647	13019
100	224	40	1409	1547	1726	1932	2161	2409	2678	300	671	40	2769	2960	3241	3564	3916	4291	4687
100	224	60	1901	2093	2362	2672	3019	3399	3814	300	671	60	3677	3980	4385	4848	5354	5895	6472
100	224	80	2335	2620	2988	3414	3891	4420	5001	300	671	80	4545	4942	5475	6082	6747	7463	8228
100	224	100	2773	3141	3616	4167	4789	5481	6248	300	671	100	5374	5869	6533	7290	8121	9020	9986
100	224	120	3203	3660	4251	4938	5716	6589	7560	300	671	120	6177	6772	7572	8485	9489	10581	11759
100	224	140	3628	4180	4895	5728	6677	7745	8937	300	671	140	6959	7659	8600	9674	10859	12153	13555
125	280	40	1611	1755	1948	2171	2417	2683	2969	325	727	40	2919	3112	3403	3739	4104	4493	4904
125	280	60	2148	2368	2657	2988	3356	3758	4193	325	727	60	3871	4183	4603	5083	5606	6166	6761
125	280	80	2662	2958	3349	3798	4300	4851	5454	325	727	80	4786	5193	5743	6371	7058	7796	8584
125	280	100	3159	3537	4037	4614	5261	5977	6765	325	727	100	5659	6164	6849	7630	8486	9411	10402
125	280	120	3647	4110	4728	5441	6245	7141	8131	325	727	120	6505	7109	7933	8873	9905	11025	12232
125	280	140	4092	4682	5424	6284	7257	8346	9556	325	727	140	7329	8036	9004	10108	11324	12649	14081
150	335	40	1806	1950	2157	2396	2658	2941	3244	350	783	40	3066	3260	3562	3910	4288	4691	5116
150	335	60	2398	2628	2934	3287	3676	4098	4555	350	783	60	4061	4382	4816	5312	5853	6431	7044
150	335	80	2972	3277	3689	4163	4689	5264	5890	350	783	80	5022	5437	6006	6654	7362	8122	8932
150	335	100	3528	3911	4436	5039	5712	6454	7266	350	783	100	5940	6452	7158	7963	8843	9793	10810
150	335	120	4029	4536	5181	5923	6754	7675	8689	350	783	120	6828	7439	8287	9253	10313	11461	12696
150	335	140	4548	5157	5928	6818	7819	8933	10164	350	783	140	7693	8405	9400	10533	11780	13135	14597
175	391	40	1996	2135	2355	2609	2888	3187	3507	375	839	40	3213	3405	3717	4077	4468	4884	5323
175	391	60	2637	2875	3199	3571	3981	4424	4901	375	839	60	4248	4576	5024	5537	6095	6691	7322
175	391	80	3271	3581	4014	4511	5061	5661	6310	375	839	80	5254	5677	6263	6931	7660	8441	9273
175	391	100	3839	4267	4816	5447	6146	6914	7751	375	839	100	6215	6734	7461	8289	9194	10168	11211
175	391	120	4417	4942	5614	6385	7245	8193	9233	375	839	120	7147	7762	8633	9625	10713	11889	13151
175	391	140	4982	5611	6411	7331	8362	9504	10761	375	839	140	7974	8766	9788	10950	12227	13612	15105
200	447	40	2137	2311	2544	2813	3108	3423	3759	400	894	40	3308	3546	3870	4242	4645	5074	5526
200	447	60	2867	3111	3452	3844	4273	4738	5235	400	894	60	4431	4766	5229	5757	6332	6945	7594
200	447	80	3521	3871	4326	4846	5420	6043	6716	400	894	80	5482	5912	6515	7203	7952	8755	9608
200	447	100	4165	4609	5182	5839	6565	7359	8222	400	894	100	6488	7011	7758	8609	9538	10537	11604
200	447	120	4790	5332	6030	6830	7720	8696	9762	400	894	120	7389	8077	8973	9991	11106	12309	13599
200	447	140	5400	6047	6876	7827	8888	10060	11344	400	894	140	8311	9121	10168	11359	12666	14081	15604

NOTE: Detector time constant at a reference velocity of 5 ft/sec (1.5m/sec).
For SI units: 1 ft = 0.305 m; 1000 Btu/sec = 1055 kW.

Table B-4(gg)   Installed Spacing of Heat Detector: 30ft t_g: 300 seconds to 1000 Btu/sec α: 0.011 Btu/sec^3

τ	RTI	ΔT	Ceiling Height (ft) Fire Size at Detector Response (Btu/sec) 4.0	8.0	12.0	16.0	20.0	24.0	28.0	τ	RTI	ΔT	Ceiling Height (ft) Fire Size at Detector Response (Btu/sec) 4.0	8.0	12.0	16.0	20.0	24.0	28.0
25	56	40	422	511	625	761	916	1093	1290	225	503	40	1367	1495	1666	1862	2077	2311	2562
25	56	60	594	753	952	1187	1456	1761	2101	225	503	60	1849	2051	2311	2608	2938	3299	3692
25	56	80	772	1018	1316	1666	2068	2520	3026	225	503	80	2310	2589	2946	3356	3813	4318	4871
25	56	100	958	1302	1717	2194	2743	3361	4052	225	503	100	2759	3121	3583	4115	4712	5375	6108
25	56	120	1154	1606	2147	2766	3476	4275	5169	225	503	120	3200	3651	4226	4889	5639	6477	7407
25	56	140	1359	1930	2605	3378	4262	5257	6369	225	503	140	3636	4181	4877	5683	6598	7625	8770
50	112	40	587	675	793	931	1089	1267	1466	250	559	40	1460	1592	1770	1974	2197	2439	2698
50	112	60	805	961	1159	1392	1661	1965	2307	250	559	60	1973	2181	2450	2758	3098	3469	3872
50	112	80	1023	1259	1552	1898	2297	2749	3256	250	559	80	2464	2749	3117	3539	4007	4523	5088
50	112	100	1249	1572	1974	2447	2993	3611	4304	250	559	100	2941	3309	3783	4328	4939	5613	6357
50	112	120	1478	1901	2424	3038	3745	4545	5441	250	559	120	3409	3865	4453	5131	5893	6743	7684
50	112	140	1714	2251	2901	3668	4549	5545	6660	250	559	140	3872	4421	5131	5951	6878	7916	9070
75	168	40	723	818	943	1089	1253	1437	1640	275	615	40	1552	1686	1871	2082	2314	2564	2831
75	168	60	985	1148	1352	1591	1864	2171	2515	275	615	60	2093	2307	2585	2903	3253	3635	4048
75	168	80	1248	1484	1780	2128	2528	2982	3491	275	615	80	2613	2905	3284	3717	4197	4725	5300
75	168	100	1510	1830	2230	2703	3248	3867	4561	275	615	100	3118	3492	3979	4536	5159	5846	6601
75	168	120	1775	2187	2705	3315	4020	4820	5717	275	615	120	3614	4074	4676	5367	6143	7005	7956
75	168	140	2036	2558	3203	3964	4842	5838	6955	275	615	140	4074	4654	5378	6213	7153	8204	9368
100	224	40	846	949	1081	1235	1408	1599	1809	300	671	40	1642	1778	1969	2188	2427	2685	2961
100	224	60	1152	1320	1533	1780	2060	2374	2723	300	671	60	2211	2430	2717	3044	3405	3796	4219
100	224	80	1453	1693	1997	2352	2757	3215	3728	300	671	80	2760	3057	3446	3890	4383	4922	5509
100	224	100	1753	2072	2478	2955	3503	4124	4821	300	671	100	3293	3671	4169	4740	5375	6075	6842
100	224	120	2041	2459	2979	3592	4297	5098	5998	300	671	120	3789	4278	4893	5598	6388	7263	8226
100	224	140	2337	2856	3501	4261	5138	6135	7253	300	671	140	4292	4882	5621	6470	7424	8487	9663
125	280	40	959	1069	1210	1373	1554	1753	1971	325	727	40	1732	1867	2065	2291	2538	2804	3088
125	280	60	1307	1481	1703	1960	2249	2571	2927	325	727	60	2325	2550	2845	3182	3553	3955	4388
125	280	80	1646	1889	2203	2567	2981	3446	3963	325	727	80	2903	3205	3604	4061	4565	5116	5713
125	280	100	1968	2300	2716	3201	3755	4381	5082	325	727	100	3464	3845	4356	4939	5588	6300	7079
125	280	120	2294	2717	3245	3864	4574	5378	6280	325	727	120	3979	4477	5105	5825	6628	7516	8492
125	280	140	2619	3142	3792	4556	5436	6434	7555	325	727	140	4505	5105	5858	6722	7691	8767	9955
150	335	40	1068	1183	1332	1503	1693	1901	2126	350	783	40	1801	1954	2158	2392	2646	2920	3212
150	335	60	1454	1633	1864	2131	2430	2761	3126	350	783	60	2438	2667	2971	3318	3698	4110	4553
150	335	80	1831	2075	2399	2774	3198	3671	4196	350	783	80	3044	3350	3760	4227	4743	5305	5915
150	335	100	2178	2518	2944	3439	4003	4636	5342	350	783	100	3605	4016	4538	5135	5796	6522	7313
150	335	120	2534	2964	3502	4130	4847	5657	6564	350	783	120	4165	4672	5314	6047	6865	7767	8755
150	335	140	2888	3416	4075	4847	5732	6734	7859	350	783	140	4714	5324	6091	6970	7954	9044	10245
175	391	40	1171	1292	1448	1627	1826	2042	2276	375	839	40	1882	2039	2250	2490	2753	3034	3333
175	391	60	1595	1778	2019	2296	2605	2946	3319	375	839	60	2549	2782	3094	3450	3840	4262	4715
175	391	80	1989	2253	2588	2974	3409	3892	4425	375	839	80	3183	3492	3912	4390	4918	5492	6113
175	391	100	2379	2726	3164	3671	4245	4887	5601	375	839	100	3764	4183	4717	5326	6001	6740	7543
175	391	120	2763	3201	3751	4389	5116	5934	6847	375	839	120	4347	4864	5518	6266	7098	8013	9014
175	391	140	3145	3679	4350	5131	6025	7034	8163	375	839	140	4918	5538	6319	7213	8212	9317	10531
200	447	40	1270	1395	1559	1747	1954	2179	2421	400	894	40	1962	2123	2340	2587	2857	3146	3453
200	447	60	1733	1917	2167	2455	2774	3125	3508	400	894	60	2657	2894	3215	3580	3980	4412	4875
200	447	80	2152	2424	2770	3168	3614	4107	4650	400	894	80	3320	3631	4061	4551	5090	5676	6308
200	447	100	2572	2927	3377	3896	4481	5133	5856	400	894	100	3920	4347	4893	5515	6203	6954	7770
200	447	120	2985	3429	3991	4642	5380	6207	7129	400	894	120	4526	5051	5719	6481	7327	8256	9271
200	447	140	3394	3934	4617	5410	6313	7331	8467	400	894	140	5119	5748	6544	7453	8467	9586	10814

NOTE: Detector time constant at a reference velocity of 5 ft/sec (1.5m/sec).
For SI units: 1 ft = 0.305 m; 1000 Btu/sec = 1055 kW.

Table B-4(hh) Installed Spacing of Heat Detector: 30ft t_g: 500 seconds to 1000 Btu/sec α: 0.004 Btu/sec^3

τ	RTI	ΔT	Ceiling Height (ft)							τ	RTI	ΔT	Ceiling Height (ft)						
			4.0	8.0	12.0	16.0	20.0	24.0	28.0				4.0	8.0	12.0	16.0	20.0	24.0	28.0
			Fire Size at Detector Response (Btu/sec)										Fire Size at Detector Response (Btu/sec)						
25	56	40	310	397	506	634	781	947	1132	225	503	40	943	1050	1187	1344	1518	1710	1919
25	56	60	455	615	810	1034	1292	1582	1906	225	503	60	1295	1466	1682	1932	2212	2523	2867
25	56	80	609	858	1153	1490	1877	2312	2798	225	503	80	1638	1878	2185	2540	2942	3393	3894
25	56	100	777	1124	1531	1997	2527	3125	3793	225	503	100	1966	2293	2700	3173	3713	4322	5003
25	56	120	955	1413	1943	2552	3237	4013	4880	225	503	120	2296	2714	3231	3835	4527	5311	6190
25	56	140	1145	1722	2385	3145	4001	4970	6052	225	503	140	2626	3142	3779	4525	5384	6359	7453
50	112	40	416	501	608	735	881	1047	1232	250	559	40	1004	1116	1257	1419	1599	1795	2009
50	112	60	591	742	930	1153	1409	1699	2024	250	559	60	1379	1554	1776	2031	2317	2634	2983
50	112	80	770	1002	1288	1623	2008	2444	2931	250	559	80	1745	1986	2299	2661	3069	3525	4031
50	112	100	958	1285	1683	2142	2672	3270	3939	250	559	100	2088	2420	2833	3313	3858	4471	5156
50	112	120	1154	1586	2106	2705	3393	4170	5039	250	559	120	2436	2858	3381	3991	4688	5475	6357
50	112	140	1359	1906	2559	3309	4169	5138	6222	250	559	140	2783	3302	3945	4696	5559	6536	7633
75	168	40	512	595	705	834	981	1148	1334	275	615	40	1066	1179	1325	1492	1677	1878	2097
75	168	60	712	862	1050	1272	1528	1818	2144	275	615	60	1462	1639	1867	2129	2421	2743	3097
75	168	80	918	1144	1426	1759	2142	2578	3066	275	615	80	1837	2092	2411	2779	3193	3655	4166
75	168	100	1127	1443	1832	2291	2819	3417	4087	275	615	100	2206	2543	2964	3450	4001	4620	5309
75	168	120	1344	1762	2268	2866	3552	4329	5199	275	615	120	2572	2998	3529	4145	4847	5639	6525
75	168	140	1567	2095	2736	3480	4338	5308	6394	275	615	140	2935	3458	4108	4865	5732	6713	7813
100	224	40	594	682	795	928	1078	1248	1436	300	671	40	1125	1241	1392	1564	1753	1960	2183
100	224	60	823	975	1166	1390	1647	1939	2266	300	671	60	1543	1723	1956	2224	2522	2850	3210
100	224	805	1053	1280	1562	1894	2278	2714	3202	300	671	80	1935	2194	2520	2895	3316	3784	4300
100	224	100	1285	1598	1984	2441	2968	3566	4237	300	671	100	2322	2664	3091	3585	4142	4767	5460
100	224	120	1522	1930	2434	3028	3713	4490	5361	300	671	120	2704	3136	3673	4296	5005	5802	6691
100	224	140	1765	2277	2910	3653	4509	5480	6567	300	671	140	3084	3612	4268	5032	5904	6890	7993
125	280	40	671	763	881	1018	1173	1345	1537	325	727	40	1183	1302	1457	1633	1828	2039	2268
125	280	60	927	1082	1277	1505	1765	2059	2388	325	727	60	1623	1804	2043	2317	2621	2955	3320
125	280	80	1180	1409	1694	2029	2414	2850	3340	325	727	80	2031	2295	2627	3009	3436	3910	4433
125	280	100	1435	1747	2134	2591	3118	3716	4389	325	727	100	2435	2782	3217	3717	4282	4912	5611
125	280	120	1693	2096	2599	3191	3875	4652	5524	325	727	120	2834	3271	3815	4446	5161	5963	6858
125	280	140	1956	2459	3087	3828	4683	5653	6742	325	727	140	3230	3762	4426	5197	6075	7065	8172
150	335	40	744	839	962	1104	1263	1441	1636	350	783	40	1239	1361	1520	1701	1901	2117	2351
150	335	60	1023	1184	1384	1616	1881	2178	2509	350	783	60	1701	1884	2128	2408	2718	3059	3430
150	335	80	1301	1533	1822	2161	2549	2987	3479	350	783	80	2125	2393	2732	3121	3555	4035	4564
150	335	100	1578	1890	2281	2740	3268	3868	4541	350	783	100	2546	2898	3340	3848	4419	5056	5761
150	335	120	1857	2258	2761	3354	4038	4816	5689	350	783	120	2962	3403	3955	4593	5315	6124	7023
150	335	140	2128	2637	3264	4004	4858	5828	6918	350	783	140	3374	3910	4581	5360	6244	7240	8352
175	391	40	813	912	1040	1187	1351	1533	1733	375	839	40	1295	1419	1582	1768	1973	2194	2433
175	391	60	1118	1281	1487	1725	1994	2295	2630	375	839	60	1766	1962	2212	2498	2814	3160	3527
175	391	80	1417	1652	1947	2290	2682	3124	3618	375	839	80	2217	2490	2835	3230	3672	4159	4693
175	391	100	1715	2029	2424	2887	3418	4020	4695	375	839	100	2655	3011	3460	3976	4555	5198	5909
175	391	120	2004	2414	2921	3516	4202	4981	5855	375	839	120	3087	3532	4092	4738	5467	6283	7188
175	391	140	2300	2809	3438	4179	5033	6004	7095	375	839	140	3515	4054	4734	5520	6412	7414	8531
200	447	40	883	982	1115	1267	1436	1623	1827	400	894	40	1349	1476	1643	1834	2043	2270	2513
200	447	60	1208	1375	1586	1830	2104	2410	2749	400	894	60	1839	2038	2294	2586	2908	3260	3643
200	447	80	1529	1767	2067	2416	2813	3259	3757	400	894	80	2307	2584	2936	3339	3787	4280	4821
200	447	100	1840	2163	2564	3031	3566	4171	4849	400	894	100	2762	3123	3579	4102	4688	5339	6056
200	447	120	2153	2566	3077	3677	4365	5146	6022	400	894	120	3210	3660	4228	4881	5618	6440	7351
200	447	140	2466	2977	3610	4353	5209	6181	7274	400	894	140	3654	4197	4885	5679	6578	7586	8709

NOTE: Detector time constant at a reference velocity of 5 ft/sec (1.5m/sec).
For SI units: 1 ft = 0.305 m; 1000 Btu/sec = 1055 kW.

Table B-4(ii)   Installed Spacing of Heat Detector: 30ft t_g: 600 seconds to 1000 Btu/sec α: 0.003 Btu/sec^3

τ	RTI	ΔT	\\ 4.0	8.0	12.0	16.0	20.0	24.0	28.0
			Ceiling Height (ft) — Fire Size at Detector Response (Btu/sec)						
25	56	40	281	369	476	602	747	909	1091
25	56	60	419	579	773	995	1249	1535	1855
25	56	80	568	817	1110	1445	1826	2257	2737
25	56	100	730	1079	1484	1948	2470	3062	3723
25	56	120	906	1364	1890	2494	3174	3943	4802
25	56	140	1091	1669	2328	3082	3932	4893	5967
50	112	40	372	455	561	685	828	991	1173
50	112	60	534	685	872	1092	1345	1631	1952
50	112	80	703	938	1221	1554	1935	2365	2847
50	112	100	883	1211	1608	2065	2589	3181	3844
50	112	120	1070	1506	2024	2620	3303	4072	4933
50	112	140	1268	1820	2471	3217	4070	5032	6107
75	168	40	451	536	642	767	911	1074	1257
75	168	60	639	787	972	1191	1443	1729	2050
75	168	80	829	1056	1335	1665	2045	2475	2957
75	168	100	1026	1345	1731	2187	2710	3302	3966
75	168	120	1230	1651	2161	2752	3433	4203	5066
75	168	140	1443	1975	2616	3358	4210	5172	6249
100	224	40	527	610	719	847	993	1157	1341
100	224	60	735	884	1070	1289	1542	1828	2150
100	224	80	946	1171	1449	1777	2156	2587	3070
100	224	100	1162	1473	1857	2310	2833	3425	4089
100	224	120	1383	1792	2295	2886	3565	4336	5199
100	224	140	1611	2131	2760	3500	4351	5314	6392
125	280	40	594	680	793	924	1072	1239	1425
125	280	60	825	976	1165	1386	1640	1928	2250
125	280	80	1056	1281	1561	1890	2269	2700	3183
125	280	100	1290	1600	1983	2435	2956	3549	4214
125	280	120	1529	1933	2432	3020	3699	4469	5334
125	280	140	1773	2281	2907	3644	4494	5456	6536
150	335	40	657	747	863	998	1150	1320	1508
150	335	60	911	1064	1256	1481	1737	2027	2351
150	335	80	1161	1388	1670	2001	2381	2813	3298
150	335	100	1413	1723	2107	2559	3081	3674	4339
150	335	120	1669	2070	2568	3156	3834	4604	5469
150	335	140	1930	2431	3054	3789	4637	5600	6681
175	391	40	717	810	930	1069	1225	1398	1590
175	391	60	990	1148	1345	1573	1833	2126	2452
175	391	80	1261	1491	1776	2110	2493	2926	3413
175	391	100	1532	1842	2229	2683	3205	3799	4466
175	391	120	1805	2204	2703	3291	3969	4740	5606
175	391	140	2073	2578	3201	3935	4782	5745	6827
200	447	40	774	871	995	1138	1298	1475	1670
200	447	60	1069	1229	1430	1663	1927	2223	2552
200	447	80	1358	1590	1880	2217	2603	3039	3528
200	447	100	1646	1958	2348	2805	3330	3925	4593
200	447	120	1929	2335	2836	3426	4105	4877	5744
200	447	140	2217	2723	3346	4080	4928	5891	6974

τ	RTI	ΔT	\\ 4.0	8.0	12.0	16.0	20.0	24.0	28.0
			Ceiling Height (ft) — Fire Size at Detector Response (Btu/sec)						
225	503	40	830	930	1058	1205	1369	1550	1749
225	503	60	1144	1308	1514	1751	2019	2319	2651
225	503	80	1452	1686	1981	2323	2713	3152	3643
225	503	100	1758	2070	2465	2925	3453	4051	4721
225	503	120	2053	2462	2967	3559	4240	5013	5882
225	503	140	2356	2863	3489	4225	5073	6038	7122
250	559	40	887	987	1119	1270	1438	1623	1826
250	559	60	1217	1384	1595	1837	2110	2413	2750
250	559	80	1543	1780	2080	2427	2821	3263	3757
250	559	100	1858	2180	2579	3044	3575	4176	4848
250	559	120	2175	2587	3096	3691	4375	5150	6020
250	559	140	2492	3001	3631	4369	5219	6185	7270
275	615	40	937	1042	1178	1333	1506	1695	1901
275	615	60	1289	1459	1674	1921	2198	2507	2847
275	615	80	1632	1872	2176	2528	2927	3374	3871
275	615	100	1961	2287	2691	3161	3697	4301	4976
275	615	120	2293	2708	3222	3822	4509	5287	6159
275	615	140	2624	3137	3770	4512	5364	6332	7419
300	671	40	987	1096	1236	1395	1572	1765	1976
300	671	60	1358	1531	1751	2003	2285	2598	2943
300	671	80	1720	1961	2271	2628	3032	3483	3984
300	671	100	2062	2392	2802	3277	3817	4424	5103
300	671	120	2408	2827	3347	3951	4642	5423	6298
300	671	140	2753	3269	3907	4653	5509	6479	7568
325	727	40	1037	1149	1292	1456	1637	1834	2049
325	727	60	1427	1602	1827	2084	2371	2688	3037
325	727	80	1796	2048	2364	2726	3135	3591	4096
325	727	100	2160	2495	2910	3391	3936	4548	5229
325	727	120	2521	2944	3469	4079	4775	5559	6437
325	727	140	2879	3399	4043	4794	5653	6626	7717
350	783	40	1087	1201	1348	1515	1700	1902	2120
350	783	60	1494	1671	1901	2163	2455	2777	3131
350	783	80	1878	2134	2454	2823	3237	3698	4207
350	783	100	2257	2595	3016	3503	4053	4670	5355
350	783	120	2631	3059	3590	4205	4906	5694	6576
350	783	140	3003	3527	4177	4932	5796	6773	7867
375	839	40	1135	1251	1402	1573	1762	1968	2191
375	839	60	1560	1740	1973	2241	2538	2865	3223
375	839	80	1958	2218	2544	2918	3338	3804	4318
375	839	100	2351	2694	3121	3613	4169	4791	5481
375	839	120	2740	3171	3708	4330	5036	5829	6714
375	839	140	3125	3653	4308	5070	5939	6919	8016
400	894	40	1182	1301	1455	1630	1824	2034	2260
400	894	60	1625	1806	2045	2317	2619	2951	3314
400	894	80	2037	2300	2632	3012	3437	3908	4427
400	894	100	2444	2791	3224	3722	4284	4911	5605
400	894	120	2846	3282	3825	4453	5164	5962	6851
400	894	140	3245	3777	4438	5206	6080	7065	8165

NOTE: Detector time constant at a reference velocity of 5 ft/sec (1.5m/sec).
For SI units: 1 ft = 0.305 m; 1000 Btu/sec = 1055 kW.

Table B-4(jj)   Installed Spacing of Heat Detector: 50ft t_g: 50 seconds to 1000 Btu/sec α: 0.400 Btu/sec^3

τ	RTI	ΔT	4.0	8.0	12.0	16.0	20.0	24.0	28.0
			Fire Size at Detector Response (Btu/sec)						
25	56	40	3499	3611	3846	4155	4512	4907	5336
25	56	60	4555	4723	5115	5604	6162	6780	7454
25	56	80	5425	5806	6382	7078	7867	8740	9695
25	56	100	6348	6890	7670	8597	9645	10804	12073
25	56	120	7255	7981	8987	10169	11502	12976	14589
25	56	140	8154	9087	10339	11799	13441	15255	17239
50	112	40	4909	5030	5295	5650	6060	6511	6996
50	112	60	6378	6566	6990	7530	8145	8819	9549
50	112	80	7744	8031	8633	9376	10217	11141	12143
50	112	100	9044	9459	10257	11224	12313	13511	14814
50	112	120	10300	10869	11881	13089	14449	15946	17578
50	112	140	11383	12270	13513	14982	16634	18454	20442
75	168	40	6159	6216	6507	6906	7368	7874	8417
75	168	60	7858	8112	8577	9173	9851	10593	11390
75	168	80	9536	9909	10558	11365	12277	13273	14347
75	168	100	11130	11646	12493	13527	14690	15963	17337
75	168	120	12669	13346	14407	15683	17116	18685	20385
75	168	140	14167	15023	16312	17847	19567	21454	23502
100	224	40	7107	7258	7585	8027	8537	9096	9692
100	224	60	9264	9497	9999	10649	11389	12196	13060
100	224	80	11254	11597	12290	13161	14144	15216	16366
100	224	100	13136	13616	14514	15619	16861	18215	19671
100	224	120	14944	15582	16698	18051	19567	21222	23005
100	224	140	16700	17513	18859	20474	22281	24255	26387
125	280	40	8093	8224	8576	9057	9613	10220	10868
125	280	60	10578	10773	11309	12011	12809	13679	14607
125	280	80	12863	13156	13891	14823	15876	17021	18247
125	280	100	15016	15438	16384	17559	18880	20317	21856
125	280	120	17078	17654	18823	20254	21856	23600	25472
125	280	140	19076	19822	21227	22925	24823	26891	29115
150	335	40	9028	9128	9503	10022	10620	11273	11969
150	335	60	11831	11970	12538	13288	14142	15072	16063
150	335	80	14401	14620	15394	16385	17505	18722	20021
150	335	100	16567	17143	18143	19387	20784	22302	23924
150	335	120	18854	19594	20824	22331	24018	25851	27814
150	335	140	21069	21988	23459	25240	27230	29393	31714
175	391	40	9927	9984	10381	10935	11574	12271	13013
175	391	60	12795	13093	13702	14500	15407	16394	17445
175	391	80	15570	15997	16820	17869	19053	20339	21709
175	391	100	18190	18766	19814	21124	22596	24192	25896
175	391	120	20701	21442	22727	24308	26078	27999	30052
175	391	140	23131	24052	25583	27445	29526	31784	34201
200	447	40	10597	10789	11218	11806	12484	13223	14009
200	447	60	13882	14175	14815	15658	16617	17659	18767
200	447	80	16902	17322	18185	19288	20535	21888	23327
200	447	100	19748	20319	21414	22788	24332	26005	27789
200	447	120	22474	23212	24549	26202	28054	30060	32202
200	447	140	25110	26028	27618	29560	31730	34082	36596

τ	RTI	ΔT	4.0	8.0	12.0	16.0	20.0	24.0	28.0
			Fire Size at Detector Response (Btu/sec)						
225	503	40	11385	11571	12023	12643	13359	14138	14966
225	503	60	14931	15215	15886	16771	17780	18875	20039
225	503	80	18187	18598	19498	20654	21962	23379	24886
225	503	100	21253	21815	22954	24389	26004	27752	29613
225	503	120	24186	24916	26304	28028	29959	32049	34277
225	503	140	27020	27932	29579	31599	33857	36301	38909
250	559	40	12148	12327	12799	13451	14203	15021	15890
250	559	60	15948	16220	16919	17847	18904	20050	21268
250	559	80	19433	19830	20766	21974	23341	24820	26393
250	559	100	22713	23260	24443	25938	27621	29442	31379
250	559	120	25848	26564	28002	29794	31802	33974	36287
250	559	140	28875	29774	31477	33573	35915	38450	41151
275	615	40	12890	13059	13552	14234	15020	15876	16785
275	615	60	16939	17194	17922	18890	19993	21189	22460
275	615	80	20648	21026	21997	23255	24678	26219	27855
275	615	100	24135	24663	25887	27442	29191	31083	33094
275	615	120	27466	28163	29649	31508	33591	35844	38240
275	615	140	30681	31562	33318	35489	37915	40539	43332
300	671	40	13614	13770	14283	14994	15815	16707	17655
300	671	60	17906	18141	18896	19903	21052	22297	23618
300	671	80	21835	22189	23194	24501	25979	27579	29278
300	671	100	25525	26028	27293	28904	30719	32680	34763
300	671	120	29048	29720	31253	33176	35334	37665	40143
300	671	140	32446	33302	35111	37354	39863	42573	45456
325	727	40	14322	14463	14995	15735	16589	17517	18502
325	727	60	18854	19065	19846	20892	22084	23377	24748
325	727	80	22998	23323	24361	25715	27248	28906	30665
325	727	100	26887	27359	28663	30331	32208	34238	36392
325	727	120	30598	31238	32817	34804	37034	39442	42000
325	727	140	34176	35000	36861	39175	41763	44559	47531
350	783	40	15017	15141	15691	16459	17344	18307	19329
350	783	60	19786	19968	20774	21857	23093	24432	25851
350	783	80	24142	24431	25501	26902	28487	30203	32021
350	783	100	28226	28660	30003	31725	33665	35761	37984
350	783	120	32121	32723	34346	36395	38695	41179	43815
350	783	140	35876	36660	38571	40954	43622	46502	49561
375	839	40	15700	15803	16371	17166	18083	19080	20137
375	839	60	20703	20851	21681	22801	24079	25463	26930
375	839	80	25267	25516	26617	28063	29701	31471	33348
375	839	100	29544	29934	31314	33089	35090	37251	39543
375	839	120	33621	34176	35842	37953	40322	42880	45594
375	839	140	37550	38285	40245	42697	45442	48404	51549
400	894	40	16373	16452	17038	17859	18807	19837	20929
400	894	60	21607	21717	22571	23727	25046	26475	27988
400	894	80	26378	26580	27711	29201	30890	32715	34649
400	894	100	30845	31184	32600	34427	36488	38713	41071
400	894	120	35101	35602	37310	39480	41918	44549	47338
400	894	140	39201	39880	41887	44406	47227	50270	53499

NOTE: Detector time constant at a reference velocity of 5 ft/sec (1.5m/sec).
For SI units: 1 ft = 0.305 m; 1000 Btu/sec = 1055 kW.

Table B-4(kk)   Installed Spacing of Heat Detector: 50ft t_g: 150 seconds to 1000 Btu/sec α: 0.044 Btu/sec^3

τ	RTI	ΔT	\multicolumn{7}{c}{Ceiling Height (ft) — Fire Size at Detector Response (Btu/sec)}						
			4.0	8.0	12.0	16.0	20.0	24.0	28.0
25	56	40	1494	1637	1837	2075	2342	2639	2963
25	56	60	2041	2305	2663	3076	3539	4049	4605
25	56	80	2568	3009	3555	4177	4868	5625	6449
25	56	100	3120	3754	4516	5373	6319	7353	8475
25	56	120	3687	4541	5542	6658	7883	9219	10667
25	56	140	4272	5369	6630	8025	9552	11213	13012
50	112	40	2075	2207	2409	2651	2924	3225	3553
50	112	60	2794	3039	3386	3792	4248	4753	5304
50	112	80	3496	3880	4398	4998	5672	6417	7232
50	112	100	4193	4741	5456	6279	7200	8217	9328
50	112	120	4893	5628	6564	7635	8831	10147	11583
50	112	140	5571	6544	7724	9066	10559	12199	13986
75	168	40	2572	2696	2907	3163	3450	3765	4105
75	168	60	3460	3681	4034	4450	4917	5431	5990
75	168	80	4274	4656	5174	5778	6455	7202	8018
75	168	100	5097	5637	6343	7160	8076	9087	10194
75	168	120	5914	6633	7549	8603	9784	11089	12517
75	168	140	6730	7647	8794	10109	11580	13204	14980
100	224	40	3026	3136	3358	3628	3932	4263	4621
100	224	60	4020	4262	4627	5059	5543	6072	6647
100	224	80	4991	5364	5892	6510	7201	7960	8787
100	224	100	5937	6462	7174	8000	8924	9942	11054
100	224	120	6869	7564	8480	9538	10721	12026	13453
100	224	140	7793	8678	9817	11127	12593	14211	15983
125	280	40	3409	3542	3775	4060	4381	4730	5105
125	280	60	4563	4802	5179	5629	6132	6681	7274
125	280	80	5658	6024	6566	7201	7911	8688	9531
125	280	100	6721	7233	7957	8800	9740	10773	11898
125	280	120	7762	8439	9364	10434	11630	12945	14381
125	280	140	8790	9650	10794	12112	13585	15208	16983
150	335	40	3791	3924	4167	4468	4805	5172	5564
150	335	60	5076	5310	5701	6169	6692	7261	7874
150	335	80	6292	6648	7204	7859	8590	9388	10251
150	335	100	7466	7962	8701	9564	10525	11577	12720
150	335	120	8612	9268	10207	11296	12510	13842	15293
150	335	140	9741	10574	11729	13063	14551	16188	17973
175	391	40	4156	4286	4540	4855	5209	5593	6003
175	391	60	5567	5794	6198	6684	7227	7817	8451
175	391	80	6900	7241	7812	8489	9241	10062	10947
175	391	100	8183	8658	9414	10298	11282	12357	13520
175	391	120	9367	10060	11016	12126	13362	14716	16185
175	391	140	10583	11457	12628	13983	15491	17146	18948
200	447	40	4507	4633	4897	5226	5596	5996	6424
200	447	60	6041	6257	6674	7179	7741	8352	9007
200	447	80	7427	7810	8397	9094	9870	10713	11621
200	447	100	8795	9326	10099	11005	12014	13113	14300
200	447	120	10128	10821	11795	12929	14189	15567	17059
200	447	140	11435	12307	13496	14874	16406	18084	19906
225	503	40	4846	4967	5241	5584	5969	6386	6830
225	503	60	6504	6703	7133	7655	8238	8869	9545
225	503	80	7974	8358	8961	9679	10478	11345	12276
225	503	100	9439	9970	10761	11690	12723	13848	15060
225	503	120	10862	11556	12549	13707	14993	16396	17913
225	503	140	12256	13128	14337	15739	17297	19000	20846
250	559	40	5177	5289	5573	5929	6329	6762	7224
250	559	60	6877	7133	7577	8117	8719	9371	10068
250	559	80	8503	8889	9508	10246	11068	11958	12914
250	559	100	10062	10594	11402	12355	13413	14563	15801
250	559	120	11574	12268	13280	14463	15776	17206	18749
250	559	140	13053	13924	15153	16582	18167	19897	21768
275	615	40	5500	5602	5895	6264	6679	7128	7606
275	615	60	7295	7552	8009	8565	9187	9859	10577
275	615	80	9018	9404	10039	10798	11642	12556	13535
275	615	100	10668	11200	12026	13002	14086	15262	16525
275	615	120	12266	12960	13992	15200	16540	17997	19567
275	615	140	13828	14698	15948	17403	19017	20775	22674
300	671	40	5818	5907	6209	6590	7020	7484	7977
300	671	60	7702	7959	8428	9002	9642	10334	11072
300	671	80	9521	9906	10556	11335	12201	13139	14143
300	671	100	11259	11790	12634	13633	14742	15944	17234
300	671	120	12942	13635	14685	15919	17286	18772	20370
300	671	140	14584	15453	16724	18205	19848	21635	23563
325	727	40	6055	6202	6514	6908	7352	7831	8340
325	727	60	8100	8356	8838	9428	10087	10799	11557
325	727	80	10011	10396	11061	11860	12748	13710	14737
325	727	100	11837	12367	13228	14250	15384	16612	17928
325	727	120	13602	14293	15363	16622	18017	19531	21157
325	727	140	15323	16190	17482	18991	20663	22480	24437
350	783	40	6344	6491	6813	7219	7676	8170	8695
350	783	60	8489	8745	9238	9845	10522	11253	12031
350	783	80	10492	10875	11554	12373	13284	14268	15319
350	783	100	12403	12930	13808	14853	16012	17267	18610
350	783	120	14250	14937	16027	17310	18733	20276	21930
350	783	140	16048	16911	18224	19760	21462	23309	25296
375	839	40	6628	6775	7105	7523	7994	8503	9042
375	839	60	8871	9125	9631	10253	10948	11698	12496
375	839	80	10964	11344	12038	12876	13808	14815	15890
375	839	100	12960	13482	14377	15445	16629	17909	19278
375	839	120	14886	15568	16677	17986	19436	21007	22690
375	839	140	16760	17618	18951	20515	22246	24124	26141
400	894	40	6907	7053	7391	7822	8306	8828	9383
400	894	60	9246	9498	10015	10653	11365	12134	12952
400	894	80	11429	11804	12512	13369	14323	15353	16450
400	894	100	13507	14024	14936	16025	17234	18540	19936
400	894	120	15512	16187	17315	18649	20126	21725	23438
400	894	140	17461	18311	19666	21256	23017	24926	26973

NOTE: Detector time constant at a reference velocity of 5 ft/sec (1.5m/sec).
For SI units: 1 ft = 0.305 m; 1000 Btu/sec = 1055 kW.

Table B-4(II) Installed Spacing of Heat Detector: 50ft t_g: 300 seconds to 1000 Btu/sec α: 0.011 Btu/sec^3

| τ | RTI | ΔT | Ceiling Height (ft) Fire Size at Detector Response (Btu/sec) | | | | | | | τ | RTI | ΔT | Ceiling Height (ft) Fire Size at Detector Response (Btu/sec) | | | | | | |
			4.0	8.0	12.0	16.0	20.0	24.0	28.0				4.0	8.0	12.0	16.0	20.0	24.0	28.0
25	56	40	942	1100	1296	1520	1769	2043	2340	225	503	40	2892	3014	3230	3488	3775	4087	4422
25	56	60	1344	1649	2008	2409	2849	3328	3848	225	503	60	3910	4155	4510	4924	5385	5888	6432
25	56	80	1772	2256	2807	3414	4075	4794	5572	225	503	80	4898	5273	5786	6380	7041	7765	8551
25	56	100	2227	2923	3686	4523	5433	6420	7487	225	503	100	5862	6386	7077	7873	8759	9733	10794
25	56	120	2703	3634	4636	5726	6908	8189	9573	225	503	120	6813	7503	8392	9411	10547	11798	13165
25	56	140	3209	4392	5653	7016	8493	10091	11818	225	503	140	7757	8631	9736	10999	12408	13963	15663
50	112	40	1272	1412	1598	1815	2058	2326	2620	250	559	40	3087	3203	3425	3690	3985	4306	4649
50	112	60	1775	2039	2374	2758	3187	3660	4175	250	559	60	4164	4409	4770	5194	5664	6177	6730
50	112	80	2275	2704	3221	3806	4454	5165	5939	250	559	80	5211	5585	6106	6709	7380	8113	8908
50	112	100	2795	3413	4138	4951	5847	6826	7889	250	559	100	6231	6752	7451	8255	9150	10132	11201
50	112	120	3333	4165	5123	6187	7355	8628	10008	250	559	120	7235	7920	8815	9842	10985	12242	13614
50	112	140	3890	4959	6171	7508	8970	10560	12283	250	559	140	8229	9096	10205	11474	12888	14446	16150
75	168	40	1559	1690	1876	2094	2333	2607	2901	275	615	40	3251	3387	3614	3887	4190	4519	4871
75	168	60	2143	2398	2726	3104	3528	3996	4508	275	615	60	4410	4655	5024	5456	5936	6459	7021
75	168	80	2728	3131	3630	4201	4840	5544	6313	275	615	80	5516	5889	6417	7030	7711	8454	9259
75	168	100	3320	3895	4593	5386	6269	7240	8298	275	615	100	6591	7109	7815	8629	9534	10525	11602
75	168	120	3922	4695	5617	6657	7810	9074	10449	275	615	120	7646	8327	9229	10264	11416	12681	14060
75	168	140	4537	5531	6699	8008	9454	11036	12754	275	615	140	8690	9549	10664	11940	13362	14926	16634
100	224	40	1811	1944	2133	2356	2605	2878	3176	300	671	40	3429	3566	3799	4078	4390	4727	5088
100	224	60	2484	2731	3058	3438	3862	4330	4842	300	671	60	4651	4896	5271	5712	6202	6734	7307
100	224	80	3145	3532	4025	4591	5225	5925	6691	300	671	80	5814	6185	6721	7344	8035	8789	9604
100	224	100	3807	4356	5040	5822	6695	7659	8713	300	671	100	6942	7457	8171	8996	9911	10912	11999
100	224	120	4474	5208	6107	7130	8271	9526	10896	300	671	120	8049	8724	9633	10679	11841	13115	14502
100	224	140	5150	6091	7228	8515	9946	11518	13231	300	671	140	9141	9993	11114	12399	13829	15401	17115
125	280	40	2048	2180	2374	2603	2858	3138	3442	325	727	40	3603	3741	3979	4265	4585	4930	5300
125	280	60	2802	3043	3374	3758	4186	4658	5173	325	727	60	4886	5130	5512	5963	6462	7005	7587
125	280	80	3538	3911	4404	4971	5605	6305	7070	325	727	80	6105	6474	7018	7651	8353	9118	9943
125	280	100	4269	4795	5474	6251	7120	8080	9131	325	727	100	7287	7797	8520	9355	10282	11294	12390
125	280	120	4977	5701	6589	7602	8733	9982	11348	325	727	120	8443	9112	10030	11086	12259	13543	14939
125	280	140	5706	6632	7750	9023	10441	12005	13712	325	727	140	9583	10426	11556	12851	14291	15871	17593
150	335	40	2272	2403	2602	2638	3101	3388	3699	350	783	40	3774	3911	4154	4448	4775	5129	5507
150	335	60	3104	3339	3675	4065	4500	4978	5498	350	783	60	5116	5359	5748	6208	6717	7270	7863
150	335	80	3892	4272	4768	5339	5977	6680	7447	350	783	80	6391	6757	7309	7952	8665	9441	10277
150	335	100	4680	5215	5893	6671	7541	8500	9550	350	783	100	7625	8130	8861	9707	10646	11669	12776
150	335	120	5464	6175	7058	8067	9194	10439	11802	350	783	120	8831	9492	10419	11486	12670	13965	15371
150	335	140	6249	7156	8264	9527	10938	12495	14198	350	783	140	10018	10851	11990	13295	14746	16336	18067
175	391	40	2486	2615	2820	3063	3334	3629	3948	375	839	40	3941	4079	4327	4628	4962	5324	5710
175	391	60	3395	3622	3963	4361	4804	5289	5816	375	839	60	5342	5584	5980	6448	6967	7530	8133
175	391	80	4241	4618	5118	5696	6340	7048	7821	375	839	80	6672	7034	7594	8248	8972	9759	10607
175	391	100	5089	5619	6300	7082	7954	8916	9968	375	839	100	7958	8456	9196	10054	11004	12039	13158
175	391	120	5930	6632	7514	8524	9651	10896	12258	375	839	120	9214	9865	10801	11880	13076	14382	15799
175	391	140	6770	7663	8765	10025	11432	12985	14686	375	839	140	10397	11268	12416	13733	15195	16796	18537
200	447	40	2692	2818	3028	3279	3558	3861	4188	400	894	40	4105	4242	4496	4803	5146	5516	5910
200	447	60	3648	3893	4241	4647	5099	5593	6128	400	894	60	5564	5805	6207	6684	7213	7786	8399
200	447	80	4675	4951	5457	6043	6695	7410	8189	400	894	80	6949	7306	7874	8538	9274	10072	10931
200	447	100	5482	6009	6694	7482	8361	9327	10383	400	894	100	8287	8777	9526	10395	11357	12404	13534
200	447	120	6379	7075	7959	8972	10102	11349	12712	400	894	120	9536	10231	11177	12267	13475	14794	16222
200	447	140	7271	8154	9256	10516	11923	13475	15175	400	894	140	10806	11678	12835	14164	15638	17251	19002

NOTE: Detector time constant at a reference velocity of 5 ft/sec (1.5m/sec).
For SI units: 1 ft = 0.305 m; 1000 Btu/sec = 1055 kW.

Table B-4(mm)   Installed Spacing of Heat Detector: 50ft t_g: 500 seconds to 1000 Btu/sec α: 0.004 Btu/sec³

τ	RTI	ΔT	4.0	8.0	12.0	16.0	20.0	24.0	28.0
			Fire Size at Detector Response (Btu/sec)						
25	56	40	706	875	1073	1292	1532	1793	2076
25	56	60	1053	1381	1738	2130	2557	3020	3520
25	56	80	1435	1948	2496	3091	3735	4432	5186
25	56	100	1851	2574	3337	4159	5047	6008	7047
25	56	120	2299	3253	4258	5323	6481	7732	9083
25	56	140	2777	3981	5241	6577	8025	9590	11280
50	112	40	920	1068	1252	1462	1697	1955	2236
50	112	60	1322	1610	1950	2331	2751	3210	3708
50	112	80	1748	2208	2734	3316	3953	4647	5399
50	112	100	2200	2859	3597	4406	5287	6245	7281
50	112	120	2672	3564	4532	5590	6740	7988	9337
50	112	140	3173	4311	5535	6863	8303	9864	11552
75	168	40	1107	1247	1427	1633	1864	2119	2398
75	168	60	1568	1836	2163	2534	2948	3403	3899
75	168	80	2039	2469	2975	3545	4175	4864	5614
75	168	100	2529	3148	3862	4657	5531	6484	7518
75	168	120	3040	3874	4818	5862	7003	8246	9594
75	168	140	3572	4644	5840	7152	8584	10141	11827
100	224	40	1278	1415	1593	1799	2029	2284	2562
100	224	60	1798	2051	2372	2738	3147	3599	4093
100	224	80	2312	2724	3217	3777	4400	5085	5832
100	224	100	2845	3435	4130	4912	5777	6726	7757
100	224	120	3393	4188	5108	6136	7270	8508	9853
100	224	140	3959	4981	6149	7446	8869	10420	12104
125	280	40	1439	1572	1751	1958	2190	2446	2725
125	280	60	2004	2257	2574	2939	3346	3796	4288
125	280	80	2571	2969	3455	4009	4626	5308	6052
125	280	100	3146	3715	4396	5168	6026	6970	7999
125	280	120	3732	4498	5399	6414	7539	8772	10114
125	280	140	4333	5317	6461	7742	9156	10702	12383
150	335	40	1592	1721	1902	2112	2347	2605	2886
150	335	60	2204	2453	2771	3135	3542	3992	4484
150	335	80	2817	3206	3688	4238	4853	5532	6274
150	335	100	3433	3988	4660	5425	6277	7216	8243
150	335	120	4058	4801	5689	6693	7810	9038	10377
150	335	140	4694	5648	6773	8041	9445	10986	12664
175	391	40	1740	1863	2047	2260	2499	2760	3044
175	391	60	2396	2642	2960	3327	3736	4186	4679
175	391	80	3053	3436	3916	4465	5078	5755	6497
175	391	100	3710	4253	4920	5680	6528	7464	8488
175	391	120	4373	5098	5976	6972	8082	9306	10642
175	391	140	5045	5973	7084	8340	9737	11273	12948
200	447	40	1867	2000	2187	2404	2646	2911	3199
200	447	60	2580	2824	3145	3514	3925	4378	4873
200	447	80	3280	3658	4138	4687	5301	5979	6720
200	447	100	3978	4511	5175	5932	6778	7712	8735
200	447	120	4679	5388	6259	7250	8356	9575	10909
200	447	140	5369	6292	7392	8640	10030	11561	13232

τ	RTI	ΔT	4.0	8.0	12.0	16.0	20.0	24.0	28.0
			Fire Size at Detector Response (Btu/sec)						
225	503	40	2000	2132	2322	2544	2790	3059	3351
225	503	60	2759	3000	3324	3696	4112	4568	5065
225	503	80	3501	3873	4355	4906	5522	6200	6943
225	503	100	4240	4762	5425	6182	7027	7960	8982
225	503	120	4958	5671	6538	7526	8628	9845	11177
225	503	140	5695	6605	7697	8939	10323	11850	13519
250	559	40	2128	2260	2453	2679	2930	3204	3500
250	559	60	2932	3172	3498	3875	4294	4754	5255
250	559	80	3717	4083	4567	5121	5740	6420	7164
250	559	100	4474	5006	5670	6428	7274	8208	9229
250	559	120	5241	5948	6813	7799	8900	10115	11445
250	559	140	6011	6911	7999	9236	10617	12140	13806
275	615	40	2253	2385	2581	2811	3067	3345	3646
275	615	60	3101	3338	3668	4050	4474	4938	5442
275	615	80	3910	4287	4774	5332	5954	6638	7384
275	615	100	4716	5245	5911	6671	7519	8453	9476
275	615	120	5517	6219	7084	8070	9170	10385	11715
275	615	140	6319	7212	8297	9531	10909	12430	14095
300	671	40	2374	2506	2706	2940	3200	3484	3789
300	671	60	3267	3501	3835	4221	4650	5119	5627
300	671	80	4111	4487	4977	5540	6166	6853	7603
300	671	100	4952	5479	6147	6910	7761	8698	9722
300	671	120	5787	6485	7350	8337	9439	10654	11984
300	671	140	6621	7507	8590	9824	11201	12720	14384
325	727	40	2493	2624	2828	3066	3331	3619	3929
325	727	60	3430	3660	3997	4388	4822	5296	5810
325	727	80	4307	4683	5176	5744	6375	7066	7820
325	727	100	5182	5709	6379	7146	8001	8940	9967
325	727	120	6050	6745	7612	8602	9705	10922	12252
325	727	140	6916	7797	8880	10114	11491	13010	14673
350	783	40	2610	2740	2947	3190	3460	3753	4068
350	783	60	3570	3815	4157	4553	4992	5472	5990
350	783	80	4498	4874	5372	5944	6580	7277	8035
350	783	100	5408	5933	6607	7379	8237	9181	10211
350	783	120	6309	7001	7871	8864	9970	11188	12521
350	783	140	7205	8082	9166	10402	11779	13299	14963
375	839	40	2724	2854	3063	3311	3586	3884	4204
375	839	60	3722	3968	4313	4715	5160	5644	6168
375	839	80	4686	5062	5564	6142	6783	7485	8247
375	839	100	5630	6154	6832	7609	8472	9420	10453
375	839	120	6562	7252	8125	9122	10232	11453	12788
375	839	140	7489	8362	9448	10686	12066	13587	15252
400	894	40	2837	2965	3178	3430	3710	4012	4337
400	894	60	3871	4117	4467	4874	5324	5815	6344
400	894	80	4871	5247	5753	6336	6983	7691	8458
400	894	100	5848	6371	7053	7835	8703	9656	10694
400	894	120	6812	7500	8376	9377	10491	11716	13054
400	894	140	7768	8638	9727	10968	12351	13874	15540

NOTE: Detector time constant at a reference velocity of 5 ft/sec (1.5m/sec).
For SI units: 1 ft = 0.305 m; 1000 Btu/sec = 1055 kW.

Table B-4(nn)   Installed Spacing of Heat Detector: 50ft t_g: 600 seconds to 1000 Btu/sec α: 0.003 Btu/sec^3

τ	RTI	ΔT	Ceiling Height (ft)							τ	RTI	ΔT	Ceiling Height (ft)						
			4.0	8.0	12.0	16.0	20.0	24.0	28.0				4.0	8.0	12.0	16.0	20.0	24.0	28.0
			Fire Size at Detector Response (Btu/sec)										Fire Size at Detector Response (Btu/sec)						
25	56	40	645	818	1016	1233	1470	1728	2007	225	503	40	1762	1895	2078	2291	2528	2787	3068
25	56	60	979	1311	1669	2058	2480	2938	3432	225	503	60	2445	2690	3007	3370	3776	4222	4710
25	56	80	1350	1868	2416	3006	3645	4336	5082	225	503	80	3117	3497	3974	4518	5126	5797	6531
25	56	100	1757	2484	3250	4063	4945	5898	6928	225	503	100	3790	4327	4988	5740	6581	7508	8523
25	56	120	2195	3154	4156	5217	6366	7608	8950	225	503	120	4467	5183	6052	7038	8138	9351	10676
25	56	140	2665	3874	5130	6460	7899	9454	11133	225	503	140	5152	6068	7167	8411	9795	11318	12981
50	112	40	826	978	1163	1373	1605	1860	2137	250	559	40	1874	2006	2192	2408	2648	2911	3195
50	112	60	1204	1500	1842	2222	2639	3094	3587	250	559	60	2595	2838	3157	3523	3932	4380	4870
50	112	80	1610	2081	2611	3191	3824	4513	5258	250	559	80	3303	3679	4156	4701	5310	5981	6717
50	112	100	2044	2721	3460	4266	5142	6093	7122	250	559	100	4009	4538	5197	5949	6788	7714	8728
50	112	120	2503	3409	4382	5437	6581	7820	9160	250	559	120	4718	5421	6285	7268	8365	9574	10898
50	112	140	2991	4145	5372	6696	8129	9681	11359	250	559	140	5414	6330	7421	8659	10038	11557	13218
75	168	40	987	1131	1309	1514	1743	1995	2271	275	615	40	1982	2115	2303	2522	2766	3032	3320
75	168	60	1414	1689	2018	2390	2801	3253	3744	275	615	60	2741	2983	3304	3674	4085	4536	5029
75	168	80	1860	2298	2809	3379	4007	4692	5435	275	615	80	3484	3856	4334	4882	5492	6164	6900
75	168	100	2322	2958	3677	4473	5343	6290	7317	275	615	100	4223	4745	5403	6155	6994	7919	8933
75	168	120	2811	3665	4617	5660	6798	8034	9372	275	615	120	4944	5654	6516	7496	8591	9798	11120
75	168	140	3323	4423	5623	6935	8362	9910	11587	275	615	140	5681	6588	7674	8907	10281	11798	13456
100	224	40	1136	1274	1450	1653	1880	2131	2406	300	671	40	2087	2220	2411	2634	2882	3151	3443
100	224	60	1610	1872	2193	2558	2966	3414	3903	300	671	60	2884	3124	3448	3821	4236	4691	5186
100	224	80	2092	2512	3010	3570	4192	4873	5614	300	671	80	3662	4029	4509	5059	5672	6346	7083
100	224	100	2591	3197	3898	4682	5546	6489	7514	300	671	100	4415	4947	5606	6359	7198	8124	9138
100	224	120	3110	3926	4856	5886	7017	8249	9586	300	671	120	5178	5883	6743	7723	8816	10023	11343
100	224	140	3649	4698	5878	7176	8596	10141	11815	300	671	140	5943	6842	7924	9153	10525	12038	13694
125	280	40	1274	1410	1586	1789	2016	2266	2541	325	727	40	2190	2323	2517	2744	2995	3269	3564
125	280	60	1797	2048	2364	2726	3130	3576	4064	325	727	60	3024	3262	3589	3966	4384	4843	5341
125	280	80	2314	2721	3209	3762	4378	5056	5795	325	727	80	3837	4199	4682	5234	5850	6527	7266
125	280	100	2849	3433	4120	4894	5751	6690	7713	325	727	100	4616	5145	5806	6560	7401	8328	9342
125	280	120	3399	4185	5096	6115	7238	8466	9801	325	727	120	5407	6108	6968	7947	9040	10246	11567
125	280	140	3967	4978	6135	7420	8832	10373	12046	325	727	140	6199	7091	8171	9398	10768	12279	13934
150	335	40	1406	1538	1715	1920	2148	2400	2675	350	783	40	2291	2424	2621	2851	3106	3384	3683
150	335	60	1965	2217	2531	2892	3294	3739	4226	350	783	60	3161	3397	3728	4108	4531	4993	5495
150	335	80	2526	2924	3405	3954	4566	5240	5978	350	783	80	3989	4365	4851	5407	6026	6706	7448
150	335	100	3096	3664	4340	5106	5957	6893	7914	350	783	100	4812	5340	6003	6759	7602	8531	9546
150	335	120	3678	4441	5337	6345	7461	8685	10018	350	783	120	5631	6330	7190	8170	9263	10470	11790
150	335	140	4275	5256	6393	7666	9071	10607	12277	350	783	140	6450	7337	8415	9641	11010	12520	14174
175	391	40	1532	1662	1840	2047	2278	2532	2808	375	839	40	2390	2523	2722	2956	3215	3497	3800
175	391	60	2131	2380	2694	3054	3457	3901	4388	375	839	60	3296	3530	3863	4248	4675	5141	5647
175	391	80	2730	3120	3598	4144	4753	5426	6162	375	839	80	4152	4528	5017	5577	6200	6883	7628
175	391	100	3335	3890	4559	5319	6165	7097	8116	375	839	100	5005	5532	6197	6956	7802	8733	9750
175	391	120	3948	4693	5577	6576	7686	8906	10236	375	839	120	5852	6547	7409	8391	9485	10692	12013
175	391	140	4574	5530	6652	7914	9311	10843	12510	375	839	140	6697	7579	8657	9883	11251	12761	14414
200	447	40	1654	1781	1961	2171	2404	2660	2939	400	894	40	2487	2620	2822	3059	3323	3608	3916
200	447	60	2291	2537	2852	3214	3618	4063	4549	400	894	60	3429	3660	3997	4386	4817	5288	5797
200	447	80	2927	3311	3788	4332	4940	5611	6346	400	894	80	4314	4689	5181	5745	6372	7059	7807
200	447	100	3565	4111	4775	5530	6373	7302	8319	400	894	100	5195	5720	6388	7151	8000	8933	9953
200	447	120	4211	4941	5816	6807	7912	9128	10456	400	894	120	6068	6762	7625	8609	9706	10914	12236
200	447	140	4866	5801	6910	8162	9552	11080	12745	400	894	140	6939	7818	8896	10123	11491	13001	14654

NOTE: Detector time constant at a reference velocity of 5 ft/sec (1.5m/sec).
For SI units: 1 ft = 0.305 m; 1000 Btu/sec = 1055 kW.

B-5   Smoke Detector Spacing for Flaming Fires.

B-5.1   Ideally, the placement of smoke detectors should be based on a knowledge of fire plume and ceiling jet flows, smoke production rates, particulate changes due to aging, and the unique operating characteristics of the detector being used. Knowledge of plume and jet flows enabled the heat detector spacing information presented in Section B-3 to be developed. Unfortunately, that knowledge does not apply to smoke originating from smoldering fires. Understanding of smoke production and aging lags considerably behind that of heat production. The operating characteristics of smoke detectors in specific fire environments are not often measured or made generally available for other than a very few combustible materials. Therefore, the existing data base precludes the development of complete engineering design information for smoke detector location and spacing.

Manufacturers provide some information regarding the response of smoke detectors in their specifications, and on the information it labels on the backs of the detectors. However, this response information indicates only the detectors nominal response values with respect to gray smoke, not to black, and may be provided as a response range, instead of an exact response value. This range is in accordance with UL 268, *Standard for Safety, Smoke Detectors for Fire Protective Signaling Systems.*

The response ranges allowable for gray versus black smoke are shown below in Figure B.3:

Figure B.3   UL 268 Smoke Detector Test Acceptance Criteria for Different Colored Smoke (Table courtesy of Christopher Marrion, FireTech, Meilen, Switzerland)

Color of Smoke	Acceptable Response Range	Max/Min
Gray Smoke	1.6 %/m (0.5 %/ft) to 12.5 %/m (4.0 %/ft)	7.8
Black Smoke	1.6 %/m (0.5 %/ft) to 29.2 %/m (10.0 %/ft)	18.25

Note that the response levels are different for black and gray smoke. This is due to the different optical density levels created by different fuels and different types of smoke. Examples of this are shown by Heskestad and

Delichatsios[10] in tests they performed as follows in Figure B.4:

Figure B.4   Values of Optical Density at Response (for flaming fires only). (Table courtesy of Christopher Marrion, FireTech, Meilen, Switzerland)

Material	10^2 Dur		Relative Smoke Color
	Ionization	Scattering	
Wood	0.5	1.5	Light
Polyurethane	5.0	5.0	Dark
Cotton	0.05	0.8	Light
PVC	10.0	10.0	Dark
Variation	200:1	12.5:1	

Note the large variations in response not only to materials producing relatively the same color of smoke, but also to smoke of different color which is much more pronounced.

B-5.2   In a flaming fire, smoke detector response is affected by ceiling height and the size and rate of growth of the fire in much the same way as heat detector response. The thermal energy of the flaming fire transports smoke particles to the smoke sensor just as it does heat to a heat sensor. While the relationship between the amount of smoke and the amount of heat produced by a fire is highly dependent upon the fuel and the way it is burning, research has shown that the relationship between temperature and the optical density of smoke remains essentially constant within the fire plume and on the ceiling in the proximity of the plume.

These results were based upon the work by Heskestad and Delichatsios[10] and are indicated below. Note that for a given fuel, the optical density to temperature rise ratio between the maximum and minimum levels is 10 or less. See Figure B.5.

B-5.3   In smoldering fires, thermal energy also provides a force for transporting smoke particles to the smoke sensor. However, because the rate of energy release is usually small and the rate of growth of the fire is slow, other factors such as airflow can have a stronger influence on the transport of smoke particles to the smoke sensor. Additionally, for smoldering fires, the relationship between temperature and the optical density of smoke is not constant and, therefore, not useful.

Figure B.5   Ratio of Optical Density to Temperature Rise For Various Fuels. (Table courtesy of Christopher Marrion, FireTech, Meilen, Switzerland)

Material	$\underline{10^2\ D_u}$ (1/ft°F) ΔT	Value Range	Maximum/Minimum
Wood	0.02	0.015-0.055	3.6
Cotton	0.01/0.02	0.005-0.03	6
Paper	0.03	Data not available	-
Polyurethane	0.4	0.2-0.55	2.8
Polyester	0.3	Data not available	-
PVC	0.5/1.0	0.1-1.0	10
Foam Rubber PU	1.3	Data not available	-
Average	0.4	0.005-1.3	260

B-5.4 Smoke detectors, regardless of whether they detect by sensing scattered light, loss of light transmission (light extinction), or reduction of ion current, are particle detectors. Particle concentration, size, color, and size distribution affect each sensing technology differently. It is generally accepted that the concentration of submicron-diameter particles produced by a flaming fire is greater than that produced by a smoldering fire. Conversely, the concentration of larger particles is greater from a smoldering fire. It is also known that the smaller particles agglomerate and form larger ones as they age and are carried away from the fire source. More research is necessary to provide sufficient data, first, to predict particle concentration and behavior and, second, to predict the response of a particular detector.

B-5.5 Unlike heat detectors, listed smoke detectors are not given a listed spacing. It has become general practice to install smoke detectors on 30-ft (9.1-m) centers on smooth ceilings, with reductions made empirically to that spacing for beamed or joisted ceilings and for areas having high rates of air movement. Spacing adjustments for ceiling height are also necessary as discussed herein.

B-5.5.1 Figures B-5.5.1(a), (b), and (c) are based on the assumption that smoke transport to the detector is entirely from fire plume dynamics. It assumes that the ratio of the gas temperature rise to the optical density of the smoke is a constant and that the detector will actuate at a constant value of temperature rise equal to 20°F (–6.7°C), which is considered indicative of concentrations of smoke from a number of common fuels that would cause detection by a relatively sensitive detector. It is cautioned that many fuel/detector combinations can cause operation at a higher temperature rise. In addition, it is assumed that the detector design does not significantly impair smoke entry. The data presented in Figures B-5.5.1(a), (b), and (c) clearly indicate that spacings consid-

erably greater than 30 ft (9.1 m) are acceptable for detecting geometrically growing flaming fires when Q_d is 1000 Btu/sec (1055 kW) or more.

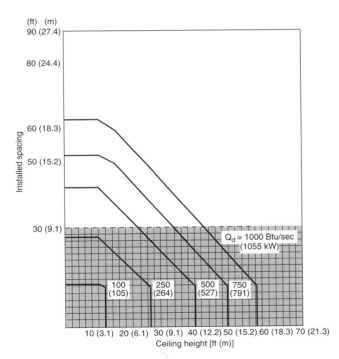

Figure B-5.5.1(a)   Smoke detector — fast fire.

B-5.5.2 In the early stages of development of a growing fire, where the heat release rate is approximately 250 Btu/sec (264 kW) or less, the environmental effects in spaces having high ceilings can dominate the transportation of smoke. Examples of such environmental effects are heating, cooling, humidity,

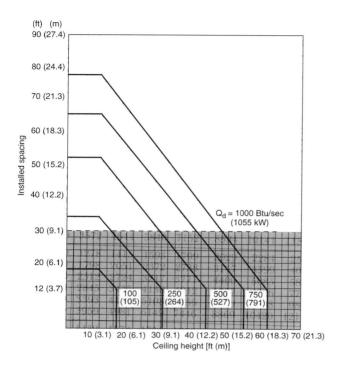

Figure B-5.5.1(b)   *Smoke detector — medium fire.*

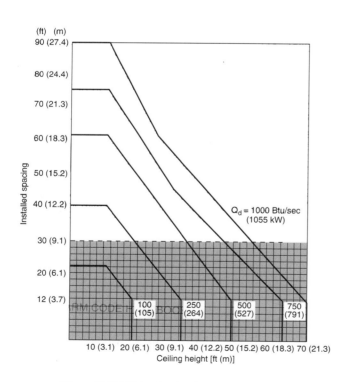

Figure B-5.5.1(c)   *Smoke detector — slow fire.*

and ventilation. Greater thermal energy release from the fire might be necessary to overcome such environmental effects. Until the growing fire reaches a sufficiently high level of heat release, closer spacing of smoke detectors on the ceiling might not significantly improve the response of the detectors to the fire. Therefore, where considering ceiling height alone, smoke detector spacing closer than 30 ft (9.1 m) might not be warranted, except in instances where an engineering analysis indicates additional benefit will result. Other construction characteristics also should be considered. (*See the appropriate sections of Chapter 5 dealing with smoke detectors and smoke detectors for the control of smoke spread.*)

B-5.6   The method used to determine the spacing of smoke detectors is similar to that used for heat detectors and is based on fire size, fire growth rate, and ceiling height.

B-5.6.1   In order to use Figure B-5.5.1(a), (b), or (c) to determine the installed spacing of a smoke detector, the designer first selects Q_d, the threshold flaming fire size at which detection is desired.

See also References 5 & 11 in Appendix C.

B-5.6.2   **Fire Growth Rate.**

See also Reference 11 in Appendix C.

(a) In addition to threshold flaming fire size, Q_d, the designer needs to consider the expected fire growth rate. Figures B-5.5.1(a), (b), and (c) are used for fast-, medium-, and slow-growing flaming fires, respectively. (*See B-2.2.2 for heat release rate and fire growth rate information.*)

(b) Figures B-5.5.1(a), (b), and (c) use the following values for t_g:

Fast fire growth rate, $t_g = 150$ sec

Medium fire growth rate, $t_g = 300$ sec

Slow fire growth rate, $t_g = 600$ sec.

B-5.6.3   **Example 1.**   To determine the installed spacing of a smoke detector on a 30-ft (9.1-m) ceiling required to detect a 750 Btu/sec (791 kW) fire that is growing at a medium rate, Figure B-5.5.1(b) should be used.

Input:

Ceiling height: 30 ft (9.1 m)

Q_d: 750 Btu/sec (791 kW)

Fire growth rate: medium; $t_g = 300$ sec

Required installed spacing:

From Figure B-5.5.1(b), using the 750 Btu/sec (791 kW) curve, installed spacing is 41 ft (12.8 m).

B-5.6.4 **Example 2.** The following example considers a 20-ft (6.1-m) ceiling with a threshold fire size of 250 Btu/sec (264 kW) growing at a medium rate.

Input:

Ceiling height: 20 ft (6.1 m)

Q_d: 250 Btu/sec (264 kW)

Fire growth rate: medium; t_g = 300 sec

Required installed spacing:

From Figure B-5.5.1(b), using the 250 Btu/sec (264 kW) curve, installed spacing for the smoke detector is 30 ft (9.1 m), since the intersection of a vertical line at 20 ft (6.1 m) and the Q_d = 250 curve fall within the shaded area below 30 ft (9.1 m) spacing.

NOTE: Both a slow and a fast rate-of-growth fire would result in the same 30-ft (9.1-m) spacing using Figures B-5.5.1(a) and (c). Engineering judgment and consideration of all factors affecting smoke movement and transport should be exercised in deciding whether a reduced spacing is warranted.

B-5.6.5 Smoke detector spacings of less than 30 ft (9.1 m) may be permitted to be used for detection of flaming fires where no other detector type is suitable and where environmental conditions allow the use of a smoke detector.

B-6   Theoretical Considerations.

B-6.1 **Introduction.** The design methods of this guide are the joint result of extensive experimental work and of mathematical modeling of the heat and mass transfer processes involved. This section outlines models and data correlations used to generate the design data presented in this guide. Only the general principles are described. More detailed information can be obtained from references 4, 9, 10, and 16 in Appendix C.

B-6.2 **Temperature and Velocity Correlations.** In order to predict the operation of any detector, it is necessary to characterize the local environment created by the fire at the detector location. For a heat detector, the important variables are the temperature and velocity of the gases at the detector. Through a program of full-scale tests and using mathematical

modeling techniques, general expressions for temperature and velocity at a detector location have been developed (*see references 4, 9, 10, and 16 in Appendix C*). The expressions are valid for fires that grow according to:

$$Q = \alpha t^2$$

where:

Q = the theoretical fire heat release rate

α = the fire intensity coefficient characteristic of a particular fuel and configuration

t = time.

The calculations used to produce the spacing curves assume that the ratio of the actual convective heat release to the theoretical heat release for all fuels is equal to that ratio for wood crib fires.

B-6.3 **Heat Detector Model.** The heating of a heat detector is given by the equation:

$$\Delta T_s/dt = (T_g (T_d)/\tau$$

where:

T_s = the temperature rating or set point of the detector

T_g = the gas temperature at the detector

τ = the detector time constant.

The time constant is a measure of the detector's sensitivity and is given by:

$$\tau = (mc)/(hA)$$

where:

m = the detector element's mass

c = the detector element's specific heat

h = the convection heat transfer coefficient for the detector

A = the surface area of the detector's element.

The value for h varies approximately as the square root of the gas velocity, u.

It is customary to refer to the time constant using a reference velocity of u_o = 5 ft/sec (1.5 m/sec) as follows:

$$\tau = \tau_o(u_o/u)^{1/2}$$

τ_o can be measured most easily by a plunge test. It can also be related to the listed spacing of a detector through a calculation; Table B-3.2.2 results from these calculations. This model uses the temperature and velocity of the gases at the detector to predict the temperature rise of the detector

I.   Input Data for Design and Analysis (*B-2*)

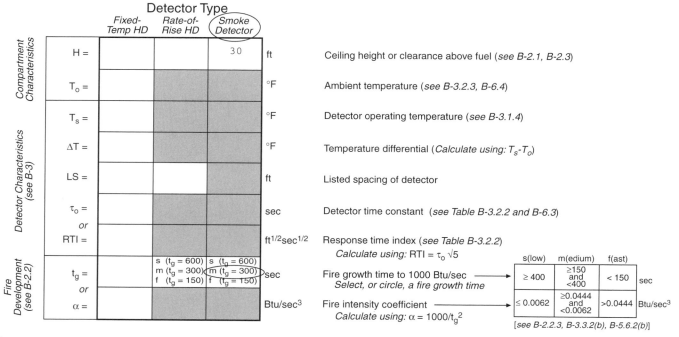

IIa. Input Data for *Design* (B-3)

IIa.1 Establish Design Objectives

Determine the size of the fire at which detector response is desired using either 1 or 2:
1. *Select* $Q_d = \underline{\quad 750 \quad}$ Btu/sec   [see B-2.2.4, Table B-2.2.2.1(a), (b), or B-2.2.2.3], or

2. *Calculate* Q_d using the time after ignition at which detector response is desired, t_d, using: $Q_d = \alpha t_d^2 = \underline{\qquad}$ Btu/sec

IIa.2 Calculate Detector Response

Fixed-Temperature HD (B-3.2)	*Rate-of-Rise HD (B-3.3)*	*Smoke Detector (B-5)*
1. Fill in the variables: $Q_d = \underline{\qquad}$ Btu/sec $t_g = \underline{\qquad}$ sec, or $\alpha = \underline{\qquad}$ Btu/sec³ $\tau_o = \underline{\qquad}$ sec, or RTI = $\underline{\qquad}$ ft¹/² sec¹/² $\Delta T = \underline{\qquad}$ °F and H = $\underline{\qquad}$ ft 2. Using Q_d and t_g (or α), select the appropriate design table (a) through (y) from Table B-3.2.4: $\underline{\qquad}$ 3. Using τ_o (or RTI), ΔT, and H, determine the installed spacing: $\underline{\qquad}$ ft	1. Fill in the variables: Fire growth (s,m,f) = $\underline{\qquad}$ H = $\underline{\qquad}$ ft $Q_d = \underline{\qquad}$ Btu/sec 2. Select installed spacing from Table B-3.3.2(a): $\underline{\qquad}$ ft 3. Select spacing modifier from Table B-3.3.2(b): x $\underline{\qquad}$ 4. Calculate installed spacing: $\underline{\qquad}$ ft	1. Fill in the variables: Fire growth (s,m,f) = $\underline{\quad m \quad}$ H = $\underline{\quad 30 \quad}$ ft $Q_d = \underline{\quad 750 \quad}$ Btu/sec 2. Using fire growth (s, m, or f), select Figure B-5.5.1(a), (b), or (c): $\underline{\quad b \quad}$ 3. Using H and Q_d, determine the installed spacing: $\underline{\quad 41 \quad}$ ft

IIb. Input Data for *Analysis* of an Existing Heat Detection System (B-4)

1. Fill in the variables:
    S = $\underline{\qquad}$ ft (installed spacing of the existing heat detector)
    $t_g = \underline{\qquad}$ sec, or $\alpha = \underline{\qquad}$ Btu/sec³
    $\tau_o = \underline{\qquad}$ sec, or RTI = $\underline{\qquad}$ ft¹/²sec¹/²
    $\Delta T = \underline{\qquad}$ °F and H = $\underline{\qquad}$ ft
2. Using S and t_g (or α), select an analysis table (a) - (nn) from Table B-4: $\underline{\qquad}$
3. Using τ_o (or RTI), ΔT, and H, determine the fire size at detector response $Q_d = \underline{\qquad}$ Btu/sec
4. Calculate the time to detector response using: $t_d = \sqrt{\dfrac{Q_d}{\alpha}} = \sqrt{\dfrac{\text{Btu/sec}}{\text{Btu/sec}^3}} = \underline{\qquad}$ sec

Figure B-5.6.3   *Completed worksheet for B-5.6.3, Example 1.*

element. Detector operation occurs when the preset conditions are reached.

The detector's sensitivity can also be expressed in units that are independent of the air velocity used in the test to determine the time constant. This is known as the response time index (RTI) and is expanded as follows:

$$RTI = \tau\sqrt{u}$$

The RTI value can therefore be obtained by multiplying τ_o values by $\sqrt{u_o}$ for example, where $u_o = 5$ ft/sec (1.5 m/sec), a τ_o of 30 sec corresponds to an RTI of 67 $sec^{1/2}$ $ft^{1/2}$ or 36 $sec^{1/2}$ $m^{1/2}$.

A detector having an RTI of 67 $sec^{1/2}$ $ft^{1/2}$ would have a τ_o of 23.7 sec, if measured in an air velocity of 8 ft/sec (2.4 m/sec).

B-6.4 Ambient Temperature Considerations. (*See also B-3.2.3.*) The maximum ambient temperature expected to occur at the ceiling dictates the choice of temperature rating for a fixed-temperature heat detector application. However, the minimum ambient temperature likely at the ceiling constitutes the worst-case condition for response to a fire by that detector.

The mass, specific heat, heat transfer coefficient, and surface area of a detector's sensing element characterize that detector's time constant. The response time of a given detector to a given fire depends only on the detector's time constant and the difference between the detector's temperature rating and the ambient temperature at the detector where the fire starts. When ambient temperature at the ceiling decreases, more heat from a fire is needed to bring the air surrounding the detector's sensing element up to its rated (operating) temperature; this translates into slower response and, in the case of a growing fire, a larger fire size at the time of detection. In a room or work area that has central heating, the minimum ambient temperature would usually be about 68°F (20°C). Certain warehouse occupancies might be heated only enough to prevent water pipe from freezing; in this case, the minimum ambient should be considered to be 35°F (2°C), even though, during many months of the year, the actual ambient is much higher. An unheated building in northern states and in Canada should be presumed to have a minimum ambient of –40°F (–40°C) or lower.

B-6.5 Heat and Smoke Analogy—Smoke Detector Model.
For smoke detectors, the temperature of the gases at the detector is not directly relevant to detection, but the mass concentration and size distribution of particulates is relevant. For many types of smoke, the mass concentration of particles is directly proportional to the optical density of the smoke, D_o. A general correlation for flaming fires has been shown to exist

between the temperature rise of the fire gases at a given location and the optical density.

Where the optical density at which a detector responds, D_o, is known and is independent of particle size distribution, the response of the detector can be approximated as a function of the heat release rate of the burning fuel, rate of fire growth, and ceiling height, assuming that the above correlation exists.

However, the more popular ionization and light-scattering detectors exhibit widely different D_o where particle size distribution is changed; hence, where D_o for these detectors is measured in order to predict response, the test aerosol used needs to be very carefully controlled so that particle size distribution is constant.

B-7 Computer Fire Models.

B-7.1 Several special application computer models are available to assist in the design and analysis of both heat detectors (e.g., fixed temperature, rate-of-rise, sprinklers, fusible links) and smoke detectors. These computer models typically run on personal computers and are available from the NIST Center for Fire Research computer bulletin board.

B-7.1.1 DETACT - T^2. DETACT - T^2 (DETector ACTuation - Time Squared) calculates the actuation time of heat detectors (fixed temperature and rate-of-rise) and sprinklers to user-specified fires that grow with the square of time. DETACT - T^2 assumes the detector is located in a large compartment with unconfined ceiling, where there is no accumulation of hot gases at the ceiling. Thus, heating of the detector is only from the flow of hot gases along the ceiling. Input data includes H, τ_o, RTI, T_s, S, and a. The program calculates the heat release rate at detector activation, as well as the time to activation. The response of a smoke detector can also be modeled by assuming the smoke detector to be a low temperature, zero lag time heat detector.

DETACT T2 uses the same methodology as that used to make the tables in this Appendix.

B-7.1.2 DETACT - QS. DETACT - QS (DETector ACTuation - Quasi-Steady) calculates the actuation time of heat detectors and sprinklers in response to fires that grow according to a user-defined fire. DETACT - QS assumes the detector is located in a large compartment with unconfined ceilings, where there is no accumulation of hot gases at the ceiling. Thus, heating of the detector is only from the flow of hot gases along the ceiling. Input data includes H, τ_o, RTI, T_s, the distance of the detector from the fire's axis, and heat release rates at user-specified times. The program calculates the

heat release rate at detector activation, the time to activation, and the ceiling jet temperature. The response of a smoke detector can also be modeled by assuming the smoke detector to be a low temperature, zero lag time heat detector.

DETACT - QS can also be found in HAZARD I, FIRE-FORM, FPETOOL.

B-7.1.3 LAVENT. LAVENT (Link Actuated VENT) calculates the actuation time of sprinklers and fusible link-actuated ceiling vents in compartment fires with draft curtains. Inputs include the ambient temperature, compartment size, thermophysical properties of the ceiling, fire location, size and growth rate, ceiling vent area and location, RTI, and temperature rating of the fusible links. Outputs of the model include the temperatures and release times of the links, the areas of the vents that have opened, the radial temperature distribution at the ceiling, and the temperature and height of the upper layer.

C

Referenced Publications

C-1

The following documents or portions thereof are referenced within this code for informational purposes only and thus are not considered part of the requirements of this document. The edition indicated for each reference is the current edition as of the date of the NFPA issuance of this document.

C-1.1 **NFPA Publications.** National Fire Protection Association, 1 Batterymarch Park, P.O. Box 9101, Quincy, MA 02269-9101.

NFPA 11, *Standard for Low-Expansion Foam and Combined Agent Systems*, 1994 edition.

NFPA 11A, *Standard for Medium- and High-Expansion Foam Systems*, 1994 edition.

NFPA 12, *Standard on Carbon Dioxide Extinguishing Systems,* 1993 edition.

NFPA 12A, *Standard on Halon 1301 Fire Extinguishing Systems,* 1992 edition.

NFPA 13, *Standard for the Installation of Sprinkler Systems,* 1996 edition.

NFPA 14, *Standard for the Installation of Standpipe and Hose Systems*, 1996 edition.

NFPA 15, *Standard for Water Spray Fixed Systems for Fire Protection*, 1996 edition.

NFPA 17, *Standard for Dry Chemical Extinguishing Systems*, 1994 edition.

NFPA 70, *National Electrical Code®,* 1996 edition.

NFPA 80, *Standard for Fire Doors and Fire Windows*, 1995 edition.

NFPA 90A, *Standard for the Installation of Air Conditioning and Ventilating Systems,* 1996 edition.

NFPA 90B, *Standard for the Installation of Warm Air Heating and Air Conditioning Systems*, 1996 edition.

NFPA 92A, *Recommended Practice for Smoke-Control Systems,* 1996 edition.

NFPA 92B, *Guide for Smoke Management Systems in Malls, Atria, and Large Areas,* 1995 edition.

NFPA *101, Life Safety Code®,* 1994 edition.

NFPA 170, *Standard for Fire Safety Symbols,* 1996 edition.

NFPA 231C, *Standard for Rack Storage of Materials,* 1995 edition.

NFPA 1221, *Standard for the Installation, Maintenance, and Use of Public Fire Service Communication Systems*, 1994 edition.

C-1.2   Other Publications.

C-1.2.1   ANSI Publications. American National Standards Institute, 11 West 42nd Street, New York, NY 10036.

ANSI A17.1, *Safety Code for Elevators and Escalators, 1993.*

ANSI S3.41, *Audible Emergency Evacuation Signal, 1990.*

C-1.2.2   IES Publication. Illuminating Engineering Society of North America, 120 Wall Street, New York, NY 10005.

Lighting Handbook Reference and Application, 1993.

C-1.2.3   UL Publication. United Laboratories Inc., 333 Pfingsten Road, Northbrook, Il 60062.

UL 1971, *Standard for Safety Devices for the Hearing Impaired, 1992.*

C-1.2.4   U.S. Government Publication. U.S. Government Printing Office, Superintendent of Documents, Washington, DC 20402.

FCC Rules and Regulations, Volume V, Part 90, March 1979.

C-2   Bibliography.

This part of the appendix lists other publications pertinent to the subject of this NFPA document that might or might not be referenced.

1. Alpert, R. "Ceiling Jets." *Fire Technology,* Aug. 1972.

2. Evaluating Unsprinklered Fire Hazards, SFPE *Technology Report* 83-2.

3. Babrauskas, V.; Lawson, J. R.; Walton, W. D.; and Twilley, W. H. "Upholstered Furniture Heat Release Rates Measured with a Furniture Calorimeter," (NBSIR 82-2604) (Dec. 1982). National Institute of Standards and Technology (formerly National Bureau of Standards), Center for Fire Research, Gaithersburg, MD 20889.

4. Beyler, C. "A Design Method for Flaming Fire Detection." *Fire Technology,* vol. 20, No. 4, Nov. 1984.

5. DiNenno, P., ed. Chapter 31, SFPE Handbook of Fire Protection Engineering, by R. Schifiliti, Sept. 1988.

6. Evans, D. D. and Stroup, D. W. "Methods to Calculate Response Time of Heat and Smoke Detectors Installed Below Large Unobstructed Ceilings," (NBSIR 85-3167) (Feb. 1985, issued Jul. 1986). National Institute of Standards and Technology (formerly National Bureau of Standards), Center for Fire Research, Gaithersburg, MD 20889.

7. Heskestad, G. "Characterization of Smoke Entry and Response for Products-of-Combustion Detectors." Proceedings, 7th International Conference on Problems of Automatic Fire Detection, Rheinish-Westfalischen Technischen Hochschule Aachen (Mar. 1975).

8. Heskestad, G. "Investigation of a New Sprinkler Sensitivity Approval Test: The Plunge Test." FMRC Tech. Report 22485, Factory Mutual Research Corporation, 1151 Providence Turnpike, Norwood, MA 02062.

9. Heskestad, G. "The Initial Convective Flow in Fire: Seventeenth Symposium on Combustion." The Combustion Institute, Pittsburgh, PA (1979).

10. Heskestad, G. and Delichatsios, M. A. "Environments of Fire Detectors — Phase 1: Effect of Fire Size, Ceiling Height and Material," Measurements vol. I (NBS-GCR-77-86), Analysis vol. II (NBS-GCR-77-95). National Technical Information Service (NTIS), Springfield, VA 22151.

11. Heskestad, G. and Delichatsios, M. A. "Update: The Initial Convective Flow in Fire." *Fire Safety Journal,* vol. 15, No. 5, 1989.

12. International Organization for Standardization, *Audible Emergency Evacuation Signal,* ISO 8201,1987.

13. Klote, J. and Milke, J. "Design of Smoke Management Systems," American Society of Heating, Refrigerating and Air Conditioning Engineers, Atlanta, GA 1992.

14. Lawson, J. R.; Walton, W. D.; and Twilley, W. H. "Fire Performance of Furnishings as Measured in the NBS Furniture Calorimeter, Part 1," (NBSIR 83-2787) (Aug. 1983). National Institute of Standards and Technology (formerly National Bureau of Standards), Center for Fire Research, Gaithersburg, MD 20889.

15. Morton, B. R.; Taylor, Sir Geoffrey; and Turner, J.S. "Turbulent Gravitational Convection from Maintained and Instantaneous Sources," Proc. Royal Society A, 234, 1-23, 1956.

16. Schifiliti, R. "Use of Fire Plume Theory in the Design and Analysis of Fire Detector and Sprinkler Response." Master's thesis, Worcester Polytechnic Institute, Center for Firesafety Studies, Worcester, MA, 1986.

17. Title 47, Code of Federal Regulations, Communications Act of 1934 Amended.

Index